Introduction to
Plant Tissue Culture

"...this book is a good starting text, especially for students who wish to have a general overview of the various applications of plant tissue culture".

Geert-Jan de Klerk
Plant Cell, Tissue and Organ Culture 77:117, 2004

Introduction to
Plant Tissue Culture

Third Edition

MK Razdan

University of Delhi
Delhi, India

Oxford & IBH Publishing Co. Pvt. Ltd.

New Delhi

(*A Unit of* CBS Publishers & Distributors Pvt Ltd)

CBSPD

CBS Publishers & Distributors Pvt Ltd

New Delhi • Bengaluru • Chennai • Kochi • Kolkata • Lucknow • Mumbai
Gujarat • Hyderabad • Jharkhand • Nagpur • Patna • Pune • Uttarakhand

Introduction to
Plant Tissue Culture

Third Edition

ISBN-13: 978-81-204-1793-9
ISBN-10: 81-204-1793-3

OXFORD & IBH
New Delhi
(A Unit of CBS Publishers & Distributors Pvt Ltd)

Published by **Satish Kumar Jain** and produced by **Varun Jain** for

CBS Publishers & Distributors Pvt Ltd

4819/XI Prahlad Street, 24 Ansari Road, Daryaganj, New Delhi 110 002, India.
Ph: 011-23266838, 23289259   Website: www.cbspd.com
                                  e-mail: delhi@cbspd.com

Corporate Office: 204 FIE, Industrial Area, Patparganj, Delhi 110 092
Ph: 011-4934 4934   Fax: 011-4934 4935
                        e-mail: publishing@cbspd.com; publicity@cbspd.com

Branches

- **Bengaluru:** Seema House 2975, 17th Cross, KR Road, Banasankari 2nd Stage, Bengaluru 560 070, Karnataka, India
  Ph: +91-80-26771678/79   Fax: +91-80-26771680   e-mail: bangalore@cbspd.com
- **Chennai:** 18/8B, Subbaraya Street, Shenoy Nagar, Chennai 600 030, Tamil Nadu, India
  Ph: +91-044-42032115, 044-26681266   e-mail: chennai@cbspd.com
- **Kochi:** 42/1325, 1326, Power House Road, Opp KSEB, Power House, Ernakulum Kochi 682 018, Kerala, India
  Ph: +91-484-4059061-65,67   Fax: +91-484-4059065   e-mail: kochi@cbspd.com
- **Kolkata:** 147, Hind Ceramics Compound, 1st Floor, Nilgunj Road, Belghoria, Kolkata-700056, West Bengal, India
  Ph: +033-25633055, 033-25633056   e-mail: kolkata@cbspd.com
- **Lucknow:** Basement, Khushnuma Complex, 7 Meerabai Marg (Behind Jawahar Bhawan), Lucknow-226001, UP, India
  Ph: +0522-4000032   e-mail: tiwari.lucknow@cbspd.com
- **Mumbai:** PWD Shed, Gala no 25/26, Ramchandra Bhatt Marg, Next to JJ Hospital Gate no. 2, Opp. Union Bank of India, Noorbaug, Mumbai-400009, Maharashtra, India
  Ph: 022-66661880/89   e-mail: mumbai@cbspd.com

Representatives

- Gujarat   0-9879558667
- Nagpur   0-8692091830
- Uttarakhand 0-9716462459
- Hyderabad   0-9885175004
- Patna   0-9334159340
- Jharkhand   0-9811541605
- Pune   0-9664372571

Printed at Mudrak, Noida, UP, India

Foreword

In this book Dr Razdan has combined his extensive research experience using plant tissue culture with his extensive knowledge of the needs of students for a subject that increasingly pervades most aspects of the plant sciences. Moreover, Dr Razdan's book avoids the fragmentation often present in multi-authored compilations. Students need to have at hand a clearly written, well-documented text that puts into perspective the plant tissue culture requirements for particular applications, yet which also enables students to undertake experiments themselves with the minimum of additional guidance. Students will benefit greatly from this clearly presented and illustrated text, and from the analysis of the historical development of a subject with its beginnings in the 18th Century,

in studies of callus formation associated with wound healing in trees, to its use in cell and molecular biological studies of plant genetic manipulations. This book is particularly well referenced, enabling the reader to follow up in greater detail points of interest, either theoretical or practical. Plant tissue culture, mainly because of the great number of interacting factors involved, tends to be largely empirical. In future years the applications of plant tissue culture in clonal propagation, plant breeding and plant genetic manipulation will increasingly depend on a better understanding of these interacting factors. Thereby, hopefully in future years the subject will become less empirical. This book by Dr Razdan will undoubtedly contribute significantly by inspiring students in this direction.

EDWARD C. COCKING, FRS
University of Nottingham
UK

Preface to Third Edition

The concept of plant cell and tissue culture actually developed in early twentieth century although its historical development had begining in eighteenth century. Initial studies were based on experiments with main focus on understanding the interactions among various physical and chemical factors that could induce successful regenerative response *in vitro* of cells, organs and tissues. With subsequent advances made in the development of protocols that enable organogenesis and whole plant regeneration from various explants of different plant species, the plant tissue culture *per se* has evolved as a subject having both theoretical and practical value. It is being increasingly recognized that plant tissue culture has potential applications in clonal propagation, breeding, genetic transformation programmes, production of plants free of pathogens, pharmaceuticals, and other secondary metabolites of industrial value. Further it has proven to have substantial role in conservation of biodiversity. Notwithstanding these developments the plant tissue culture is now an integral component of agriculture biotechnology.

This third edition is a comprehensive text book, like earlier two editions, covering fundamental as well as applied aspects of the subject mainly aimed as learning *cum* teaching aid to undergraduate and postgraduate students. More importantly, it is overwhelming to read review of the previous edition of this book stating "...It is well edited and highly recommended to senior graduate students, professors, and researchers in the fields of biology... This is a must for anyone working in the field of plant tissue culture. Readers will gain valuable knowledge from this eye-opening publication" (The Quarterly Review of Biology, Vol. 79, pp. 208-209, 2004). Another review also stated "this is a good starting text, especially for students who wish to have a general overview of the various applications of plant tissue culture" (Plant Cell, Tissue and Organ Culture 77: 117-2004).

This book in effect is an update of the previous edition published in 2003 giving insight of the major technological advancements from that period onwards on various chapters dealt within the book. During the past two decades a rapid advancement has taken place in the production of new improved plants, products or processes, using biotechnology via application of plant tissue culture and gene transformation methods. Further, agriculturally important new plant varieties have also been obtained through conventional selection, crossings, mutation and polyploidisation. Such new plant varieties, products and processes, obtained by biotechnological and conventional procedures are mostly of commercial or industrial value. As a consequence, issues like legal characterisation of Intellectual Property (IP) and Intellectual Property Protection (IPP) have arisen; this can be addressed by giving 'Rights' such as Intellectual Property Rights (IPR). A new chapter on IPR, therefore, has been added elucidating guiding principles of patenting plants, biotechnological products or processes, which have commercial/industrial value. These guidelines can provide protection rights to patent holders for giving permission to use their patents in research activities at universities and other institutions, including the industries for manufacture or production purposes.

The literature available on data regarding plant tissue culture being vast there are likely to be some omissions as it is impossible to cover all the information in one book. Yet, this edition should continue to be valuable to students, academia, researchers, plant propagators, and industrialists. Any suggestions for its improvement in future are most welcome.

The encouragement of the Publishers to bring out the third edition of this title is greatly appreciated. I am equally indebted to Delhi University Library System for providing me access to source information that helped in preparation of the book. This book is dedicated to my wife, Saroj, who has always been a great inspiration and support to me in all academic endeavours.

December 2015

M.K. Razdan
Delhi, India

Contents

Part I: Introduction and Techniques

PART I
Introduction and Techniques

Introductory History

Development of the science of tissue culture is historically linked to the discovery of the cell and subsequent propounding of the cell theory. More than 260 years ago, Henri-Louis Duhamel du Monccau's (1756) pioneering experiments on wound healing in plants demonstrated spontaneous' callus formation on the decorticated region of elm plants. His studies, according to noted biologist Gautheret (1985), could be considered a 'foreword' for the discovery of plant tissue culture. Further contributions to plant tissue culture could be attributed to the cell doctrine, which implicitly admitted that a cell is capable of autonomy and even demonstrated the potential for totipotency. The development of the multicellular or multiorganed body of a higher organism from a single-celled zygote supports the totipotent behaviour of a cell. Germination of spores into complete individuals is an obvious feature of cells in haplobiontic plants; similarly, epidermal cells of *Begonia* are transformed into new begonia plants. In the animal kingdom, a cell from any part of a *Hydra* may give rise to a new individual. Although these aspects of cellular behaviour point to the totipotent nature of some plant or animal cells, many somatic cells do not produce complete organisms since they form multicellular tissues or organs and are highly differentiated. They are mostly irreversible and lose the meristematic activity.

One needs, therefore, to understand more about the interrelationships between the various cells of a tissue and the various organs of an organism.

There must be some factor(s) superimposed on theoretically genetically identical cells which brings about certain cellular or subcellular changes leading to morphological differentiation. Trécul (1853) observed callus formation in a number of decorticated trees. He published excellent pictures of callus sections and established that apart from the cambium, the medullary rays, phloem, and youngest xylem elements also contribute as raw materials in the development of tissue culture. An interesting observation made by Vöchting (1878) suggested 'polarity' as a characteristic feature guiding the development of plant fragments. In his classical experiments on stem cuttings, Vöchting observed that the upper portion of a piece of stem always produced buds and the basal region callus or roots. Further, the grafting experiments which he undertook among species of *Opuntia, Salix, Beta* and other trees, demonstrated that behaviour of a tissue is not altered by contact with other tissue because the dependence of morphogenetic capacity on hereditary internal factors is very strict. Wiesner (1884) proposed a general theory that suggested the existence of organ-forming substances distributed in a polar fashion. Minimal size of explants could be another factor that determined the potential for differentiation. Small pieces of explants isolated from buds, roots and stems were placed by Rechinger (1893) on a surface of moistened sand. Pieces thicker than 20 mm were able to produce buds or even plants whereas smaller explants (with vascular elements) proliferated without organisation and those less than 1.5 mm (representing no more than 21 cell layers) failed to grow.

These experiments were really outlines that ultimately paved the way for studying the role of *in vitro* cultures for understanding the theoretical and practical aspects of plant tissue culture.

1.1 Concept of Cell Culture

German botanist Gottlieb Haberlandt (1902) developed the concept of *in vitro* cell culture. He was the first to culture isolated, fully differentiated cells in a nutrient medium containing glucose, peptone, and Knop's salt solution. Using palisade cells of *Lamium purpureum*, pith cells from petioles of *Eicchornia crassipes*, glandular hair of *Pulmonaria* and *Utrica*, stamen hair cells of *Tradescantia*, stomatal guard cells of *Ornithogalum*, and other plant materials, Haberlandt realised that asepsis was necessary when culture media are enriched with organic substances metabolised by micro-organisms. In his cultures, free from microcontamination, cells that were able to synthesise starch as well as increase in size survived for several weeks. However, Haberlandt failed in his goal to induce these cells to divide. Despite drawbacks, this plant physiologist made several predictions about the requirements for cell division under experimental conditions in 1902, which have been confirmed with the passage of time. Haberlandt is thus regarded as the father of tissue culture.

1.2 Development of Tissue Culture

Many plant physiologists have tried to obtain multiplication of isolated cells but failed. Hannig (1904) initiated a new line of investigation involving the culture of embryogenic tissue, which later became an important applied area of investigation, using *in vitro* techniques. He excised nearly mature embryos of some crucifers (*Raphanus sativus, R. landra, R. caudatus,* and *Cochlearia donica*) and successfully grew them to maturity on mineral salts and sugar solution. Winkler (1908) cultivated segments of string bean and observed some cell divisions but no proliferation. More promising results were obtained in the same year by Simon, who achieved success in the regeneration of a bulky callus, buds and roots from poplar stem segments. In effect, Simon may be credited with having established the basis for callus culture and to some extent even micropropagation.

1.2.1 Root-tip Culture

A new approach to tissue culture was conceived simultaneously in 1922 by Kotte (Germany) and Robbins (USA). They postulated that a true *in vitro* culture could be made easier by using meristematic cells, such as those that operate in the root tip or bud. Small excised root tips of pea and maize cultivated by Kotte in a variety of nutrients containing salts of Knop's solution, glucose and several nitrogen compounds (such as asparagine, alanine, and yeast extract) grew successfully for two weeks. Robbins, working independently, maintained his maize root tip cultures for a longer period by subculturing them, but the growth gradually diminished and the cultures were ultimately lost. An important breakthrough for continuously growing root tip cultures came from White (1934, 1937), who initially used yeast extract in a medium containing inorganic salts and sucrose but later replaced yeast extract by three B vitamins, namely, pyridoxine, thiamine and nicotinic acid. White's synthetic medium later proved to be one of the basic media for a variety of cell and tissue cultures.

The experiments on root cultures raised an interesting question on the chemical nature of yeast extract. Bonner (1937) demonstrated the importance of thiamine in yeast extract; eventually it was found that thiamine could be replaced by its two components, thiazole and pyrimidine. Many workers continued efforts to develop a complete medium for ensuring prolonged growth of root cultures. In this regard, mention may be made of the contributions by Bürstrom (1953), Street and associates (1951, 1952, 1954: see Gautheret (1985), who demonstrated the importance of copper, manganese, iodine and chelating agents on root metabolism, and Sheat et al. (1959) who obtained information on the effect of NH_4^+ and amino acids.

Although the root culture technique helped in solving many morphological, physiological and pathological problems, it was soon realised that cultures should lead to organogenesis in order to fulfil Haberlandt's objectives.

1.2.2 Embryo Culture

Very early in the history of tissue culture, Laibach (1925, 1929) demonstrated the practical applications of zygotic embryo culture in the field of plant breeding, although Hannig was the first to initiate work along these lines. Laibach raised zygotic embryos isolated from non-viable seeds of *Linum perenne* × *L. austriacum* to maturity on a culture medium and obtained hybrids which in natural course died out due to embryo abortion. Later, Van Overbeek et al. (1941) used coconut milk (embryo sac fluid) for embryo development and callus formation in *Datura*, which proved a turning point in the field of embryo culture. These studies gave impetus to further work in the area and to date several hybrids have been raised through embryo culture which could not otherwise be obtained due to embryo abortion. Further embryo culture has significant application in plant biotechnology and clonal propagation. Feeder cell and "double medium" culture methods in particular improve the survival of zygotes and proembryos thus proving as invaluable tools in embryo rescue method (Haslam and Yeung 2011).

1.2.3 Stem-tip Culture

Like embryo culture, stem-tip cultures yielded success in producing whole plants. Loo (1945)

obtained excellent cultures from stem tips of dodder and *Asparagus,* while the following year an ingenious method was devised by Ball (1946) to identify the exact part of the shoot meristem that gives rise to a whole plant. Thus shoot-tip culture is now used extensively in plant propagation industries and production of disease free plants throughout the world (*see* Chapter 15).

1.3 Role of Auxin

Successes achieved in tissue culture would be regarded as incomplete without an understanding of the role of auxins since earlier attempts involving the culture of isolated cells, root tips, or stem tips ended in the development of calluses. There remained two goals to fulfil the predictions of Haberlandt. First, the callus obtained from explants had to be perpetuated for unlimited periods and, second, the regenerated calluses had to undergo organogenesis to form whole plants. Gautheret (1934) cultured cambium cells of some tree species (*Acer pseudoplatanus, Ulmus campestre, Robinia pseudoacacia,* and *Salix caprea*) on the surface of a medium (Knop's solution containing glucose and cysteine hydrochloride) solidified with agar. After two months he observed the proliferation of callus from these cells which, however, ceased six months later due to distinct nutrient deficiency. Meanwhile, Snow (1935) demonstrated that indoleacetic acid (IAA—a growth substance discovered by Went in 1926) and occurring naturally in plant tissues stimulated cambial activity, and Gaurheret, following these observations, found that the addition of this auxin enhanced the proliferation of his cambial cultures, making it possible to prepare subcultures. White (1939) reported similar results in the cultures from tumour tissues of the hybrid *Nicotiana glauca* × *N. langsdorffii* and Nobécourt (1939) established continuously growing cultures of carrot slices. Finally, the possibility of cultivating plant tissues for an unlimited period, using media enriched with auxins, was announced independently by Gautheret, White, and Nobécourt in 1939. In plant tissue culture medium presence of auxin is now must for growth and differentiation.

1.4 Discovery of Cytokinin

Caplin and Steward (1948) reported for carrot explants that coconut milk enhanced more proliferation of callus than did auxin. In a medium enriched with this natural extract, Morel (1950) obtained indefinite growth with respect to monocotyledonous tissues of *Amorphophallus rivieri, Sauromatum guttatum,* gladiolus, iris and lily. Even the cultures of royal fern tissue evinced a similar response with coconut milk. Van Overbeek et al. (1941) had suggested earlier that liquid endosperm such as coconut milk would be a good medium for embryo culture. The fact that in coconut milk the proliferation of a tumoural type of callus was more enhanced than in auxin indicated that the milk contained a stimulating substance that was not auxin. This prompted researchers to ascertain the chemical nature of this non-auxinic substance found in plant tissues. Skoog (1944), Skoog and Tsui (1951) demonstrated that adenine stimulates cell division and induces bud formation in tobacco tissue even in the presence of IAA (which normally acts as a bud inhibitor). This convinced Skoog and collaborators that nucleic acids which contain substances such as adenine influence tissue proliferation. With the collaboration of Miller and associates, he undertook experiments with nucleic acids extracted from herring sperm, calf thymus and yeast. In 1955, Skoog and collaborators (Miller et al. 1955) finally isolated from autoclaved yeast extract, a derivative of adenine (6-furyl arninopurine), named *kinetin*. A substance with kinetin-like properties was also detected in young maize endosperm (Miller 1961), which was isolated by Letham (1963) and named *zeatin*. It was also verified that a similar substance called *ribosylzeatin* occurred in coconut milk (Letham 1974). Now many synthetic as well as natural compounds with kinetin-like activity are known which show bud-promoting properties. These substances are collectively called *cytokinins* and are used to induce divisions in cells of highly mature and differentiated tissues (such as mesophyll or endosperm from dried seeds) to induce shoot differentiation even in the presence of auxin in cultures. Topolins are naturally occurring aromatic cytokinins and becoming popular in tissue culture studies because of positive effects of organogenic differentiation (Adeyami et al. 2012).

1.5 Hormonal Control of Organ Formation

Skoog and Miller (1957) proposed the concept of hormonal control of organ formation. Their classic experiments on tobacco pith cultures showed that root and bud initiation were conditioned by a balance between auxin and kinetin. High concentration of auxin promoted rooting whereas proportionately more kinetin initiated bud or shoot formation. Unequal proportion of auxin and cytokinin led to unorganised growth of the tissue. The discoveries of gibberellin and abscisic acid relatively altered this concept as a multiplicity of factors are responsible for control of organ formation. The determination of organogenesis also depends upon the source of plant tissue, environmental factors, composition of media, polarity, growth substances and other factors and, therefore, may not be restricted to hormonal balance only. These aspects are discussed in Chapters 5 and 6.

1.6 Improvement of Media

The advances made in tissue culture technology are directly related to the mineral content and overall composition of the culture media. Initially, tissue culturists used Knop's mineral solution (Gautheret 1942, Nobécourt 1937) and White's medium supplemented with various trace elements. Gradually, other workers tried to develop media differing essentially in mineral content. For example, Murashige and Skoog (1962) proposed a solution for tobacco tissues with a concentration of salts 25 times higher than in Knop's solution. This medium allowed five to seven times more active growth than other media. Following this conclusive improvement, other researchers suggested new media for very special purposes. Increase in NO_3 content was proposed to facilitate flower initiation (Margara et al. 1967), whereas Zn and B were considered important in the formation of whole plants. The various media generally used in tissue culture of plants are described in Chapter 3 and in some other chapters, where considering their subject of topic media with special nutrient requirements have been mentioned.

For large-scale cultures and germplasm cryopreservation alternatives to gelling agent agar, mostly used in media preparation, have been suggested for reduction of culture media costs (Saad and Elshahed 2012, Lalitha et al. 2014)

1.7 Preparation and Cloning of Single Cell Cultures

During the period 1934-1954, the main focus remained on the growth and development of callus cultures from explants under *in vitro* conditions. However, the principles of tissue culture envisaged by Haberlandt on the basis of cell theory and the related concept of totipotency were sidetracked. Two aspects of these principles required experimental verification: first, the multiplication of a single cell *in vitro* and, second, the development of a whole plant from proliferated tissue of this cell.

Sanford et al. (1948) initiated studies on single cell cultures by demonstrating divisions in animal cells using conditioned media (media in which tissue has been growing for sometime). The isolated cell was placed in a very thin glass tube and the tube plunged into a medium containing numerous living cells. Some unknown substance excreted by the surrounding cells in the medium influenced the isolated cell in the tube to divide. Muir (1953) demonstrated that on transfer of callus tissues of *Tagetes erecta* and *Nicotiana tabacum* to a liquid medium and subsequent agitation in a shaking machine it was possible to break down the callus tissue into single cells and small cell aggregates. The following year Muir and associates (1954) applied the 'conditioning

principle' to induce divisions in isolated single cells of *Tagetes* and tobacco growing in shake (suspension) cultures. A single isolated cell was placed on a filter paper positioned on top of a callus mass. About 8% of these isolated single cells on the filter paper multiplied and formed colonies (Fig. 1.1). Earlier, Steward et al. (1952) had designed a shaking apparatus which allowed dissociation of tissues to form cell suspension cultures. In these cultures crowds of isolated single cells or cell clumps were formed which could further multiply while remaining suspended in the medium under constant shaking. With improvement in culture vessel designs like bioreactors cells from suspension cultures are cultivated on mass scale which happens to be important basic requisite for industrial production of secondary metabolites (Chapter 17). Plant regeneration in species which earlier did not respond to tissue cultures has also been possible using suspension cultures. Additionally cell suspensions have increasing role in protoplast isolation, somatic hybridization /cybridization and genetic transformation.

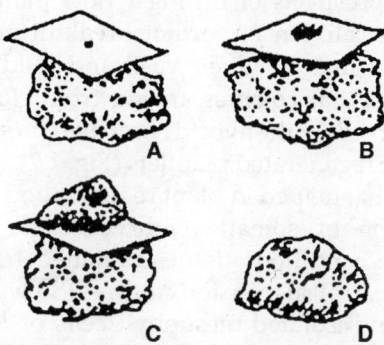

Fig. 1.1 First example of single-cell culture technique diagrammed from the work of Muir et at. (1954). A: Single cell placed on a small piece of filter paper positioned on top of a callus tissue; B: Isolated cell multiplies; C: Division of cells leads to the development of a callus tissue; D: Callus regenerated from single-cell culture.

The plating technique for cloning a large number of isolated single cells was developed by Bergmann in 1960. He filtered suspension cultures of *Nicotiana tabacum* and *Phaseolus vulgaris* and obtained a population of nearly 90% free cells. This was followed by blending these free cells in a medium containing 0.6% agar, not completely cold, and finally plating in a petri dish where the medium solidified. In this experiment some of the single plated cells divided and formed visible colonies. This technique was later widely used, particularly for cloning isolated single protoplasts.

Jones et al. (1960) designed a microculture method using hanging drops of free cells in a conditioned medium. This facilitated continuous observation of cells growing in cultures. Using the hanging drop method, Vasil and Hildebrandt (1965) substituted fresh medium enriched with coconut milk and NAA (1-naphthaleneacetic acid) for 'conditioning' and observed divisions in isolated tobacco hybrid cells.

1.8   Regeneration of Single Cell to Whole Plant

The success achieved thus far in single cell cloning created great enthusiasm for research on the prospects of raising whole plants from a single cell. An important breakthrough was achieved in 1965 when Vasil and Hildebrandt observed that colonies arising from cloning of isolated cells of the hybrid *Nicotiana glutinosa × N. tabacum* regenerated plantlets (Fig. 1.2). Another aspect that helped in plant regeneration was the induction of somatic embryogenesis. Under suitable conditions tissues obtained from free cells in carrot suspension cultures (Steward et al. 1966) and isolated mesophyll cells of *Macleaya cordata* (Kohlenbach 1966) differentiated somatic embryos. These embryos on subsequent culture developed into normal plants. The phenomenon of somatic embryogenesis leading to plantlet formation in cultures was later reported in many species and its potential application in synseed technology presently is well recognized (Nower 2014, Siew et al. 2014, *see* also Chapter 7). All these discoveries contributed to the establishment of totipotency of somatic cells under experimental conditions, thereby accomplishing the goals set by Haberlandt.

1.9   Practical Applications and Recent Advances

Soon after working out the theoretical aspects of *in vitro* cultures, plant biologists made efforts to realise the practical applications of plant tissue culture technology. Success achieved has been spectacular since tissue culture techniques can now be gainfully employed to a range of plant groups, e.g., cereals and grasses, legumes, medicinals, vegetable crops, oil seeds, fruit crops, floriculture, clonal forestry, etc. (see Vasil and Thorpe 1994, Bhojwani and Razdan 1996, Smith 2000, 2013). From a survey of the literature, it is apparent that principal applications are based on advancements made in the areas of morphology, biochemistry, pathology and genetics and agricultural biotechnology.

1.9.1   Morphological Aspects

Tissue culture, to begin with, would be a good means for understanding the factors responsible for cell differentiation and organ formation. Based on histological observations of callus development, Trécul (1853) pointed out that establishment of a tissue culture involves the process of cell dedifferentiation. A differentiated cell is induced *in vitro* to divide, thereby providing a method for dedifferentiation. Buvat (1944, 1945) devised a scheme according to which dedifferentiation followed two steps: (1) regression to the cambial stage and (2) return to the cytological structure of primary meristematic cells. The first stage affects highly differentiated cells while the second is accompanied by the organisation of bud or root meristems.

Cell cultures provide valuable information on morphogenesis and plant development. Studies on molecular, physiological and biochemical aspects of cells in culture have contributed to an in-depth understanding of cytodifferentiation, organogenesis and somatic embryogenesis (Fukuda 1997, Thomson and Thorpe 1997, Zhou et al. 2000, Kondo et al. 2015).

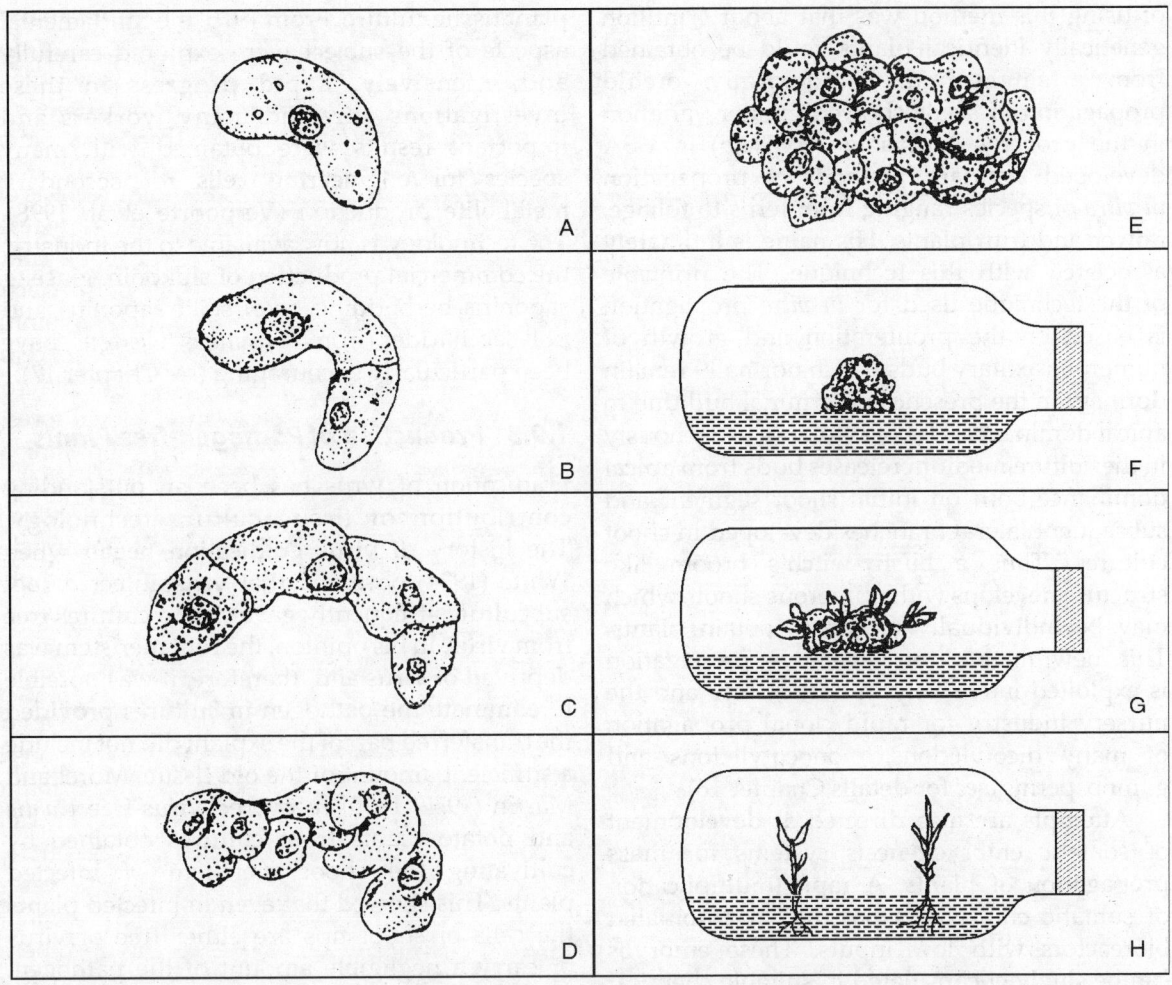

Fig 1.2 Regeneration of a single cell to whole plant. A: Isolated single cell cultured *in vitro*; B-E: Multiplication of this cell leading to the development of a callus; F: Subculture of the callus on a nutrient medium solidified with agar; G: Proliferation of shoots in the callus cultured on a medium containing a high amount of cytokinin; H: Shoots transferred onto a medium without growth substances to induce root formation; I: *In vitro* regenerated plant transplanted to soil inside a pot (diagrammed from the work of Vasil and Hildebrandt 1965).

Several theories have been proposed for the organ formation in tissue cultures. Of these, hormonal control of organogenesis, proposed by Skoog and Miller in 1957, is the most popular. This has already been described (Section 1.4). The regeneration of whole plants in tissue cultures may occur via shoot or root differentiation. Alternatively, the cells may undergo embryogenic development to give rise to bipolar embryos, also called *embryoids*.

Another important morphological application of plant tissue culture is micropropagation. Small amounts of tissue can be used to raise hundreds or thousands of plants in a continuous process. Ball (1946) successfully raised transplantable whole plants of *Lupinus* and *Tropaeolum* by culturing their shoot meristem (tip). The potential for this work was soon realised by Morel (1963) for rapid propagation of orchids *Cymbidium* and *Odontoglossum*. The advantage of using this method was that about 4 million genetically identical plants could be obtained from a single bud. Paradoxically, orchid propagation of seeds showed greater variation in the progeny. Murashige (1978a,b) in USA developed standard methods of propagation *in vitro* of species ranging from ferns to foliage, flower and fruit plants, His name is intimately associated with this technique. The principle of the technique used for *in vitro* propagation is based on the proliferation and growth of numerous axillary buds which normally remain dormant in the presence of terminal bud due to apical dominance. Use of cytokinin exogenously in the culture medium releases buds from apical dominance both on initial shoot segment and subsequent lateral branches developed in shoot cultures. Thus a bushy-witch's broom like structure develops with numerous shoots which may be individually rooted to obtain plants. This new method of vegetative propagation is exploited intensively in horticulture and the nursery industry for rapid clonal propagation of many dicotyledons, monocotyledons and gymnosperms (*see* for details Chapter 16).

Attempts are also directed at development of somatic embryogenesis systems for mass propagation of plants. A rapid multiplication of somatic embryos is possible in automated bioreactors with low inputs. These embryos can be singly encapsulated in suitable chemical compound for use as 'synthetic seeds' (*see* Chapter 7). Different types of bioreactors have been tested and successfully used to scale-up embryogenesis, e.g., carrot, poinsettia (Redenbaugh 1993), or other plant cell and tissue cultures (Eibl and Eibl 2008).

1.9.2   Production of Secondary Metabolites

The industrial production of secondary metabolites using cell cultures was initiated during the period 1950-1960 by the Pfizer Company with the assistance of G.N. Nickell, a distinguished expert on plant tissue culture. These early attempts were disappointing and little progress was made on this aspect of applied plant tissue culture. From 1975, the fundamental aspects of the subject were explored carefully and intensively. Rapid progress in these investigations attracted many workers and important results were obtained with many species for engineering cells for secondary metabolite production (Verpoorte et al. 1998). The technology is now available to the industry; the commercial production of shikonin, ginseng saponins, berberidine, taxol, seiko saponins, and polysaccharides (from *Polyanthes tuberosus*) have been particularly encouraging (*see* Chapter 17).

1.9.3   Production of Pathogen-free Plants

Eradication of virus has been an outstanding contribution of tissue culture technology. The history of virus eradication began when White (1934) observed that virus-infected root subcultures frequently gave rise to cultures free from virus. In his opinion, the root meristem was deprived of virus and, therefore, it was possible to eliminate the pathogen in cultures provided the transferred part of the explant did not include a sufficient amount of the old tissue. Morel and Martin (1952, 1955) recovered virus-free *Dahlia* and potato plants from cultures obtained by cultivating the shoot meristem of infected plants. This showed that even in infected plants the cells of shoot tips are either free of virus or carry a negligible amount of the pathogen. This technique is economical and used very frequently. Meristem-tip culture is often coupled with thermotherapy or chemotherapy for virus eradication (*see* Chapter 15). Cryotherapy and electrotherapy are also emerging as potential techniques of virus elimination (Wang et al. 2009, Falah et al. 2009)

1.9.4   Germplasm Conservation

Plant germplasm is traditionally stored in the form of seeds, tubers, roots, bulbs, corms, rhizomes, buds, cuttings, etc. Field plots, nurseries and greenhouses are other sources of maintaining germplasm or gene banks. A number of difficulties are, however, encountered while applying these modes of conservation. Important crop species produce recalcitrant seeds with early embryo degeneration (Reed and Chang 1997, Pritchard et al. 2000). Besides, the maintenance of field collections, nurseries or greenhouses are expensive with the further disadvantage of plants being vulnerable to insects, pathogens and various climatic hazards. About 2000-60,000 species of higher plants are believed endangered, rare and threatened with extinction (Bajaj 1995). According to FAO (2009) most of these are crop plants. Plant tissue culture is being developed as an effective means of germplasm conservation since a low maintenance *in vitro* germplasm storage collection provides a cost effective alternative to growing plants under field collections, nurseries, or greenhouses (Taylor 1997, Razdan and Cocking 1997, 2000; Engelmann 2012). Conservation of plant genetic resources *in vitro* is dealt comprehensively in Chapter 18.

1.9.5 Genetic Manipulations

The role of cell and tissue culture in plant genetic manipulations has been increasingly recognised. Heredity variations can be observed in cell colonies or plants regenerated *in vitro*, which may later express at the time of vegetative multiplication or sexual reproduction. The principal aspects of genetic manipulation are outlined below.

(I)   Genetic Variability

Pioneering investigations (Gautheret 1955, Nobécourt 1955) on tissue cultures showed that cells in long-term cultures are genetically unstable. Many workers have confirmed that a tissue colony is a mosaic of different kinds of cells. Some of these, when cultivated separately, produce unorganised colonies while others regenerate plants frequently affected by morphological variations. Thus, tissue cultures are a direct source of genetic variability. Caryological investigations of cultured cells have revealed abnormal mitosis. The unstable

chromosome numbers ultimately result in the variability of tissue cultures leading to somaclonal and gametoclonal variant selection (*see* Chapter 14) and commercial release of some crop varieties.

(II)   *In Vitro* Pollination and Fertilisation

In the 1960s, the Botany School at the University of Delhi became actively engaged with the *in vitro* culture of reproductive organs of flowering plants. The initial success with intraovarian pollination led to the subsequent development of test-tube fertilization or *in vitro* Fertilization (IVF). This technique, developed by Kanta et al. (1962), involves culturing excised ovules and pollen grains in the same medium to make it easier for the germinating pollen to fertilise the ovule under *in vitro* conditions. Through the application of this technique the incompatibility barriers existing at the sexual level can be overcome, particularly those for which reaction occurs at the stigma or style. This method has been successfully employed to overcome self-incompatibility in *Petunia axillaris*. Similarly, interspecific (*Melandrium album* × M. *rubrum*) and intergeneric (*M. album* × *Silene schafta*) hybrids, hitherto unknown in nature, were obtained by Zenkteler and co-workers in 1975. Obtention of germinable embryos from 'naked zygote' developed *in vitro* by fusion of male and female gametes (Kranz and Lörz 1993) has been a landmark of this technique. IVF system for rice using electrofusion is also reported to be successful (Okamoto 2011).

(III)   Induction of Haploidy

Another major contribution by the Botany School of Delhi University that attracted worldwide attention is the application of tissue culture to synthesise haploid plants. Normally, somatic cells of higher plants have a diploid chromosome number while reproductive cells (gametes) are haploid. In 1966, Guha and Maheswari cultured immature anthers of *Datura innoxia* (Solanaceae family) and were able to raise embryoids and plantlets. These plants turned out to be haploid; apparently they had been released from microspores within the anthers. This opened the field of androgenesis. In the following year Bourgin and Nitsch (1967) confirmed the totipotency cf pollen grains by raising full haploid plants of tobacco, rice and wheat. Haploids by anther culture are now reported to have been raised in 250 species belonging to 40 families (Mishra and Goswami 2014). Experimental evidence points out that the pollen populations are basically dimorphic and only 'smaller' pollen are capable of forming haploids. These pollen occur in low frequency and appear to be different from the majority destined to form gametes (Rashid 1983, Cho and Zapata 1990).

Induction of haploid plants from unpollinated ovaries and ovules (gynogenesis) is another subsequent advance in plant tissue culture and experimental embryology. San Noeum (1976) reported her first result on *in vitro* culture of ovary isolated from *Hordeum vulgare* and, thereafter, many other workers have obtained gynogenic haploid plants from cultured ovaries and ovules of tobacco, wheat, rice, lilium, maize, mulberry and other plants (*see* Chapter 8). This demonstrates that not only the microspore, but also the megaspore or female gametophyte of angiosperms can be triggered *in vitro* to sporophytic development, thus making way for an alternative approach to haploid plant breeding. It may be noted, however, that in some cases anther culture has given rise to diploid, polyploid and aneuploid plants. Due to these erratic and sporadic results, this technique cannot supplant the conventional methods of plant breeding.

(IV) Somatic Hybridisation

Isolation, regeneration and fusion of plant protoplasts *in vitro* is another area with potential for genetic manipulation. Somatic hybridisation, an asexual hybridisation procedure using isolated somatic protoplasts (wall-less cells), is now being developed as a new tool in plant breeding. Very early,

Klercker (1892) isolated protoplasts and the first fusion was achieved in 1909 by Küster, a noted cytologist, who established that some salt solutions facilitated this process. This gave an indication that hybridisation by fusion of protoplasts could be a distinct possibility in future. Products of fusion of two protoplasts (heterokaryon) could be cultured to regenerate a somatic hybrid plant of desired genotype. New genetic variations might also be introduced as a consequence of cytoplasmic interactions during the process of fusion in protoplasts. The tempo of somatic hybridisation as a technique in higher plants accelerated after the pioneering work of Professor Edward C. Cocking (1960) in the UK, who demonstrated the feasibility of isolating a large number of protoplasts by incubating a small amount of tissue with the enzyme cellulase. Soon divisions were reported in isolated protoplasts leading to regeneration of plants (Nagata and Takebe 1971). Having established totipotency in plant protoplasts, the next endeavour was to fuse the regenerating protoplasts *in vitro* to produce hybrid plants. Carlson et al. (1972) obtained the first somatic hybrid by fusion of protoplasts isolated from *Nicotiana glauca* with *N. langsdorffii*. This attracted many people to this field of research and attempts were made to obtain somatic hybrid plants from species between which hybrids could not be produced by means of sexual crosses. Some success has been achieved in obtaining somatic hybrid plants between sexually compatible and incompatible species; the relative merits of this technique are discussed in Chapter 12. However, the role of cybridisation in bringing about subtle genetic variations by partial genome transfer in plants has been increasingly recognised.

(V) Genetic Transformation

The current importance of genetic transformation of cells by uptake of exogenous DNA has generated enormous interest in harnessing the advantages offered by plant tissue and cell culture technology. This consists of four steps: insertion, integration, expression and replication of foreign DNA inside the host cell. The principle of transformation is based on the mechanism of infection by viruses, but cell function is so greatly disturbed by this procedure that the host dies subsequently. Since 1976, several researchers have reported that various eukaryotic genes introduced in bacteria are able to express their specific activity in the host and that it may be possible to obtain a similar transformation of eukaryotic cells. In nature, *Agrobacterium tumefaciens* can transfer genetic information to plant cells in the form of a plasmid that promotes tumoural formation (crown gall). Using recombinant DNA technology it was possible to introduce the gene for kanamycin resistance to the plasmid DNA of *A. tumefaciens* and the modified plasmid was later incorporated into tobacco protoplasts. Plants from the transformed tobacco protoplasts ultimately expressed for kanamycin resistance.

Attempts have now been made to genetically modify plants by introducing other useful genes for crop improvement, such as resistance to insects, weeds, and plant diseases, or in study of mechanism of gene action. Genetically modified (GM) plants are obtained by DNA transfer via vector-independent or vector-dependent means. Direct DNA transfer by electroporation, electrofusion, microinjection, microprojectile bombardment are vector independent and used for transformation of protoplasts and bacterial cells. Some of these methods are also used in callus cultures. In vector-mediated gene transfer, the genes of interest are inserted on plant expression binary vectors (Schardl et al. 1987) in 'sense' or 'antisense' orientation and these gene constructions mobilised into agrobacterium for transformation of cells, tissues, organ cultures or whole plant parts (floral dip: Clough and Bent 1998). Vectors for RNAi interference as well as the expression of polygenic traits have been constructed in genetic manipulation of crops aimed at their improvement (Stewart 2008).

The development of tissue culture techniques has thus a long and complicated history. Initial efforts concentrated on understanding the various aspects of plant growth and development. Success regarding the plant regeneration *in vitro* was achieved only during the first quarter of previous century when basic methods for plant tissue culture were established. Soon it was discovered that this new area of plant biology had practical value for commercial and agricultural plant propagators as well as in conservation of valuable germplasm (Razdan and Cocking 1997, 2000). Plant tissue culture is currently finding increased applications in study of molecular approaches followed in biotechnology. Success achieved in commercial plantings of GM crops (corn, soybean, cotton, canola and potato) and production additives (enzymes from genetically engineered microorganisms) for use in processed products like soft drinks, cakes, cheese, bread, meat, fish, etc. account for 90% common foodstuffs likely to contain GM components in USA (see Trigiano and Gray 2000, 2011). The production of GM crops has, however, been a matter of concern in certain parts of the world and issues involved are debatable. Yet plant tissue culture methods play a vital role in biotechnological improvement of plants particularly the agricultural crops. GM cotton (Bt cotton) introduced in India is estimated to result in increase of the yield by 50% (300kg/ha in 2002-2003 to 567kg/ha in 2007-2008), whereas in China GM crops with reduced (70%) application of pesticides have been produced (see Kumar and Loh 2012). These examples and other examples mentioned in Chapter 13 are claimed to have contributed significantly to socio-economic benefits for farmers in some countries.

Laboratory Organisation

The laboratory set-up for tissue culture depends on the nature of the research undertaken and the availability of funds. Regardless of the fact that the techniques employed may be for simple plant propagation, synthesis of secondary metabolites, or genetic manipulations, the basic facilities that should be available to an individual or group for tissue culture usually include;

(a) Washing and storage facilities.
(b) Media preparation, sterilisation and storage room.
(c) Transfer area for aseptic manipulations.
(d) Culture rooms or incubators for maintenance of cultures under controlled conditions of temperature, light and humidity.
(e) Observation or data collection area.

The layout for organisation of a tissue culture laboratory is shown in Fig. 2.1.

2.1 Washing and Storage Facilities

The main requirement for washing and storage is an area sufficient to accommodate large sinks (some lead-lined to resist acids or alkalis) with provision for running water, draining-boards or racks and ready access to a de-ionised, distilled and double-distilled water apparatus. Space should also be available to set up drying ovens, washing machines, plastic or steel buckets for soaking labware, acid or detergent baths, pipette washers, driers and cleaning brushes. For storage of washed and dried labware, the washing area may be provided with dustproof cupboards or storage cabinets.

A tissue culture facility requires good quality running cold and hot water as well as provision for waste water disposal. The disposal of any waste should be governed by local municipal codes and international norms to minimize health risks and environmental hazards. It is mandatory that maximum cleanliness is maintained in all areas within the organization of a tissue culture set-up.

2.1.1 Cleaning Glassware

In tissue culture, glassware should be resistant to heat. Generally, good-quality borosilicate glass (preferably Pyrex, Corning, Kimax, or Kimble) is recommended as it also appears to be less prone to breakage and scratching. The conventional method of cleaning laboratory glassware is to soak it for 4 hr in sodium dichromate-sulphuric acid (concentrated), then wash vigorously under a jet of tap water. Given the corrosive nature of this acid and the consequent need to adopt measures for protection of the skin and eyes, detergents are presently preferred. The glassware is soaked in a detergent solution for 16 hr, then rinsed first in tap water followed by a second rinse in distilled water. If washing is done mechanically, in a

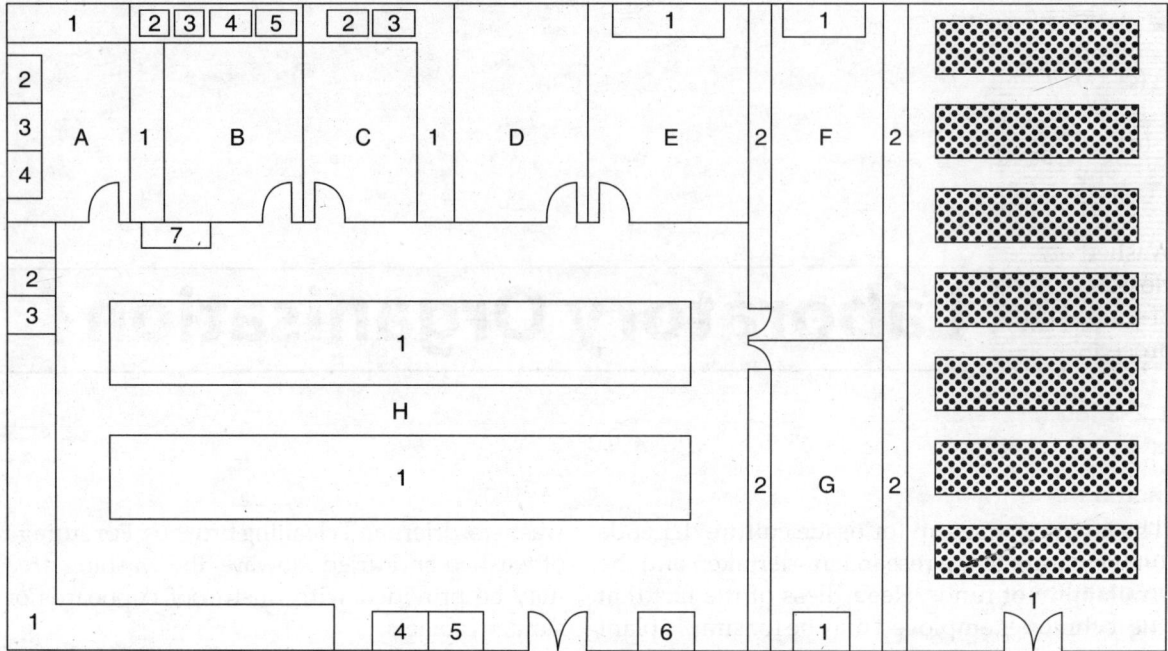

Fig. 2.1 Layout for organisation of a tissue culture laboratory. A: Media preparation and sterilisation room with lab bench (1), space for autoclave (2), water bath (3), heat-dry sterilisation unit (4); B: Washing area with lab bench (1), double-basin, lead-lined sink with provision for running water taps (2), space for detergent baths, brushes etc. (3), draining-boards/racks (4), drying ovens (5); C: Dark room with lab bench (1), refrigerator and deep-freeze space (2), double-basin, lead-lined sink (3); D: Storage room for chemicals and glassware; E: Transfer room fitted with overhead UV and HEPA filter ventilation unit, laminar air-flow hood (1) for aseptic manipulations; F–G: Temperature-controlled culture room with shaking machines installed (1), culture racks (2) fitted with light-adjustable fluorescent tubes; H: Main laboratory with lab bench and tables (1) for microscopy work, data collection etc., centrifuge (2), pH meter (3), hot plate/stirrers (4), balances (5), space for buckets and used labware (6), double-distillation unit (7) within close access to A and B; I: Transplantation area with controls for light, temperature and humidity (modified from Torres 1989).

dishwasher, the glassware may be treated with detergent at 70°C for 5-10 min. Further cleansing involves a 5 min rinse in hot water, 3 min rinse in de-ionised water, and a final hand rinse in distilled water.

To dry the cleansed glassware, it is placed in an oven at high temperature or else air-dried. Sometimes dry agar sticks to the sides of used glass culture vessels. It may be removed by melting at low temperature in an autoclave before the treatment with detergent.

2.1.2 Using Plastic Labware

A wide range of presterilised disposable polystyrene culture containers (e.g., Falcon, Corning, Nalgene and Nalge Co.) are now available and used in place of glassware in order to dispense with washing. However, the increased costs of the disposable plastic containers have led to the introduction of reusable plastic labware. These plastics may be washed with a mild non-abrasive detergent and rinsed with tap and distilled water. The procedure for washing must be matched to the type of plastic under consideration, which may include:

(a) Rinsing with an organic solvent such as alcohol. Acetone, chloroform, or methylene chloride cannot be used for acrylic.

(b) Boiling in dilute sodium bicarbonate. This treatment is not generally recommended for polycarbonate, polyethylene, acrylic and polystyrene.

(c) Soaking for less than 4 hr in a chromic acid-sulphuric acid bath.

(d) Soaking in 1 N HCl for 8 hr or less followed by another soak in 1 N HNO$_3$ and rinsing in distilled water.

Washed and dried labware is finally stored in a closed cupboard or cabinet to avoid the collection of dust. Additional information about cleaning procedures is summarised in Appendix 2.1.

2.2 Media Preparation Room

The area earmarked for media preparation should have ample storage and bench space for chemicals, labware, culture vessels, closures and miscellaneous equipment required for media preparation and dispensing. There should be provision for placing hot plates or stirrers, pH meters, balances (e.g., top-loading balance, 0.001-400.00 g; semi-microbalance, 0.00001-160.0 g), water baths and bunsen burners with a gas source. Space can also be provided for them in the main laboratory. A microwave oven, autoclave, or domestic pressure cooker for sterilising media, culture vessels and instruments are appliances most needed for media preparation. Other requirements to be included in the media room are vacuum sources, refrigerator and freezer for safe storage of chemicals and media stocks.

2.3 Transfer Area

Tissue culture techniques can be successfully performed on an open laboratory bench under very clean and dry atmospheric conditions. It is very essential, however, that all precautions be taken to prevent entry of any contaminant into the culture vial during the process of subculture or inoculation.

The simplest type of transfer area is an enclosed plastic box (glove box) which can be sterilised with an ultraviolet (UV) light and by swabbing the floor surface with 95% ethyl alcohol when in operation. Ordinarily a small wooden hood may also be used for tissue culture work. The hood has a glass or plastic door either sliding or hinged. The door should be dustproof. The culture hoods can be conveniently placed on a bench in a quiet corner of the laboratory. To carry out the operations, the worker inserts his hands through two holes in front of the chamber. The transparent glass or plastic doors enable him to see inside and make the necessary manipulations.

The glove box has limited applications as relatively few transfers can be effected in it. When a large number of cultures or transfers are being manipulated, large equipment is required and the most desirable arrangement is a dust-free room equipped with an overhead UV and positive-pressure ventilation unit. The ventilation should possess a high-efficiency particulate air (HEPA) filter. It has been observed that a 0.3μm HEPA filter shows 99.97-99.99% efficiency in not allowing the bacteria to pass through. All surfaces in the room are so constructed and designed that no dust or micro-organisms can accumulate. Further, the surfaces can be thoroughly cleansed and disinfected regularly. The room is air-conditioned and provided with sources of electricity, gas, compressed air and vacuum.

Another type of transfer area found convenient and currently used in most of the laboratories is a laminar airflow cabinet (*see* Fig. 4.1). A small motor blows air into the unit first through a coarse filter, where large dust particles are separated, and subsequently passes through a 0.3 μm HEPA filter. The air is directed either downward (vertical flow unit) or outward (horizontal flow unit) over the working surface. The air coming out of the fine filter is ultraclean (free from fungal or bacterial contaminant) and its velocity (27 ± 3 m/min) adequately prevents the microcontamination of the working area by a worker sitting in front of the cabinet. Various other contaminants (hair, salts, flakes) are also blown away by the ultraclean airflow. In this way an aseptic environment is maintained as long as the cabinet is switched on.

Laminar airflow cabinets are commercially available in various sizes and shapes. The principal advantage of working in these cabinets is that the flow of air does not hamper the use of a spirit lamp or bunsen burner and each cabinet occupies a relatively small space within an ordinary laboratory. Under climatic conditions (tropical or subtropical) where atmospheric dust is very high, it may, however, be advisable to house the airflow cabinet in a culture room fitted with double doors, in order to prolong the effective life of the filter. An important precaution to note is that a laminar airflow cabinet should never face a window or door that is frequently used.

2.4 Culture Room

Environmental factors greatly influence the process of growth and differentiation of tissues in cultures. All types of plant tissues are therefore incubated under conditions of well-controlled temperature, humidity, illumination and air circulation. Anther, low density suspension, and protoplast cultures are affected significantly under *in vitro* environmental conditions. A typical culture room should have both light and temperature programmable for a 24-hr period. Usually air-conditioners and heaters are used to maintain the temperature around 25 ± 2°C. Lighting is adjusted in terms of quantity and photoperiod duration by using automatic clocks, Cultures are generally grown in diffused light (less than 1 klx*) although provision should also be made for maintaining cultures in higher light intensity as well as total darkness. Further prerequisites are a humidity range of 20-98%, controllable to ± 3%, and uniform forced-air ventilation.

Specially designed shelves made of glass, rigid wire mesh, or wood are provided in the culture room for storing cultures. Each shelf is illuminated by a separate set of fluorescent tubes**. Insulation between the lamps and shelves above ensures a more even temperature around the cultures. Individual shelves may also be ventilated by fitting a small fan at one end of the shelf and blowing air through a plastic pipe running along the length of the shelf. Holes drilled along the sides of the pipe at appropriate distances will further allow an even airflow and thus prevent build-up of hot air in the shelves due to lamps.

The culture vessels (flasks, jars and petri dishes) can be placed directly on the shelf or trays of suitable size whereas culture tubes require metallic racks to hold them. A label giving details of the experiment (name of the plant, explant, medium, date of culture and other information) is stuck on each tray or rack to ensure identity and for recording monitored results. Installation of a shaking machine (horizontal or rotary) in the culture room is also advisable. Shakers with controlled temperature and light are commercially available.

Incubators, large plant growth chambers, and walk-in environmental rooms have generally the range of control and flexibility desirable for growth and development of plant tissues under *in vitro* conditions. Besides, they occupy less space and are readily available in the market. The general characteristics of incubators and growth chambers are summarised in Appendix 2.2.

It is most essential that precautionary measures be taken against a major power breakdown. Failure of electricity may ruin important experiments and suspension cultures

---

* The light can be measured in foot-candles (fc; full sunlight measures 10,000 fc), or microeinsteins (μE) per second per square metre (1 μE·sec^{-1}·m^{-2} = 6.02×10^{17} photons^{-1}·m^{-2} = μmol·sec^{-1}·m^{-2}); the range of light readings can be 40-200 fc or 20-100 μE·sec^{-1}·m^{-2}. A meter to measure foot-candles and a quantum radiometer-photometer light meter to measure in nucroeinstein may be used to measure the light level (see Smith 2000).

** As alternative to fluorescent tubes, LED (light emitting diode) source may be tested for growing cultures considering that power consumption by LED panels is comparably less and these can be configured to deliver lights of specific wavelengths (blue 460nm, red 630nm) to plants growing *in vitro* for chlorophyll absorption (Kumar and Loh 2012).

may die due to stoppage of shakers. A generator should be kept as a stand-by and emergency power points in the culture room, incubators, or growth chambers attached to it in order to maintain the necessary light and temperature conditions. The equipment most commonly used for tissue culture work is listed in Appendix 2.3.

2.5 Data Collection Area and Specialised Facilities

The growth and development of tissues cultured *in vitro* are generally monitored by observing cultures at regular intervals in the culture room or incubators where they have been maintained under controlled environmental conditions. Data based on the observations under aseptic conditions may be collected using a laminar airflow cabinet or suitable transfer area designed for the purpose, whereas for the microscopic examination of cultures the usual laboratory space is used. For germplasm conservation, it is essential to have lower temperature rooms and installation of cryopreservation units is a must (Engeimann 1994). Low temperature rooms can also be used for cold storage of temperate (2-4°C) and tropical (Ca 15°C) plants for commercial purpose.

2.6 Transplantation Area

Plants regenerated from *in vitro* tissue cultures are transplanted to soil in pots. The potted plants are ultimately transferred to greenhouses or growth cabinets and maintained for further observations under controlled conditions of light, temperature and humidity.

Prior to transfer of potted plants to greenhouse the acclimatisation of these plants may begin in growth chambers under mist benches, or in a standard greenhouse bay (Gamborg and Phillips 1995). Even if funds are a deterrent for providing greenhouse facilities, it is, however, essential that minimum facility (a small chamber in potting room) for maintaining high humidity by fogging, misting, or a fan system for cooling in summer, and heating system for winter, are provided in a potting room adjacent to the tissue culture laboratory (*see* Fig. 2.1).

Care needs to be taken that potting mix does not get too wet during fogging or misting and water drops do not accumulate on plantlets. To prevent this individual pots with plants are covered with clear plastic bags or groups of potted plants kept under the cover of plastic sheets. The humidity is then gradually decreased to ambient level and after 7-15 days light intensity increased. The potted plants are finally exposed to greenhouse conditions (Fasella and Hussain 2014).

APPENDIX 2.1

Additional Guidelines for Washing and Handling of Glassware

General

1. Reusable glassware for tissue culture should be emptied immediately after use and soaked. Never allow media or agar to dry on the glassware.
2. Segregate all glassware containing corrosive chemicals or fixatives from the rest of the tissue culture glassware.
3. All glassware contaminated or coming into contact with micro-organisms should be decontaminated before washing.
4. Remove labels and marking ink before washing. Marking ink may be wiped out by abrasive cleanser or acetone and immediate rinsing.
5. Melted agar in the culture vessels should be poured into a collecting sieve and discarded.
6. Rubber-lined screw caps should be soaked only in distilled water. Never soak caps in detergent or soap solution.
7. Glassware for growing cells should be acid-cleaned if proteinaceous deposits are not removed by conventional washing.
8. Silicone-treated glassware should be permanently labelled and placed separately from the rest of the glassware.

Glass-washing

1. Tissue culture glassware should be soaked in 5% detergent solution for a minimum of 1 hr.
2. The automatic washer is used only as a rinsing unit and no detergent is added to the washer. Glassware should be treated with detergent prior to its transfer into the washing unit.
3. The automatic washer is programmed for no less than six tap water rinses and a 1 min distilled water rinse.
4. The glassware that does not fit into an automatic unit is rinsed by hand. Hand-rinsing requires a minimum of six thorough tap water rinses and four rinses in quality water.
5. If an automatic dish-washing facility is not available, a simple washing machine with brushes of different sizes maybe used for manual washing.

Acid-cleaning

1. Personnel concerned should wear a full-face mask and acid-resistant apron and gloves.
2. Handling of acid and processing of glassware in the acid should be done in a fume hood.
3. Never add water to acid. To dilute acid, add acid slowly to water while stirring.
4. After acid-cleaning, if the solution in the acid bath becomes dark coloured, it should be discarded.

APPENDIX 2.2

General Characteristics of Commercially Available Incubators or Growth Chambers

1. Temperature range, 2–40°C.
2. Temperature control, ± 0.5°C.
3. Safety high- and low-temperature limits.
4. Continuous temperature recorder.
5. Twenty-four-hour temperature and light programming.
6. Adjustable fluorescent lighting up to 10,000 lx.
7. Relative humidity range, 20-98%.
8. Relative humidity control, ±3%.
9. Uniform forced-air distribution.
10. Capacity up to $0.7 m^3$ of $0.5 m^2$ shelf space.

APPENDIX 2.3

Equipment Commonly Used in a Laboratory for Many Tissue Culture Techniques[a]

Preparation of Media

1. Gas, water, and electricity supplies.
2. Provision for compressed air* and vacuum lines.*
3. Water heater or small stove.
4. Hot plate with magnetic stirrer, vortex.
5. Glass or stainless steel containers for heating and dissolving media.
6. Autoclave or pressure-steam steriliser,
7. pH meter.
8. Centigram balances (see Section 2.2).
9. Graduated measuring cylinders, flasks, beakers, petri dishes and pipettes with teats.
10. Culture tubes, bottles and other glassware with suitable closures (such as cotton plugs, aluminium foil, plastic film, or metal caps).
11. Small transfer instruments such as spatulas, knife, scalpels, forceps, or dissecting needles.
12. Hot-air oven for rapid heating of media and agar mixtures (microwave oven will also do rapid thawing of frozen items).
13. Water de-ioniser, distilled or double-distilled water units.
14. Filter membranes with holders and hypodermic syringes to filter-sterilise solutions.
15. Chemicals for preparing culture media or commercially available powdered culture media, growth hormones and other organic constituents.
16. Detergents, disinfectants.
17. Pipette washers (acidproof).
18. Drying and draining racks.
19. Storage space for chemicals, glassware, nutrient media, sterile water and other items.

---

[a]Additional requirements for single cell and/or protoplast work have not been included.
*For work on a small scale these items may not be absolutely necessary.

Isolation of Cultures

1. Laminar airflow cabinet.
2. Spirit lamp or bunsen burner in the inoculation cabinet.
3. Stereo-microscope.
4. Ethyl alcohol (96%) for sterilisation and flaming of small metal instruments.
5. Tiles or glass plates for use during sterile cutting.
6. Hypochlorite solution for sterilisation of plant material.

General Equipment

1. Refrigerators (temp, range as low as 4°C), deep-freeze (-80°C).
2. Automatic dish-washer.*
3. Glassatomiser.
4. Dispensing devices (e.g., wire-mesh baskets, trolleys with trays and metal racks for holding test-tubes or culture vials in the autoclave).
5. Acidproof baths for cleansing glassware.
6. Microscopes (e.g., compound, inverted) with microphotographic equipment.
7. Markers, labels and vitafilm (or similar material for wrapping culture vessels, glassware, tiles and other labware).
8. Light meter for measurement of light intensity.

Culture Room

1. Temperature control (17-27°C).
2. Electricity supply essential for lighting, cooling and heating.
3. Shelves for culture racks.
4. Fluorescent tubes for lighting.
5. Timer for regulating day-length.
6. Racks for culture vials.
7. Rotary shaker.
8. Observations table.

Transplantation Facility

A small area where high humidity, light and temperature can be controlled in the form of a greenhouse or a commercially available growth chamber.

---

* For work on a small scale these items may not be absolutely necessary.

3 Media

Growth and morphogenesis of plant tissues *in vitro* are largely governed by the composition of the culture media. Although the basic requirements of cultured plant tissues are similar to those of whole plants, in practice, nutritional components promoting optimal growth of a tissue under laboratory conditions may vary with respect to the particular species. Media compositions are therefore formulated considering specific requirements of a particular culture system. For example, some tissues show better response on solid medium while others prefer a liquid medium. Considerable progress has been made on the development of media for growing plant cells, tissues and organs aseptically in culture. The present chapter deals with aspects of the plant tissue culture media that influence the growth of various types of cells or explants and further help to achieve the regeneration of plants.

3.1 Media Composition

The principal components of most plant tissue culture media arc inorganic nutrients (macronutrients and micronutrients), carbon source(s), organic supplements, growth regulators and a gelling agent. Some tissues grow on simple media containing only inorganic salts and utilisable carbon (sugar) source, but for most others it is essential to supplement the medium with vitamins, amino acids and growth substances. Very often, complex nutritive substances are added to the medium. Such a medium, composed of 'chemically defined' compounds, is referred to as a 'synthetic' medium.

A number of media have been devised for specific tissues and organs. These are described in relevant chapters of this book. White's medium is one of the earliest plant tissue culture media originally formulated for root culture. To induce organogenesis and regeneration of plants in cultured tissues, MS (Murashige and Skoog 1962) and LS (Linsmaier and Skoog 1965) media contain the desired salt composition and are widely used. B5 medium, originally designed for cell suspension or callus cultures, has, with modifications, proved valuable for protoplast culture. It may also be useful in regeneration of protoplast-derived plants. Another medium, designated N6 has been developed especially for cereal anther culture and other tissues, though in experiments on anther culture the medium devised by Nitsch and Nitsch (1969) is frequently used. Similarly, E1 medium found use for the culture of soybean, red clover and other legume species in addition to other media. Nutritional components of these media support rapid growth of cells for embryogenesis and in protoplast culture. Success in employing these various media in all probability lies in the fact that the ratios as well as the concentrations of nutrients (Table 3.1) nearly match the optimum requirement with regard to the growth and differentiation of respective cell or tissue systems.

The concentration of inorganic and organic constituents in the plant tissue culture media is expressed either in mass values (mg/1 or ppm, though mg 1^{-1} is now acceptable) or mole* values (1 M or 1 mol l^{-1}). According to the recommendations of the International Association for Plant Physiology, mmol l^{-1} should be used for expressing the concentration of macronutrients and organic nutrients and μmol l^{-1} for micronutrients, hormones, vitamins and other organic constituents. The advantage in using mole values is that the number of molecules per mole is constant for each compound and while following a particular recipe, during the preparation of a medium, the original mole value can be used irrespective of the number of water molecules in the salt sample. This cannot be done when concentrations are expressed in mass values.

3.1.1 Inorganic Nutrients

A variety of mineral elements (salts) supply the needed macro- and micronutrients required in the life of a plant. Elements required in concentrations greater than 0.5 mmol l^{-1} are referred to as macronutrients and those less than 0.05 mmol l^{-1} as micronutrients.* When mineral salts are dissolved in water they undergo dissociation and ionisation. In a medium one type of ion may be contributed by more than one salt. For example, NO_3 ions are contributed by NH_4NO_3 and KNO_3 in MS medium and K^+ ions by KNO_3 and KH_2PO_4. Hence it is useful to

compare various media by evaluating the total concentration of ions present in them, to obtain some idea about the suitability of a medium towards a particular tissue or culture system. A comparative statement about the concentration of ions for the plant tissue culture media (inorganic constituents) listed in Table 3.1 is presented in Table 3.2.

(I) Macronutrients

The macronutrients include six major elements; nitrogen (N), phosphorus (P), potassium (K), calcium (Ca), magnesium (Mg) and sulphur (S) present as salts that constitute various media. All are essential for plant cell and tissue growth. Culture media should contain at least 25 m mol l^{-1} nitrate and potassium. However, considerably better results are obtained if the source for nitrogen in media is contributed by both nitrates and ammonium (2-20 mmol l^{-1}) or any other reduced nitrogen source. In case ammonium only is used, there is need to add one or more tricarboxylic acid (TCA) cycle acids (e.g., citrate, succinate, or malate) so that any deleterious effect due to ammonium concentrations in excess of 8 mmol l^{-1} in the medium is diluted. When nitrate and ammonia ions are present together in the culture medium, the latter are used more rapidly. Other major elements, Ca, P, S and Mg, at concentrations in the range of 1-3 mmol l^{-1}, appear adequate provided other requirements are satisfied.

(II) Micronutrients

Inorganic elements required in small quantity but essential for plant cell and tissue growth constitute microelements. These are iron (Fe),

---

* 1 mole = 1 gram molecular weight (i.e., formula weight of a substance in grams). The formula weight is equal to the sum of the weights of the atoms in the formula of a substance.
1 molar (1 M or 1 mol l^{-1}) = one litre of solution containing 1 mole of a substance or 1 mol. wt. in g/l. It may also be expressed as 1000 or 10^3 mrnol l^{-1} = 1,000,000 or 10^{-6} μmol l^{-1}.
1 millimolar (1 mM) = one litre of solution contains 0.001 mole of a substance or 0.001 mol l^{-1} or mol.wt. in mg/l.
1 micromolar (1 μM) = one litre of solution contains 0.00001 mole of a substance or 0.00001 mol l^{-1} or 10^{-3} mg in mol.wt./l.
1 microgram (1 μg) = 0.001 or 10^{-3} mg = 0.00001 or 10^{-t6} g.
parts per million (ppm) = 1 mg l^{-1}.

Table 3.1 Nutritional components of some plant tissue culture media[a]

Components	Amount (mg l^{-1})[b]					
	White's	MS	B5	Nitsch's	N6	E1
Macronutrients						
$MgSO_4 \cdot 7H_2O$	750	370	250	185	185	400
KH_2PO_4	—	170	—	68	400	250
$NaH_2PO_4 \cdot H_2O$	19	—	150	—	—	—
KNO_3	80	1900	2500	950	2830	2100
NH_4NO_3	—	1650	—	720	—	600
$CaCl_2 \cdot 2H_2O$	—	440	150	—	166	450
$(NH_4)_2 \cdot SO_4$	—	—	134	—	463	—
Micronutrients						
H_3BO_3	1.5	6.2	3	—	1.6	3
$MnSO_4 \cdot 4H_2O$	5	22.3	—	25	4.4	—
$MnSO_4 \cdot H_2O$	—	—	10	—	3.3	10
$ZnSO_4 \cdot 7H_2O$	3	8.6	2	10	1.5	2
$Na_2MoO_4 \cdot 2H_2$	—	0.25	0.25	0.25	—	0.25
$CuSO_4 \cdot 5H_2O$	0.01	0.025	0.025	0.025	—	0.025
$CoCl_2 \cdot 6H_2O$	—	0.025	0.025	0.025	—	0.025
KI	0.75	0.83	0.75	—	0.8	0.8
$FeSO_4 \cdot 7H_2O$	—	27.8	—	27.8	27.8	—
$Na_2EDTA \cdot 2H_2O$	—	37.3	—	37.3	37.3	—
EDTA Na ferric salt	—	—	43	—	—	43
Sucrose (g)	20	30	20	20	50	25
Organic supplements						
Vitamins						
Thiamine HCl	0.01	0.5	10	0.5	1	10
Pyridoxine HCl	0.01	0.5	1	0.5	0.5	1
Nicotinic acid	0.05	0.5	1	5	0.5	1
Myoinositol	—	100	100	100	—	250
Others						
Glycine	3	2	—	2	—	—
Folic acid	—	—	—	0.5	—	—
Biotin	—	—	—	0.05	—	—
pH	5.8	5.8	5.5	5.8	5.8	5.5

[a] Growth regulators and complex nutrient mixtures used by various workers are not included here as the quantity and constituents of these compounds vary for specific tissue and organ.

[b] Abbreviations and references:

White's (1953; *Am. J. Bot.* 40: 517-524, or 1963; The Cultivation of Plant and Animal Cells, 2nd Edition, The Ronald Press, NY, 228 pp.).

MS (Murashige and Skoog 1962; *Physiol. Plant.* 15: 473).

B5 (Gamborg et al. 1968; *Exp. Cell Res.* 50: 151).

Nitsch's (Nitsch and Nitsch 1969; *Science*, N.Y., 163: 85).

N6 (Chu 1978; *Proc. Symp. Plant Tissue Culture*, Science Press, Peking, p. 43).

El (Gamborg et al. 1983; *Plant Cell Rep.* 2: 209).

Note: LS (Linsmaier and Skoog 1965; *Physiol. Plant.* 18: 100-127) medium has same composition as MS, but differs in concentrationof thiamine (6.4 mg l^{-1}); inositol (100 mg l^{-1}) is used instead of other vitamins and glycine.

Table 3.2 Concentration of ions for the media listed in Table 3.1

Ions	Media					
	White's	MS	B5	Nitsch's	N6	E1
Macronutrients (mmol l^{-1})						
Ca	1.27	2.99	1.02	1.49	1.1	3.0
Cl	0.87	5.98	2.04	2.99	—	—
K	1.66	20.05	25.00	9.90	31.00	0.20
NH$_4$	—	20.62	2.00	9.00	6.6	7.50
NO$_3$	3.33	39.41	25.00	18.40	28.0	27.50
PO$_4$	0.14	1.25	1.1	0.50	2.94	1.80
SO$_4$	—	1.8	—	—	0.2	1.6
M	3.04	1.5	1.0	0.75	0.75	1.6
Micronutients (μmol l^{-1})						
B	24.20	100.00	48.50	161.80	26.00	50.00
Co	—	0.10	0.10	—	0.11	0.10
Cu	0.04	0.10	0.10	0.10	—	0.10
Fe	12.50	100.00	100.00	100.00	100.00	100.00
L	4.50	5.00	4.50	—	4.80	4.50
Mn	22.40	92.50	60.00	112.00	19.50	60.00
Mo	0.08	1.00	1.00	1.00	—	1.00
Zn	10.40	30.00	7.00	34.70	5.2	7.00

manganese (Mn), zinc (Zn), boron (B), copper (Cu) and molybdenum (Mo). Of these, iron seems more critical since chelated forms of iron and zinc are commonly used in preparing culture media. Iron tartrate and citrate are difficult to dissolve and frequently precipitate in the medium. This problem may be overcome by using diaminetetraacetic acid (EDTA)-iron chelate in place of iron citrate, particularly for embryo induction. EDTA chelates are not entirely stable in liquid culture and the procedure developed by Steiner and van Winden (1970) may be followed for preparing an iron chelate solution that will not precipitate.

Certain media are enriched with cobalt (Co), iodine (I) and sodium (Na), but strict cell growth requirements of these elements have not been established. Generally 0.1 μmol l^{-1} Cu and Co, 1 μmol l^{-1} Fe and Mo, 5 μmol l^{-1} I, 5-30 μmol l^{-1} Zn, 20-90 μmol l^{-1} Mn and 2-5100 μmol l^{-1} B are added to culture media depending upon the requirement of the experiment.

3.1.2 *Carbon and Energy Source*

The most preferred carbon source in plant tissue culture is sucrose. Glucose supports equally good growth, while fructose is less efficient. Sucrose, while autoclaving the medium, is converted into glucose and fructose, In the process, first glucose is used and then fructose. Other carbohydrates, such as lactose, galactose, rafinose, maltose, cellobiose, melibiose and trehalose, generally yield inferior results. Tissue cultures of *Sequoia* and maize endosperm can even metabolise starch as the sole carbon source.

Plant cells and tissues in the culture medium lack autotrophic ability and therefore need external carbon for energy. Even tissues which are initially green or acquire green pigments under special conditions during the culture period are not autotrophs for carbon. The addition of an external carbon source to the medium enhances proliferation of cells and regeneration of green shoots.

A partial hydrolysis of sucrose occurs when media are autoclaved. Further, cultures grow better on media with autoclaved sucrose than on media with filter-sterilised sucrose. This suggests that cells benefit from the ready supply of glucose and fructose brought about by hydrolysis of autoclaved sucrose. Use of autoclaved fructose is not recommended as it could be detrimental to the growth of tissue. It is interesting to note that dextrose (3-4%) as carbon source has been found most effective for *in vitro* propagation of radish (*Raphanus sativus L.*) var. Beeralu Rabu (Manawadu and Dahanayake 2015).

3.1.3 Organic Supplements

(I) Vitamins

Plants synthesise vitamins endogenously and these are used as catalysts in various metabolic processes. When plant cells and tissues are grown *in vitro*, some essential vitamins are synthesised but only in suboptimal quantities. Hence it is necessary to supplement the medium with required vitamins and amino acids to achieve the best growth of the tissue. Thiamine (B_1), nicotinic acid (B_3), pyridoxine (B_6), calcium pentothenate (B_5), and myoinositol are used more often; of these, thiamine is the basic vitamin required by all cells and tissues. Nicotinic acid and pyridoxine are usually added to culture media but may not be essential for cell growth in many species. Other vitamins such as biotin, folic acid, ascorbic acid, pantothenic acid, vitamin E (tocophenol), riboflavin and *p*-aminobenzoic acid are also invariably used, particularly when cells are grown at very low population densities, although their requirement by plant cell or tissue cultures is apparently negligible. Generally, these vitamins are added in the range of 0.1 to 10.0 mg l^{-1}.

(II) Amino Acids

Cultured tissues are normally capable of synthesising the amino acids necessary for various metabolic processes. In spite of this, the addition of amino acids to media is important for stimulating cell growth, protoplast culture, and for establishing cell lines. Unlike inorganic nitrogen, amino acids are taken up more rapidly by plant cells. Casein hydrolysate (0.05-0.1 %), L-glutamine (8 mmol l^{-1}), L-asparagine (100 mmol l^{-1}), L-glycine (2 mmol l^{-1}), L-arginine and L-cysteine (10 mmol l^{-1}) are common sources of organic nitrogen used in culture media. Tyrosine (100 mmol l^{-1}) should be used only when agar is added to the medium. Amino acids added singly prove inhibitory to cell growth while their mixtures are frequently beneficial. Supplementing the medium with adenine sulphate can stimulate cell growth or enhance shoot formation. In similar manner, vitamin myoinosital in culture media is likely to stimulate cell growth because of its breakdown to ascorbic and pectin for incorporation into phosphoinositides and phosphotidyl-inositol (Saad and Elshahed 2012).

(III) Other Organic Supplements

Culture media are often supplemented with a variety of organic extracts which have constituents of an undefined nature. These include protein (casein) hydrolysates, coconut milk, yeast and malt extracts, ground banana, orange juice and tomato juice. In tissue culture the success achieved with the use of coconut milk (5 to 20%) and protein (casein) hydrolysates (0.05 to 1.0%) has been significant. Similarly, potato extract has been found a suitable medium for anther culture. Generally, the use of natural extracts is avoided because the quality and quantity of growth-promoting constituents vary with age of the tissue from which the extract has been obtained, thereby affecting the reproducibility of results. Moreover, it may be desirable to replace these substances with a defined organic substance, such as a single amino acid. For example, L-asparagine could effectively replace yeast extract and tomato juice in the medium for callus culture of maize endosperm. L-glutamine alone has demonstrated favourable tissue responses in several species and enrichment by a mixture of amino acids or fruit extracts was found unnecessary. Coconut milk from six month old fruits in MS medium, in place of growth regulators, reportedly increases shoot length and produces multiple shoots in cultures of the potato variety 'Desiree' and hence found useful for micropropagation of this crop (Muhammad et al. 2015).

(IV) Activated Charcoal

The addition of activated charcoal (AC) to culture media is reported to stimulate growth and differentiation in orchids, carrot, ivy and tomato. Paradoxically, its effect on tobacco, soybean and *Camellia* has proved inhibitory. Inhibition of growth is attributed to the adsorption of phytohormones to AC whereas stimulation could be due to any one of the factors, namely, adsorption of inhibitory compounds to AC and darkening of the medium. AC is generally acid-washed and neutralised before its addition at concentrations of 0.5-3% to the culture medium. It also helps to reduce toxicity by removing toxic compounds (e.g., phenols) produced during the culture and permits unhindered cell growth (see Mohamed-Yaseen et al. 1995).

(V) Antibiotics

Addition of antibiotics to culture media is generally avoided because their presence in the medium retards the cell or tissue growth. However, some plant cells have a systemic infection of micro-organisms. To prevent the growth of these microbes it is essential to enrich the media with antibiotics. Streptomycin or kanamycin at low concentration effectively controls systemic infection and media supplemented with these antibiotics do not adversely inhibit the growth of cell cultures.

For gene cloning and transformation experiments using *E. coli* and *Agrobacterium*, besides these antibiotics, rifampicin, tetracycline, carbenicillin, gentamycin sulfate, hygromycin B, cefotaxime, chloramphenicol, augmentin, etc. are used at varying concentrations depending on the vector and selection of the transformant (Gamborg and Phillips 1995, Smith 2000).

3.1.4 Growth Regulators

Four broad classes of growth regulators, namely auxins, cytokinins, gibberellins, and abscisic acid, are important in tissue culture. The growth, differentiation and organogenesis of tissues become feasible only on the addition of one or more of these classes of hormones to a medium. The ratio of hormones required for root or shoot induction varies considerably with the tissue, which seems directly correlated to the quantum of hormones synthesised at endogenous levels within the cells of the explant. The joint action of auxins and cytokinins seems to regulate cell cycle. Auxing exert an effect on DNA replication, while cytokinins exercise some control on events leading to mitosis (Pasternak et al. 2000).

(I) Auxins

Media are supplemented with various auxins: 1 H indole-3-acetic acid (IAA), 1-naphthalene-acetic acid (NAA), 1 H-indole-3-butyric acid (IBA), 2,4-dichlorophenoxyacetic acid (2,4-D) and naphthoxyacetic acid (NAA). IAA occurs naturally in the plant tissues. There are other auxins which have proven particularly effective in plant cell culture. These include 4-chlorophen-oxyacetic acid (4-CPA) or *p*-chlorophenoxyacetic acid (PCPA), 2,4,6-trichlorophenoxyacetic acid (2,4,6-T), 2-methyl 4-chlorophenoxyacetic acid (MCPA), 4-amino-3,5,6-trichloropicolinic acid (picloram), and 3,6-dichloro-2-methoxybenzoic acid (dicamba). A common feature of auxins is the property of inducing cell division. In nature the hormones of this group are involved with such activities as elongation of stem, internodes, tropism, apical dominance, abscission and rooting.

Various auxins differ in their physiological activity and the extent to which they move through tissue or are bound to the cells. Based on stem curvature essays, 2,4-D has 8 to 12 times the activity of IAA compared to 2,4,6-T, PCPA, and picloram, which are four times or twice as effective as IAA. 2,4-D and 2,4,6-T are mostly employed to induce callus production. Auxins are generally dissolved either in ethanol or dilute NaOH. Auxin transduction pathway depicts that each target plant cell is presumed to possess receptors (auxin-binding proteins) which are able to detect hormonal signals initiating molecular events chain leading to physiological response (Hagan et al. 2004). Topolins, naturally occurring cytokinins, particularly metatopolin [mT or 6-(3-hydroxybenzylamins-purine)], have shown positive effects like high rate of shoot multiplication, better rooting, and acclimatization of plants raised in tissue cultures (Adayemi et al. 2012). They are likely to gain increasingly the application for the growth and morphogenesis of cell and tissues in cultures.

(II)   Cytokinins

Cytokinins are adenine derivatives which are mainly concerned with cell division, modification of apical dominance and shoot differentiation of the tissue in culture. The most frequently used cytokinins are 6-benzylaminopurine (BAP) or 6-benzyladenine (BA), 6-γ-γ-dimethylamino-purine (2-ip),** N-(2-furfurylamino)l-H-purine-6-amine (kinetin), and 6-(4-hydroxy-3-methyl-trans-2-butanylamino) purine (zeatin). Zeatin and 2-ip are naturally occurring cytokinins while BA and kinetin are synthetically derived cytokinins. They are generally dissolved in dilute HC1 or NaOH.

It was subsequently discovered that N,N′-diphenylurea (DPU), thidiaziron (TD2), N-2-chloro-4-puridyl-N-phenyl urea (CPPU) and other derivatives of diphenyl urea show the cytokinin-type activity (Pierik 1989, Murthy et al. 1998). The specific effect of these urea-based compounds on the growth of plant tissues calls for further investigation.

The ratio of auxins and cytokinins is important with respect to morphogenesis in the culture system. For embryogenesis, callus

induction, and root initiation the requisite ratio of auxins to cytokinin is high, while the reverse leads to axillary and shoot proliferation. Equally important is the concentration of these two groups of growth regulators. It has been observed that 2,4-D and BA together at a concentration of 5.0 mg l^{-1} promote callus formation in *Agrostis* but if used at 0.1 mg l^{-1} they promote shoot formation, although in both cases the ratio of auxin to cytokinin is one. The mechanism of cytokinin action is not clearly understood although there are reports about the presence of compounds with cytokinin activity in transfer RNA (tRNA). Cytokinins have also been shown to activate RNA synthesis and to stimulate protein and enzyme activity in certain tissues.

(III)   Gibberellins and Abscisic Acid

These growth compounds are occasionally used in tissue culture. In some species these hormones are required to enhance and in others to inhibit growth. GA_3 is the most common gibberellin used of the over 20 known gibberellins. It promotes the growth of cell cultures at low density, enhances callus growth and induces dwarf or stunted plantlets to elongate. Stock solutions of GA_3 should be made up fresh and filter-sterilised before addition to the medium. Abscisic acid (ABA) in the culture medium either stimulates or inhibits the callus growth depending on the species. Reportedly useful in embryo culture, ABA is heat stable but light sensitive (Smith 2000).

3.1.5 Solidifying Agents

Gelling or solidifying agents are commonly used for preparing semisolid or solid tissue culture media. In static liquid cultures the tissue or cells become submerged and die due to lack of oxygen. Gels provide a support to tissues growing in static conditions. Agar, a polysaccharide obtained from seaweeds, has several advantages over other gelling agents. First agar gels do not react with media constituents. Secondly, they

---

** Or N^6-(β^2-isopentenyl) adenine.

are not digested by plant enzymes and remain stable at all feasible incubation temperatures. Normally, 0.5 to 1% agar is used in the medium to form a firm gel at the pH typical of plant cell culture media. In nutritional studies the use of agar is avoided because commercially available agar contains impurities in the form of Ca, Mg, K, Na and trace elements. Consequently, a change in the agar concentration also affects the nutrients present in it as well as the overall nutrient concentration in the experiment. Impurities can be removed, however, for such critical experiments by washing agar in doubled-distilled water for at least 24 hr, rinsing in ethanol and drying at 60°C for 24 hr.

Gelatin at high concentrations (10%) has been tried as a gelling agent but has limited use because it melts at low temperature (25°C). Other compounds successfully tested include methacel, alginate, phytagel and Gelrite. Use of guargum or isabgol as gelling agent (Jain and Babbar 2005) and market sugar as carbon source (Agrawal et al. 2010) in plant tissue culture medium is reported to be cost effective. Other plant derived gelling agents namely cornflour (*Zea mays* var. *amylacea*), arrow root (*Maranta arundinacea*) and cassava powder (*Manihot esculenta*) have all been used as cost effective substitute for agar (Lalitha et al. 2014). The search for more cheap alternatives provide that wheat (white) flour, potato starch, rice powder may be as good gelling agents as agar. A combination of laundry starch, potato starch and semolina in a ratio of 2:1:1 reduces costs by more than 70% in comparison to gelling agents normally used (Saad and Elshahed 2012). The FMC Corp. has developed a highly purified agarose called Sea Plaque *(k)*, which can be used in the recovery of single protoplasts from cultures. Perforated cellophane, filter paperbridges, filter paper wicks, polyurethane foam and polyester fleece are alternative methods of support used in the medium for cell or tissue growth.

The advantage of working with synthetic gelling compounds is that they form clear gels at relatively low concentrations (1.25-2.5 g l^{-1}) and are valuable aids for detecting contamination that may develop during the span of cultures. Whether explants grow better on agar or other supporting agents, depends on the tissue and the species.

3.1.6 pH

Plant cells and tissues require optimum pH for growth and development in cultures. While preparing a medium, the pH can be adjusted to the requirement of an experiment. The pH affects uptake of ions and for most of the culture media pH 5.0 to 6.0 before sterilisation is considered optimal. Higher pH is likely to give a hard medium while a low pH results in unsatisfactory solidification of the agar. Most tissue culture media are poorly buffered, resulting in pH fluctuations which could prove detrimental to long-term survival and growth of single cells or cell populations at low density. For single cell cultures of *Datura innoxia* and protoplast-derived colonies of *Santalum album,* the media buffered with 1 mM 2-(N-morpholino) ethane sulphonic acid (MES) or 1% PB74 in the normal pH range proved superior to unbuffered media.

3.2 Media Preparation

Media-making can be time-consuming. Nowadays the plant tissue culture media most commonly used are available in the market*** as dry powders. The simplest method of preparing media is to dissolve these powders containing inorganic and organic nutrients in some quantity of distilled water. After the contents have been thoroughly mixed in water, sugar and agar (melted) other organic supplements are added. Finally, the volume is made up to one litre. The pH is adjusted and the medium autodaved.

*** Standard plant tissue culture media as dry powders can be obtained from Flow Laboratories Ltd. (UK), Gibco (USA) and Hi-Media (India).

Powdered media are useful for propagation of plant species requiring nutrients according to the recipe of standard media. In experiments in which changes in the quantity and quality of media constituents become necessary, it is desirable to weigh and dissolve each ingredient separately before mixing them together. Another convenient procedure is to prepare stock solutions (Table 3.3) which, when mixed together in appropriate quantities, constitute a basal medium. Four stock solutions are prepared, consisting of (1) major salts (20 × concentrated), (2) minor salts (200 × concentrated), (3) iron (200 × concentrated) and (4) organic nutrients except sucrose (200 × concentrated). For each growth regulator a separate stock solution is prepared by dissolving it in a minute quantity of appropriate solvent (*see* Section 3.1.4) and then adjusted with distilled water to the desired volume to make an overall concentration of 1 mmol l^{-1} or 10 mmol l^{-1}. All the stock solutions are stored in proper plastic or glass containers at low temperatures. Iron stocks should be stored in coloured (amber) bottles. It is obligatory to shake the containers before use and any container with a contaminant or suspension in the form of precipitate must be discarded. Extra care is needed in storing coconut milk. The liquid extract (endosperm) from the fruit is boiled to deproteinise it, filtered and stored in plastic bottles in a deep-freeze at –20°C.

The sequence of events involved in preparing a medium is given in Appendix 3.1.

3.3 Selection of New Medium

In order to formulate a suitable medium for new systems, a broad spectrum approach may be desirable. To design broad spectrum experiments all components of the medium are divided into four categories: minerals, auxins, cytokinins and organic nutrients (such as sucrose, amino acids, or inositol). Each group of substances is prepared in three concentrations: low (L), medium (M), and high (H). According to De Fossard et al. (1974), various combinations

Table 3.3   Stock solutions for MS basal medium[a]

Constituents	Amount (mg l^{-1})
Stock solution I	
$MgSO_4 \cdot 7H_2O$	7400
KH_2PO_4	3400
KNO_3	38000
NH_4NO_3	33000
$CaCl_2 \cdot 2H_2O$	8800
Stock solution II	
H_3BO_3	1240
$MnSO_4 \cdot 4H_2O$	4460
$ZnSO_4 \cdot 7H_2O$	1720
$Na_2MoO_4 \cdot 2H_2O$	50
$CuSO_4 \cdot 5H_2O$	5
$CoCl_2 \cdot 6H_2O$	5
Stock solution III[b]	
$FeSO_4 \cdot 7H_2O$	5560
$NA_2 \cdot EDTA \cdot 2H_2O$	7460
Stock solution IV	
Inositol	20000
Thiamine HCl	100
Pyridoxine HCl	100
Nicotinic acid	100
Glycine	400

[a] To prepare one litre of medium take 50 ml of stock solution I, 5 ml each stock solutions II-IV.
[b] Dissolve $FeSO_4 \cdot 7H_2O$ and $Na_2 \cdot EDTA \cdot 2H_2O$ separately in 450 ml distilled water by heating and constant stirring. Mix the two solutions, adjust the pH to 5.5 and add distilled water to make up the final volume to one litre.

of these four categories of components at three different concentrations lead to an experiment with 81 treatments (Table 3.4). The best of these treatments is denoted by a four-letter code, e.g., MLMH (treatment composed of medium salts, low auxin, medium cytokinin and high organic nutrients). This may thus be selected as a new medium suitable for an untested system.

Mistakes in the preparation of media can do greater harm than any other fault in the tissue culture technique. It is absolutely essential that all steps in the media preparation and composition be followed carefully. To minimise error it is beneficial to prepare a reference sheet

Table 3.4 Constituents and their concentrations used in the broad spectrum experiment[a]

Constituents	Concentration range (mmol l^{-1})		
	Low	Medium	High
Minerals			
NH_4NO_3	5	10	20
KNO_3	—	10	20
KH_2PO_4	0.1	—	—
NaH_2PO_4	—	1	2
KCl	1.9	—	—
$CaCl_2$	1	2	3
$MgSO_4$	0.5	1.5	3
H_3BO_3	0.01	0.05	0.15
$MnSO_4$	0.01	0.05	0.1
$ZnSO_4$	0.001	0.02	0.04
$CuSO_4$	0.00001	0.0001	0.0015
Na_2MoO_4	0.00001	0.0001	0.001
$CoCl_2$	0.0001	0.0005	0.001
KI	0.0005	0.0025	0.005
$FeSO_4$	0.01	0.05	0.1
$Na_2 \cdot$ EDTA	0.01	0.05	0.1
Auxin	0.0001	0.001	0.01
Cytokinin	0.0001	0.001	0.01
Organic nutrients			
Inositol	0.1	0.3	0.6
Nicotinic acid	0.004	0.02	0.04
Pyridoxine HCl	0.0006	0.003	0.006
Thiamine HCl	0.0001	0.002	0.04
Biotin	0.00004	0.0002	0.001
Folic acid	0.0005	0.001	0.002
D-Ca-Pantothenate	0.0002	0.001	0.005
Riboflavin	0.0001	0.001	0.01
Ascorbic acid	0.0001	0.001	0.01
Choline chloride	0.0001	0.001	0.01
L-cysteine HCl	0.01	0.06	0.12
Glycine	0.0005	0.005	0.05
Sucrose	6	60	120

[a] De Fossard et al. (1974).

giving details of constituents and the sequence of steps to be followed during the preparation or composition of the particular medium. After a component is added or a step is completed, it should be cancelled on the sheet. While selecting a new medium the components in the standard tissue culture media may be tried.

APPENDIX 3.1

Sequence of Some Important Steps Involved in Preparing a Medium

Materials

1. Water should be glass distilled or demineralised to remove impurities.
2. Inorganic and organic compounds required must be of the highest grade (AnalaR Grade). Purification of growth substances may require charcoal decolorisation and recrystallisation using water-ethanol mixtures. Amino acids used have to be the L isomer.
3. To prepare semi-solid or solidified media Bacto agar from Difco is generally used.
4. The required quantity of agar and sucrose is dissolved in water (three-fourths of the final volume) by heating on a water bath or autoclaving at low temperature. This step is not necessary with a liquid medium because sucrose can dissolve even in lukewarm water.

Procedures

1. Vitamins and auxins may be added after autoclaving. Their solutions are adjusted to the desired pH and then sterilised by filtering through a microfilter of pore size 0.22-0.45 μm (manufactured by Millipore Intertech Inc., P.O. Box 255, Redford, MA 01730, USA).
2. Adjusting of pH is done by using 0.1 NaOH or 0.1 NHCl.
3. Only the medium in a uniformly liquid state is poured into the desired vessels (250-500 ml flasks) with suitable closures. If the medium starts to gel, the container with medium should be heated.
4. The medium should occupy three-fourths of the volume of the space inside the vessel. The vessels containing medium are then covered with aluminium foil and autoclaved at 120°C and 15 psi for 15-40 min. or less depending on the volume.
5. The autoclaved medium is allowed to cool to approximately 60°C and finally dispensed into sterilised vials aseptically.
6. The liquid medium may also be directly dispensed into culture vials occupying one-fifth the volume of its space, covered and then autoclaved.
7. Finally, media are brought down to room temperature and stored at 4°–10°C.

APPENDIX 3.2

Molecular Weights of Nutritional Components and Growth Regulators Commonly Used in Plant Cell and Tissue Culture Media

Compound	Chemical formula	Molecular weight
Macronutrients		
Ammonium nitrate	NH_4NO_3	80.04
Ammonium sulphate	$(NH_4)_2SO_4$	132.15
Calcium chloride	$CaCl_2 \cdot 2H_2O$	147.02
Calcium nitrate	$Ca(NO_3)_2 \cdot 4H_2O$	236.16
Magnesium sulphate	$MgSO_4 \cdot 7H_2O$	246.47
Potassium chloride	KCl	74.55
Potassium nitrate	KNO_3	101.11
Potassium dihydrogen *ortho*-phosphate	KH_2PO_4	136.09
Sodium dihydrogen *ortho*-phosphate	$NaH_2PO_4 \cdot 2H_2O$	156.01
Micronutrients		
Boric acid	H_3BO_3	61.83
Cobalt chloride	$CoCl_2 \cdot 6H_2O$	237.93
Cupric sulphate	$CuSO_4 \cdot 5H_2O$	249.68
Manganese sulphate	$MnSO_4 \cdot 4H_2O$	223.01
Potassium iodide	KI	166.01
Sodium molybdate	$Na_2MoO_4 \cdot 2H_2O$	241.95
Zinc sulphate	$ZnSO_4 \cdot 7H_2O$	287.54
Sodium EDTA	$Na_2 \cdot EDTA \cdot 2H_2O$ $(C_{10}H_{14}N_2O_8Na_2 \cdot 2H_2O)$	372.25
Ferrous sulphate	$FeSO_4 \cdot 7H_2O$	278.03
Ferric-sodium EDTA	$FeNa \cdot EDTA$ $(C_{10}H_{12}FeN_2NaO_8)$	367.07
Sugars and sugar alcohols		
Fructose	$C_6H_{12}O_6$	180.15
Glucose	$C_6H_{12}O_6$	180.15
Mannitol	$C_6H_{14}O_6$	182.17
Sorbitol	$C_6H_{14}O_6$	182.17
Sucrose	$C_{12}H_{22}O_{11}$	342.31
Vitamins and amino acids		
Ascorbic acid (vitamin C)	$C_6H_8O_6$	176.12
Biotin (vitamin H)	$C_{10}H_{16}N_2O_3S$	244.31
Calcium pantothenate (Ca salt of vitamin B_5)	$(C_9H_{16}NO_5)_2Ca$	476.53
Cyanocobalamine (vitamin B_{12})	$C_{63}H_{90}CoN_{14}O_{14}P$	1357.64

Contd...

Appendix 3.2 Contd.

Compound	Chemical formula	Molecular weight
L-Cysteine-HCl	$C_3H_7NO_2S \cdot HCl$	157.63
Folic and (vitamin B_c, vitamin M)	$C_{19}H_{19}N_7O_6$	441.40
Inositol	$C_6H_{12}O_6$	180.16
Nicotinic acid or niacin (vitamin B_3)	$C_6H_5NO_2$	123.11
Pyridoxine HCl (vitamin B_6)	$C_8H_{11}NO_3 \cdot HCl$	205.64
Thiamine HCl (vitamin B_1)	$C_{12}H_{17}ClN_4OS \cdot HCl$	337.29
Glycine	$C_2H_5NO_2$	75.07
L-Glutamine	$C_5H_{10}N_2O_3$	146.15
Glutathione	$C_{10}H_{17}N_3O_6S$	307.33

Hormones

Auxins

p-Chlorophenoxyacetic acid (p-CPA)	$C_8H_7O_3Cl$	186.59
3,6-Dichloro-o-anisic acid (Dicamba)	$C_8H_6Cl_2O_3$	221.04
2,4-Dichlorophenoxyacetic acid (2,4-D)	$C_8H_6O_3Cl_2$	221.04
Indole-3-acetic acid (IAA)	$C_{10}H_9NO_2$	175.18
3-Indolebutyric acid (IBA)	$C_{12}H_{13}NO_2$	203.23
2-Methyl-4-chlorophenoxyacetic acid (MCPA)	$C_9H_9ClO_3$	200.62
α-Naphthaleneacetic acid (NAA)	$C_{12}H_{10}O_2$	186.20
β-Naphthoxyacetic acid (NOA)	$C_{12}H_{10}O_3$	202.20
4-Amino-3,5,6-Trichloropicolinic acid (Picloram)	$C_6H_3Cl_3N_2O_2$	241.46
2,4,5-Trichlorophenoxyacetic acid (2,4,5-T)	$C_8H_4Cl_3O_3$	255.49

Cytokinins

Adenine (Ad)	$C_5H_5N_5 \cdot 3H_2O$	189.13
Adenine sulphate (AdSO$_4$)	$(C_5H_5N_5)_2 \cdot H_2SO_4 \cdot 2H_2O$	404.37
6-Benzyladenine or 6-benzylamino purine or (BA or BAP)	$C_{12}H_{11}N_5$	225.20
6-γ,γ- Dimethylallylamino purine or N-isopentenylamino purine (2-ip)	$C_{10}H_{13}N_5$	203.3
6-Furfurylamino purine (kinetin)	$C_{10}H_9N_5O$	215.21
6-(Benzylamino)-9-(2-tetrahydropyranyl)-9H-purine (SD8339)	$C_{17}H_{19}N_5O$	309.40
n-Phenyl-N-l,2,3-thiadiazol-5-urea (thidiazuron)	$C_9H_8N_4OS$	220.2
6-(4-Hydroxy-3methylbut-2-enylamino)-purine (zeatin)	$C_{10}H_{13}N_5O$	219.20

Gibberellins

Gibberellic acid (GA$_3$)	$C_{19}H_{22}O_5$	346.37

Contd...

Appendix 3.2 Contd.

Compound	Chemical formula	Molecular weight
Other compounds		
Abscisic acid	$C_{15}H_{20}O_4$	264.31
2′-Isopropyl-4′-(trimethylammonium chloride)-5′methylphenyl piperidine carboxylate (Amo 1618)		
α-Cyclopropyl-α-4-methoxyphenyl (ancymidol)	$C_{15}H_{16}N_2O_2$	256.3
β-Chloroethyltrimethyl ammonium chloride (CCC)	$C_5H_{13}Cl_2N$	158.07
Colchicine	$C_{22}H_{25}NO_6$	399.43
N-Dimethylaminosuccinamic acid (paclobutrazol)	$C_{15}H_{20}ClN_3O$	293.80
Phloroglucinol	$C_6H_6O_3$	126.11
1,4-Diaminobutane; tetramethylene-diamine (putrescine dihydrochloride)	$C_4H_{12}N_2 \cdot 2HCl$	161.1
N-(3-Aminopropyl)-1,4-butanediamine (spermidine)	$C_7H_{19}N_3$	145.2
N,N′-(Bis 3-aminopropyl)-1,4-butane-diamine (spermine)	$C_{10}H_{26}N_4$	202.3
2,3,5-Tri-iodobenzoic acid (TIBA)	$C_7H_3I_3O_2$	499.81

Aseptic Manipulation

The most important and rather difficult aspect of the *in vitro* techniques is the requirement to carry out various operations under aseptic conditions. Bacteria and fungi are the most common contaminants observed in cultures. These microbes are omnipresent in our environment and on coming into contact with the culture media they find conditions optimal and grow much faster than the cultured tissue. Consequently, the tissue is killed d ne to contamination. The contaminants may also give out metabolic wastes which are toxic to plant tissues. In view of this the maintenance of a completely aseptic environment inside the culture vial where tissue grows is absolutely essential. This chapter, therefore, deals with those measures that must be taken in order to prevent the contamination of plant tissue cultures.

4.1 Sterilising the Culture Vessels and Instruments

Glass culture vials, metal instruments and aluminium foil can all be sterilised by exposure to hot dry air (160°-180°C) for 2-4 hr in a hot-air oven. All items should be properly sealed before sterilisation. Sealing with paper is not advisable as it decomposes at high temperatures. Disadvantages of dry-heat sterilisation are poor circulation of air and slow penetration of

heat; therefore, sterilisation by *autoclaving* is recommended.

Autoclaving is a method of sterilising with water vapour under high pressure. Nearly all microbes are killed on exposure to the superheated steam of an autoclave. It is normal practice now to sterilise glassware and other accessories, such as cotton plugs, gauze, plastic caps, filters or pipettes in a commercially available autoclave at 121°C and 15psi for 15-20 min. Some types of plastic labware can also be repeatedly autoclaved, such as propylene, polymethyl pentene, polly-allomer, Tfzel, ETPE, and Teflon FEP. However, some polycarbonates show slow loss of mechanical strength with repeated autoclaving.

The instruments used for aseptic manipulations are usually made of metal. Normally, instruments such as forceps, scalpels, needles and spatulas are sterilised by dipping in 95% ethanol, followed by flaming and cooling, This technique is called *flame sterilisation*. Autoclaving of metallic instruments is generally avoided as they may rust and become blunt. During flame sterilisation the major concern should be the safety of alcohol because it is inflammable and if spilled near a flame wiil cause an instant fire. Alternatively, in place of flame sterilisation, dry sterilisation of instruments using steri pots may be practised in order to avoid instant fires.

4.2 Sterilisation of Nutrient Media

The nutrient media used in tissue cultures are commonly sterilised by *autoclaving* and *filter sterilisation*. Distilled water, micro- and macronutrients, and other stable mixtures are autoclaved, whereas the solutions that contain thermolabile compounds are filter-sterilised.

Culture media in glass containers sealed with cotton plugs, aluminium foils or plastic closures are autoclaved at 15 psi and 121°C for 15-40 min from the time the medium reaches the required temperature. Exposure time depends on the volume of the liquid to be sterilised (Table 4.1). The pressure should not exceed 20 psi as higher pressure may lead to the decomposition of carbohydrates and other components of a medium.

Table 4.1 Minimum time necessary for autoclaving nutrient media

Volume (ml)	Sterilisation time^a (min)
1-200	15
200-1000	30
1000-2000	40

^{a} At 121°C and 15 psi.

Vitamins, amino acids, plant extracts, hormones and carbohydrates are thermolabile and may decompose during autoclaving. The solutions of these compounds require filter sterilisation. In this process the solutions are passed through a bacteria proof membrane-filter under positive pressure. A Millipore or Seitz filter with a pore size of not more than 0.2 µm is generally used in filter sterilisation. Filter-sterilised solutions are then combined with other nutrient substances sterilised in the autoclave to give a complete medium. The procedure for filter sterilisation is given in Appendix 4.1.

4.3 Sterilising Culture Rooms and Transfer Area

Initially, the culture rooms are cleaned by gently washing all floors and walls with a detergent soap. This is followed by carefully wiping them with 2% sodium hypochlorite solution or 95% ethyl alcohol. Extreme care is needed to avoid stirring up any contamination that may have settled on their surfaces. Commercially available Lysol, Zephiran and Roccal are equally effective disinfectants used in the sterilisation of work surfaces and culture rooms. The process of sterilisation of culture rooms should be repeated at regular intervals.

The transfer area is also sterilised once or twice a month by washing with a commercial brand of antifungal spirocyte. However, larger transfer rooms are best sterilised by exposure to UV light. Since UV radiation is harmful to the eyes, sterilisation should be done when there are no experiments in progress. The time of sterilisation varies according to the size of the room. It may also be convenient to maintain an aseptic transfer area by installing an HEPA filter ventilation unit (*see* Chapter 2). Laminar airflow hoods are usually sterilised by switching on the hood and wiping the working surface with 95% ethyl alcohol 15-30 min before initiating any operation under the hood.

4.4 Aseptic Culture Technique

4.4.1 Sterilising Plant Material

All tissue cultures are likely to end up contaminated if the inoculum or explant used is not obtained from properly disinfected plant material. To obtain sterile plant material is difficult because in the process of sterilisation living materials should not lose their biological activity; only bacterial or fungal contaminants should be eliminated. Plant organs or tissues are, therefore, only surface-sterilised by treatment with a disinfectant solution at suitable concentrations for a specified period. The disinfectants most widely used and their concentrations in the solution are given in Table 4.2.

Fairly hard explants are treated directly with disinfectants. For example, in the culture of mature seeds or mature endosperm (euphor-

Table 4.2 Disinfectants used for sterilising plant material[a]

Disinfectant	Concen-tration	Duration of treatment[b] (min)
Benzalkonium chloride[c]	0.01-0.1%	5-20
Bromine water	1-2%	2-10
Calcium hypochlorite[d]	9-10%	5-30
Ethyl alcohol[d]	75-95%	—[e]
Hydrogen peroxide	3-12%	5-15
Mercuric chloride	0.1-1.0%	2-10
Silver nitrate	1%	5-30
Sodium hypochlorite[d]	0.5-5%	5-30

[a] In plants in which the infection of micro-organisms is systemic, the plant tissues or cells after treatment with disinfectant may be immersed in an antibiotic (40-50 mg l^{-1}) for 30-60 min. Explants may also be drenched in antifungal solutions to remove any fungal contaminants.

[b] Duration of sterilisation is inversely related to the concentration of the sterilant.

[c] BTC, Roccal, or Zephiran.

[d] Most widely used. Commercial bleach contains about 5% sodium hypochlorite and may be used at 10-20% v/v concentration.

[e] Several seconds to several minutes.

biaceous plant), whole seeds or decoated seeds are surface-sterilised. An explant that carries a heavy load of micro-organisms needs to be washed in running tap-water for 1-2 hr prior to its treatment with disinfectant solution. Ethyl alcohol or isopropyl alcohol is used to surface-sterilise delicate tissues such as shoot apices, pollen grains, and shoot or flower buds. Such explants are given a rinse in 70% ethanol for a few seconds and then left exposed in the sterile hood until the alcohol evaporates. Usually shoot apices or pollen grains are free from microcontaminants and may be used for inoculation without surface sterilisation. Addition of a few

drops of a surfactant (Triton-X or Tween-20) to the solution or treating the plant material in a solution of Cetavlon for 2 min before exposing to sterilant may enhance sterilisation efficiency.

4.4.2 Transfer of the Explant

Successful control of contamination largely depends upon the operator's technique while transferring the sterilised explant into the sterile culture vial containing nutrient media under aseptic conditions. During this process dust, hair, hands and clothes are potential sources of contamination, against which it is essential to take precautions. This is particularly important for commercial tissue culture for which the contamination problem could be hazardous. To minimise the risk of contamination in practice, it is advisable that the operator change into sterilised clothes and wear sterilised headgear before entering the sterile area. The most important and easy precautionary measure is to wash the hands with 95% ethyl alcohol before starting an experiment. Talking or sneezing while the culture material is being inoculated or transferred from one container to the other should be avoided. A further point to note is that cultured tissue should never touch inside the edges of culture vessel. Other measures as well as the steps involved in aseptic culture of plant tissues are summarised in Appendices 4.2 to 4.4.

The aseptic manipulation of plant tissues or cells *in vitro* is governed by a number of factors. Cleanliness and proper care of equipment are vital to the operation of a tissue culture laboratory besides controlling the contamination problem. Equally important are commonsense safety practices to reduce the possibility of accidents or injuries in a laboratory. This warrants following the rules on all aspects of aseptic manipulation rigorously.

APPENDIX 4.1

Procedure for Filter Sterilisation

1. The choice and size of membrane filter is governed by the volume of fluid to be filtered.
2. For small volumes (10–100 ml), the filters available in three sizes from Millipore are suitable.
3. The membranes are fitted into filter assembly holders of appropriate size and needle. The entire assembly along with filter is wrapped in aluminium foil and autoclaved.
4. Sterilisation temperature for filters is critical and should not exceed 121°C.
5. A graduated syringe (it need not be sterilised) carrying the liquid is fixed to one end of the sterilised filter assembly and the solution gradually pushed through the membrane present in the middle of the assembly.
6. The filtered sterile solution dripping out from the other end (needle) is either directly added to the medium (autoclaved) or collected in a sterilised jar. The filter-sterilised solution is transferred from the jar to the medium using a sterilised pipette. All operations are carried out in a laminar airflow hood.
7. Large filter assemblies may be used for sterilisation of a large volume of solution. For liquids in the range of 500 ml to one litre, stainless steel (47 µm) pressure filter holders are available; when more than one litre must be filtered, 142 µm filter holders have been found most efficient.
8. Membrane filter units manufactured by Nalgene, Corning and Bioquest are disposable and hence expensive because after use they are discarded. Usually small volumes (upto 500 ml) can be handled. The advantage of using these units is that they are completely assembled, sterilised and individually sealed; however, filtration is by negative pressure, which can cause foaming.
9. To facilitate passage of viscous materials (e.g., serum extract, embryo extract), the material is first filtered sequentially through a series of fibreglass filters (Dacron separators) of pore size 0.8, 0.45 and 0.22 µmn respectively.
10. Minimum pressure should be applied to obtain the desired flow rate. The pressure should never exceed 15 psi and normally the pressure exerted by air coming out from a 5% CO_2 cylinder is used to begin filtration.
11. Most membrane filters are impregnated with some quantity of detergent during the manufacturing process. The detergent is removed by passing distilled water at 80°C through the filter. This is followed by a cold saline rinse. The filter is then air-dried and used as per prescribed schedule.
12. Membrane filters should never be allowed to accumulate dust or other particles during air-drying. Used filter assemblies should be properly cleansed according to the manufacturer's instructions.

APPENDIX 4.2

Precautions Observed during Aseptic Manipulations in Tissue Culture Laboratory

General

1. Enter the laboratory wearing shoes and laboratory coat.
2. Avoid handling alcohols around open flames.
3. Never pipette by mouth.
4. Handle strong acids and alkalis with extreme caution.

5. Wash (preferably with alcohol) and bandage all cuts immediately.
6. Ensure before opening an autoclave that the pressure is reduced to zero and the temperature is below 100°C.
7. Switch off the electrical appliances (e.g., stirrers, pH meters, balances, hood) when not in use, particularly overnight.
8. Cover all equipment to avoid contact with dust and other contaminants.
9. Clean benches and work surfaces regularly; take care that dirty glassware or other items are kept away while work is in progress.
10. Mop the floor in the laboratory and the culture room with an approved disinfectant.
11. Separate space should be allocated for storage of items such as chemicals, photographic equipment, cleaned labware and other solutions.

APPENDIX 4.3
Operation of Laminar Airflow Hood (Fig. 4.1)

1. Switch on light and blower motor 15-30 min before making transfers.
2. Check to make sure that no paper or other object blocks the air intake at the bottom of the unit and that the sterile air coming out of the filter flows uniformly in the hood.
3. Use 95% ethanol to wipe the working surface of the hood.
4. Place only sterile instruments, appliances and culture tubes or vessels inside the hood.
5. During flame sterilisation of the small instruments keep spirit lamp or bunsen burner at a sufficient distance from the bottle containing alcohol to avoid fire.
6. After using the hood, rewipe its surface with ethanol, remove all appliances, dishes, or containers, and switch off the light and blower motor. Cover the hood with an airtight screen.

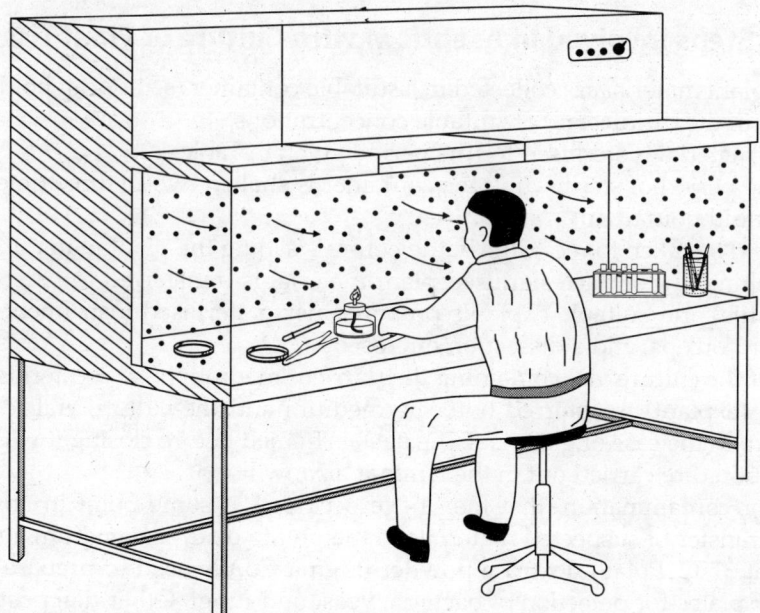

Fig. 4.1a Laminar airflow hood used in tissue culture.

Fig. 4.1b Removing lower epidermal peel from leaves under aseptic conditions in laminar airflow hood.

APPENDIX 4.4

Steps Involved in Aseptic *in vitro* Culture of Plant Tissues

1. Pieces of plant material are collected in a suitable container (screw-cap bottle) and immersed in a solution of the disinfectant at suitable concentrations.
2. The plant material is sterilised for the period given in Table 4.2.
3. During the period of sterilisation the container is shaken two to three times, the cap removed and the liquid poured out.
4. Care should be taken to recap the bottle before washing the plant material.
5. After washing, the surface-sterilised plant material is transferred to presterilised petri dishes (or china tile) and suitable explants prepared using sterilised instruments (scalpels, needles, cork-borer, forceps) and a dissection microscope.
6. The cap of the culture vial containing the nutrient medium (also presterilised) is removed, the inoculum (explant) transferred onto the medium and the culture vial recapped all in quick succession. In the case of glass vial, the neck is flamed before closing it with the cap.
7. All operations are carried out in the laminar airflow hood.
8. Testing for contamination, that may have occurred at some point in any operation, can be done by transfer of suspected materials to the potato dextrose medium and made to incubate for 24 h at 37°C. Potato dextrose powder (Sigma Co.) is used to prepare the medium liquid or as agar plates for detection of bacteria, yeast and fungi. Other diagnostic media for testing contamination are also available (PhytaSource 1994).

PART II
Basic Aspects

Cell Culture

Establishment of single cell cultures provides an excellent opportunity to investigate the properties and potentialities of plant cells. Such systems contribute to our understanding of the interrelationships and complementary influences of cells in multicellular organisms. The pioneering attempts made by Haberlandt (*see* Chapter 1) failed to achieve divisions in free cells but his detailed paper in 1902 stimulated further studies in this area. Subsequently, several workers reported spectacular success in achieving isolated single cell divisions and even raised complete plants from single cell cultures. This generated much interest among plant biotechnologists who recognised the merits of applying cell cultures over an intact organ or whole plant cultures to synthesise natural products. Using cell cultures in studies designed to describe the pathways of cellular metabolism was another aspect that initially attracted the attention of plant biologists. It was soon realised that single cell systems have a great potential for crop improvement. Free cells in cultures permit quick administration and withdrawal of diverse chemicals/substances, thereby making them easy targets for mutant selection. Moreover, the individual cells within a population of cultured cells invariably show cytogenetical and metabolic variations depending on the stage of the growth cycle and culture conditions.

Such variability, termed 'spatial heterogeneity' (Lindsey and Yeoman 1985), has been the subject of much interest since differences between cells in their karyotype and the ability to accumulate secondary metabolites would manifest during morphogenesis in the clones regenerated from single cells. In this way the cell line selection technique can be usefully applied to produce high-yielding cultures as well as plants with superior agronomic traits. Cell cultures are now important tool in the study of Molecular biology and agricultural biotechnology (Thorpe 2007) and use of a 3-dimensional fibre (3-D scaffold) in recent has proved an investigative aid in studying morphogenesis among isolated cells in cultures leading to plant development (R. Wightman: BMC Plant Biology Blog 14.09.2015).

5.1 Isolation of Single Cells

5.1.1 From Plant Organs

The most suitable material for the isolation of single cells is the leaf tissue since a more or less homogeneous population of cells in the leaves offer good candidates for raising defined and controlled large-scale cell cultures. From such intact plant organs (as leaf tissue) single cells can be isolated using *mechanical* or *enzymatic* methods.

(I)   Mechanical Method

Gnanarn and Kulandaivelu (1969) developed a procedure which has since been successfully used to isolate mesophyll cells, active in photosynthesis and respiration, from mature leaves of several species of dicots and monocots induding the grasses. Even metabolically active single cells from the bundle sheath of crab grass (*Digitaria sanguinalis*) can be isolated using a similar procedure. The procedure involves mild maceration of 10 g leaves in 40 ml of the grinding medium (20 μ mol sucrose, 10 μ mol $MgCl_2$, 20 μ mol tris-HCl buffer, pH 7.8) with a mortar and pestle. The homogenate is passed through two layers of muslin cloth and the cells thus released are washed by centrifugation at low speed using the same medium.

The mechanical isolation of free parenchymatous cells can also be achieved on a large scale. The details of this procedure are given in Appendix 5.1.

(II)   Enzymatic Method

In 1968, Takebe and his co-workers treated tobacco leaf tissue with the enzyme pectinase and obtained a large number of metabolically active cells. A point to note is that potassium dextran sulphate in the enzyme mixture improved the yield of free ceils (for details *see* Appendix 5.2).

Isolation of single cells by the enzymatic method has been found convenient as it is possible to obtain high yields from preparations of spongy parenchyma with minimum damage or injury to the cells. This can be accomplished by providing osmotic protection to the cells while the enzyme macerozyme degrades the middle lamella and cell wall of the parenchymatous tissue. Applying the enzymatic method to cereals (*Hordeum vulgare, Zea mays*) has proven rather difficult since the mesophyll cells of these plants are apparently elongated with a number of interlocking constrictions, thereby preventing their isolation.

5.1.2   *From Cultured Tissues*

The most widely applied approach is to obtain a single cell system from cultured tissues. Freshly cut pieces from surface-sterilised plant organs are simply placed on a nutrient medium (solidified) consisting of a suitable proportion of auxins and cytokinins to initiate cultures. Explants on such a medium exhibit callusing at the cut ends, which gradually extends to the entire surface of the tissue. The callus is separated from an explant and transferred to a fresh medium of the same composition to enable it to build up a mass of tissue. Repeated subculture on an agar medium improves the friability of the callus, a prerequisite for raising a fine cell suspension in a liquid medium. The pieces of undifferentiated and friable callus are transferred in a continuously agitated liquid medium dispensed in autoclaved flasks or other suitable vials. Agitation is done by placing the culture flasks/vials on an *orbital-platform shaker* or suitable device. Movement of the culture medium exerts mild pressure on small pieces of tissue, breaking them into free cells and small cell aggregates. Further, it augments the gaseous exchange between the culture medium and the culture air, and also ensures uniform distribution of cells as well as cell clumps in the medium.

5.2   Growth and Subculture of Suspension Cultures

Cell suspensions are clonally maintained by the routine transfer (subculture) of cells in the early stationary phase to a fresh medium. During the incubation period the biomass of the suspension cultures increases due to cell division and cell enlargement. This continues for a limited period since the viability of cells in suspension after the stationary phase decreases due to the exhaustion of some factors or the accumulation of toxic substances in the medium. At this stage an aliquot of the cell suspension with uniformly dispersed free cells and cell aggregates is transferred to a fresh liquid medium of the original composition. The timing of subcultures is very important. The incubation period from culture initiation to the stationary phase is determined primarily by: (a) initial cell density, (b) duration of lag phase and (c) growth rate of cell line. The cell density used to subculture is critical and depends largely on the type of suspension culture to be maintained. Low initial cell densities will prolong the lag phase and exponential phases of growth. While initiating new suspension culture it is necessary to determine optimal cell density, proportionate to the volume of the culture medium, in order to achieve maximum growth. At an initial cell density of $9\text{-}15 \times 10^3$ ml^{-1}, the cells will generally undergo an eightfold increase in cell number before entering the stationary phase. Subcultures established with a high inoculum rate $(0.5\text{-}2.5 \times 10^5$ cells $ml^{-1})$ show an increase in cell number during the incubation period to a range $1\text{-}4 \times 10^6$ ml^{-1} before entering the stationary phase. The normal incubation time of stock cultures is 21-28 days between subcultures although cloning may occur within 18-25 days. In cases in which the cells are in a very active state of division, the passage length (periods between subcultures) may be reduced to 6-9 days. Cell cultures initiated at very low cell densities will not grow unless the medium is enriched with the metabolites necessary to grow single cells or a small population of cells. The presence of multicellular structures in the inoculums with single cells seems essential to maintain optimum density of single cells in suspension cultures of wheat and barley (Dong et al. 2010).

5.3 Types of Suspension Cultures

5.3.1 Batch Cultures

These cultures are maintained continuously by propagating a small aliquot of the inoculum in the moving liquid medium and transferring it to a fresh medium (ca. 5 × dilution) at regular intervals. Generally, cell suspensions are grown in flasks (100-250 ml) containing 20-75 ml of the culture medium. The biomass growth in batch cultures follows a fixed pattern (Fig. 5.1).

When the cell number in suspension cultures is plotted against the time of incubation, a growth curve is obtained depicting that initially the culture passes through a lag phase, followed by a brief exponential growth phase—the most fertile period for active cell division. The growth declines after three to four cell generations, signalling that the culture has entered the stationary phase.

For a subculture, the flask containing the suspension culture is allowed to stand still for a few seconds to enable the large colonies to settle down. A pipette or a syringe with an orifice fine enough to hold aggregates of two to four cells or only single cells is used. The suspension is taken from the upper part of the culture and transferred to a fresh medium.

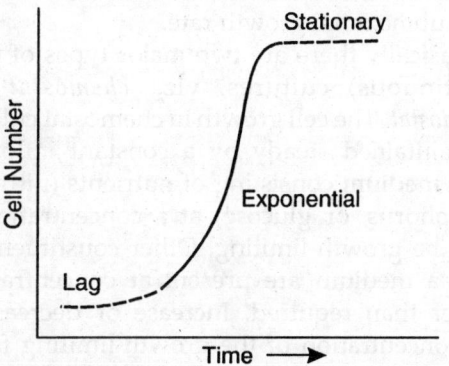

Fig. 5.1 Model curve showing different growth phases in batch cultures.

Batch cultures are characterised by a constant change in the pattern of cell growth and metabolism. As a result, batch cultures are not ideal systems for studies related to various aspects of cellular behaviour. In these cultures exponential growth with constancy of cell doubling time may be achieved, but there is no period of steady-state growth in which relative concentrations of metabolites and enzymes are constant. The drawbacks of batch cultures are overcome, to a certain extent, in *continuous cultures*.

5.3.2 Continuous Cultures

The large-scale cultures grown under steady state for long periods by adding fresh medium and draining out the used medium in a number of specially designed culture vessels are known as continuous or *mass* cultures. Continuous cultures are of the *closed* or *open* type. In the closed type, the addition of fresh medium is balanced by the outflow of old medium. The cells passing through the outflowing medium are separated mechanically and reintroduced in the culture. Cell biomass continues to increase as the growth proceeds. Paradoxically, the inflow of medium in the open type is accompanied by a balancing harvest of an equal volume of the culture medium and cells. This allows the indefinite maintenance of cultures at a constant and submaximai growth rate.

Basically there are two major types of open (continuous) cultures, viz., *chemostat* and *turbidostat*. The cell growth in chemostat cultures is maintained steady by a constant inflow of fresh medium consisting of nutrients (nitrogen, phosphorus, or glucose) at a concentration so as to be growth-limiting. Other constituents of such a medium are present at concentrations higher than required. Increase or decrease in the concentration of the growth-limiting factor is correspondingly expressed by increase or decrease in the growth rate of cells. Thus, the desired rate of cell growth can be maintained by adjusting the level of concentrations with respect to the growth-limiting factor and other constituents. In turbidostat cultures, on the contrary, the input of medium is intermittent as it is mainly required to control the rise in *turbidity* due to cell growth. The turbidity is preselected on the basis of biomass density in cultures and can be maintained by intermittent flow of medium and washout of cells.

A variety of rig configurations ranging from conventional stirred-tank reactors to bubble column and airlift systems (*see* Chapter 17) have been tested to achieve good mixing and homogeneous growth conditions that will not be deleterious to cell growth. Culture vessels are generally home made and manufactured principally of glass and/or stainless steel. A wide range of bioreactor configurations and sizes have been designed for continuous cultures depending on the variety of plant cells. These are described in detail by Martin (1980), Fowler (1982), Panda et al. (1989) and Scragg (1994).

Continuous cultures, besides commercial applications, offer certain other advantages: (a) ease of maintaining sterility over a long period of time, (b) less detrimental effects during mechanical failures, (c) a degree of automation and (d) versatility with regard to growth conditions such as temperature, aeration, stirring speed, illumination, nutrient and growth regulator levels. In spite of these advantages, plant tissue culturists refrain from using continuous cultures, probably because they require constant attention and specially designed equipment which pose practical rather than conceptual problems.

5.4 Culture Medium for Cell Suspensions

To obtain a fine suspension culture it is of prime importance that the callus used initially be friable. Moreover, the texture of a callus is genetically controlled and very often one experiences difficulty in achieving a good dispersion of cells. Manipulations in the media constituents and subculture routine may help in tissue dissociation although the addition to the medium of 2,4-D, small amounts of hydrolytic enzymes (cellulase and pectinase), or substances such as yeast extract, appears to have promotory effect on cell dispersion.

A good cell dissociation may also be achieved by permanently maintaining the cultures in the late lag phase by adding fresh medium every other day in a proportion that the biomass/medium volume is kept at 2. Sometimes it may be necessary to transfer small callus pieces or cell aggregates back to the agar or semi-solid medium. After two to three passages these pieces develop into a friable callus tissue which, on transfer to liquid medium, gives rise to a fine suspension.

Theoretically, the medium used for raising a fast-growing friable callus should prove suitable for initiating the cell suspension cultures of that particular species in a liquid medium. In practice, the requirements for rapidly growing cell suspensions differ from those for tissue or callus cultures. For example, the culture medium for a tobacco cell suspension requires an increase in the concentration of 2,4-D from 0.3 mg l^{-1} to 2 mg l^{-1}, followed by supplementing the callus medium with additional vitamins and casein hydrolysate (Table 5.1). Furthermore, the inorganic phosphate is rapidly utilised in actively growing suspension cultures and, consequently, becomes a limiting factor.

Table 5.1 Culture medium for tobacco suspension cultures[a]

Constituents	Amount (mg l^{-1})
Inorganic nutrients	As in MS medium
Thriamine HCl	10
Pyridoxine HCl	10
Nicotinic acid	5
Myoinositol	100
Casein hydrolysate	1000
2,4-D	2
Kinetin	0.1
Sucrose	30000
pH	5.7

[a] After Reynolds and Murashige (1979).

Many media have very little buffering capacity and the pH can change with an increase in the cell biomass. This necessitates monitoring and adjustment of pH in suspension culture. B5 (see Chapter 3) and ER (Eriksson 1965) media are specially recommended for suspension culture of higher plants. These and other synthetic media are used for initial population density 5 × 10 cells ml^{-1} or more. With lower cell density, the medium requires to be *conditioned* or enriched with various other compounds.

5.4.1 Conditioning of Medium

In initiating cell cultures at low inoculum density a conditioned medium is used. A simple method

is to filter out cells growing at high density from 4-6-week-old liquid cultures and to use this medium in drops or as thin layers to culture single cell/cells at low population density. The principle of conditioning followed by Torres (1989) involves the separation of a high-density cell culture from a low-density culture medium by a barrier that permits the diffusion of solutes and air. A high-density cell suspension (the nurse culture) kept inside a dialysis tube (Fig. 5.2) is suspended by means of a thread or rod in the flask containing the culture medium with low cell density (low-density medium). The metabolites produced by the nurse culture diffuse into the low-density medium, thereby increasing the latter's growth-promoting activity. This meets the conditions of growth for low cell populations since the necessary substances that may not be found in the low-density medium are released into it by the biosynthetic activity of the nurse cells.

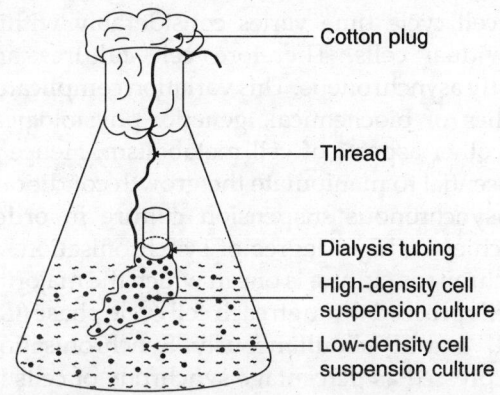

Fig. 5.2 Apparatus designed for conditioning of a low-density cell culture medium (modified from Torres 1989).

5.4.2 Agitation of the Medium

Suspension cultures require constant agitation of the medium for adequate aeration. This also facilitates dispersion of cells. It can be achieved using a shaker and suitable flasks. Muir (1953) was the first to introduce the orbital-platform shaker for growing suspension cultures of tobacco and *Tagetes erecta*. The platform of the shaker is fitted with interchangeable clips of appropriate size for holding the flasks. A shaking speed of 30-150 rpm is optimum for most tissues. Rotary shakers are also used which have a disc that can be rotated at slow speed (1-2 rpm) by a shaft. About 10 ml medium is dispensed in a specially designed tube (12.5 cm long and 3.5 cm diameter) having a wide neck (1.7 cm diameter) and each tube mounted near the margin of the disc. The inoculum is introduced from the neck and its mouth closed with a cotton plug. When the disc rotates, the cells and tissues are alternately bathed in the culture medium and thereby exposed to both nutrients and the culture air. Sometimes the tubes are substituted by special *nipple* flasks. Both orbital and rotary shakers have a control for regulating the speed.

5.5 Synchronisation of Suspension Cultures

Cells in suspension cultures vary greatly in size, shape, DNA and nuclear content. Moreover, the cell cycle time varies considerably within individual cells. Therefore cell cultures are mostly asynchronous. This variation complicates studies of biochemical, genetic, physiological and other aspects of cell metabolism. Hence it is essential to manipulate the growth conditions in asynchronous suspension culture in order to achieve a high degree of synchronisation. A synchronous culture is one in which the majority of cells proceed through each cell cycle phase (G_1, S, G_2 and M) simultaneously. Synchronisation is expressed as percentage synchrony of cells in suspension cultures. The methods employed to achieve synchronisation of suspension cultures may be grouped under two categories: *physical* and *chemical*.

5.5.1 *Physical Methods*

Physical properties of cells (e.g., size of individual cell/small cell aggregates) or environmental growth conditions (light, temperature) can be monitored successfully to achieve a high degree of synchronisation. Some of the physical methods helpful in achieving synchrony are:

(I) Selection by Volume

Synchronisation may be achieved on the basis of selecting the size of cell aggregates present even in the finest possible suspension cultures. This approach proved successful for carrot suspension cultures to the extent that 90% cell aggregates isolated were in early embryogenic stages. The procedure followed for selection of cell aggregates using the cell fractionation technique is given in Appendix 5.3.

(II) Temperature Shock

Low temperature shocks are reported to induce synchronisation of suspension cultures. This approach (*see* detailed procedure in Appendix 5.4) is now widely followed to increase the degree of cell synchronisation.

5.5.2 *Chemical Methods*

Cell cultures are *starved* of a nutrient or supplied with a *biochemical inhibitor* to prevent cells from completing a cell cycle. Through this approach the cells are first arrested at a particular stage of the cell cycle and, subsequently, allowed to undergo simultaneous divisions either by supplementing the starved chemical or withdrawing the inhibition.

(I) Starvation

The principle of starvation is based on depriving suspension cultures of an essential growth compound leading to a stationary growth phase. Resupplying the missing compound is expected to induce resumption of cell growth synchronously. This procedure has been *very* effective for sycamore (*Acer pseudoplatanus*) suspension cultures. Cultures starved of nitrogen, phosphorus or carbonate, result in the arrest of cell growth during the G_1 or G_2 phase of the cell cycle. After a period of starvation, when these growth-limiting compounds are supplied to the medium, the stationary cells enter divisions synchronously (*see* Appendix

5.5). Growth hormone starvation is also reported to induce synchronisation of cell cultures.

(II)   Inhibition

Synchronisation is achieved by temporarily blocking the progression of events in the cell cycle and accumulating cells in a specific stage using a biochemical inhibitor. On release the block cells will synchronously enter the next stage. Inhibitors of DNA synthesis (5-aminouracil, FudR or 5-fluorodeoxypurine, hydroxyurea, TdR or excess thymidine) in cell cultures accumulate cells at the G_1/S boundary. Removal of the inhibitor is followed by synchronous division of cells. The procedure for using, a biochemical inhibitor to obtain synchronisation of cell cultures of *Haplopappus gracilis*, soybean, tobacco and tomato is described in Appendix 5.6.

(III)   Mitotic Arrest

Colchicine has been widely used to arrest cells at metaphase. Suspension cultures in exponential growth are supplied with 0.02% (w/v) colchicine for 4-8 hr in order to inhibit spindle formation, Longer colchicine treatment leads to an increased frequency of abnormal mitoses and chromosome stickiness. Colchicine should be filter-sterilised and only shorter duration treatment is recommended. This technique has been used for synchronisation of *Zea mays* suspension cultures but the likelihood of colchicine inducing genomic changes raises the possibility of obtaining asynchronous cultures.

5.6 Measurement of Growth in Suspension Cultures

Assessment of the growth in suspension cultures can be accomplished by following selected parameters at regular intervals. These include: (a) *cell counting*, (b) *packed cell volume* and (c) *fresh/ dry weight increase of cells and cell colonies*.

5.6.1   Cell Counting

Cell count is a relatively more accurate measure adopted to determine the growth of cultures. Increase in cell number depends on the rate of mitotic index (MI) of cells in suspension cultures. Determination of cell number is a simple but tedious procedure since suspension cultures invariably carry cell colonies of various sizes. Therefore, it becomes essential to first disrupt cell aggregates by treating them with 5-15% chromic acid or pectinase (0.1% w/v, pH 3.5). The procedure described to count sycamore cells has been found suitable for cell counting. To 1 volume of cell suspension culture may be added 2 volumes of 8% chromic acid (trioxide) solution and the mixture heated at 70°C for 2-15 min. The mixture is cooled and then agitated vigorously for 10 min on a shaking machine. The suspension is now centrifuged, the chromic acid poured off and the pellet resuspended in 8% saline (NaCl) solution. After 10-15 min free cells are counted on a haemocytometer. Heating is avoided if an enzyme is used to disrupt cell aggregates.

5.6.2   Packed Cell Volume (PCV)

For PCV determination a small sample (10 ml) is removed from the uniformly disposed suspension culture aseptically and centrifuged in 15 ml graduated conical centrifuge tubes at $1000 \times g$ for 5 min. The packed cell volume is expressed as ml pellet ml^{-1}culture.

5.6.3   Cell Fresh Weight

The cells are collected on a preweighed (wet) circular filter of nylon fabric supported in a Hartley funnel, washed with distilled water under vacuum and the filter discs reweighed along with the cells.

5.6.4   Cell Dry Weight

A procedure similar to that for fresh weights is followed for determining cell dry weight except that the filter discs are dried in an oven for 12 hr at 60°C. After cooling in a desiccator containing silica gel, the dried filter is reweighed and the cell weight expressed as g ml^{-1} of culture or per 10^6 cells.

5.7 Viability of Cultured Cells

The growth of cultures is largely dependent on the viability of the cells, which can be assessed by microscopic examination of untreated cells or after staining them with substitute chemicals. These include:

5.7.1 Phase Contrast Microscopy

Cytoplasmic streaming and the presence of a healthy nucleus indicate that the cells are viable. Phase contrast microscopy is recommended as it is difficult to observe these aspects in unstained cells under a bright field.

5.7.2 Reduction of Tetrazolium Salts

This test is used to measure respiratory efficiency of cells by reduction of 2,3,5-triphenyltetrazolium chloride (TTC) to the red dye formazon. Formazon can be extracted and measured spectrophotometrically.

5.7.3 Fluorescein Diacetate (FDA) Method

The FDA method offers a quick visual assessment of the viability of cells. Stock solution of FDA (0.5% w/v) is prepared in acetone and stored at 0°C. Viability is tested by adding this solution to the cell or protoplast suspension at a final concentration of 0.01%. For protoplasts an appropriate osmotic stabiliser is added to the FDA solution. After 5 min incubation the cells are examined under a microscope with a suitable excitation or suppression filter. FDA, though non-fluorescing and non-polar, is cleaved by esterase activity inside the living cell, resulting in release of the polar portion of fluorescein, which fluoresces under UV. Since fluorescein is not freely permeable across the plasma membrane, it accumulates mainly in the cytoplasm of intact cells and thus becomes distinctly visible. In a dead cell the fluorescein is lost and remains invisible. Under UV light fluorescein gives a green fluorescence.

5.7.4 Evan's Blue Staining

This simple procedure is usually used as a complement to FDA. A dilute solution (0.025% w/v) of Evan's blue dye stains the dead or damaged cells while the living or viable cells repel the dye and remain unstained.

5.8 Culture of Isolated Single Cells

Free cells isolated either from plant organs (mesophyll tissue) or cell suspensions are grown as single cells under *in vitro* conditions using a suitable medium. This process, called *plating,* is of particular importance when attempting to obtain single-cell clones. Success in the culture of single cells, therefore, depends on the technique and various factors affecting cell plating.

5.8.1 Plating Technique

The technique developed by Bergmann (1960) is the most popular one for plating of single cells (Fig. 5.3). Free cells are suspended in liquid medium at a density twice the finally desired plating cell density. Equal volumes of the media containing single cells and a melted (30° to 35°C) agar medium (0.6-1%, w/v) are mixed and rapidly spread out in petri dishes in such a manner that the cells become fixed in an evenly distributed thin layer (ca 1 mm thick) after the agar has cooled and solidified. The dishes are sealed with parafilm and incubated in the dark or diffused light at 25°C. Free cells can also be plated in the liquid medium but follow-up of individual cells or their derivatives is difficult in this procedure because the cells do not remain in a fixed position. It is important to note that suspension cultures are filtered aseptically through a sieve that allows only the single cells required for culture to pass through; the cell aggregates are discarded.

5.8.2 Plating Efficiency (PE)

The plates or culture dishes may be observed under an inverted microscope and single cells marked on the outside of the plate with a fine

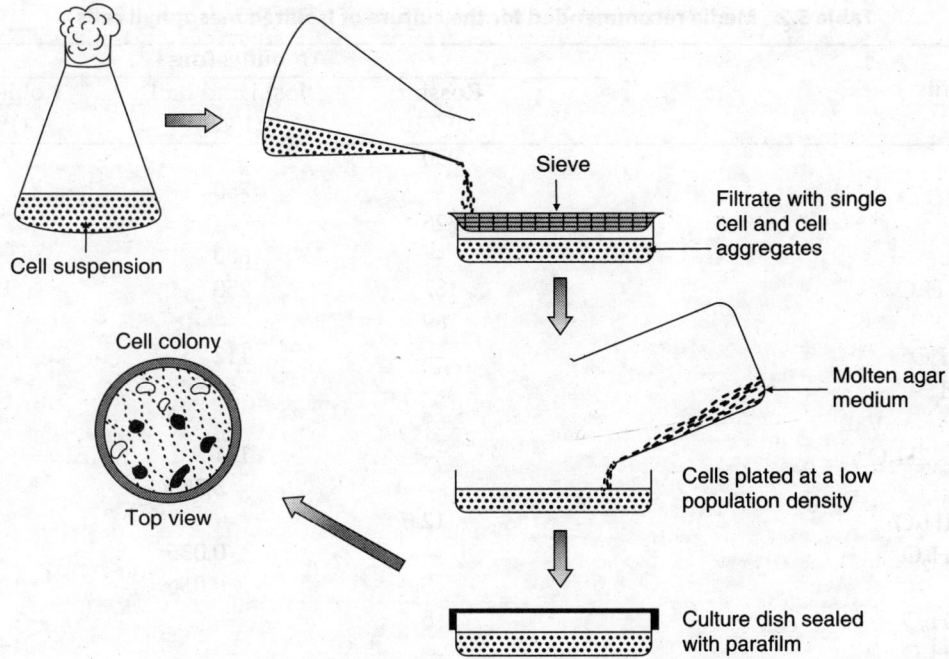

Fig. 5.3 Steps involved in Bergmann's technique of cell plating.

marker to keep track of their regeneration potential. This ensures the isolation of pure single-cell clones. In preparing cultures, if a known volume of suspension is transferred to each plating dish, it should be possible to assess the PE quantitatively using the formula:

$$PE = \frac{\text{Final number of colonies/plate}}{\text{Initial number of cell units/plate}} \times 100$$

Usually, plating at cell densities of 10^3–10^5 cells ml^{-1} or more yields a high plating efficiency. Other parameters for obtaining a high PE are:

(a) Using a conditioned medium or synthetic medium designed to permit growth from a low initial density.

(b) Avoid plating cells held too long in stationary phase.

(c) Harvesting cells during exponential growth phase.

(d) Exposure of cells to temperature should never exceed 35°C.

(e) Incubating the plates in diffused light or darkness.

5.8.3 *Medium and Technique of Low-Density Cell Cultures*

(I)   Medium

Efforts have been made to develop a synthetic medium for cells plated at low density. Cells plated in a culture medium synthesise necessary metabolites; when their concentrations reach a threshold value, a cell divides. This process of cells releasing metabolites into the medium continues until an equilibrium is reached between the cells and the medium. At initial high cell density the equilibrium is reached much earlier than at low ceil density. Below a critical cell density the equilibrium is never reached and the cells fail to divide. This impediment to cell division, i.e., the population effect, can be overcome by supplementing the minimal medium with such undefined factors as coconut milk, casein hydrolysate and yeast extract. Various media developed for culturing isolated mesophyll cells are summarised in Table 5.2.

Table 5.2 Media recommended for the culture of Isolated mesophyll cells

Constituents	Amounts (mg l^{-1})		
	Rossini (1972)	Joshi and Ball (1968)	Kohlenbach (1984)
KNO_3	950	—	950
KCl	—	750	—
NH_4NO_3	725	—	720
$NaNO_3$	—	600	—
$MgSO_4 \cdot 7H_2O$	187	250	185
$CaCl_2$	169	—	—
$CaCl_2 \cdot 6H_2O$	—	112	—
$CaCl_2 \cdot 2H_2$	—	—	16.6
KH_2PO_4	69	—	68
$NaH_2PO_4 \cdot 2H_2O$	—	141	—
NH_4Cl	—	5.35	—
$MnSO_4 \cdot 4H_2O$	12.5	—	25
$MnCl_2 \cdot 4H_2O$	—	0.036	—
H_3BO_3	5	0.056	10
$ZnSO_4 \cdot 4H_2O$	5	—	—
$ZnSO_4 \cdot 7H_2O$	—	—	10
$ZnCl_2$	—	00.15	—
$Na_2MoO_4 \cdot 2H_2O$	0.125	0.025	0.25
$CuSO_4 \cdot 5H_2O$	0.0125	—	0.025
$CuCl_2 \cdot 2H_2O$	—	0.054	—
$CoCl_2$	—	0.02	—
$FeSO_4 \cdot 7H_aO$	13.9	—	27.85
$FeCl_3 \cdot 6H_2O$	—	0.5	—
Na · EDTA	18.6	—	37.25
Disodium salt of ethylene dinitritotetraacetic acid	—	0.8	—
Glutamine	—	—	14.7
Glycine	2	—	2
Nicotinic acid	5	—	5
Pyridoxine HCl	0.5	—	0.5
Thiamine HCl	0.5	—	0.5
Biotin	0.05	—	0.05
Folic acid	0.5	—	0.5
Casein hydrolysate (acid hydrolysate, acid and vitamin free)	—	400	—
Zn-Inositol	100	—	100
Adenine	—	—	20.25
BAP	0.1	—	—
Kinetin	—	0.1	1
2,4-D	1	1	1
Sucrose	10,000	20,000	10,000
pH	5.0	?	5.5

(II) Technique

(a) *The filter-paper raft nurse technique.* The principle of this technique is similar to the conditioning of the culture medium described under Section 5.5.1. The basic difference is that a callus is used in place of the liquid medium to nurse the culture of an isolated single cell. Individual cells from a suspension culture or callus tissue (e.g., tobacco, marigold) are placed by means of a micropipette or spatula on top of an actively growing callus but separated by a filter-paper raft (*see* Fig. 1.1). After some days a cell, which normally fails to divide in the culture medium, is able to grow under the nurse effect of the callus. The whole operation is done aseptically and cell transfer should be rapid in order to avoid excessive drying of the cell and raft. Once a macroscopic colony develops from the cell on the filter-paper raft, it is transferred to an agar medium for further growth and maintenance under aseptic conditions. This method is now widely used to clone isolated single cells (*see* Chapter 1).

(b) *The microchamber technique.* De Ropp (1955) made the first attempt to culture single cells in a liquid medium using hanging drops. Success in obtaining divisions was limited up to the formation of aggregates of 10 cells or more, which could not meet the ultimate objective of raising clones from isolated single cells. Jones et al. (1960) accomplished this goal by developing the microchamber technique. In this method (Fig. 5.4), a drop of the medium carrying a single cell is placed on a sterile microscope slide and ringed with sterile mineral oil. Again one drop of mineral oil is placed on either side of the ringed culture drop and a cover glass placed on each oil drop. A third cover glass is then placed on the culture drop bridging the two cover glasses. As a result, a microchamber is formed enclosing the single cell aseptically within the mineral oil. The oil prevents water loss from the chamber but permits gaseous exchange. The microchamber slide is now incubated by placing it in a petri dish. The cover glass is removed as soon as the cell colony becomes visible to the naked eye and the tissue subcultured by transferring to fresh liquid or a semi-solid medium.

The microchamber technique enables visual monitoring of the divisions in an isolated cell. This method has been applied to raise a complete flowering plant of tobacco from a single cell in a culture medium containing mineral salts, sucrose, vitamins, Ca-pentothenate and coconut milk (*see* Chapter 1).

5.8.4 Bioreactor for Large Scale Culture

Mostly large scale cultivation of plant cells has been achieved as continuous cultures, which requires bioreactor configurations of various sizes (*see* Section 5.3.2). A bioreactor is a glass or steel vessel fitted with probes to monitor the pH, temperature and dissolved oxygen in the cell culture under aseptic conditions. Bioreactors have found increasing applications in industrial production of valuable compounds (Scragg 1994, 1999, Eibl and Eibl 2008, Ruffoni et al. 2010). The details of design regarding bioreactors used in cell cultures are discussed in Chapter 17.

In conclusion, reasonable progress has been made to develop the methods of cell culture from most plant tissues. It is now possible to nurse isolated free cells *in vitro* at increasingly low plating densities under defined conditions. Due to the occurrence of a high degree of spontaneous variability in cultures, cloning of individual cells has the potential for application in mutant selection and synthesis of natural plant products. These aspects are detailed later in Parts III and IV.

The plant cell cultures currently have increasing role in large-scale production of pharmaceutical proteins. Protalix Biotherapeutics develops recombinant proteins and produces them in plant cell culture. Teliglucerase alfa is reported as the first biotherapeutic protein expressed in plant cells and is now under approval for commercial use in the world. Other therapeutic proteins being developed for production in plant cell cultures and major milestones reached by protalix biotherapeutics are summarised in a review by Tekoah et al. (2015).

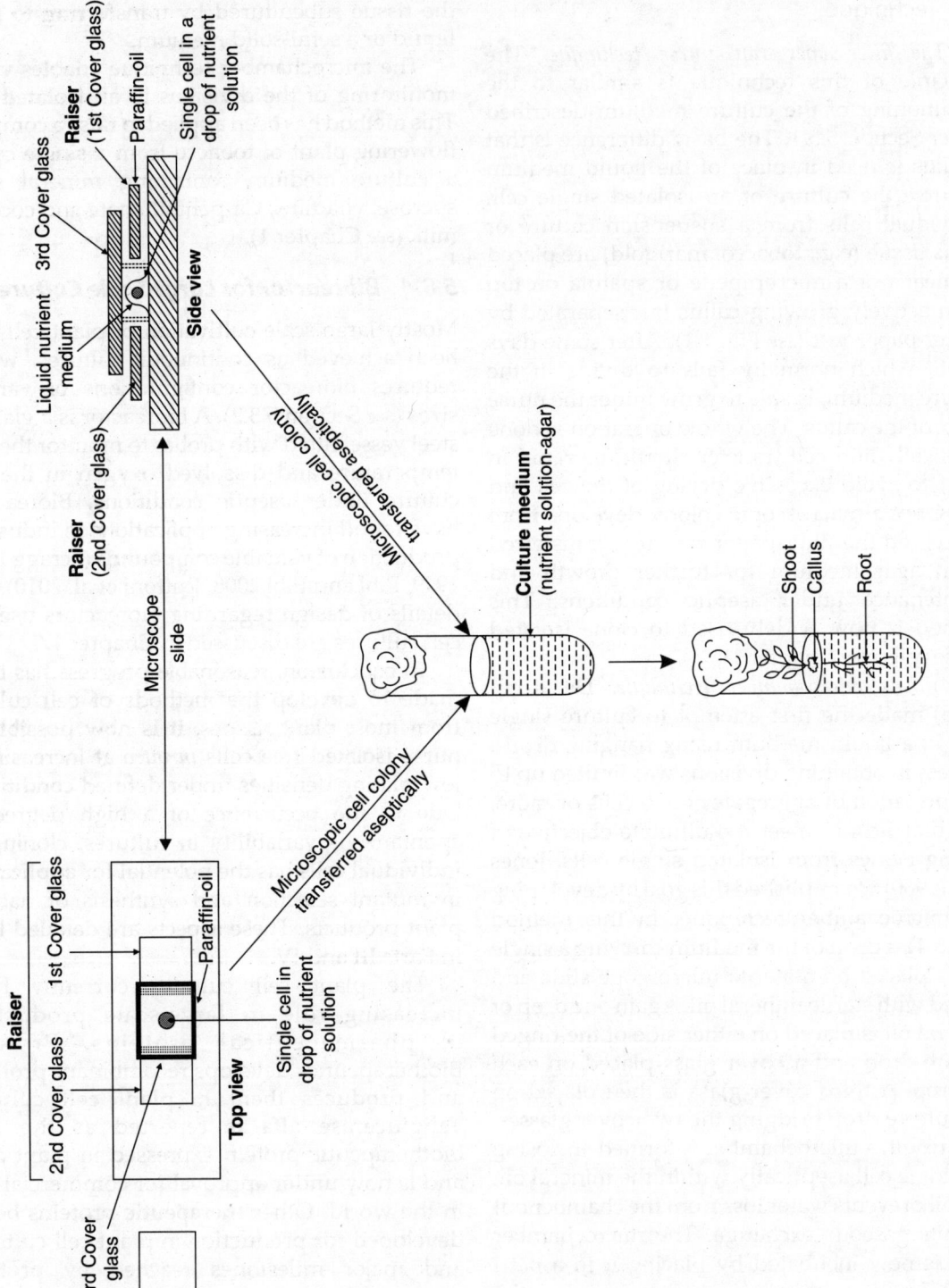

Fig. 5.4 Microchamber technique of cell plating.

APPENDIX 5.1

Large-scale Mechanical Isolation of Mesophyll Cells from the Leaves of *Calystegia sepium*[a]

1. Surface sterilise the leaves by rapid immersion in 95% ethanol followed by rinsing for 15 min in filter-sterilised 7% solution of calcium hypochlorite. Wash in sterile distilled water.
2. Cut the leaves into small pieces of less than 1 cm^2.
3. Transfer and homogenise 1.5 g of leaf pieces with 10 ml of culture medium (*see* Table 5.2) in a Potter-Elvehhem glass homogeniser tube.
4. Filter the homogenate through sterile metal Tyler filters[b] of mesh diameters 61 µm (of upper filter) and 38 µm (of lower filters) respectively.
5. Remove the fine debris from the filtrate by slow-speed centrifugation in order to sediment the free cells. Pipette out the supernatant and suspend the cells in a volume of the medium sufficient to achieve the required cell density.
6. Plate the cells in a thin layer of culture medium.

APPENDIX 5.2

Enzymatic Isolation of Mesophyll Cells from Tobacco Leaves[c]

1. Immerse fully expanded leaves taken from 60- to 80-day-old plants in 70% ethanol for 30 s and transfer the leaves to 3% sodium hypochlorite solution containing 0.05% Teepol or cetavlon.
2. Wash the leaves after 30 min surface sterilisation (by hypochlorite solution) with sterile distilled water. Place the leaves on a sterile tile and peel off the lower epidermis with sterile forceps.
3. Excise peeled areas as 4 cm^2 pieces using a sterile scalpel blade.
4. Transfer 2 g of leaf pieces to 100 ml Erlenmeyer flasks containing 20 ml filter-sterilised enzyme macerozyme (0.5%) solution, 0.8% mannitol and 1% potassium dextran sulphate (MW 560, S content 17.3%. Meito Sangyo Co. Ltd., Japan).
5. Evacuate the flasks with a vacuum pump to infiltrate the enzyme into the leaf tissue.
6. Incubate the flasks at 25°C for 2 hr on a reciprocating shaker with a stroke of 4-5 cm at the rate of 120 cycles/min.
7. Change enzyme solution every 30 min. The enzyme solution after the second 30 min should contain largely spongy parenchyma cells and after the third and fourth 30 min periods predominantly palisade cells.
8. Wash the cells twice with culture medium and plate them in culture dishes.

---

[a] After Rossini (1972).
[b] W.S. Tyler Co., Cleveland, OH, U.S.A.
[c] After Takebe et al. (1968).

APPENDIX 5.3

Synchronisation of Cultures by Cell Fractionation[a]

1. Subculture suspensions for 7 days in basal medium supplemented with 0.5 µM 2,4-D.
2. Pass the cells through a 47 µm nylon sieve and then filter using a 31 µm sieve.
3. Collect cells and cell aggregates retained on the 31 µM sieve in a sterile tube and add .*m* equal volume of liquid medium.
4. Gently transfer 1 ml of cell aggregate suspension in a 10 ml centrifuge tube containing 8 ml Ficoll (10-18%, w/v) discontinuous density gradient prepared in 2% sucrose solution.
5. Centrifuge the contents of the tube at 180 g for 5 min.
6. Collect 1.5 ml fractions and resuspend the heaviest fraction in 8 ml basal medium.
7. Centrifuge again at 50 g for 5 s; repeat this step four times.
8. Transfer the cells in the lower portion to an embryo-inducing medium. Embryo formation can be detected in 4 or 5 days.

APPENDIX 5.4

Protocol for Employing Temperature Shocks to Increase Degree of Cell Synchronization[b]

1. Transfer 10 ml cultured cells to 100ml fresh medium.
2. Maintain cultures on a shaker (155 strokes/min) at 27°C until cell number plateaus. Keep the shaker on for another 40 hr.
3. Place the culture in a cold room (4°C) and allow it to stand for 3 days.
4. Add warm (27°C) fresh medium to the culture 10 times to its volume and allow the cells to grow for 24 hr at 27°C.
5. Repeat the cold treatment for another 3 days.
6. Incubate the culture again at 27°C and observe the increase in cell number and frequency of synchronised divisions after 2 days.

APPENDIX 5.5

General Procedure for Cell Synchronization of Suspension Cultures by Starvation[c]

1. Grow suspension cultures in complete medium[d] until cells enter the stationary phase.
2. Pass cells through a sterile 350 µm nylon sieve.
3. Transfer cell aggregates larger than 350 µM to basal medium.
4. Adjust cell density 25-30 × 10^3 cells ml^{-1}. Most cell aggregates should consist of approximately 25 cells.
5. Add growth-limiting compound after 30-35 hr. Growth should resume synchronously, reaching a MI peak within 60-70 hr.

---

[a] After Fujimura and Komamine (1979).
[b] After Okamura et al. (1973).
[c] Devised according to the procedure developed by Jouanneau (1971).
[d] See King et al. (1974).

APPENDIX 5.6

Synchronisation of Suspension Cultures through Biochemical Inhibition[a, b]

1. Add FudR (2 μg ml^{-1}) and uridine (1 μg ml^{-1}) to suspension cultures 1 day after subculture. The cell concentration should be between 5×10^3 and 3×10^6 cells ml^{-1} of the medium.
2. Incubate for 12-24 hr.
3. Remove FudR and uridine by washing three times with fresh medium.
4. Subculture in fresh medium. Add thymidine (2 μg ml^{-1}) for 12-24 hr.
5. Wash the culture by removing the medium containing thymidine.
6. Add fresh medium containing colchicine* (0.005%, w/v) to the culture.

---

[a] See Wang and Phillips (1984).

[b] To synchronise suspension cultures of *Haplopappus gracilis*, only excess TdR (6 mM) may be added for 24 hr, washed to remove medium containing TdR and cultured finally in fresh medium. A peak in the MI may be observed in 16 hr.

* This may increase the MI 5- to 10-fold in one cell cycle.

Cellular Totipotency

The technique of cloning isolated single cells *in vitro* has demonstrated the fact that somatic cells, under appropriate conditions, can differentiate to a whole plant. This potential of a cell to grow and develop a multicelluiar or multiorganed higher organism is termed cellular totipotency. Since the potential lies mainly in cellular differentiation, this indicates that all genes responsible for differentiation are present within individual cells and many of them that remain inactive in differentiated tissues or organs are able to express only under adequate culture conditions. The development of an adult organism from a single cell (zygote) is the result of the integration of cell division and cell differentiation (Fig. 6.1). Isolated cells from differentiated tissues are generally non-dividing and quiescent; to express totipotency the differentiated cell first undergoes *dedifferentiation* and then *redifferentiation*. The phenomenon of a mature cell reverting to a meristematic state and forming undifferentiated callus tissue is termed dedifferentiation, whereas the ability of a dedifferentiated cell to form a whole plant or plant organs is termed redifferentiation. Differentiation from one organ directly from another organ may be called transdifferentiation or just differentiation (Sugimoto et al. 2011). Thus, cell differentiation is the basic event of development in higher organisms and conveniently referred to as *cytodifferentiation*.

In animals, the differentiation is irreversible. This is in contrast to plants where even highly mature or differentiated cells have the ability to regress to a meristematic state as long as they are viable. Tissue culture techniques offer an excellent opportunity to study factors responsible for differentiation of cells. These factors control cellular totipotency through cytological, histological and organogenic differentiation.

6.1 Vascular Differentiation

One of the efficient systems for the study of cytodifferentiation *in vitro* is *xylogenesis*. Xylogenesis is the differentiation of parenchyma into cells that have localised secondary wall thickenings as seen in the xylem of vascular plants. These xylem-like cells have been variously named as wound vessel members, vessel elements, tracheids and tracheary elements. In this chapter the general term tracheary element has been used for such differentiated cells.

Knowing how tracheary elements form helps in understanding the mechanism of differentiation in higher plant cells because (1) the morphological characteristics of tracheary elements such as annular, spiral, reticulate and pitted secondary wall thickenings enable us to distinguish differentiated cells from undifferentiated cells; (2) the formation of

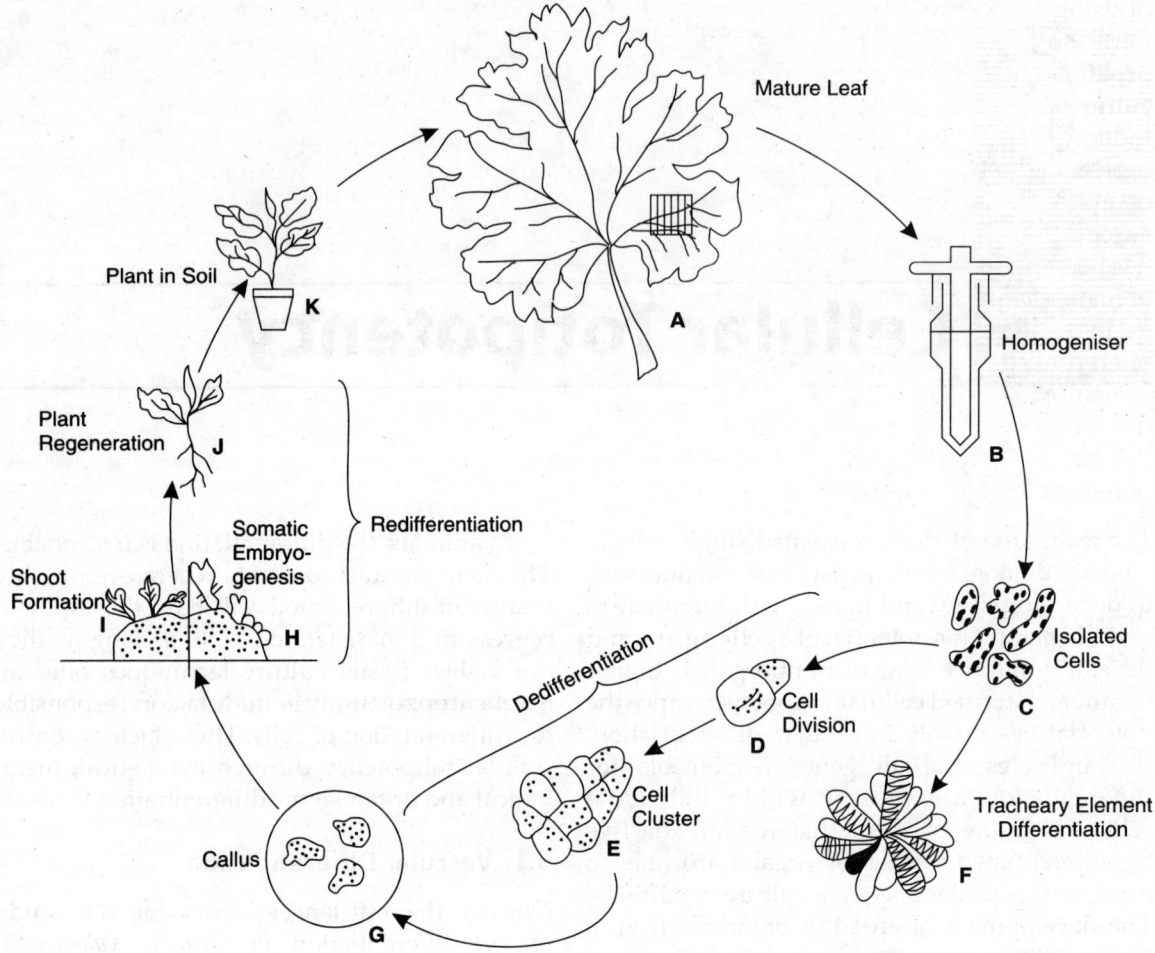

Fig. 6.1 Scheme showing cytodifferentiation in plant ceils (modified from Kohlenbach 1984).

tracheary elements can be induced in tissue and cell cultures of many species; (3) specific biochemical events leading to deposition of cell wall polysaccharides and lignin make it possible to trace marker proteins associated with the process of cytodifferentiation, and (4) due to loss of nucleus the tracheary element loses its capacity for de- or redifferentiation.

6.1.1 Culture Systems

Wounding the internodes of intact plants has often been used to induce tracheary element formation in cultures. Explants without pre-existing xylem tissue, phloem tissue, pith and tubers can also be used for this purpose. It is desirable that explants have cells in a uniform physiological state so that differentiation occurs synchronously.

(I) Callus Culture

Callus, which shows stable characteristics under specific conditions after subculture through many successive passages, is a suitable material for cytodifferentiation. The advantage of using such callus is that it is composed of a fairly homogeneous mass of cells and can be proliferated in large amounts under known culture conditions. Wetmore and Sorokin (1955) induced vascular strands in *Syringa* callus derived from the cambial region of the stem or grafted apices of shoots. Since then (1950s) vascular or tracheary element differentiation has been induced in callus derived from tissues of many plant species.

In a callus system, tracheary elements very often pre-exist and this necessitates selection of a maintenance medium that does not support the formation of these cells while subculturing the callus. However, many cultures lose their potential for differentiation during continual subculture due to epigenetic changes, such as those in the hormonal requirements of cells necessary for differentiation. Another disadvantage in using callus as experimental material is that tracheary elements are formed as discrete nests or nodules, giving the appearance of a tissue, rather than as scattered groups or single elements. Therefore, it is difficult to analyse the process of cytodifferentiation as distinct from tissue formation.

(II) Suspension Culture

Cells in suspension cultures receive more homogeneous stimuli in a defined medium supplemented with the requisite amount of inducers such as sugar or auxin. Theoretically, the percentage and synchrony of differentiation may be expected to be higher in cells of a suspension than in a callus culture system. Also, it may be possible to analyse biochemical pathways related to differentiation of cells in suspension cultures. For example, enzymes regulating lignin synthesis in tobacco suspension cultures showed that enzymes in the shikimate and cinnamate pathways were co-ordinately enhanced during tracheary element differentiation (Kuboi and Yamada 1978a,b). Similarly, phenylalanine ammonia lyase and peroxidase were found associated with differentiation in bean and pea cultures.

(III) Single-Cell Culture

Single-cell cultures are ideal systems for investigating cellular differentiation. Mechanically isolated mesophyll cells of *Zinnia elegans* are able to differentiate (preceding cell division) into tracheary elements within 3 days of culture in a medium containing 1 mg l^{-1} 2,4-D and 1 mg l^{-1} kinetin. The frequency and synchrony in differentiation further improved when mesophyll cells were isolated from the first leaves of seedlings and cultured in a medium containing a low concentration of ammonium, 0.1 mg l^{-1} NAA and 1 mg l^{-1} BA (Fig. 6.2). Cells isolated by enzymatic maceration can also differentiate but only at low frequencies.

(IV) Protoplast Culture

Vascular element formation has been reported in protoplasts isolated from mesophyll tissue of *Zinnia elegans* and cultured haploid cells of *Brassica napus* (Kohlenbach et al. 1982). The protoplasts differentiated without cell division in the presence of auxin and cytokinin. These systems need to be improved because of the low rate of tracheary element formation and their asynchroneity; however, protoplast culture may offer the chance to examine the fundamental process of cytodifferentiation, particularly after regeneration of cell walls, fusion and transfer of exogenous DNA.

(V) Leaf-Disc Culture

During vascular development procambial and cambial cells give rise to xylem and phloem cells. Because the vascular tissue is deeply embedded it has remained difficult to analyze the processes of vascular development in detail. Kando et al. (2015) established a novel *in vitro* experimental system in which vascular system is induced in *Arabidopsis thaliana* leaf-disc cultures using bikinin (inhibitor of glycogen synthase kinase 3 proteins). Transcriptome analysis reveals that mesophyll cell in leaf discs synchronously turn into procambial cells and then plants expressing the procambial markers TDR$_{pro}$, GUS and YFP

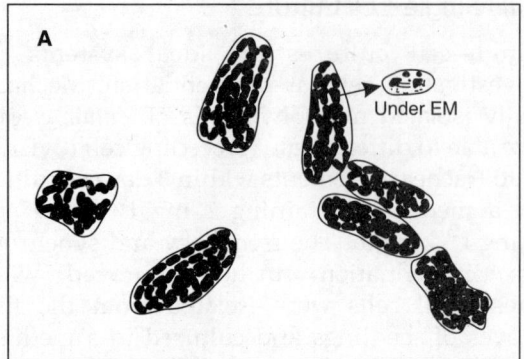

Fig. 6.2 A: Drawing of single *Zinnia* mesophyll cells isolated from first leaves of seedlings. The arrow shows one of the chloroplasts observed under EM.

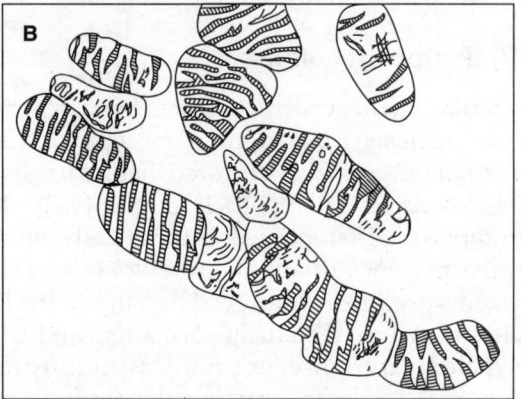

Fig 6.2 B: Isolated *Zinnia* mesophyll cells after 72 hr in culture medium containing NAA and BA show tracheary element formation at high frequency (based on the work of Fukuda and Komamine 1985).

can be used for spatiotemporal visualization of procambial cells. Further analysis with the TDR mutant and TDIF (tracheary element differentiation inhibitory Factor) indicates that key signalling factor TDIF-TDR-GSK3s regulates xylem differentiation in leaf-disc cultures. This new *in vitro* experimental system has potential as a tool for analyzing xylem cell differentiation in other plant vascular systems.

6.1.2 Cytological and Cytochemical Aspects

Electron microscope (EM) studies during various stages of cell differentiation reveal some early cytological changes. Mostly the cytoplasm in isolated *Zinnia* mesophyll protoplasts is occupied by lens-shaped chloroplasts and a large vacuole (Fig. 6.3A). Other organelles are hardly prominent. However, the volume of cytoplasm and number of organelles other than chloroplast increases when cells are cultured in a differentiation-inducing medium, but the arrangement of chloroplasts gradually falls into disorder, leaving large spaces between them and the plasma membrane (Fig. 6.3B-C). Soon cells elongate due to synthesis of metabolites and thickenings appear in the cell wall, which are mostly reticulate or helical in pattern (Fig. 6.3D). Secondary wall thickenings include cellulose microfibrils (arranged parallel to one another), xylan, lignin, hemicellulose, pectin and protein (Fig. 6.3E). The presence of ER, dictyosomes and microtubules between thickenings give the impression that these organelles blanket those parts of the wall that remain unthickened (Fig. 6.3F).

Detection of lignin and its quantitative measurement is done by the phloroglucinol test (*see* Morrison 1972). Lignification is mediated by the enzymes peroxidases, laccase, dehydrogenase, CoA reductase and phenylalanine-ammonia-lyase, which can remain bound to the cell wall (Whetten and Sederoff 1995).

Cell autolysis is the general feature after secondary wall thickenings due to loss of nucleus and cytoplasmic contents in cultured cells. This process, leading towards cell death, seems to be programmed at the beginning of secondary wall thickening. Vacuoles in the tracheary elements may function as lysosomes at some stage and release hydrolytic enzymes, resulting in disruption of the intracellular structures (Church 1993). Twin phenomenon of cytoquiescence and cytosenescence of cells are exemplified by *Dianthus* tissue cultures (c.f. Fasella and Hussain 2014). Degradative changes associated with RNA, chlorophyll, ER/dictyosomes, mitochondria (loss of grana, stroma), chloroplasts (loss of thylakoid bodies) and alterations in membranes of organelles

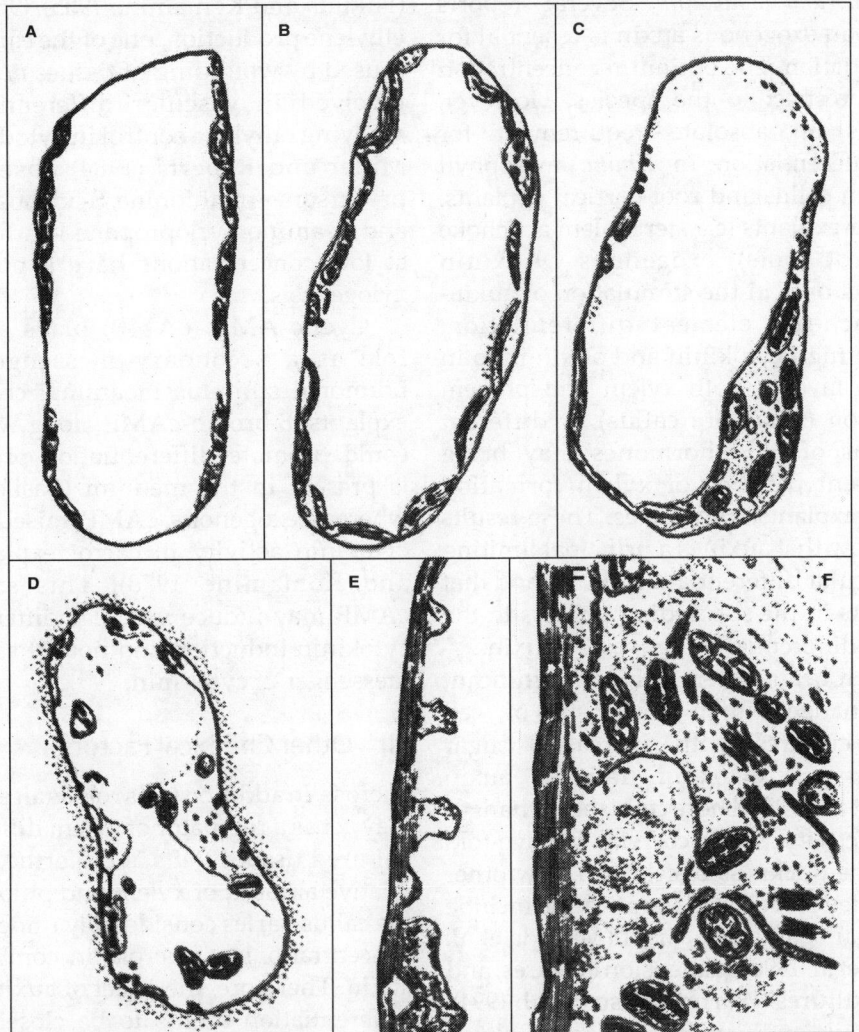

Fig. 6.3 Drawings based on EM observations by Fukuda and Komamine (1985) from longitudinal sections of *Zinnia* cells. For explanation, see the text.

result in cytoquiescence that ultimately lead to cytosenescence showing no recovery of cells. Cytoquiescense, however, seems to have a functional role in cytodifferentiation. Cells that are senescent do not show dedifferentiation, yet quiescent cells are dependant on autolytic activity of compartmentalised hydrolytic enzymes of lysomal system located in vacuoles of these cells resulting either in cell proliferation (callus formation) or xylogenesis.

6.1.3 *Physiological Aspects*

(I)   Hormonal Control

Phytohormones, particularly auxins, are reported to effect vascular differentiation quantitatively as well as qualitatively. Some evidence also points towards the involvement of cytokinins and gibberellins in the process of xylogenesis.

(a) Auxin and cytokinin: Several reports suggest that an exogenous auxin is essential for cytodifferentiation. The optimum concentration varies with respect to the species. However, cytokinin is an absolute requirement for vascular differentiation in *Zinnia* mesophyll cells, soybean callus and root cortical explants. In most other explants (e.g., Jerusalem artichoke tubers, carrot roots), exogenous cytokinin appears to act only in the stimulation of auxin-induced tracheary element differentiation. Sometimes a high cytokinin and auxin ratio in the medium favours both xylem and phloem differentiation (sycamore callus) or different combinations of phytohormones may bring about different patterns of xylem formation (lettuce pith explants) in cultures. These results appear to show that auxin is a principal limiting factor of vascular differentiation *in vitro* and that cytokinin acts in the succeeding process to the onset of cytodifferentiation caused by auxin.

The french bean system exhibits significant metabolic changes upon treatment of cell suspension cultures with increased sugar concentration and cytokinin ratio to auxin (Robertson et al. 1995). These cells show changes in enzyme activity which correlate to five-fold increase in the thickness of cell wall. Few other systems studied for cell wall lignification include tobacco (Nagai et al. 1994), carrot (Masuda et al. 1984, Suzuki et al. 1990) suspension cultures, and lettuce pith cultures (Warren-Wilson et al. 1994). For more details on the subject refer to Bolwell and Robertson (1997).

(b) Gibberellic acid, abscisic acid, ethylene and cAMP: Application of gibberellic acid (GA) in the presence of auxin and cytokinin reportedly has both promotory and inhibitory effects on cytodifferentiation (Gautheret 1985). This discrepancy may be due to the difference in the endogenous level of GA, which changes during various stages of cultured tissues. The requirement of GA in the medium for cellular differentiation is not absolutely necessary.

Abscisic acid (AA) inhibits cytodifferentiation in Jerusalem artichoke explants and bean callus

(Fukuda and Komamine 1985). On the contrary, ethylene production, one of the earliest reactions caused by wounding of tissues *in situ*, is closely involved in vascular differentiation. While studying ethylene control in cytodifferentiation, Miller and Roberts (1984) observed that its precursors—methionine, S-adenosylmethionine and 1-amino-cyclopropane-l-carboxylic acid—at low concentrations have a positive role in xylogenesis.

Cyclic AMP (cAMP) plays an important role as a secondary messenger following hormonal stimulus in animal cells. In lettuce explants, 8-bromo-cAMP along with cytokinin could promote differentiation provided auxin is present in the medium (Basile et al. 1973), whereas exogenous cAMP raised endogenous cytokinin activity in carrot explants (Mizuno and Komamine 1978). This suggests that cAMP may induce vascular differentiation via cytokinin induction and not act as a secondary messenger of cytokinin.

(II) Other Chemical Factors

Sucrose, in addition to its role as an energy source, may act as a regulator of xylem differentiation in cultured tissues (Jeffs and Northcote 1967). The relative amount of xylem and phloem tissues in the callus varies considerably under the varying concentrations of sucrose in combination with auxin. Therefore, the effect of auxin on vascular differentiation seems to be closely dependent on the presence of sugar. Other carbohydrates used as substitutes for sucrose are α-glucosyl disaccharides, glucose, maltose and trehalose, glycerol, and myoinositol (Bolwell and Robertson 1997).

Observations on various experiments deduce that a reduction in the total nitrogen (NH_4^+; and NO_3^-) content of the medium in cultures contributes to an increase in the percentage of vascularisation without affecting cell division in tissues and cells. In view of the fact that a low level of NH_4^+ favours the induction of embryogenesis in carrot suspension cultures (Fujimura and Kumamine 1979), intracellular changes in the nitrogen source in cultured cells might play an important role in switching over to a differentiation state from a quiescent state. Similarly calcium uptake has a significant role in tracheary element differentiation of cell suspension cultures (Robertson and Haigler 1990):

(III) Physical Factors

The role of light, temperature and pH has also been assessed for vascular differentiation in cell cultures. All these physical factors are effective only when provided at optimum levels and vary with the species. In *Helianthus,* vascular elements differentiated within the range of 17-31°C. A difference in the optimum temperature between that for cytodifferentiation and cell replication has been reported in cultured explants of Jerusalem artichoke tubers. For cytodifferentiation of *Zinnia* cells, the optimum pH in the medium was found to be 5.0 after autoclaving. A few reports have shown that increase or decrease in the optimum pH results either in delay in tracheary element formation or a decrease in their number.

6.1.4 *Cell Division and Cytodifferentiation*

Several earlier workers suggested that cell replication precedes the differentiation of xylary elements. The suggestion was based on an *in vivo* situation wherein the procambium exhibited continual divisions to produce xylem and phloem. The *in vitro* system is a good tool for studying the relationship between cell division and differentiation. Since the chemical factors (auxin, cytokinin, sugar) involved in vascular differentiation are generally the same as those regulating cell division, it appears that cell division and cytodifferentiation are two sides of the same coin.

The first report suggesting a close relationship between tracheary element differentiation and the cell cycle was published by Fosket (1968). He found that inhibitors of DNA synthesis (mitomycin C, fluorodeoxyuridine and colchicine) prevented cytodifferentiation in cultured stem segments of *Coleus* in which cell division preceded the differentiation. Such observations gave the impression that cell division is essential for cell differentiation. On the contrary, there are several reports suggesting that tracheary differentiation occurs without cell division. Torrey (1975) observed that some of the parenchymatous cells differentiated directly in single-cell cultures of *Centaurea cyanus* and cell division was unnecessary. In support of this Shininger (1975) found bromodeoxyuridine blocked cytodifferentiation but allowed cells to divide. Using the *Zinnia* system, Fukuda and Komamine (1985) made several observations on tracheary element differentiation by plating single cells isolated from mesophyll tissue, and confirmed that 60% differentiation originated from undivided cells.

From the point of view of cytodifferentiation, it may be interesting to consider the type of cell cycle preceding the origin of a new cell. The cell cycle may be either of the *quantal* or *proliferative* type. In a quantal cell cycle the daughter cells that form differ from the mother cell. Paradoxically, the proliferative cell cycle generates copies of the mother cell. Cells originating from the quantal cycle are unlikely to differentiate, whereas a cell formed by the proliferative cycle need not undergo further division for differentiation. The cytodifferentiation of tracheary elements is an irreversible specialisation of cells accompanied by the loss of nuclei and cell contents at maturity. This rules out dedifferentiation or redifferentiation of a cell that has once differentiated to a tracheary element or is in the progression of cytodifferentiation (Fukuda and Komamine 1985). However, the *Macleaya* mesophyll tissue system seems an exception wherein even differentiated single cells are reported to demonstrate totipotency (*see* Kohlenbach 1984).

6.1.5 *Vascular Differentiation by DNA Transformation*

Application of DNA technology to understand

the mechanism of vascular differentiation using various cell culture systems has led to identification of a number of vascular-specific genes. The lignification pathway is known in its entirety and considerable progress made towards cloning cell-specific proteins and regulation of polysaccharide synthesis. The promoter analysis of cloned genes of vascularization would identify transcription factors and elucidate signal transduction pathways coupled with biochemical studies using modulators of signalling components (Whetten and Sederoff 1995).

A number of molecular markers for tracheary element differentiation have been isolated for *Zinnia* cell culture system. Based on these markers three stages are involved in tracheary element differentiation. The stage I is characterized by dedifferentiation due to loss of photosynthesis process in cells. cDNA clones of 12 genes isolated control these events (Fukuda 1997). Dedifferentiated cells redifferentiated in stage II. Demura and Fukuda (1994) isolated 3 cDNA clones of tracheary element genes (TED2, TED3, TED4) expressing preferentially in mesophyll cells of *Zinnia* during their redifferentiation into tracheary elements at stage II. Xylem parenchyma and patternal secondary wall formation including programmed cell death occurs at stage III, the expression of which is controlled by several genes (Fukuda 2004). Involvement of phenylalanine-ammonia-lyase (PAL) promoter in conferring vascular-specific expression in addition to stress-related expressions has also been extensively analysed (Fukuda 1992). The promoters of the *Eucalyptus* CAD (Feuillet et al. 1995) and CCR genes (Boudet 1995), and PRP-like proteins in loblolly pine have been investigated for xylem-specific gene expression (Whetten and Sederoff 1995).

6.2 Phloem Differentiation

Induction of sieve elements of phloem in cell cultures can also be used to monitor differentiation and cellular potency as well as a source for the acquisitions of molecular probes.

Toth et al. (1994) induced *Stephanthus tortuosus* cultures to form sieve elements by substituting 0.1 mg l^{-1} NAA in the general maintenance medium with 5 l^{-1} 2,4-D in the presence of 2% sucrose and 1 mg l^{-1} kinetin. Thus a phloem enriched preparation could be obtained by preferential digestion of parenchymatous cell walls and collection of undigested sieve elements by centrifugation on percoll gradients. The advantage in phloem differentiation is that many phloem specific gene products are likely to give some insight into novel regulatory aspects of protein synthesis (Sakuth et al. 1993).

6.3 Organogenic Differentiation

Organogenic differentiation is an outcome of the process of dedifferentiation followed by redifferentiation of ceils. Dedifferentiation favours unorganised cell growth and the resultant developed callus has meristems randomly divided. Most of these meristems, if provided appropriate *in vitro* conditions, would redifferentiate shoot buds and roots. Mostly, the whole plant regeneration from cultured cells may occur either through shoot-bud differentiation or somatic embryogenesis. The formation of organs *in vitro* via a callus phase is termed indirect organogenesis and without callus phase as direct organogenesis. This establishes the totipotency of somatic cells which, under natural conditions, can also be observed to some extent during the vegetative reproduction of plant species. The cell(s) of stem, leaf and root cuttings of several taxa are able to directly differentiate shoots and roots, leading to the establishment of new individuals under *in vivo* conditions.

6.3.1 Factors Affecting Shoot-Bud Differentiation

(I)   Chemical Factors

Shoot-bud differentiation in cultured tissues besides genotypic influence is dependent on the auxin/cytokinin ratio in the medium. Skoog's group (USA) has done detailed and comprehensive work on this subject. Skoog and Miller (1957) rejected the concept of organ-forming substances (rhizocaulines and caulocaulines) proposed by Went (1938) and suggested instead that organ formation is controlled by quantitative interaction (ratio rather than absolute concentration) of substances in growth and development.

Cytokinin (adenine or kinetin) in the medium leads to the promotion of bud differentiation and development. Kinetin is 30,000 times more potent than adenine. The shoot-forming effect is modified by auxin (IAA and NAA), which at lower concentrations (5 µM IAA) suppresses the differentiation of bud in the tobacco callus. About 15,000 molecules of adenine or 2 molecules of kinetin are required to neutralise the inhibitory effect of 1 molecule of IAA on shoot-bud differentiation. A relatively high concentration of auxin favours cell proliferation and root differentiation while higher levels of cytokinin promote bud differentiation. Thus, root-shoot differentiation is a function of quantitative interaction between auxin and cytokinin (*see* Fig. 6.4).

Increased levels of phosphate (PO_4^-) in the medium is reported to counteract the inhibitory effect of auxin and promotes bud formation in the absence of cytokinin. Casein hydrolysate or tyrosine also induces kinetin type bud formation even in the presence of higher levels of IAA in the medium. Components of tobacco smoke, such as benzoapyrene, benzoepyrene, may be used as a substitute for kinetin and IAA in shoot-bud differentiation, but strangely these compounds seem to be effective for haploid tissues or callus.

The requirement for exogenous auxin and cytokinin in the process of bud differentiation varies with the tissue system and apparently depends on the endogenous levels of the two hormones in the tissue, viz. an auxin and a cytokinin. Endosperm cultures require cytokinin

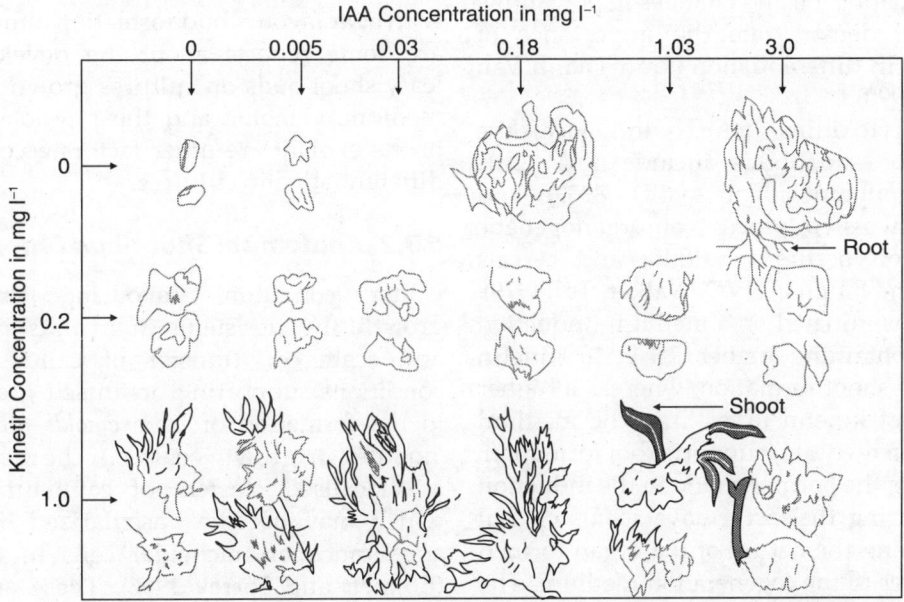

Fig. 6.4 Effect of increasing IAA concentrations at different kinetin levels plus casein hydrolysate (3 mg l⁻¹) at all levels of combinations on organogenesis in tobacco callus cultured on semi-solid White's medium. Drawings based on observations by Skoog and Miller (1957) from 62-day-old cultures. Note root formation in the absence of kinetin and in the presence of 0.18-3.0 mg l⁻¹ IAA; shoot formation in the presence of 1.0 mg l⁻¹ kinetin, especially with IAA concentrations, in the range 0.005-0.18 mg l⁻¹.

alone or in combination with a very low level of auxin for bud differentiation. Three plasmid genes, *tms* 1, *tms* 2 (encoding a tryptophan mono-oxygenase and indoleacetamide hydrolase respectively) and *ipt* (encoding an isopentyl transferase) ensure the biosynthesis of auxin and cytokinin in *Agrobacterium*. The portion of plasmid DNA (T-DNA) from which auxin or cytokinin gene have been deleted can be inserted under different promoters to plant cells. The *ipt* gene stimulates bud differentiation and inhibits root formation in transformed tobacco and plant tissues (Ooms et al. 1985).

Polyamines have been shown to be associated with induction of cell division, growth and differentiation of bacterial, animal and plant cells. However, their mechanism of action is not known. Abnormal development of flowers in a tobacco mutant with a high polyamine content, accumulation of conjugated polyamines in flowering (not vegetative) plants and correlation between the expression of Ri TL-DNA (*see* Chapter 13) and changes in polyamine metabolism, demonstrate the importance of polyamines in differentiation (Tran Thanh Van and Trinh 1990).

Other cytokinins which influence the induction of shoot-buds include BAP, 2-iP, 6-tetrahy-dropyrane-adenine (SD 8339) and zeatin. A two-step process of organogenetic differentiation occurs in alfalfa and cereals (Saunders and Bingham 1972, Walker et al. 1979). Alfalfa callus initiated on a medium (induction medium) containing higher 2,4-D to kinetin also favours shoot formation, whereas a higher proportion of kinetin to 2,4-D in the medium (regeneration medium) supports root formation. Interestingly, the hormone ratio in the induction medium during the last 4 days seems critical in determining the nature of the organ formed upon transfer to the regeneration medium. The cereal callus, on the other hand, is initiated in the induction medium containing 2,4-D and kinetin but organogenesis occurs only when pieces of callus are transferred to a hormone-free regeneration medium.

(II) Physical Factors

Light intensity plays an important role in organogenesis. *Pelargonium* callus maintained under continuous light or high light intensity remains whitish and does not exhibit organogenesis. High Sight intensity has been shown to be inhibitory for shoot-bud formation in tobacco. The quality of light also influences organogenic differentiation. Blue light promotes shoot-bud differentiation in tobacco callus while red light stimulates rooting (*see* Bhojwani and Razdan 1983). In general, maintenance of callus under alternating light and dark periods (15-16 hr) may prove satisfactory for differentiation of shoots.

Temperature also affects the callus growth and differentiation. Increase in temperature up to 33°C may be associated with rise in the growth of tobacco callus but for shoot-bud differentiation a lower temperature (18°C) may be optimal. The physical state of the medium also influences the shoot-bud differentiation. Medium solidified with agar favours bud formation although there are some reports about the development of leafy shoot-buds on cultures grown in a liquid medium. Genome and the physiological state of the explant are other factors accounting for differentiation in cultures.

6.3.2 Anatomy of Shoot-bud Differentiation

Under conditions favouring unorganised growth, the meristems in a callus are random and scattered. Transfer of callus pieces to conditions supporting organised growth leads to the formation of *meristemoids* (also termed 'nodules' or 'growth centres'). The meristemoids are localised clusters of cambium-like cells which may become vascularised due to the appearance of tracheidal cells in the centre (Bonnett and Torrey 1966). These are the site for organ formation in the callus and can form roots or shoots. The origin of roots or shoots is endogenous despite their later sometimes appearing to have grown exogenously.

6.3.3 Explant and Electric Stimulation

Regenerability of explants is generally influenced by their physiological state, size and orientation on the medium. In some plants (e.g., tobacco) all parts are amenable to regeneration *in vitro*, whereas potential for other plants may be restricted to only certain tissues. Application of weak electric current is reported to markedly increase the organogenic and embryogenic potential in tissue cultures (Rathore and Goldsworthy 1985). Bagga et al. (1985) suggested the role of phytochrome in shoot induction.

6.4 Totipotency of Epidermal Cells

Development of shoot-buds from cultured single epidermal cells of lax *(Linum usitatissimum)* hypocotyl is the classic example of cellular totipotency. Several other species have also been reported to form shoot buds/embryos from superficial cell layers of stems in cultures, namely, *Daucus carota, Ranunculus sceleratus, Exocarpus cupressiformis, Torenia fournieri* and *Nicotiana tabacum*. The potential of forming shoot-buds in flax epidermal peels depends on the age of the hypocotyl. Epidermal cells peeled all along the length of the hypocotyl in very young seedlings form shoot-buds, while in peels taken from seedlings older than 15 days this potential is restricted to the basal half of the hypocotyl. The formation of shoots from *Pinus radiata* cotyledons in culture depends on the developmental state of the latter at the time of excision. Epidermal and subepidermal layers of cotyledons (younger) predisposed to form shoots are active in cell division and possess less-developed stomatal complexes, thinner cell walls, absence of epicuticular wax, partially reduced protein reserves and unhydrolysed lipid reserves. On the other hand, cells from cotyledons (older) that did not form shoots have more developed stomatal complexes, thicker cell walls with epicuticular wax, depleted protein reserves and nearly depleted lipid reserves (Aitken-Christie et al. 1985). These cells do not actively participate in cell division.

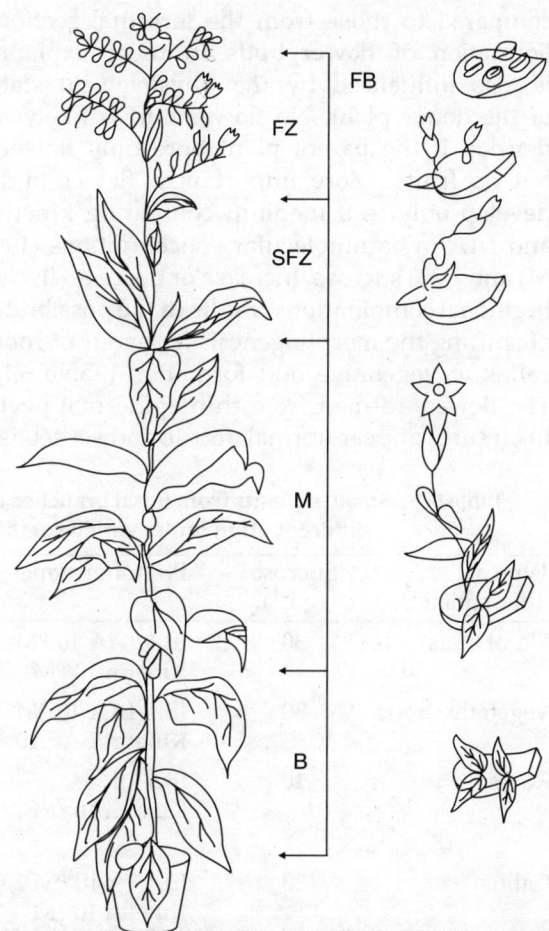

Fig. 6.5 Diagram showing capacity of peels excised from different regions of a flowering branch from tobacco plant for vegetative or floral bud formation in culture (B—base; M—middle zone; SFZ—subfloral zone; FZ—floral zone; FB—floral branches: after Tran Thanh Van 1977).

In tobacco, the type of bud formed in cultures varies with the region from which the peel is derived. Under certain culturing conditions the peels from floral branches produce only flower buds (Fig. 6.5) and those from the basal part of the plant produce vegetative buds. Peels derived from the middle section bear both types of buds in different proportions according to their distance from the base of the plant. Similar explants taken from the basal region of the inflorescence branch give a better response compared to those from the terminal section. Formation of flower buds in these explants is also influenced by the physiological state of the donor plant. No flower bud is likely to develop if the parent plant bore only flowers but no fruits. More importantly, flower buds develop only in a medium containing kinetin and IAA in equimolecular concentrations (10^{-6} M) and 2-3% sucrose. Increase or decrease in the hormonal combinations results in the possibility of shifting the morphogenesis in favour of root, callus or vegetative bud formation (Table 6.1). The flowers formed from thin epidermal peels in cultures appear normal, fertile and set seeds.

technique (*see* Section 5.9.1) to clone single cells grown in shake cultures from pieces of a crown gall and obtained calli differentiating into shoots. Tips (3-5 mm long) excised from these abnormal teratoma shoots were grafted onto the cut end of the stems of normal tobacco plants from which axillary buds had been removed. The successful grafts were also observed to have abnormal shoots. Braun and Wood (1976) continued to graft these shoot tips onto normal tobacco plants until after successive generations of grafting they obtained shoots which appeared structurally, histologically and functionally normal. Thus, whole plants were recovered in

Table 6.1   Small explants from floral branches of tobacco expressing different types of organogenic differentiation under varied combinations of hormones and other factors[a]

Type of differentiation	Sucrose ($g\ l^{-1}$)	Phytohormones	Ratio auxin/ cytokinin	Other favourable conditions required
Floral buds	30	IBA/IAA 10^{-6}M Kinetin 10^{-6}M	1.0	Light; terminal bud in green fruit stage
Vegetative buds	30	IBA/IAA 10^{-6}M Kinetin/BAP 10^{-5}M	0.1	Light
Roots	10	IBA 10^{-5}M Kinetin 10^{-7}M	100.0	Darkness; terminal bud in mature fruit stage
Callus	30	IBA 3×10^{-6}M/2,4-D 5×10^{-6}M Kinetin 10^{-7}M	50.0	—

[a] After Tran Thanh Van et al. (1974) and Tran Thanh Van and Trinh (1990).

6.5 Totipotency of Crown-gall Cells

Typical crown-gall cells show a complete lack of organogenic differentiation but have a capacity for unlimited growth independent of exogenous hormones in cultures. The bacterium *Agrobacterium tumefaciens* infects some plant species and induces inside the host tissue a special type of tumour (teratoma) called a crown gall. Cells of these tumours possess a pronounced capacity to differentiate shoot-buds and leaves when cultured *in vitro* for unlimited periods. However, shoots derived from a teratoma are abnormal in growth and morphology. Braun (1959) used the nurse culture

cultures from crown gall tissue, which under natural conditions grows as an unorganised mass of cells.

The crown-gall system has been extensively used in plant genetic manipulations (Chapter 13) since regenerated normal plants have been found to be fertile. Yang et al. (1980) demonstrated that explants from vegetative or reproductive organs of crown-gall cells derived normal plants when cultured reverted to tumour characteristics (such as auxin autonomy); but the F_1 progeny raised from seeds of such plants had lost the tumour trait. These studies point out that temporary loss of totipotency can be restored in crown-gall cells under *in vitro* conditions.

Somatic Embryogenesis

Somatic embryogenesis is the process of a single cell or a group of cells initiating the developmental pathway that leads to reproducible regeneration of non-zygotic embryos capable of germinating to form complete plants. Under natural conditions this pathway is not normally followed, but from tissue cultures *somatic embryogenesis* occurs most frequently and as an alternative to organogenesis for regeneration of whole plants. Adherence to this pattern of morphogenesis depends on co-ordinated behaviour of a cell or cells to establish polarity as a unit and thereby initiate gene action sequentially specific to emerging tissue regions. According to Sharp et al. (1982), somatic embryogenesis is initiated either by 'pre-embryogenic determined cells' (PEDCs) or by 'induced embryogenic determined cells' (IEDCs). In PEDCs, the embryogenic pathway is predetermined and the cells appear to only wait for the synthesis of an inducer (or removal of an inhibitor) to resume independent mitotic divisions in order to express their potential. Such cells are found in embryonic tissues (including scutellum of cereals), certain tissues of young *in vitro* grown plantlets, the nucellus and the embryo sac (within ovules of mature plants). IEDCs, on the other hand, require redetermination to the embryogenic state by exposure to specific growth regulators such as 2,4-D. These cells are differentiated generally in microspore (anther) cultures and callus cultures. Once the embryogenic state has been reached both cell types proliferate in a similar manner as *embryogenic determined cells* (EDCs). Plantlets are then produced directly by following the full embryogenic pathway as a co-ordinated group of EDCs. Sometimes individual cell or cells from the group may escape and give rise to either *embryoids* or nodular embryogenic callus (e.g., scutellar callus) consisting of *proembryoids* (Williams 1987). These embryo-like structures are bipolar units and germinate into full plantlets under suitable culture conditions. Thus, a complete sexual apparatus is not an essential prerequisite for embryogeny in tissue cultures. Many species that were once considered recalcitrant have been induced to respond *in vitro* and a significant body of information has been gathered to establish the embryogenic potential of somatic plant cells (*see* Reviews: Zimmerman 1993, Thorpe 1995, Zhou et al. 2000, Thorpe and Stasolla 2001). In plantation forestry somatic embryogenesis has potential for rapid production of seedlings in at least 115-120 known *Pinus* species at very low costs (Pullman and Bucalo 2011).

Embryos formed in cultures have been variously designated as accessory embryos, adventive embryos, embryoids and supernumerary embryos. Kohlenbach (1978) has proposed the following classification of embryos.

1. Zygotic embryos—those formed by fertilised egg or the zygote.
2. Non-zygotic embryos—those formed by cells other than the zygote.
   - (I) Somatic embryos—those formed by the sporophytic cells (except zygote) either *in vitro* or *in vivo*. Such somatic embryos arising directly from other embryos or organs (stem embryos in carrot and buttercup) are termed *adventive embryos*.
   - (II) Parthenogenetic embryos—those formed by unfertilised egg.
   - (III) Androgenic embryos—those formed by the male gametophyte (microspore pollen grains).

7.1 Somatic Embryogenesis in Dicotyledonous Cultures

7.1.1 *Explant for Initiation of Embryogenic Callus*

Totipotent embryogenic cells have been most commonly obtained from explants of embryonic or young seedling tissues. Excised small tissues from young inflorescences (before maturation of floral primordia) are equally effective for the induction of somatic embryogenesis in cultures. Other explants used are the scutellum, young roots, petioles, immature leaf, and immature hypocotyl. Corredoira et al. (2015a) successfully developed protocol for induction of somatic embryogenesis in explants of shoot cultures established from adult trees of *Eucalyptus globulus* and the hybrid *E. saligna* × *E. maidenii*. In *Ranunculus sceleratus* various floral (including anthers) and vegetative tissues proliferate to form a callus on a medium containing coconut milk (10%) with or without IAA. Within 3 weeks, numerous embryos appear from the peripheral and deep-seated cells of the callus. A high yield of embryogenic calli can also be obtained from isolated fully differentiated mesophyll cells or protoplasts in a defined culture medium (*see* Bhojwani and Razdan 1983). *Citrus* nucellar cells have a natural potential for somatic embryogenesis, which is also manifested in their cultures. Even the nucellus from mono-embryonate (seeds bearing only single zygotic embryo) species of *Citrus* exhibit somatic embryo formation in cultures. Interestingly, the addition of malt extract (500 mg l^{-1}) to a basal medium has proven most promotive for embryogeny in nucellus cultures.

Somatic embryos germinate *in situ* or when they are excised and cultured individually on a fresh semi-solid medium. A special and noteworthy feature may be the development of a fresh crop of adventive embryos (numbering 5-50) which originate from single epidermal cells on the stem surface of the plantlets obtained from germinating embryos (Fig. 7.1).

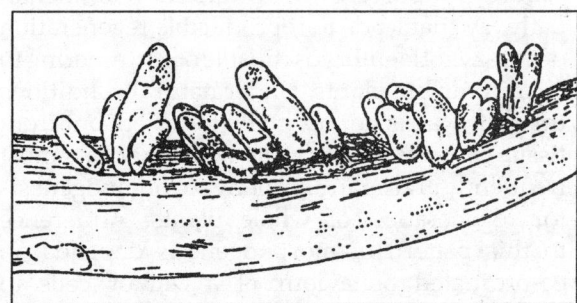

Fig. 7.1 A large number of adventive embryos originating from single epidermal cells on the stem surface *in vitro* (drawing based on observations made by Konar and Nataraja 1969).

Age, physiological state, genotype and orientations of the explant, while in contact with the medium, influence the induction of somatic embryogenesis. These aspects govern the disruption of explant tissue integrity, callus friability, isolation of cells and other requirements in order to enhance somatic embryogenesis in various species (Merkle et al. 1995).

7.1.2 *Basic Requirements*

The essential requirements for the induction and promotion of somatic embryos are primarily established by the use of suspension cultures.

The *in vitro* development of somatic embryos was first observed in carrot (*Daucus carota*) suspension cultures by Steward and associates (1958) although Reinert (1958, 1959) was also able to induce somatic embryogenesis in a callus cultured on a semi-solid medium.

(I) Auxin Supply

The presence of auxin in the medium is generally essential for embryo initiation. Tissues or calli maintained continuously in an auxin-free medium generally do not form embryos; therefore, somatic embryogenesis is achieved in two steps: First, the callus is initiated and multiplied on a medium rich in auxin (2,4-D, 0.5 mg l^{-1}) which induces the differentiation of localised groups of meristematic cells called *'embryogenic clumps'* (ECs). Second, the ECs develop into mature embryos when transferred to a medium with a very low level of auxin (0.01-0.1 mg l^{-1}) or no auxin at all. Consequently, the medium with auxin is called a *proliferation medium* (PM) and without auxin an *embryo development medium* (EDM). Mature embryos develop in EDM. In this respect the PM could be regarded as the 'induction medium' for somatic embryogenesis and each EC a disorganised embryo.

Mostly 2,4-D has been used to induce somatic embryogenesis. In some systems, other auxins have also been used for this purpose. For example, the cultures of pumpkin require NAA and IBA in the PM, whereas in nucellus cultures of *Vitis* the presence of β-naphthoxyacetic acid and BAP proved inductive for somatic embryogenesis. The fact that the auxin-grown cultures of carrot tissue produce more ethylene than the auxin-free cultures, suggests that 2,4-D-induced embryo suppression may be mediated through ethylene production. In support of this conjecture it may be pointed out that 2-chlorophosphonic acid (Ethephon), which releases ethylene in plant tissues, suppresses the development of mature somatic embryos without an appreciable reduction in growth and multiplication of the ECs in suspension cultures of carrot. Thus, in a medium containing 2,4-D

the tissue multiplication goes on but mature embryos do not appear. 2-(2,4-Dichlorophenoxy) propionic acid, however, appeared most effective for auxin inducing somatic embryogenesis in alfalfa (cf. Redenbaugh and Walker 1990). There are some reports about the promotive effect of zeatin on embryogenesis (Fujimura and Komamine 1980), which is contrary to the general inhibitory effect of cytokinins and PCIB (Fujimura and Komamine 1979a,b) on the embryogenic potential of tissue cultures. In anther cultures of *Hordeum* (Claphaum 1973) and *Zea mays* (Genovesi and Collins 1982) TIBA was more effective than auxins normally used for callus induction and embryogenesis.

(II) Nitrogen Source

A substantial amount of nitrogen, usually in reduced form such as ammonium salts, is required for both embryo *initiation* and *maturation* (Merkle et al. 1995). In carrot, somatic embryos are formed when high concentrations of inorganic nitrogen (nitrate) is supplied to the culture medium. However, a combination of reduced nitrogen and nitrate ($NH_4Cl + NO_3$) in the medium appears beneficial for a number of culture systems. The source of nitrogen can be in the form of complex addenda such as coconut milk, casein hydrolysate, a mixture of amino acids or single amino acid (*L*-glutamine, *L*-alanine) and ammonium ions (NH_4Cl, NH_4NO_3). An MS medium contains high levels of nitrogen in the form of ammonium nitrate.

(III) Other Constituents

An essential feature for embryogenesis in wild carrot is the presence of a high concentration of potassium (20 m mol^{-1}) in the medium. Similarly, the amount of dissolved oxygen (DO) in the medium is critical and should be below 1.5 mg l^{-1} (Kessell et al. 1977). A low amount of DO seems to result in the synthesis of higher levels of cellular ATP while a high amount of DO favours rooting. However, Nishimura et al. (1993) observed a high frequency of somatic embryo formation (394 SE ml l^{-1}) at DO concentration as high as 88% or more. Activated charcoal is also reported to improve embryogenesis in carrot, whereas choramquat chloride (CCC) stimulated embryogenesis in citrus (Spiegel-Roy and Kochba 1980). Certain tissues release volatile and non-volatile substances in cultures which can inhibit somatic embryogenesis in the callus. For example, ovules of monoembryonate *Citrus medica* release ethanol as one of the volatile inhibitors and IAA, ABA and GA_3 as non-volatile inhibitors (Tisserat and Murashighe 1977). This necessitates manipulating the media constituents to minimise the possibility of inhibition.

Involvement of polyamines in somatic embryogenesis has been suggested in case of celery (Altman et al. 1990), and mango (Litz et al. 1993); nevertheless, the observed changes in polyamine content, biosynthesis and their role in induction of somatic embryogenesis in plants remains to be well established. Exposure of mesophyll protoplasts of alfalfa cultivar to a low voltage electric field induced direct embryogenesis in a 100% preparations as against 40% preparations of the untreated controls (Dijak et al. 1986). The electric stimulus seems to promote differentiation like shoots or somatic embryos by affecting cell polarity through changes in organization of microtubules (de Jong et al. 1993). A[15] N-NMR spectroscopic study on carrot somatic embryogenesis has revealed association of amino acid metabolism to embryo development (Joy et al. 1996). For example, globular and torpedo stages of embryo showed increasing glutamine and arginine content, whereas germinating embryos showed decreasing content of these amino acids. For some plant species including gymnosperms like *Picea abies* (Norway spruce) arabinogalactan proteins (AGPs) were essential for somatic embryo development in cultures (Majewska–Sawkas and Nathnagel (2000). Similar response was noted using brassinosteroids, jasmonate, sterols and systemins (c.f. George et al. 2008). Thus, various factors operate in tandem to promote somatic embryo formation in dicotyledonous cultures.

7.1.3 Establishment of Embryogenic Suspension Cultures

To initiate suspension culture, 2.5 to 3.5 g callus tissue is placed in 50 ml of the growth medium in Erlenmeyer flasks and the flasks are agitated on a shaker platform at 125-160 rpm. Alternatively, the auxophyton (developed by F.C. Steward and colleagues) provides a more gentle method of agitation and aeration. Tumble tubes which hold 10 ml medium and culture flasks (nipple) containing 250 ml medium are slowly rotated on a klinostat at 1 rpm. Due to the asymmetry of these two types of culture vessels, the tissue inside them alternatively lifts up to aerate and bathes the cells in the medium. A convenient procedure, however, is to transfer the primary callus tissue to tumble tubes for the first passage and then, when there are abundant cells, to the larger culture flasks. Other more elaborate methods for growing embryogenic suspension cultures, e.g., spinning cultures, stirred cultures and various continuous culture systems are discussed in detail by Street (1977). The bioreactor, a mass culture system originally developed for microbial fermentations, has been tested for somatic embryogenesis on a large scale. The higher plant cell system (alfalfa) appears to grow successfully in airlift bioreactors in contrast to stirred-tank bioreactors used in microbial cultures. Styer (1985) reviewed bioreactor designs suitable for plant cell cultures and explained that although cells could be grown as batch, semi-continuous or continuous cultures, the highest efficiency can be obtained by continuous culture (example: carrot cell suspension culture). A kinetic model was subsequently developed in which substrate utilisation, culture growth, and embryo development over the time course of an embryogenic culture can be monitored and bioreactor conditions optimised for the production of somatic embryos mature enough to grow into plants (Cazzulino et al. 1990). To obtain synchronous development of somatic embryos on mass scale, a modified stirred-tank bioreactor was designed for carrot suspension cultures (Molle et al. 1993). The detailed description of bioreactors used in large scale production of somatic embryos is given in volume on synseeds edited by Redenbaugh (1993).

Embryogenic cell suspension develops within 2-3 weeks of agitation provided cells are transferred to a fresh medium at periodic intervals. Frequent subcultures not only prevent cells from rapid senescence at the end of the growth phase, but maintain embryogenic potential by minimising the extent of chromosomal changes in cell cultures. To subculture embryogenic suspensions, the cells in the log phase are allowed to settle for several minutes so that it is possible to decant or aspirate almost all the medium. An aliquot of one-fourth to one-sixth of the entire population of cells is then transferred to a culture flask with fresh medium and agitated. For a finer suspension, cells may be sieved (using 200 μm stainless steel/ nylon mesh sieve, or single layer or cheese cloth) at the time of transfer. The protocol for inducing somatic embryogenesis in tissue cultures of some dicot species is given in Appendix 7.1.

7.2 Somatic Embryogenesis in Monocotyledonous Cultures

Many monocotyledonous plants are of agricultural and medicinal importance. Unlike dicots, the vegetative parts of a monocot plant do not readily proliferate in cultures. Therefore, explants are best taken from embryogenic or meristematic tissues (young inflorescences and leaves). Considerable success was achieved in obtaining reproducible regeneration of plants from embryogenic cultures of all major cereals and grass species (Vasil et al. 1982). The procedures have been developed to induce somatic embryogenesis in suspension cultures of other monocot species, such as *Dioscorea (D. floribunda, D. bulbifera)*.

7.2.1 Selection of Explant

(I) Zygotic Embryo

Young caryopses (10-15 days after pollination)

or seeds are sterilised by a 30 s rinse in 70% ethanol, followed by 10-20 min soak in 20% commercial bleach to which a few drops of detergent have been added as a wetting agent. Some species may require a further 30-60 s rinse in mercuric chloride (0.01-0.1%) to eliminate the contamination problem. Explants are then washed at least three times in sterile distilled water and zygotic embryos removed aseptically with the unaided eye (maize embryo) or using a dissecting microscope (for young and immature embryos). Excised embryos are now transferred to a culture vial containing MS medium supplemented with 2,4-D (0.5-2.5 mg l^{-1} in the case of cereals or 18 μM for *Dioscorea*) and sucrose (2-6%). Cultures are incubated in diffused light or complete darkness. A small and slow-growing callus will appear in 4-6 weeks (Williams and Maheshwaran 1986). In gymnosperm (conifer) species, immature zygotic embryos have embryogenic tissue arising from cells in suspensor region which is initiated into embryogenic cultures (Bornman 1993).

(II) Young Inflorescence

Premeiotic inflorescences, with the primordia of the individual florets just beginning to protrude, have been observed to be the most suitable material in some systems. The inflorescences, generally 1-2 cm in length, are sterilised according to the procedure described for zygotic embryos. Following sterilisation, each inflorescence is exposed by a vertical incision through the surrounding leaves and then cut into 1-2 mm thick segments. Individual segments are then cultured on a medium containing 2,4-D for proliferation and initiation of an embryogenic callus.

(III) Young Leaf

Leaves (unexpanded) of young seedlings obtained from seeds, germinated under aseptic conditions, are removed and cut into 1-2 mm thick transverse segments starting from the level of the shoot meristem up to the leaf apex. Six to eight explants are placed on a nutrient medium to obtain a callus.

7.2.2 Induction of Embryogenic Cell Suspension

The callus obtained from cultured explants is sliced and teased apart into small pieces which are then incubated in a liquid medium according to the procedures described in Appendix 7.1. Once a good embryogenic suspension has been established, somatic embryos can be obtained either by allowing the culture to age or by incubating the embryogenic tissue in a medium without 2,4-D. Since an unsupplemented basal medium often encourages root formation, the normal practice is to add 500 mg l^{-1} glutamine with 0.1 μM zeatin or 0.1 μM ABA to the medium to facilitate the embryo development.

7.2.3 Orchardgrass Embryogenesis

All major species of cereals/grasses regenerate plants *via* somatic embryogenesis and 2,4-D is by far most effective for producing embryogenic culture (Vasil and Vasil 1986), but orchardgrass seems an exception where 30 μm dicamba is reported to give best response. Further the addition of NH$_4^+$ (2.5-3 mM) and CH (3 g l^{-1}) to the medium seems essential for induction and proliferation of embryogenesis (Trigiano et al. 1992, Gray et al. 1993).

Whereas somatic embryos of dicots are usually bipolar, in monocotyledons like *Dioscorea floribunda* somatic embryos appear as short conical structures (Fig. 7.2A), each with a collar of cotyledonary tissue. The cotyledonary tissue initiates as a complete ring but often develops asymmetrically; when mature, these embryos

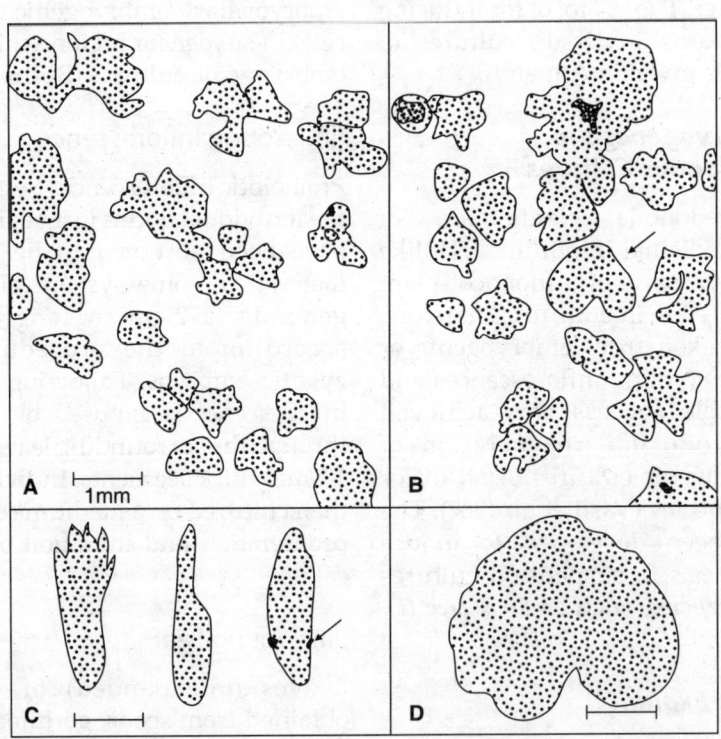

Fig. 7.2 Illustrations of somatic embryos formed in monocotyledon *Dioscorea floribunda* (*see* details in the text; designed after observations of Ammirato 1984).

have first visible leaf (Fig. 7.2B). In contrast, a typical zygotic embryo has a single fan-shaped cotyledon with a flattened apical end and round basal end (Fig. 7.2C). This cotyledon has a sheathing leaf base which encloses the shoot apex (Fig. 7.2C, arrow). The mature monocot somatic embryo (Fig. 7.2D) is capable of growing on a semi-solid MS medium supplemented with glutamine and zeatin individually, but if placed in groups of five or more the plantlet formation may proceed on an unsupplemented medium.

7.3 Embryo Maturation and Plantlet Development

Germination of the somatic embryo can occur only when it is mature enough to have functional shoot and root apices capable of meristematic growth. Somatic embryos show poor germinable quality with respect to their convertibility into plants. This is because these embryos do not go through 'embryo maturation' phase which is characteristic of seed or zygotic embryos. During this phase, accumulation of embryo-specific reserve food materials and proteins imparts desiccation tolerance to seed embryos and thereby promote their normal development for germination. Maturation of somatic embryos, however, could be accomplished with abscisic acid (ABA) which is known to increase desiccation tolerance in somatic embryos of carrot, celery, soybean, alfalfa and other plants (*see* Molle et al. l993).

High auxin levels can inhibit development and growth of the shoot meristem and often embryos mature when the embryogenic cell suspension is transferred to a medium lacking auxin. The addition of a low level of cytokinin (zeatin 0.1 µM) in combination with ABA may prove beneficial for embryogenesis of low density cell cultures. Sometimes somatic embryogenesis may be repetitive and new centres of embryogeny arise from maturing embryos. Thus, the germination phase in somatic embryos is adversely affected by their highly asynchronous development. Therefore, it becomes essential to obtain some degree of uniformity in terms of

the initial population by sieving the inoculum (*see* Nomura and Komamine 1999). A graded series of stainless steel or nylon mesh sieves is recommended for sieving the embryogenic cell suspension, followed by centrifugation in 16% Ficoll solution containing 2% sucrose. Replacing sieves with glass beads has also been effective. This ensures isolation of cell clusters, of 3-10 cells each, and as a result embryo development and maturation occurs more or less synchronously when these clusters are moved to an auxin-free medium. Alternatively, computer sorting methods could be applied to obtain synchronous embryos (Harrell and Cantliffe 1991).

Various physical factors may also affect embryo maturation depending on the requirement of the species. In such species which require cold treatment for seed germination, it may be necessary to chill the young or mature embryos for their normal development into plantlets. This procedure, which, mimics seed maturation *in vivo*, may be necessary to trigger metabolic processes needed for germination and seedling growth. Progressive increase in sucrose levels are also used to achieve maturation. Thus, embryo desiccation by a high-sucrose medium may be involved to develop plantlets. The maturation of somatic embryos proceeds more normally in complete darkness although light seems essential for somatic embryogenesis in some cultures (e.g., tomato, egg plant).

Somatic embryos germinate on agar medium without growth regulators. Single embryos may profit by the inclusion of low levels of zeatin (0.1 µM) in the medium. Application of GA_3 is required for the development of root and shoot from embryos *of Citrus*. After a number of leaves have formed, the small plantlets are transferred to jiffy pots, or vermiculite, for subsequent growth and development.

Maturation of somatic embryos in many coniferous species could be achieved by combination of activated carbon (AC) and ABA (high concentrations) in liquid or gelled medium (Pullman and Bucalo 2011). Variations included AC-coated embryos plated on ABA containing medium or plating somatic embryos on filter paper coated with AC on ABA-containing medium. Advances have also been made in optimization of palm somatic embryogenesis right from explants choice, culture methods, somatic embryo maturation and conversion to plantlets, and acclimatization of regenerated plants. These are summarised in a review by Ree and Guerra (2015).

7.4 Loss of Morphogenic Potential in Embryogenic Cultures

Callus or suspension cultures due to aging or prolonged subcultures often show a progressive decline and even complete loss of morphogenic ability. Three hypotheses have been proposed to explain this phenomenon.

7.4.1 *Genetic and Molecular Aspects*

Nuclear changes such as polyploidy, aneuploidy and chromosomal mutations in cultured cell may be responsible for the loss of organogenic or embryogenic potential in prolonged cultures. This loss is generally irreversible. Understanding the genetic mechanism of somatic embryo development, however, can be facilitated by identification of biochemical markers in the process of somatic embryogenesis. One of the approaches followed to determine biochemical markers is the use of drugs that can block specific stages of embryogenesis without affecting the proliferation of cells. For example, sensitivity of carrot cell culture to tunicamycin was used to identify the role of extracellular-secreted proteins which have to be properly glycosylated in order to act sequentially at different stages of somatic embryo development (de Vries et al. 1988). Isolation and characterisation of drug-resistant mutants also enable a search for biochemical markers. Most interesting results may be cited from the characterisation of temperature-sensitive mutants in which the embryo development process is impaired. For example, phosphorylation seems to cause a defect in one or more peptides of 'heat shock'

proteins induced at high temperature in carrot mutant cell line Ts59. From this, one can speculate that (a) heat-shock proteins are important in several steps of the development programme, (b) phosphorylation is a signal for activation of particular function and (c) kinases have strict specificity (*see* Lo Schiavo et al. 1988). Another mutant line (tsllc) at non-permissive temperatures is unable to acquire polarity because the somatic embryo reaches the globular stage and subsequently enlarges to form secondary embryos or monstrosites. However, mutants impaired in somatic embryogenesis can be rescued by the addition to the medium of conditioning factors secreted by embryogenic cultures (*see* Terzi and Lo Schiavo 1990, Kragh et al. 1996).

A number of extracellular proteins exuded by embryogenic cells in the medium are implicated in the induction and development of somatic embryos in carrot suspension cultures (de Vries et al. 1988). Some of these include EP3 (de Jong et al. 1993, van Hengel et al. 1998), AGPs (Pennell et al. 1992) and 21D7 (Smith 2000) proteins.

Somatic embryogenesis is a genetically controlled process in orchardgrass (Gavin et al. 1989) and alfalfa (Kielly and Bowley 1992), expressed as a dominant nuclear trait. Cytoplasmic factors (mitochondrial) have also been implicated in control of somatic embryogenesis (Rode et al. 1989). About 21 'embryo-specific' or 'embryo-enhanced' genes have been cloned (Zimmerman 1993, Lin et al. 1996) and it is likely that some of these genes may be useful markers for early embryo development. AGL 15-specific antibodies found to accumulate in microspore embryos in oil seed rape and somatic embryos of alfalfa (Perry et al. 1999) are found to participate in regulation of programmes active during the early stages of embryo development.

Several genes have been isolated which express during somatic embryogenesis. They are classified in three categories: (a) genes involved in cell division (21D7 and CEM1), (b) genes involved in organ formation (CAR3, CAR4 and CHB4) for hypocotyl formation; CAR5,

CAR6 and CHB involved in hypocotyl and/or root formation at globular and torpedo stage embryos), and (c) embryo-specific genes (CHB2 and CEM6). Expression and regulation of the gene action of these genes are described in detail by Komamine (2001). Some additional proteins identified at different stages of somatic-to-embryogenic transition are germin-like proteins (GLPs), embryogenic cell proteins (ECPs), Trx H proteins (Karmani et al. 2009). Similarly, leafy cotyledon (LEC) 1 and LEC 2 are two genes involved in *Arabidopsis* somatic embryogenesis (For details, *see* Bhojwani and Dantu 2013).

7.4.2 Physiological Hypothesis

Altered hormonal balance within the cells or tissues may also be associated with decline in the embryogenic potential. In such cases it may be possible to restore the potentiality of cells (differentiating organs and embryos) by modifying the growth constituents of the media. The loss of embryogenic potential could also be restored by adding 1-4% activated charcoal in the auxin-free medium (e.g., carrot cultures) or on giving cold treatment to the tissues (Ammirato 1983). The induction of embryogenic tissue can also be aided by various stress factors, such as heat, cadmium and anaerobsis (Yeung and Meinke l993).

7.4.3 Competitive Hypothesis

In the complex multicellular explant only a few cells are able to give rise to embryogenic clumps while the remaining cells are non-totipotent. According to the competitive hypothesis, the non-embryogenic cells of the explant will increase under conditions favourable to their growth, resulting in a gradual loss of embryogenic component during repeated subcultures. Restoration of embryogenesis in such cultures is impossible, but if the culture carries few embryogenic cells which are not able to express their totipotency due to the inhibitory effect of the non-embryonic cells, it should be possible to restore the morphogenic potential of these cultures by altering the composition of the medium in such way that selective proliferation of the embryogenic totipotent cells occurs.

7.5 Practical Applications of Somatic Embryogenesis

7.5.1 Clonal Propagation

Somatic embryogenesis has a potential application in plant improvement. Since both the growth of embryogenic cells and subsequent development of somatic embryos can be carried out in a liquid medium, it is possible to combine somatic embryogenesis with engineering technology to create large-scale mechanised or automated culture systems. Such systems are capable of producing propagules (somatic embryos) repetitively with low labour inputs. In this process of repetitive somatic embryogenesis (also referred to as accessory, adventive, or secondary somatic embryogenesis) a cycle is initiated whereby somatic embryos proliferate from the previously existing somatic embryo in order to produce clones. Mention may be made of four step protocol developed by Ducos et al. (2007) for mass propagation of *Coffea canephora* (Robusta Coffee) via somatic embryogenesis using an intermittent bioreactor with capacity to produce 2.5 million embryos or one million plantlet per year at 45% conversion rate of embryos. However, major impediment in clonal forestry via somatic embryogenesis in conifers seems recalcitrance of somatic embryos of high value which could be overcome by using media improved by supplementation of oxidation-reduction agents (Pullman et al. 2015).

(I) Cloning Zygotic Embryos for Repetitive Somatic Embryogenesis

A wide range of soybean genotypes have been tested for their ability to undergo auxin-stimulated somatic embryogenesis during cloning of zygotic embryos (Barwale et al. 1986, Komatsuda and Ohyama 1988). All of them are reported to form somatic embryos provided appropriate nutrients are provided in the medium. Parrot et al. (1989) evaluated a diverse group of 33 soybean genotypes and observed that the genotype with the highest regeneration capacity had an average number of somatic embryos per explanted cotyledon (SE/COT) from immature zygotic embryos equivalent to 2.09. This figure was 100 × higher than the SE/COT of the worst responder. The role of genotypes in conferring regeneration capacity is further supported by studies on zygotic embryo cloning of wheat, rice and maize. Diallel analysis of various cultivars demonstrated that the regeneration capacity of these crops was directly affected by non-additive, additive and cytoplasmic factors. However, the genotype which has the capacity to undergo repetitive somatic embryogenesis can be back-crossed to elite lines in order to transform the latter with capacity for high regeneration of somatic embryos. Such a type of transformation could play an important role in plant breeding since through *in vitro* techniques high quality somatic embryos have been produced in 80 species of tropical crops (*see* Redenbergh 1990).

(II) Raising Somaclonal Variants in Tree Species

Embryos formed directly from PEDCs appear to produce relatively uniform clonal material, whereas the indirect pathway involving IEDCs generates a high frequency of somaclonal variants. Mutation during adventive embryogenesis may give rise to a mutant embryo which on germination would form a new strain of plant. Joseph et al. (2004) reported induction of mutation in secondary embryos of cassava growing *in vitro* through γ-irradiation and obtained mutant plants with altered starch composition. Nucellar embryos, like shoot tips, are free of virus and can be used for raising virus-free (soma) clones, especially from some tree species (e.g., poly-embryonate *Citrus*) where shoot tip culture has not been successful. For clonal propagation of tree species, somatic embryogenesis from nucellar cells may offer

rapid means of obtaining juvenile plants. Clonal propagation through somatic embryogenesis has been reported in 60 species of woody trees representing 25 families (Razdan 2003). Somatic embryos with genotype of a selected elite parent are potentially convenient organs for cryopreservation and germplasm storage. Comprehensive details on these aspects are given in Chapter 18.

7.5.2 Synthesis of Synthetic/Artificial Seeds

There is considerable worldwide interest in the development of methods for encapsulation of somatic embryos to enable them to be sown under field conditions as 'synthetic' or 'artificial' seeds. Research programmes on production of artificial seeds via somatic embryogenesis in respect of commercially important crops would not only contribute to increased agricultural production, but also add to our basic knowledge of the regulatory mechanisms which control plant growth and differentiation. The concept is now extended to encapsulation of protocorm-like bodies, shoot tip, shoot buds, etc. which can be used as clonal seeds (Khor and Loh 2005).

Synthetic seeds, consisting of somatic embryos enclosed in a protective coating, have been proposed as a 'low-cost-high-volume' propagation system. The inherent advantages of synthetic seeds are the production of many somatic embryos and the use of conventional seed-handling techniques for embryo delivery. The objective is to produce clonal 'seeds' at a cost comparable to true seeds. Two types of synthetic seeds have been developed, namely, *hydrated* and *desiccated*. Redenbergh et al. (1986) developed hydrated artificial seeds by mixing somatic embryos of alfalfa, celery and cauliflower with sodium alginate, followed by dropping into a solution of calcium chloride/nitrate to form calcium-alginate beads. About 29-55% embryos encapsulated with this hydrogel germinated and formed seedlings *in vitro*. Kim and Janick (1989) applied synthetic seed coats to clumps of carrot somatic embryos to develop desiccated artificial seeds. They mixed equal volumes of embryo suspension and 5% solution of polyethylene oxide (polyox WSR N-750), a water-soluble resin, which subsequently dried to form polyembryonic desiccated wafers. The survival of encapsulated embryos was further achieved by embryo 'hardening' treatments with 12% sucrose or 10^{-6}M ABA, followed by chilling at high inoculum density.

Calcium alginate capsules tend to stick together and are difficult to handle because they lose water rapidly and dry down to a hard pellet within a few hours of exposure to the atmosphere. These problems can be offset by coating capsules with Elvax 4260 (ethylene vinyl acetate acrylic acid tetrapolymer; Du Pont, USA). From a practical sowing situation, it is necessary to produce high and uniform quality synthetic seeds at large scale. For this an automate encapsulation process has been developed. Its protocol has been described by Onishi et al. (1994) and Sakamoto et al. (1995). Processes for use of microcapsules (that release sucrose inside alginate beads), or self-breaking beads, pharmaceutical type capsules, and cellulose acetate mini-plugs, have been recently developed to promote sowing of coated somatic embryos as artificial seeds as well as their germination in non-sterile environment, such as in greenhouse or directly in the field. These techniques of embryo coating are summarised in an overview of synthetic seed germination by Dupuis et al. (1999).

Another delivery system for somatic embryos for obtaining transgenic plants is *fluid-drilling*. Embryos are suspended in a viscous-carrier gel which extrudes into the soil. The primary problem in fluid-drilling is that the sucrose level necessary to permit conversion also promotes rapid growth of contaminating micro-organisms in a non-aseptic system. Gray (1987) found that somatic embryos of orchard grass (*Dactylis glomerata*) became quiescent when desiccated in empty plastic petri dishes at 70% relative humidity at 23°C which amounted to loss of 13% water. However, after 21 days of storage, desiccated embryos when rehydrated

in vitro germinated to produce viable plantlets though limited (4%) in number. Senaratna et al. (1990) treated alfalfa somatic embryos with ABA at the torpedo to cotyledonary stages in order to increase their tolerance to desiccation. Over 60% of such desiccated embryos germinated when placed on a moist filter paper or sown directly onto sterile soil and formed plantlets. Heat-shock treatments, osmotic stress and nutrient deprivation also induce a degree of desiccation tolerance comparable to that conferred by ABA treatment and have no detrimental effect on the subsequent growth of the plantlets (Thorpe and Stasolla 2001). For efficient methods of synthetic seed production and its conversion to plantlets for clonal propagation in *Catharanthus*, *Dendrobium* and *Stevia* refer to Maqsood et al. (2012), Siew et al. (2014) and Nower (2014), respectively.

7.5.3 Source of Regenerable Protoplast System

Embryogenic callus, suspension cultures and somatic embryos have been employed as sources of protoplast isolation for a range of species. Cells or tissues in these systems have demonstrated the potentiality to regenerate in cultures and, therefore, yield protoplasts that are capable of forming whole plants. Embryogenic cultures are especially valuable in providing a source of regenerable protoplasts in the graminaceous, coniferous and citrus species. Attempts to achieve regeneration of callus or even sustained divisions in mesophyll-derived protoplasts of Gramineae proved unsuccessful until Vasil and Vasil (1986) turned to embryogenic cultures obtained from immature embryos of pearl millet (*Pennisetum purpureum*) as the source of protoplasts. Protoplasts from these cultures were induced to divide to form a cell mass from which embryoids, and even plantlets, regenerated on a suitable nutrient medium. Similar success was subsequently reported by other workers with embryogenic suspension of *Panicum maximum*, *Pennisetum purpureum*, *Oryza sativa*, *Saccharum officinarum*, *Lolium perenne*, *Festuca arundinaceae* and *Dactylis glomerata*.

Among cereals, in Gramineae, the development of a protoplast regeneration system for maize has been especially challenging. Rhodes et al. (1988) were the first to raise protoplast-derived maize plantlets from embryogenic cultures which, however, proved sterile. Later, Shillito et al. (1989) initiated embryogenic callus cultures from immature zygotic embryos of an elite inbred maize line B73, which yielded protoplasts that regenerated to form fertile plants. The protoplast-derived plant on crossing with pollen from a seed-derived plant produced viable seeds which germinated normally. Using similar techniques, Prioli and Sondahl (1989) obtained fertile maize plants from embryogenic cell suspension protoplasts of a line (Cat 100-1) adapted to tropical regions.

Embryogenic citrus suspension cultures also provide protoplasts that can be used in the production of interspecific and intergeneric somatic hybrid plants. Equal success can be obtained with protoplasts isolated from nucellus-callus cultures of this plant. Ability to isolate protoplasts from embryogenic cultures has had a large impact on their *in vitro* culture of forest trees, e.g., *Pinus taeda, Picea glauca, P. mariana, Pseudotsuga menziesii, Abies alba, Santalum album* and *Liriodendron tulipfera,* Somatic embryos induced on the protoplast-derived calli also germinate to form plantlets which finally establish in the soil.

7.5.4 Genetic Transformation

In seed embryogenesis, zygotic embryos are seated deep inside the nucellar tissue. They live in a protected environment besides being genetically heterogeneous. On the contrary, somatic embryos remain virtually unprotected and more or less give rise to genetically uniform plants. The advent of leaf-disc transformation systems has made it possible to successfully engineer species (*Nicotiana tabacum, Medicago sativa*) in which tissues are capable of regeneration via somatic embryogenesis. In these species, isolated single cells can be transformed in cultures and grown on a selection medium (nutrient medium containing an antibiotic kanamycin) to callus colonies which eventually form somatic embryos on removal of auxin from this medium (Chabaud et al. 1988). Since the callus phase seems essential in this type of indirect somatic embryogenesis, the possibility of chimeric embryos arising from transformed and non-transformed tissues cannot be ruled out. Therefore, the callus phase can be bypassed through a process of repetitive somatic embryogenesis. McGranahan et al. (1990) used repetitive embryogenesis for agrobacterium-transformed walnut (*Juglans regia* L.) cells and obtained multiple crops of somatic embryos without employing the callus phase.

There is also evidence to show that repetitive embryos originate from single epidermal or sub-epidermal cells which can be readily exposed to *Agrobacterium.* Thus, the transformation technique applied to a primary somatic embryo, instead of a zygotic embryo, should give rise to totally transgenosic somatic embryos. Repetitive embryogenesis is also ideally suited to particle gun-mediated genetic transformation. Instead of relying on *Agrobacterium* to mediate the transfer of genes into plant cells, the particle gun literally shoots DNA, that has been precipitated onto particles of a heavy metal, into the plant cells. Embryogenic suspension cultures of cotton and soybean, initiated from immature embryos, yielded an average of 30 stably transformed cell lines following each firing of the gun. This represents the stable transformation of approximately 0.7% of the cells on each bombardment (*see* McMullen and Finer 1990). The transformed cell lines can then be induced to form an unlimited number of transformed somatic embryos through repetitive embryogenesis. Corredoira et al. (2015b) developed a system of direct transfer of anti fungal candidate gene CsTL1 isolated from European chest nut cotyledons into somatic embryogenic lines of this tree which resulted in over expression of the candidate gene. Trees regenerated from transformed lines were tolerant to ink disease caused by *Phytophthora* spp.

7.5.5 Synthesis of Metabolites

The repetitive embryogenesis system is of potential use in the synthesis of metabolites such as pharmaceuticals and oils. Borage (*Borage officinalis* L.) seeds contain high levels of γ-linolenic acid, used as a precursor of postglandins, or in the treatment of atopic eczema. Somatic embryos of borage also produce this metabolite but through repetitive somatic embryogenesis a continuous supply of γ-linolenic acid is ensured, which otherwise would be limited to the growing season in the zygotic embryo (*see* Quinn et al. 1989). The same principle can be applied for production *in vitro* of industrial lubricants from jojoba (*Simmondsia chinensis*) and leo-palmi-tostearin (the major ingredient in cocoa butter) from Cacao (*Theobroma cacao*).

7.5.6 Conservation of Genetic Resources

Somatic embryos which originate from single cells and subsequently regenerate mostly genetically uniform plants are good candidates for genetic resource (germplasm) conservation. The totipotency in somatic embryos occurs because the process of somatic embryogenesis relies on reprogramming of gene expression patterns in a single cell and on triggering the cascade of structural embryogenic changes as in zygotic development (Dudits et al. 1991). Embryogenic cultures as well as somatic embryos remain viable upon storage at ambient temperature, cold storage, or cryostorage. Applications of somatic embryogenesis in plant germplasm conservation are comprehensively discussed by Zhou et al. (2000) and cryo-preservation protocols for somatic embryos of at least 38 species are now available (*see* Thorpe and Yeung (2011).

APPENDIX 7.1

Procedure for Inducing Somatic Embryogenesis in Some Tissue Cultures

A. *From root culture* (e.g., *Daucus carota;* after Smith and Street 1974)

1. Surface-sterilised (in 10% calcium hypochlorite for 15 min) seeds are washed three times in sterile distilled water and germinated on sterilised-moistened filter paper placed in petri dishes, in darkness at 25°C.

2. About 1 cm long segments of roots are excised from 7-day-old seedlings and then individually cultured on a semi-solid medium containing inorganic salts of MS medium, organic constituents of White's medium, 100 mg l^{-1} myoinositol, 0.1 mg l^{-1} 2,4-D, 2% sucrose, and 1% Difco bacto agar. Cultures are incubated in darkness.

3. Pieces (0.2 g fresh weight) of callus regenerated from a root segment after 6-8 weeks are transferred to fresh medium of original composition and the subcultures maintained now in the light at 25°C. This procedure for subcultures can be repeated every 4 weeks.

4. After some passages, when tissue is soft and friable, suspension cultures are initiated by incubating ca 0.2 g of callus tissue in a 200 ml Erlenmeyer flask containing 20-25 ml liquid medium of the same constituents used for callus growth (without agar) and rotated on a horizontal shaker at 100 rpm in light.

5. Suspensions are subcultured every 4 weeks by transferring 5 ml (sieved) of suspension to 65 ml of fresh medium (1:13).

6. Somatic embryogenesis is induced by cultivating either callus pieces or portions of suspension in the same medium without 2,4-D.

7. Within 3-4 weeks numerous embryos in different stages of development appear in culture flasks.

B. *From fruit culture* (e.g., *Citrus* sp.; after Tisserat and Murashige 1977).

1. Fruitlets (6-8 weeks old) are surface-sterilised with 1% sodium hypochlorite for 15-20 min and washed three times in sterile distilled water. The subsequent steps are followed under aseptic conditions.

2. Ovules are removed from sterilised fruitlets on a sterile tile and nucellus tissue excised from them under a dissecting microscope. Care has to be taken that the nucellus does not carry rudiments of integuments, endosperm, or embryonic tissue.

3. Excised nucellus tissue is then incubated in culture vessels using a semi-solid medium (constituents: inorganic salts of MS medium, 1 mg l^{-1} pyridoxine-HCl, 1 mg l^{-1} nicotinic acid, 4 mg l^{-1} glycine, 100 mg l^{-1} myoinositol, 0.2 mg l^{-1} thiamine-HCl, 500 mg l^{-1} malt extract, 5% sucrose, and 1% Difco bacto agar) in 16 hr light (3000 lx) at 27 ± 1°C.

4. Multiple somatic embryos develop from the nucellus-derived callus in 4-6-week-old cultures.

5. To stimulate the development of plantlets, embryos are cultured in another medium comprising the same constituents except that GA_3 (1 mg l^{-1}) is used in place of malt extract.

C. *From leaf culture* (e.g., *Coffea arabica;* after Sondahl and Sharp 1977).

1. Mature leaves taken from plagiotrophic shoots are surface-sterilised and washed following the procedure mentioned in *A* and *B*.

2. The lamina between the midrib and the margin is cut into 7 mm^2 pieces, avoiding the apical and basal portions, in saline-sugar solution made of half-strength inorganic salts

of MS medium, 10 mg l^{-1} thiamine-HCl, 86 mg l^{-1} L-cysteine-HCl, 99 mg l^{-1} myoinositol, and 3% sucrose.

3. Excised leaf pieces are now transferred to the same medium solidified with 1% agar in petri dishes and stored in the dark. The adaxial surface of the leaf pieces should always be placed facing the medium.

4. After 30 hr, the leaf pieces are cultured in jars containing 'conditioning medium' constituted by inorganic salts of MS medium, 10 mg l^{-1} thiamine-HCl, 86 mg l^{-1} cysteine-HCl, 99 mg l^{-1} myoinositol, 4% sucrose, 4.5 mg l^{-1} kinetin, 1 mg l^{-1} 2,4-D, and 1% Difco bacto agar. Cultures are stored in darkness at 25 ± 1°C.

5. The callus tissue develops from leaf tissues in 40-45 days. Small pieces of callus tissue, or the entire callus, are then transferred to an 'inducing medium' in which inorganic salts (except KNO$_3$ and sucrose have been reduced to half strength, and the hormonal combination changed to 0.6 mg l^{-1} kinetin + 0.1 mg l^{-1} NM. The cultures are maintained in 12 hr light (4600 1x) at 24-28°C.

6. Somatic embryo formation occurs within 13-15 weeks at a low frequency initially and then high frequency somatic embryos appear after another 3-6 weeks.

7. Embryos germinate into plantlets *in situ*. To stimulate embryo germination, the proembryogenic tissue (4-6 weeks old) is taken from the 'inducing medium' and grown on another medium of the same composition but lacking agar and kinetin. After maintaining the cultures in the light at 26°C for 4-6 weeks, torpedo-shaped embryos develop. These embryos are transferred to a saline-agar medium containing 0.5-1% sucrose and maintained in light.

APPENDIX 7.2

Induction of Somatic Embryogenesis from Maize[a]

1. Harvest intact ears, 12-16 days after crossing, and surface sterilise for 30 min in a 50% commercial bleach solution (or 2.5% sodium hypochlorite) containing a few drops of detergent followed by washing 5 times with sterile distilled water.

2. Insert a long forceps into the top of the ear to act as handle and cut off top-half of 20 kernels using a scalpel.

3. Scoop out endosperm inside the kernel, discard it, and then gently remove zygotic embryo in empty seed with spatula.

4. Place zygotic embryos with scutellum up (embryo axis or flat side down) on the Induction Medium (Ml) containing MS salts, iron-EDTA, B5 vitamins, 2% sucrose, 2,4-D (1 mg l^{-1}), AgNO$_3$ (10 mg l^{-1}) and 0.8% agarose. Some embryos may be placed in opposite orientation (i.e., scutellum down) on the medium. Size and orientation of embryos noted, cultures sealed and incubated.

5. Subculture embryogenic callus formed within 1-2 weeks; after 3 weeks of subculturing on MI medium transfer to Maize Development (MD) medium containing MS salts, iron-EDTA, B5 vitamins, 6% sucrose and 0.2% gelrite. Seal the cultures and incubate.

6. Somatic embryos develop within 1-4 weeks.

7. Transfer somatic embryos for maturation and germination to MS medium.

---

[a] after Finer (1995).

PART III
Applications to Plant Breeding

Haploid Production

An important aspect of plant breeding is the induction of maximum genetic variability of germplasm sources to secure a wider scope for selection and introduction of better trait qualities in existing crop species. Ever since A.D. Bergner discovered haploid plants in *Datura stramonium* in 1921, plant breeders have worked intensively to obtain haploids (sporophytes of higher plants with gametophytic chromosome constitution) either *in vivo* or *in vitro*. In nature, haploids arise as a result of parthenogenesis and these plants rarely produce characters of the male parent. The *in vivo* techniques employed (*see* Pierik 1989) to induce haploid production include:

(a) *Gynogenesis.* Production of a haploid individual by the development of an unfertilised egg-cell as a result of delayed pollination (through use of abortive pollen pre-exposed to ionising irradiation or by using an alien pollen). Gynogenesis is found in interspecific cross *Solanum tuberosum* (2n = 4x) × *S. phureja* (2n = 2x) resulting in the production of a dihaploid (2n = 2x) potato.

(b) *Ovule androgenesis.* Production of a haploid individual by development of an egg-cell containing the male nucleus. In this case elimination or inactivation of the egg nucleus occurs before fertilisation.

(c) *Genome elimination by distant hybridisation.* Arising in intergeneric and interspecific crosses due to selective elimination of one of the parental genomes during the process of development after fertilisation. Consequently, the embryo is formed with only one genome and the plant arising from such embryo is expected to be a haploid. Example; Interspecific cross between *Hordeum vulgare* and *H. bulbosum* results in the production of haploid *H. vulgare*.

(d) *Semigamy.* A cross wherein the nucleus of the egg-cell and generative nucleus of the germinated pollen grain divide independently, resulting in a haploid chimera.

(e) *Chemical treatment.* Some chemicals, such as chloramphenicol and parafluorophe-nylalanine may induce chromosome elimination in somatic cells or tissues, giving rise to haploids. Treatment with toluene blue, maleic hydrazide, nitrous oxide and colchicine may produce similar results.

(f) *Temperature shocks.* High or low temperature treatments may play a role in suppression of syngamy and induction of haploidy.

(g) *Irradiation effects.* X-rays or UV light reportedly induce breakage in chromosomes of pollen resulting in their elimination during embryo development and subsequent haploid production.

In vivo methods yield a low frequency of haploid production. On the contrary, the success achieved using *in vitro* methods has been spectacular and there are reports on haploids being induced by *anther culture* ('anther androgenesis') or *pollen culture* from about 250 species and hybrids distributed in 40 families (Mishra and Goswami 2014). Initially members of Solanaceae showed a good response though in the families of Cruciferae, Gramineae, Ranunculaceae and others, it was subsequently possible to induce haploidy either using anther androgenesis, henceforth referred to as *androgenesis*, or from cultures of individual pollen grains. Intraspecific pollination with irradiated pollen for induction of haploids has been successful in apple, blackberry, carnation, cucumber, European plum, kiwifruit, mandarin, melon, onion, pear, petunia, rose, squash, sunflower, sweet cherry, watermelon, etc. (Murovec and Bohanec 2012).

The first report showing that isolated anthers of *Datura innoxia* were able to form haploid embryos *in vitro* was published by Indian scientists, Guha and Maheshwari (1964, 1966). Subsequently, Bourgin and Nitsch (1967) obtained the first haploid plants from anther cultures of *Nicotiana*. During the last two decades much progress has been made in anther culture of rice, wheat, maize, brassica, pepper and other crop species (Dwivedi et al 2015). These advances attracted the attention of geneticists to survey the advantages offered by haploid plants over their relative diploids. This chapter outlines the technique and factors influencing haploid production. The principal applications of this technology to plant breeding are also discussed.

8.1 Technique of Androgenesis

The requirements that trigger off androgenesis are: (a) healthy plants grown under controlled environmental conditions, (b) knowledge of pollen ontogenesis in the strain to be used and (c) temperature treatment to arrest existing metabolism in order to shift it towards the new pathway of embryogenesis instead of the usual formation of mature pollen. Once the androgenic embryos are formed, their development into plants is dependent on the composition of the culture medium, light, temperature conditions, differentiation of embryo primordia as well as their growth into shoots and roots, and finally transfer of the plantlets to the greenhouse.

Young plants grown under controlled conditions of temperature, light and humidity are used as the experimental material for anther and pollen culture. Since the stage of pollen is a critical factor for *in vitro* androgenesis, it is important to use the bud of the right stage. For this a correlation is drawn between stage of pollen development and certain visible morphological features of the bud (length of the corolla tube, emergence of corolla from the calyx etc.) which may vary with the species and age of the donor plant. Some workers believe that the most productive anthers are those which contain uninucleate microspores midway between release from the tetrad and the first pollen grain mitosis (*see* Heberle-Bors 1985, Scott el al. 1991).

The selected buds are surface-sterilised and anthers removed along with their filaments. The anthers, after a quick dip in absolute ethyl alcohol, are excised under aseptic conditions and tested for the correct stage of pollen development by crushing one of them in 1 % acetocarmine. If it is found to be of the correct stage, the anthers of the remaining stamens are detached and placed horizontally on the medium. Adequate care should be taken that the anthers are not injured during the process of anther culture lest wounding cause callusing of the anther wall tissues. Another approach is to isolate pollen from anthers and prepare pollen suspension cultures. While dealing with plants having minute flowers (e.g., *Asparagus*, *Brassica* and *Trifolium*) it may be necessary to dissect anthers under a stereoscopic microscope aseptically. In such flowers only the perianth is removed and the rest of the bud with anthers intact inoculated. As regards genotypes where androgenesis occurs on simple nutrient media containing only mineral salts, vitamins and sugars, it may be possible to culture complete flower buds or inflorescences. This is due to the fact that chances of sporophytic tissue proliferating in such plants are remote on these media. However, where growth hormones are essential for androgenic development of pollen grains, the sporophytic tissue should be removed as far as possible.

The anther cultures are maintained in alternating periods of *light* (12-18 hr; 5000-10,000 1x m^2) and *darkness* (12-6 hr) regimes at 28°C depending on individual systems. Anther cultures of some species (e.g., *Brassica*) are very sensitive to light and need to be maintained in dark throughout (*see* Bhojwani and Razdan 1996). In responsive anthers, the wall tissues gradually turn brown and after 3-8 weeks burst open due to the pressure exerted by the growing pollen callus or pollen plants. Individual plants (3-5 cm in height) or shoots from callus are then transferred to a medium which supports a good root system. The rooted plants are ultimately transplanted to soil in small pots or seed trays and maintained under controlled environmental conditions in the greenhouse.

8.2 Factors Influencing Anther Cultures

Reports of haploid induction through anther culture have been steadily increasing. This has been possible by carefully monitoring a number of factors that influence androgenesis *in vitro*.

8.2.1 Genotype of Donor Plants

The genotype of the donor plant plays a significant role in determining the frequency of pollen-plant production. In cereals (wheat) the induction frequency of pollen callus and subsequent green plantlet formation is by and large controlled by many genes. This results in large genotype differences within a species. For example, in *Hordeum* each genotype differs with

respect to androgenic response in anther culture. It may therefore be worthwhile to select only highly responsive genotypes rather than direct efforts towards improving culture conditions for systems which are recalcitrant in anther culture. Improvement in the androgenic response of poorly responding lines of tobacco (Jacobsen and Sopory 1978) and barley (Foroughi-Wehr and Mix 1979) was, however, possible by following the breeding approach.

Based on RFLP analysis of the products of a number of crosses between a highly androgenic and a non-androgenic line of maize, Cowen et al. (1992) concluded involvement of two major epistatic and two recessive genes in the androgenic response of this plant. Interestingly, androgenic haploid plants of *Melandrium* (dioecious having XY mechanism of sex determination) are only obtained from pollen carrying X chromosome and these are cytologically and phenotypically female (Wu et al. 1990).

8.2.2 Anther Wall Factor(s)

Pelletier and Ilami (1972) demonstrated that pollen from one cultivar of tobacco would successfully develop into an embryo even if transferred into anthers of another cultivar. This work introduced the concept of 'wall factor' and encouraged many workers to use the nursing effects of whole anthers for androgenic development of isolated pollen of a number of species. An extract from anthers also stimulates pollen-embryo production, whereas on a conditioned medium (by culturing anthers for 7 days) a high frequency of anthers develop calli compared to a controlled medium.

Histological studies support the role of anther wall factor(s) in pollen-embryo development (Raghavan 1978). The fact that in henbane the embryogenic pollen are restricted to the periphery of the anther locule (in close proximity to tapetum), suggests that a gradient of critical substances of tapetal origin is instrumental in the induction of embryogenic division in the pollen grains. However, there are reports that glutamine alone or in combination with serine and myoinositol could replace the anther wall factor(s) for isolated pollen cultures.

8.2.3 Culture Medium and Culture Density

The available literature does not suggest an anther culture medium applicable to all systems since the requirements vary with the genotype and probably the age of the anther as well as conditions under which donor plants are grown. Androgenic development of microspores from *Nicotiana tabacum* and *Datura innoxia* can be induced on simple agar plates containing only sucrose. On such a medium the development proceeds up to the globular stage. Further growth of embryos is possible only when mineral salts are added to the medium. However, most solanaceous species exhibit androgenesis on a *complete* nutrient medium (mineral salts, vitamins and sucrose) of Nitsch or MS (Table 3.1). Of the various minerals, iron (40 μmol l^{-1} Fe·EDTA or Fe·EDDHA) seems crucial for pollen-embryo development in cultures that are 3 or 4 weeks old.

In the majority of non-solanaceous plant species known to exhibit androgenesis, pollen-embryo formation occurs only when the medium is fortified with growth adjuvants (growth regulators, complex nutrient mixtures such as yeast extract, casein hydrolysate, coconut milk) either alone or in various combinations. Several culture media developed by Chinese scientists are now widely used for anther culture of cereals. N_6 (Table 3.1) and Yu-Pei (Table 8.1) are two media proposed for the culture of maize anthers. The two media differ marginally as the former contains a low ammonium ion concentration which appears to be optimal in respect of nitrate nitrogen. The suitability of N_6 medium for promoting the efficiency of anther cultures has also been tried in other cereals, such as rice and rye. *Potato-1* medium was the other medium tested for cereal anther culture but a reproducible response could not be obtained on this medium because of the great variability in the concentration of inorganic constituents contributed by different samples of potato

tubers. This problem could be overcome through modification of the medium by incorporating six important salts at low concentrations and by reducing the concentration of potato extract from 20% (in potato-1) to 10%. This modified medium, called *potato-2* medium, produced a consistently higher number of green plants from anther cultures of wheat compared to the other three media (Table 8.1).

Whereas some species require an auxin and/ or a cytokinin to induce androgenesis, for many solanaceous species this occurs even on a basal medium (Nitsch et al. 1982). In addition, the number of plants developing from these embryos also appeared higher. Such observations on the formation of fully differentiated embryos from anthers growing in an auxin-free medium have been reported by several workers. Cultures maintained on an auxin-containing medium for prolonged periods generally develop a friable callus. Surprisingly, a compound related to auxin, namely 2,3,5-triiodobenzoic acid (TIBA), appears to give positive results at low (0.1 mg l^{-1}) concentrations in anther cultures of *Hordeum* (Clapham 1973), and *Zea mays* (Genovesi and Collins (1982)).

Sucrose is another essential constituent of the medium as long as the tissue in culture is unable to photosynthesise for itself. It is generally used at a concentration of 2-4% but some species such as *Brassica* require a higher (10%) concentration. In ornamental like *Phlox drummondii* haploid plants produced from callus through androgenesis required 9% sucrose concentration in the medium initially, but calli after induction had to be transferred to medium containing normal 3% sucrose concentration for multiplication and shoot formation (Razdan et al. 2008). Similar requirements were observed for production of androgenic haploids of neem (Chaturvedi et al. 2003a). The replacement of sucrose by the addition of glutothione, ascorbic acid and glucose is reportedly stimulatory for androgenesis in rye plant (Wenzel et al. 1977). Similarly, incorporation of activated charcoal or 2-chloroethyl-phosphate into the medium stimulates androgenesis in some systems.

Table 8.1   Constituents of the media used for anther culture of cereals in China

Constituents	Media (mg l^{-1})[a]		
	Yu-Pei[b]	Potato-1[c]	Potato-2[c]
KNO_3	2500	—	1000
NH_4NO_3	165	—	—
$(NH_4)_2SO_4$	—	—	100
$CaCl_2 \cdot 2H_2O$	176	—	—
KH_2PO_4	510	—	200
$MgSO_4 \cdot 7H_2O$	370	—	125
KCl	—	—	35
$Ca(NO_3)_2 \cdot 4H_2O$	—	—	100
$FeSO_4 \cdot 7H_2O$	27.8	—	—
Na $\cdot$ EDTA	37.3	—	—
Fe $\cdot$ EDTA	—	10^{-4} mol l^{-1}	10^{-4} mol l^{-1}
H_3BO_3	1.6	—	—
$MnSO_4 \cdot 4H_2O$	4.4	—	—
$ZnSO_4 \cdot 7H_2O$	1.5	—	—
KI	0.8	—	—
Thiamine HCl	1.0	1.0	1.0
Pyridoxine HCl	0.5	—	—
Nicotinic acid	0.5	—	—
Potato extract	—	20%	10%
Sucrose	50 g	90 g	90 g
2,4-D	2.0	1.0	1.0
Kinetin	—	0.5	0.5
Agar/agarose	10 g	6 g	6 g

[a] Constituents of N_6 medium are given in Table 3.1.
[b] From Nitsch et al. (1982).
[c] After Chuang et al. (1978).

Culture density could be a critical factor particularly with regard to isolated pollen culture. The minimum density required for embryogenesis is 3000 pollen ml^{-1} of the culture medium (example: *Brassica napus,* Huang et al. 1990). The effect of culture density in anther cultures could have varying response to the frequency of pollen embryogenesis for different species.

8.2.4   Stage of Microspore or Pollen Development

Androgenesis occurs when a microspore or pollen is induced to shift from a gametophytic pathway to a sporophytic pathway of embryo formation. The shift may occur prior to division of the microspore (premitotic) that forms the binucleate pollen grain or after microspore mitosis (postmitotic) wherein either vegetative cell or generative cell (or both) divide to undergo androgenesis.

For maximum production of pollen plants the appropriate stage of microspore development varies with plant species. As summarised by Sunderland (1982), anthers of some species (postmitotic: *Datura,* tobacco) give the best response if pollen is cultured at first mitotis or later stages, whereas in most others (premitotic: barley, wheat and rice) anthers are most productive when cultured at the uninucleate microspore stage. *Datura* buds 4.6-6.0 cm and unopened flower buds of tobacco in which calyx and corolla are at equal length (0.8-3 cm) are generally containing anthers with microspores at uninucleate stage and reported as responsive for haploid production (Smith 2013). The early bicellular stage is best for *Atropa belladonna*, *Nicotiana sylvestris* and absolutely essential for *N. knightiana*. Anthers at a very young stage (containing microspore mother cells, or tetrads) and late stage (containing binucleate, starch-filled pollen) are generally ineffective although exceptions to the rule are known. The ideal stage for androgenesis in *Arabidopsis thaliana* and *Lycopersicon esculentum* has been found when the microspore mother cells are in meiosis-1. On the other hand, the *Brassica species* show better response when mature anthers or pollen are isolated from them and cultured.

8.2.5 Effect of Temperature, Light and Stress Factors

Temperature shocks enhance the induction frequency of microspore androgenesis. The buds treated with cold temperatures at 3°C or 5°C for 72 hr induced approximately 58% anthers to yield pollen embryos in some solanaceous members *(Datura, Nicotiana)* as against 21% anthers from buds maintained at 22°C for the same period (Nitsch 1974). Excised rice panicles treated at 13°C for 10-14 days also showed the highest frequency of pollen callusing. The induction of androgenesis has been found very effective if other cereals (rye, maize, and pennisetum) are stored at low temperatures prior to anther culture. There are also reports that pretreatment of anthers at elevated temperatures (35°C) stimulate androgenesis in some *Brassica* and *Capsicum* species in early binucleate pollen grains (*see* Bhojwani and Bhatnagar 1990) and heat shock treatment at 40°C for late binucleate pollen grains (Marachin et al. 2005).

The frequency of haploids formed and the growth of plantlets is generally better in light although pollen plants from some genotypes grow both in the light and in the dark. However, isolated pollen seems to be more sensitive to light than that within the anther. For example, low intensity white light or red fluorescent light promotes faster development of embryos in isolated tobacco pollen cultures compared to high intensity white light (Nitsch 1977).

Other stress factors e.g. drought, variations in osmotic potential, reduced atmospheric pressure, γ-irradiation, presence of colchicines, heavy metals, ABA, AEC (lysine S-2 aminoethyl h-cystem), CGA (gametocide), 2-NHA (hydroxy nicotinic acid) in the culture medium induce ultra structural charges that trigger reprogramming of microspores for haploid production (Islam and Tuteja 2012).

8.2.6 Physiological Status of Donor Plant

Anthers taken from plants grown under short-day (8-hr day) conditions and high light intensity-regimes show relatively better response than long-day (16-hr day) donor plants at the same intensity. Pollen embryogenesis can be further improved if temperature under short-day conditions is maintained at 18°C. Seasonal changes, water stress, and age of the donor plant greatly influence the responding pollen. Buds from first flush of flowers produce anthers showing better response to anthers collected at the end of the flowering season (Dunwell 1985). Treatment of plants with pesticide sprays or systemic poisons should be avoided although treating the donor plants with feminizing and gametocidal compounds (etherel, auxin, antigibberllin) reportedly improve the yield of embryogenic pollen (Heberle-Bors 1985). Plants starved of nitrogen may give more responsive anthers compared to those that are well fed with nitrogenous fertilisers. Therefore, it is recommended that only materials grown under controlled environmental conditions, such as those in greenhouses, be used for microspore androgenesis.

8.3 Differentiation of Pollen into Gametophytic or Sporophytic Cell

8.3.1 Morphological

It has generally been observed that in systems where pollen plants have been obtained the yield of haploids is invariably low although thousands of pollen are present in an anther. This may be attributed to pollen dimorphism (*see* Chapter 1) as reported in anther cultures of barley, tobacco, rice and wheat (*see* Bhojwani and Soh 2001). The pollen capable of forming haploids is smaller and stains less well than most of the pollen destined to develop into male gametes. Such pollen, described as embryogenic, is usually low in quantity, which accounts for a very low frequency of haploid formation. Rashid and Reinert (1981) reported the possibility of increasing the frequency of smaller pollen by inducing tobacco plants to flower at low (15°C) temperatures and, subsequently, producing very high frequency of pollen embryos by growing the isolated microspores on a simple sucrose medium.

8.3.2 Physiological

The pollen mother cell is diploid and has a determinate structure. It undergoes cell re-organisation for differentiating into microspore (haploid) cells which are indeterminate structures. The determination into either androgenic (gametophytic) or embryogenic (sporophytic) grains is possibly a function of differential gene activity. Elaborate endoplasmic reticulum, abundance of ribosomes and normal mitochondria are visual manifestations of gametophytic pollen, whereas quiescence and repression of organelles in microspores lead to differentiation of embryogenic pollen (Fig. 8.1). Gametophytic pollen germinates in the presence of metabolic inhibitors in contrast to embryogenic pollen which does not germinate because of sensitivity to these antimetabolites.

In view of the fact that exposure of plants to low temperatures before flowering increases the frequency of embryogenic pollen in *Nicotiana tabacum,* cold treatment possibly interferes with the process of cell organisation by ensuring a continuity of sporophytic information in the form of membrane-protected enclaves of cytoplasm. The inference that can be drawn on the basis of this hypothesis is that, in nature, it is very likely that embryogenic pollen originates either from those microspore mother cells which escape the process of cell organisation or from the microspores which inherit a greater proportion of membrane-bound cytoplasm from microspore mother cells.

8.3.3 Molecular Aspects

Zarsky et al. (1990) suggested that sugar-starvation of pollen grains before exposing them to full nutrient medium causes repression of the gametophytic cytoplasm and differentiation of vegetative cell, pushing it from G1 to S phase of cell cycle, an essential requirement for embryogenic pollen development. During the starvation treatment a large fraction of pollen showed DNA replication in the vegetative nucleus. Sugar starvation is reported to induce androgenesis in tobacco and *Hordeum vulgare* microspore cultures. Most of the information on molecular basis of inducing pollen embryogenesis in microspore cultures is based on *Brassica napus* (Cordewener et al. 1994). In this system the late uninucleate microspores or early binucleate pollen, cultured at 18°C, form only gametophytes. Paradoxically at high temperature (32°C), about 90% of these microspores acquire irreversible commitment to embryogenesis within 8 hr. A comprehensive qualitative and quantitative computer analysis of 2D protein pattern in microspore cultures revealed that reduction of pollen embryogenesis correlate to synthesis of heat shock protein (HSP 70) in the nucleoplasm of the cell which formed embryo.

The use of functional genomic tools have led to identification of genes associated with the process of microspore embryogenesis (ME). In wheat, 14 genes are involved in early, middle and late stages of ME and of these 13 genes are

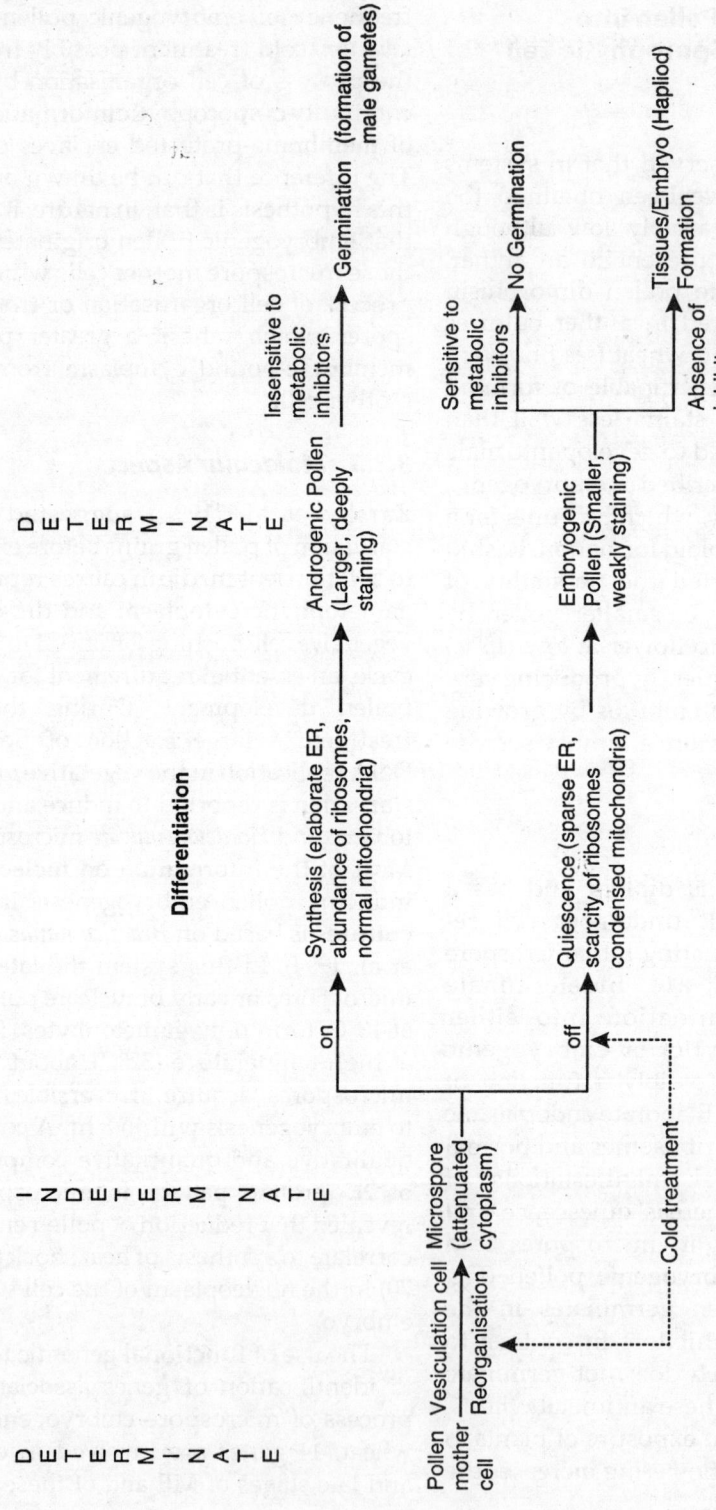

Fig. 8.1 Schematic representation of pollen differentiation into an androgenic (gametophytic) or embryogenic (sporophytic) cell (modified from Rashid 1983).

also associated with ME of Triticate lines (see for details Sanchez-diaz et al. 2013).

A specialized culture system has been developed to study ultrastructure and molecular details for barley pollen embryogenesis. Immature barley pollen are cultured as a monolayer trapped between the bottom glass-coverslip and of a live-cell chamber diaphanous PTFE membrane within liquid medium over a period of 28 days. This method allows automatic capture of images every 3 min beginning at the unicellular pollen stage till development of multicellular embryogenic structure. Daghma et al. (2014), who developed this culture system used time-lapse imaging on transgenic barley pollen expressing nuclear localised green fluorescent protein and investigated cellular dynamics during pollen embryogenesis process. They were able to identify nine distinct embryogenic and non-embryogenic types of pollen response to culture conditions. This novel technique is envisaged to assist in elucidating the still unknown molecular triggers of pollen embryogenesis.

8.3.4 Irradiation Effect

Low doses of γ-irradiation of young flower buds seems to enhance anther culture efficiency in cultures of *Brassica napus* (MacDonald et al. 1988). Direct irradiation of microspores could be deterimental for androgenesis since anther wall appears to have promotory effect on irradiated intact anthers.

8.4 Development of Androgenic Haploids

8.4.1 Pathways of Microspore Divisions

Four pathways based on the few initial divisions in the microspores have been identified as leading to *in vitro* androgenesis (Fig. 8.2).

(I) Pathway I

The microspores divide by an equal division and two identical daughter cells contribute to the sporophyte development. Vegetative and generative cells are not distinctly formed in this pathway (Example: *Datura innoxia*).

(II) Pathway II

The division of uninucleate microspores is unequal, resulting in the formation of a vegetative and a generative cell. The sporophyte arises through further divisions in the vegetative cell while the generative cell either does not divide or does so once or twice before degenerating (examples: *Nicotiana tabacum, Hordeum vulgare, Triticum aestivum, Triticale* and *Capsicum annuum*).

(III) Pathway III

The uninucleate microspore undergoes a normal unequal division but the pollen embryos are predominantly formed from the generative cell alone (example: *Hyoscyamus niger*). The generative cell either does not divide at all or does so only to a limited extent.

(IV) Pathway IV

The division of microspore is asymmetrical as in Pathway II. Both vegetative and generative cells divide further and contribute to the development of the sporophyte (examples: *Datura innoxia*, occasionally; *Datura metel, Atropa belladonna*).

8.4.2 Later Development

Irrespective of the above early pattern of microscope divisions, the embryogenic pollen grains ultimately become multicellular and burst open, gradually assuming the form of a globular embryo. This is followed by the normal stages of post-globular embryogeny until the development of a plant (e.g., *Atropa, Brassica, Datura, Hyoscyamus, Nicotiana*). Alternatively, the multicellular mass liberated from the bursting pollen grain proliferates to form a callus which may later differentiate whole plants (e.g., *Arabidopsis, Asparagus, Triticale*) either on the same medium or on a modified medium. It may sometimes be possible to obtain androgenic

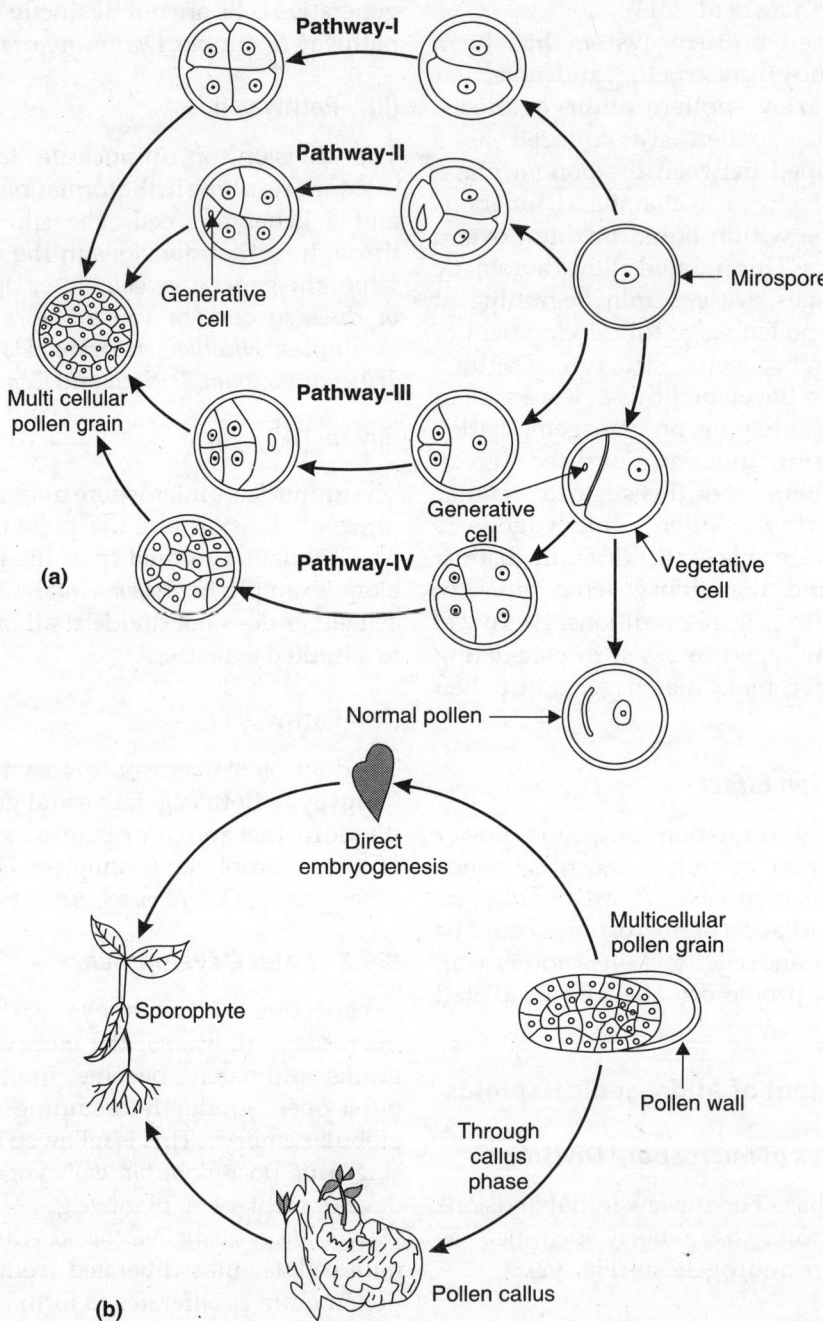

Fig. 8.2 (a) Pathways of microspore divisions leading to the formation of a multicellular pollen grain; (b) Multicellular pollen may directly form an embryo or produce sporophytes through a callus phase (after Bhojwani and Bhatnagar 1990).

haploids via embryo formation or through callusing within the same species (*Oryza sativa*) by manipulating constituents of a medium.

8.5 Haploids from Isolated Microspore or Pollen Culture

In 1953, Tulecke was able to obtain the callus from isolated pollen cultures of gymnosperms. This raised the possibility of obtaining haploids from isolated microspores or pollen cultures. The advantage of using this method was soon realised since pollen grains within an anther would constitute a heterogenous population. The first report on the tissue formation from isolated pollen of an angiosperm was from Kameya and Hinata (1970). They used hanging drops to culture mature pollen (about 80 pollen grains per drop) isolated from *Brassica oleracea* and the hybrid *B. alboglabra* × *B. oleracea* at low temperature. In the subsequent years workers found the nurse-culture technique (Fig. 8.3) suitable for raising haploid tissue clones from isolated pollen grains. Intact microspores are mounted on a filter paper disc which, in turn, is placed over an intact anther or callus derived

from the floral parts. The extract of cultured anthers has equal nursing effects on the growth and development of microspores, and may be used in place of a nurse callus or anther. A purely synthetic medium (Table 8.2) developed for raising haploid plants from isolated pollen grains of tobacco and *Datura* may also be suitable for other systems. The crucial ingredients of this medium are glutamine, *L*-serine and inositol. The details of the technique developed for raising haploids from isolated pollen are enumerated in Appendix 8.1.

The efficiency of isolated pollen cultures for haploid production may be increased by the gradient centrifugation technique. Anthers at the proper stage of development are gently macerated to obtain a microspore suspension. After removing the debris either by sieving or filtration, the suspension is layered on sucrose solution (30%, or 55% percoll and 4% sucrose solution) and centrifuged at 1200 g at 5 min. The pollen (vacuolated and starch-filled) having embryogenic potential forms a band at the top of the sucrose solution while non-embryogenic pollen (starch-filled) settles at the bottom.

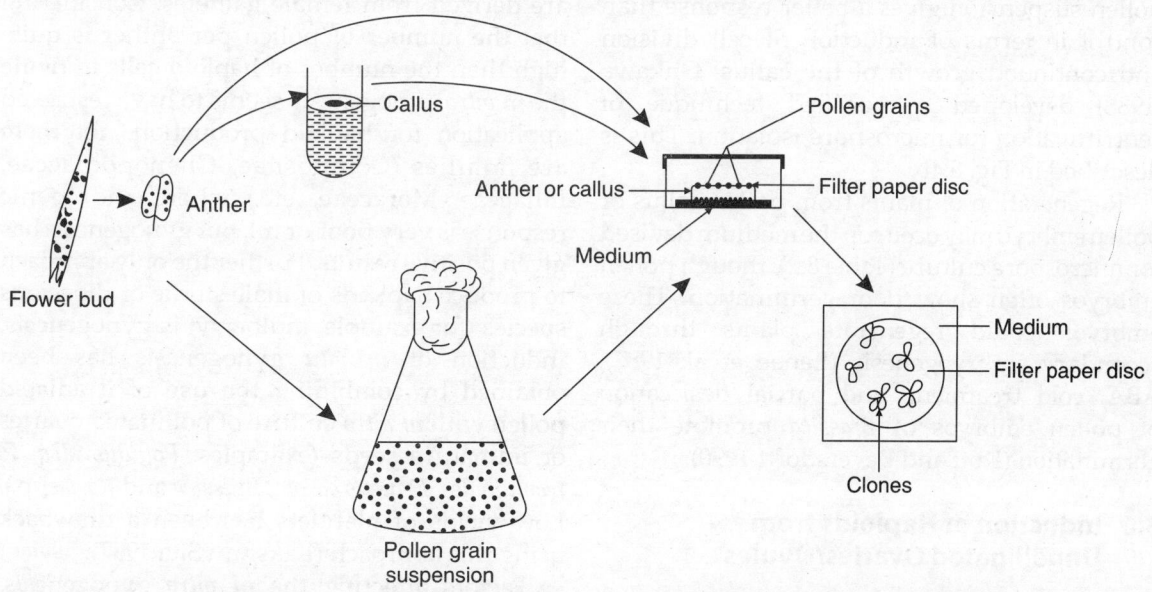

Fig. 8.3 Illustration showing the technique of nurse culture for cloning isolated pollen grains.

Table 8.2 Synthetic medium to culture isolated microspores of tobacco and *Datura*[a]

Constituents	Amount (mg l^{-1})
KNO_3	959
$NH_4 \cdot NO_3$	725
$MgSO_4 \cdot 7H_2O$	185
KH_2PO_4	68
$CaCl_2 \cdot 2H_2O$	166
Fe · EDTA	5 ml[b]
Glutamine	730
Serine	105
Inositol	4505
Sucrose	20000
pH	5.8

[a] See Nitsch (1977).
[b] The stock solution is prepared by dissolving 7.45 g of Na_2 · EDTA and 5.57 g of $FeSO_4 \cdot 7H_2O$ in 1000 ml of distilled water.

The upper band containing embryogenic or sporophytic pollen is now pipetted out and suspended in the washing medium, centrifuged again, and the supernatant removed. Finally, the pellet containing these grains is resuspended in the culture medium (Fig. 8.4a). Such a pollen suspension gives a better response than control in terms of induction of cell division and continued growth of the callus. Ohkawa (1988) developed a modified technique of centrifugation for microspore isolation. This is described in Fig. 8.4b.

Regeneration of plants from pollen callus or pollen embryo may occur on the medium devised for microspore culture (Table 8.2), though pollen embryos often show poor germination. These embryos could regenerate plants through secondary embryogenesis (Lenee et al. 1987). ABA, cold treatment, and partial desiccation of pollen embryos of *Brassica* promote their germination (Kott and Beversdorff 1990).

8.6 Induction of Haploids from Unpollinated Ovaries/Ovules

As described earlier, intensive efforts to raise parthenogenetic haploid plants using *in vivo*

techniques yielded little success. San Noeum in 1976 was the first to demonstrate that gynogenesis, an essentially *in vivo* phenomenon, can be induced under *in vitro* conditions. Sheo btained gynogenic haploids using an ovary culture of *Hordeum vulgare*. Subsequently, Zhu and Wu (1979) obtained haploid plants from cultured unpollinated ovaries of *Triticum aestivum* and *Nicotiana tabacum*. This technique has also proved successful in raising rice haploid plants. Meanwhile, haploids have also been obtained from unpollinated ovary/ovule cultures of *Gentiana triflora*, *G. scabra*, *Gerbera jamesonii*, *Lilium davidii*, *Iinum usitatissimum*, *Zea mays*, mulberry and over two dozen species (Thomas et al. 1999, Table 8.3a, Doi et al. 2013) which include ornamental, food and tree crops. These studies indicate that the induction of haploids from female gametophytes may not be inaccessible.

Some examples giving methods and results of haploid induction of unpollinated ovaries and ovules are summarised in Table 8.3b.

Gynogenesis can be obtained *in vitro* either by organogenesis or embryogenesis as long as ploidy of resulting plants is haploid and they are derived from female gametes. Considering that the number of pollen per anther is quite high than the number of haploid cells in ovule the *in vitro* gynogenesis seems to have restricted application for haploid production. Yet there are families (Compositae, Chenopodiaceae, Liliaceae, Moraceae, etc.) where androgenic response is very poor or nil, but gynogenesis has given positive results. Further the only approach to produce haploids of male sterile or dioecious species (for example, mulberry) is gynogenesis. Induction of *in situ* gynogenesis has been obtained by combining the use of irradiated pollen with *in vitro* culture of pollinated ovaries or immature seeds (examples: *Populus alba*, *P. tremuloides*, *Daucus carota*, *Brassica* and *Rosa* spp.). Low yields of plantlets has been a drawback using this approach (Lakshmi Sita 1997).

Factors affecting the *in vitro* gynogenesis, such as genotype of the explant, nutrient

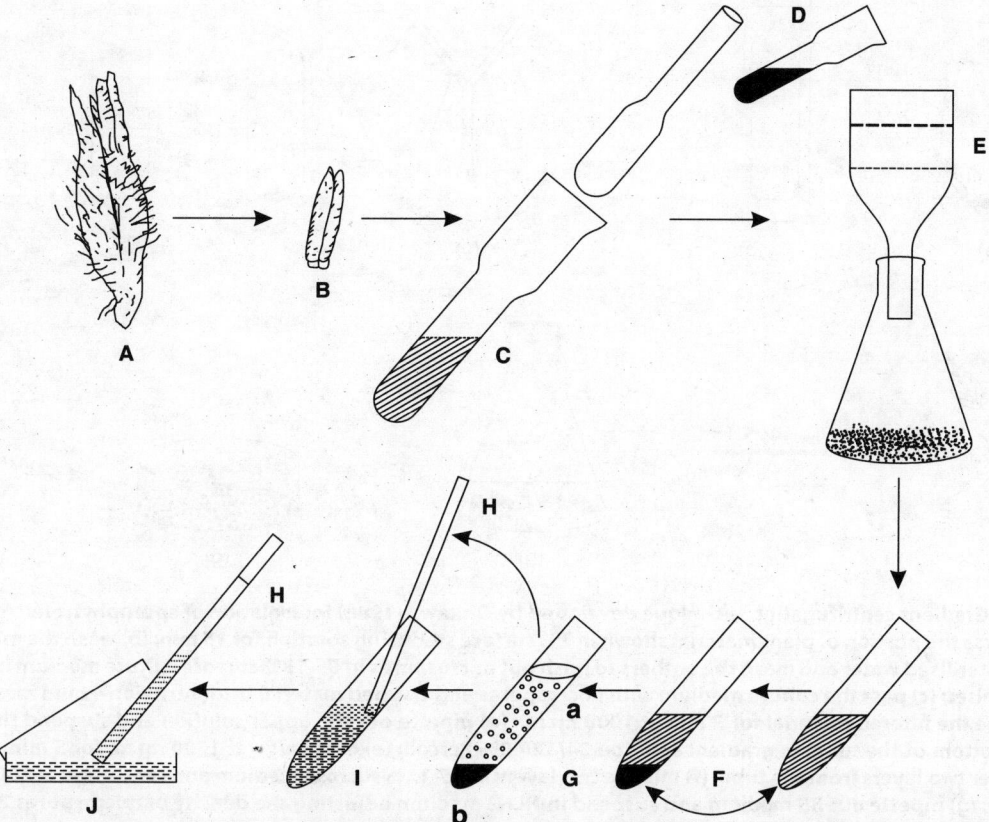

Fig. 8.4a Isolation of sporophytic pollen from gametophytic pollen by gradient centrifugation technique (after Pierik 1989). From a young flower bud (A) an anther (B) is removed and homogenised using a Potter's homogeniser (C). The crude pollen suspension (D) obtained is sieved (E), decanted (F), layered in sucrose solution and centrifuged in a tube (G). Pollen having embryogenic potential forms the top band (a) while the non-embryogenic pollen settles at the bottom (b). The upper band containing embryogenic pollen is now inoculated by pipette (H) into a culture medium (I) and finally plated in a petri dish (J).

requirements, and other culture conditions vary little to androgenesis. Details on these aspects are analysed in reviews by Lakshmi Sita (1997) and Bhojwani and Thomas (2001).

8.7 Diploidisation of Haploid Plants

Haploid plants obtained either from anther or ovule culture may grow normally under *in vitro* conditions up to the flowering stage but viable gametes are not formed due to the absence of one set of homologous chromosomes and, consequently, there is no seed set. The only mechanism of perpetuating the haploids is by duplicating the chromosome complement in order to obtain homozygous diploids. In pollen-derived haploid plants the duplication of chromosomes may occur spontaneously in cultures but due to the small percentage of such double-haploids it is necessary to diploidise the haploids by chemical means. A simple procedure designed to achieve diploidisation involves immersion of very young haploids in a filter-sterilised solution of colchicine (0.4%) for 2-4 days followed by their transfer to the culture medium for further growth. In this procedure chromosome or gene instabilities are minimum

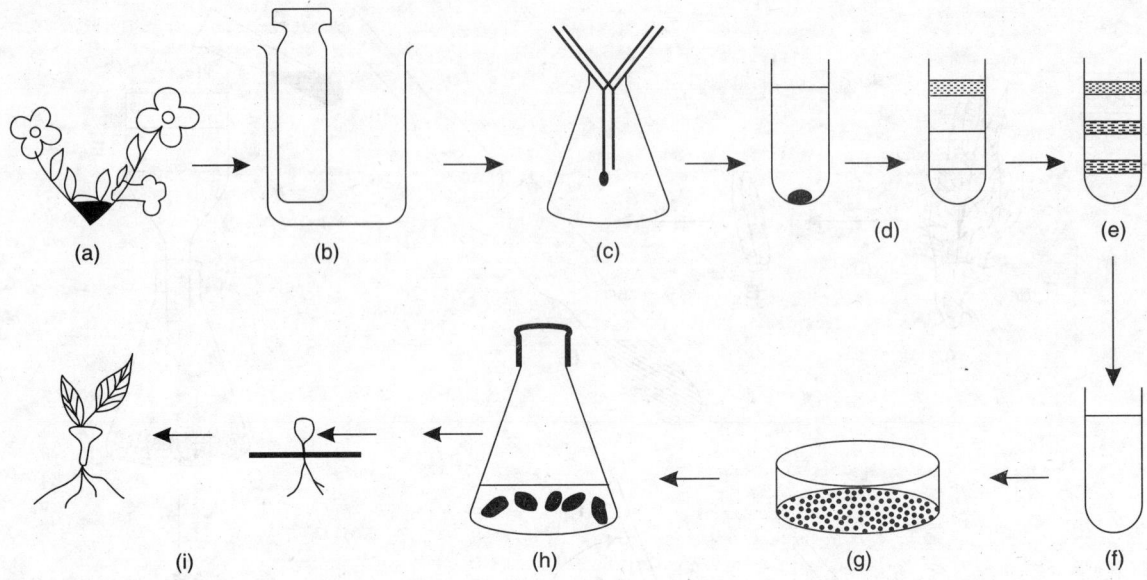

Fig. 8.4b Gradient centrifugation technique developed by Ohkawa (1988) for isolation of sporophytic microspores; (a) immerse the portion of plant material shown in 7% surface sterilising solution for 15 min; (b) wash the material 2 times in sterilised water and mash the anthers to push out microspores in B5-13% sucrose culture medium in a glass homogeniser; (c) pass the culture medium with microspores and crushed material through nylon 42 μm mesh sieve, centrifuge the filtered material for 3 min at 1000 r.p.m.; (d) pipette out the upper solution and suspend the pellet (at the bottom of the tube) in gradient solution 24/32/40% Percoll; (e) centrifuge at 1000 r.p.m. for 5 min, remove only upper two layers from the tube; (f) mix the two layers in B5-13% sucrose medium and centrifuge at 1000 r.p.m. for 3 min; (g) pipette out B5 medium and suspend in NLN* medium adjusting the density of microspores 2-5 × 10⁴ ml⁻¹, plate them by pouring 50 ml in culture dish at intervals using 2 ml each time, maintain the cultures at 32°C for 3–5 days; (h) transfer regenerated tissue/embryo to 18 ml of hormone free NLN medium in a conical flask, maintain for a week on the shaking machine at 60 r.p.m, 32°C; and (i) finally transfer the material from flask to a culture dish containing solidified B5-2% sucrose medium to develop roots and shoots.

* NLN medium (mg l⁻¹): KNO_3—62.5, $MgSO_4 \cdot 7H_2O$—62.5, $CaNO_3 \cdot 4H_2O$—250, KH_2PO_4—62.5, FeEDTA—40, $MnSO_4 \cdot 4H_2O$—25, H_3BO_3—10, $ZnSO_4 \cdot 4H_2O$—10, $Na_2MoO_4 \cdot 2H_2O$—0.25, $CuSO_4 \cdot 5H_2O$—0.025, $CoCI_2 \cdot 6H_2O$—0.025, myoinositol—100, Nicotinic acid-5, Glycine-2, pyridoxine HCl, Thiamine HCl, Folic acid—all 0.5, Glutathione—30, L-Glutamine—800, L-Serine—100, Sucrose—130,000, and pH—6.0.

compared to other methods of colchicine or chemical treatments. Oryzalin, amiprosmethyl (APM), trifluralin and pronamide are other options of chemical treatment for chromosome doubling of haploid plants (Murovec and Bohanec 2012). Treatment with nitrogen oxide (N_2O) under high pressure (600kPa) of maize seedlings (Kato and George 2002) and *in vitro* adventitious somatic bud regeneration of haploid onion lines (Jakše et al. 2010) showed equal doubling efficiency.

8.8 Application of Haploids in Plant Breeding

In vitro production of haploids can solve some problems in genetic studies since gene actions are readily manifested due to a single allelic dose present in chromosomes of an entire genome. This also makes them good material for induction of mutations and various gene manipulations, particularly gametoclonal variants which being hemizygous express even the recessive traits in R_0 plants. This is unlike somaclonal variants,

Table 8.3 (a)   Haploid plants raised through in vitro gynogenesis[a,b]

Species	Explant	Mode of regeneration	Ploidy level
Allium cepa	Ovary	Embryo	Haploid, Diploid
	Ovule	Embryo	Haploid
	Ovary, Ovule	Embryo	Haploid
	Ovary, Ovule, Flower bud	Callus	Haploid
	Flower bud	Embryo	Haploid, Diploid, Tetraploid
	Flower bud	Embryo	Haploid, Diploid, Polyploid
	Flower bud	Embryo	Haploid
	Flower bud	Embryo	Haploid
	Power bud	Embryo	Haploid, Diploid
A. tuberosum	Ovary	Embryo	Hapioid
A. porrum	Flower bud	Embryo	Diptoid, Tetraploid, Octoploid
Beta vulgaris	Ovule	Callus, Embryo	Haploid
	Ovule	Embryo	Haploid
	Ovule	Callus	Haploid, Mixoploid
	Ovule	Callus	Diploid, Tetraploid, Mixoploid
Brassica oleracea	Ovule	Embryo	Haploid
Coix lachryma-jobi	Ovary	Embryo	Haploid
Cucumis melo	Ovule	Embryo	Haploid, Diploid
Cucurbita pepo	Ovaries	Embryo	Haploid, Diploid

Contd...

Table 8.3 (a) Contd.

Species	Explant	Mode of regeneration	Ploidy level
Gerbera jamesonii	Ovule	Callus	Haploid
	Ovule	Callus	Haploid
	Ovule	Callus	Haploid
	Ovule	Callus	Haploid, Diploid, Mixoploid
Helianthus annuus	Ovary, Ovule	Embryo	Haploid, Diploid
Hevea brasiliensis	Ovule	Embryo, Callus	Haploid, Diploid, Aneuploid
Hordeum vulgare	Ovary	Embryo	Haploid
Hyoscyamus muticus	Ovary	Callus	Haploid, Diploid, Mixoploid
Lilium davidii	Ovary	Callus	Haploid, Diploid
Lilium sp.	Ovary	Embryo	Haploid, Diploid
Melandrium album	Ovule	Embryo	Haploid, Diploid, Mixoploid
Mimulus luteus	Ovary	Callus	Haploid, Diploid
Morus alba	Inflorescence segments	Embryo	Haploid, Aneuploid
Nicotiana tabacum	Ovary	Embryo	Hapioid
Oryza sativa	Ovary	Callus	Haploid
Petunia axillaris	Ovule	Embryo	Haploid, Diploid
Populus simonigra	Ovary	Embryo	Haploid, Diploid
Solanum tuberosum	Ovule	Embryo	Haploid, Diploid
Triticum aestivum	Ovary	Callus	Haploid,
T. durum	Ovary	Callus	Haploid
Zea mays	Ovary	Embryo, Callus	Haploid, Diploid, Triploid, Aneuploid

[a] See details of references for above in Bhojwani and Thomas (2001).
[b] *Guizotia abyssinica* (l.f.) cass, Bhat and Murthy (2007).

Table 8.3 (b)   Methods and results of haploid induction by culture of unpollinated ovaries and ovules[a]

Species	Stage of embryo sac	Mode of inoculation	Development pattern	Characteristics of regenerated plants
Hordeum vulgare	Nearly mature 1-2 days before anthesis	Ovary culture Vertical ovary culture	Embryoid Direct shooting	Green haploids Green haploids
	Uninucleate to mature	Vertical flower culture	Embryoid	
Triticum aestivum		Ovary culture	Direct shooting or via callus	Green haploids
Oryza sativa	Nearly mature	Ovary culture	Direct shooting or via callus	Green haploids, some mixoploids
	Uninucleate to mature	Float culture of unhusked flowers	Callus	Green haploids, some albinos and nonhaploids
	Uninucleate to mature	Ovary culture	Callus or embryoids	Green haploids, and diploid, some albinos
Nicotiana tabccum (in some cases *N. rustica* too)	Embryo sac mother cell to megaspore tetrad	Ovary culture	Embryoid	Green haploids
		Ovary culture	Direct shooting	Green haploids
	Uninucleate to mature	Ovary culture	Embryoid, some via callus	Mainly green haploids
Gerbera jamesonii	Ovule bigger than half ovary cavity	Ovary culture	Callus	Green haploids, some diploids

[a] See details of references for above in Yang and Zhou 1982.

which require selfing and further progeny analysis (Morrison and Evans 1988). Through rapid achievement of homozygous traits in double haploids, pollen-derived haploid plants have been used in breeding and improvement of crop species and more than 290 varieties have been already released (Murovec and Bohanec 2012). Some of the achievements include:

8.8.1 Releasing New Varieties Through F₁ Double-haploid (Doubled Haploid) System

Haploid breeding techniques usually involve only one cycle of meiotic recombination. However, many agronomic traits (such as the yield) are polygenically controlled. One cycle

of recombination is usually insufficient for the improvement of such quantitative traits since the linkage between polygenes will not release all potential variations available in a cross. To overcome these disadvantages, the Chinese developed a method of combining anther culture with sexual hybridisation among different genotypes of anther-derived plants. Selection of superior plants among haploids derived through anther culture of F₁ hybrids is popularly described as hybrid sorting. The anthers of the hybrid (F₁) progeny are excellent breeding material for raising pollen-derived homozygous double haploid plants called double haploids (DH) in which complementary parental characteristics are combined in one generation

(Fig. 8.5). This results in considerably shortening the breeding cycle since in a normal hybridisation programme it takes several generations of backcrossing, or pedigree breeding, to produce a stable homozygous line to release new variety (Luckett and Smithard 1991).

Double haploids are also useful in studies related to inheritance of qualitative traits. In breeding programmes the double haploids derived from pollen cultures expressed genetic variability to an extent that new varieties have been synthesised in respect of barley, brassica, rice, maize, rye, potato, pepper and asparagus. By haploid induction followed by chromosome doubling it is possible to obtain exclusively male plants in dioecious species. An important example of this is *Asparagus officinalis,* in which androgenic haploids produced from male asparagus plants are either X or Y. Chromosome doubling of Y results in supermales (YY) which can be vegetatively propagated. If XX is crossed with YY plants the progeny will be all male lines. Using this method Thevénin and Doré (1976) established high-yielding male lines of *Asparagus.*

Some notable examples of releasing new varieties via double-haploid system are tobacco cultivar 'F-211' resistant to bacterial wilt, and more importantly with mild smoking in Japan; about 100 high-yielding rice varieties in China; rye or canola (*Brassica napus*) with low euric acid and glucoside content; as well as Spanish onion of high aroma in Canada (Fayos et al. 2015). Production of doubled haploid plants using maize technique in bread wheat (*see* Section 8.8.2) and factors influencing this technique have been elaborately discussed by Xynias et al. (2015). Double haploid lines developed as a result of gametoclonal variation are described in Chapter 14.

8.8.2 Dihaploid Production

An alternative approach to production of haploids *in vitro* is the 'bulbosum technique' used successfully in barley. The technique can be applied because of improvement in methods of zygotic embryo culture (Chapter 11). The tetraploid barley, *Hordeum vulgare,* is fertilised by diploid *H. bulbosum* causing the formation of a large proportion of zygotes. In embryos, originating from such zygotes, the chromosomes of *H. bulbosum* are rapidly eliminated and require to be dissected out of the fruit for culture on a suitable nutrient medium. Nearly all plants raised from such embryos are dihaploid barley. Two varieties of barley namely 'Mingo' and 'Gwylan' are known to have been produced using bulbosum technique. Further the crosses of these dihaploids produce the progeny comprising haploids. Although having contributed to commercial varieties the limitation of bulbosum technique is that it remained restricted to only these two crops (Collins and Edwards 1998).

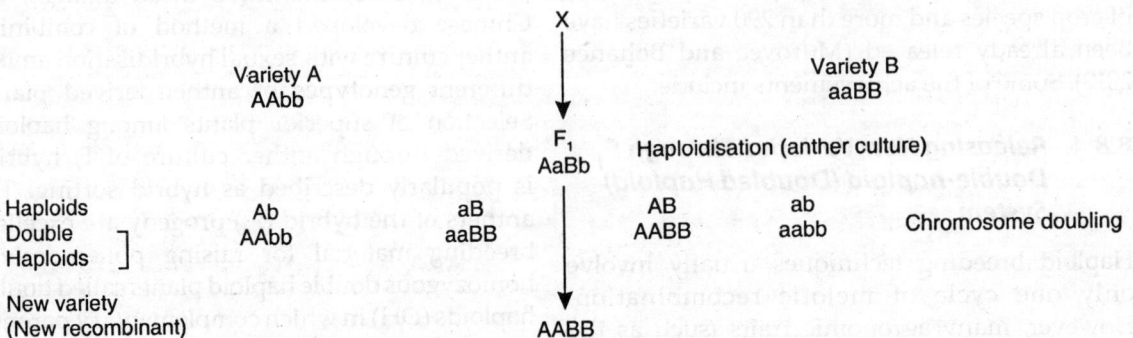

Fig. 8.5 Application of F₁ double-haploid system in releasing new recombinants (*see* Han 1987).

Chromosomes elimination has also been observed in interspecific crosses between bread wheat and maize. Hybrid embryos of these crosses in the process of further development lose maize chromosomes resulting in the production of haploid wheat plantlets. Haploid embryos have to be cultured *in vitro* because endosperm fails to develop and no seeds are formed on the plant. This maize chromosome elimination system, also called maize technique, is equally effective for induction of haploid embryos by crossing maize with barley, triticale, rye and oats (Wedzony et al. 2009).

Now there seems shift in approach toward plant breeding with emphasis on molecular mapping (Fasella and Hussain 2014).

8.8.3 Selection of Mutants Resistant to Disease

Screening mutants with resistance to disease is of prime importance in crop improvement. Haploids provide a relatively easier system for the induction of mutations; therefore, they can be employed in rapid selection of mutants having traits for disease resistance. Some examples of using anther culture technique in mutant selection successfully are tobacco mutants resistant to black shank disease, wheat lines resistant to scab *(Fusarium greminearum),* or herbicide mutants of *Brassica.*

8.8.4 Developing Asexual Lines of Trees/ Perennial Species

Chinese workers obtained a pollen-derived rubber tree taller by six metres which could then be multiplied by asexual propagation to raise several clones. Another example of using pollen-haploids in plant improvement is popular. Of particular interest in this tree species is that haploid seedlings selected for desired genotypes are naturally doubled as diploids after 7-8 years. This coincides with their flowering, enabling them to be both asexually and sexually propagated. Pollen-derived haploid plantlets have also been obtained in other perennial .woody species.,

such as *Aesulus hippocastatnum, Citrus microcarpa, Vitis vinifera, Malus prunifolia, M. pumila, Litchi chinensis, Euphoria logan, Ponicirus trifoliata, Lycium halinifolium, L. babarium, L. chinenesis* and *Camellia sinensis (see* Chen 1990).

8.8.5 Transfer of Desired Alien Genes

Chromosomal instability in haploids makes them potential tools for introduction of alien chromosomes or genes during wider crossing programmes. In rice the breeding of high-yielding varieties for resistance to blast is conventionally achieved through back-cross. This is a time-consuming process and may take more than twelve years to develop a cultivar with desired resistance. Through hybridisation and anther culture, a standard pollen-plant breeding procedure can be developed in rice to introduce genes for high yield and resistance to blast in a short span of two years (examples: cv. 'Zhonghua No. 8' and 'Zhonghua No. 9' released by the Institute of Crop Breeding and Cultivation in China). A similar approach was used to release a doubled-haploid wheat variety 'Florin' in France (de Buyser et al. 1987), in 1985. Anther culture is invariably used as an adjunct to broccoli breeding programme (Springer and Bailey 1990).

8.8.6 Establishment of Haploid and Diploid Cell Lines of Pollen Plant

The anther culture technique was used to establish both haploid and diploid somatic cell lines of pollen plants in wheat and maize. Similarly, a haploid tobacco line resistant to methionine sulfoxomide (MSO) was selected which turned out to be identical in phenotype and effect to the toxin produced by the pathogen *Pseudomonas tabaci (see* Chapter 14).

8.8.7 Production of Transgenic Plants

Pollen embryos, being haploid, can be used as tools for genetic transformation studies. Neuhaus et al. (1987) produced transgenic plants of *Brassica napus* by microprojectile bombardment of DNA into individual pollen embryos. Haploid plants regenerated from transformed embryos on diploidisation expressed the recombinant transgene homozygously. PEG-induced uptake of foreign DNA by isolated microspores of *Hordeum vulgare* also showed potential for *in vitro* androgenesis comparable to untreated controls (Kuhlmann et al. 1991). These studies suggest the possibility to produce stable transformants using pollen embryos or mirospores.

Chauhan and Khurana (2011) transformed haploid bread wheat with HVA1 gene to obtain drought tolerance and by chromosome doubling achieved stable wheat transformed of this gene. Chugh et al. (2009) transformed microspores of triticale by coupling of cell-penetrating peptides with plasmid DNA to regenerate transformed plants. The use of short peptide nanocarriers with plasmid DNA as delivery method of triticale microspore transformation opens new options for plant breeding and transgenic technology (Eudes et al. 2014).

Gametosomatic hybridization between egg-cell protoplasts and mesophyll protoplasts of *Petunia hybrida* (Sangthong et al. 2009), haploid and mixoploid protoplast fusion in cucumber (*Cucumis sativa*, Skálová et al. 2010) are other studies attempted for genetic transformation of plants. Gametosomatic hybridization refers to fusion of gametic (microspore tetrad or young pollen) and somatic protoplasts. In this process both nuclear-organelles genomic combination occurs in hybrid protoplasts.

8.8.8 *Haploid Plant Production by Centromere-Mediated Genome Elimination*

Molecular basis of genome elimination can be generated through seeds by manipulating a single centromere protein resulting in haploid plant production. Ravi and Chen (2010) obtained haploid *Arabidopsis thaliana* plants from seeds by manipulating a single centromere specific histone CENH3. They crossed cenh3 null mutant (expressing altered CENH3 protein) to wild type

and in formation of seeds of this cross there was elimination of chromosomes from mutant. The progeny generated from these seeds were haploid which spontaneously converted into fertile diploids. Application of this molecular basis of genome elimination for production of haploids in other plant species needs to be explored.

8.9 Problems Associated with Haploid Production

Although haploids have been raised mainly from anther culture of a large number of species, this technique has not proved successful in respect of all genotypes of crop species. A number of problems are encountered with the induction of haploids as well as a number of other factors which complicate the overall process of haploidisation. Some of these aspects are summarised below.

1. Often anthers fail to grow *in vitro* or the initial growth is followed by abortion of the embryos.
2. The tissue or callus developing from the anther generally comprises a chimera of diploid, tetraploid and haploid cells.
3. Selective cell division must take place in the haploid microspores, concomitantly restricting proliferation of unwanted diploid and polyploid tissues. This selective cell division is often impossible.
4. Formation of albinos in anther cultures especially with cereals can neither be avoided nor the loss of plants due to albinism reduced.
5. The technique of inducing haploids *in vitro* is not economically viable due to low success rate.
6. Callus derived from anther or pollen in a medium supplemented with growth regulators is usually detrimental for haploid production.
7. It is difficult to isolate a haploid from a mixture of haploids and higher ploidy levels since the polyploids outgrow the haploids.

8. The doubling of haploids is time consuming and may not always result in the production of a homozygote. Double haploids sometimes exhibit segregation in their progeny.

Double haploidy (DH), however, appears to emerge as potential tool for genetic studies, such as genome mapping, marker-trait association, location of QTLs and targets for transformation in addition to breeding particularly in brassica and cereals (Dwivedi et al. 2015). Microspore engineering and newer breeding technique, such as TALENs (transcription activator-like effector nucleases) are providing options to produce DH lines with desired trait(s) in one season instead of two seasons required for elucidating gene function (Chen and Gao 2013).

APPENDIX 8.1

Technique for Raising Plants from Isolated Pollen[a]

1. Surface-sterilise unopened flower buds with a suitable agent and wash thoroughly with sterile water.
2. If the pollen grains require an inductive period, float anthers in a liquid medium (medium used for anther cultures) for 2-3 days and then squeeze out the microspores.
3. Place about 50 anthers in a beaker containing about 20 ml of the liquid medium and press the anthers by means of a glass rod or syringe piston to squeeze out the microspores.
4. Filter the solution containing microspores and debris through a nylon sieve (pore size: 40 μm for *Nicotiana*, 25 μm for tomato and 100 μm for maize).
5. Centrifuge the suspension thus obtained at 500-800 rpm for about 5 min.
6. Remove the supernatant and resuspend the pellet containing pollen in fresh medium. Repeat the process described under 5 and 6 twice.
7. Finally, mix the pollen with a suitable culture medium at a density of 10^3–10^4 grains ml^{-1}.
8. Transfer the pollen suspension into the petri plate and adjust the volume in such a manner that the pollen grains are not too deeply immersed in the medium.
9. Store the cultures in diffused light (500 1x) at around 25°C.

---

[a] Modified from Nitsch (1974).

Triploid Production

Double fertilisation, a unique phenomenon, occurs in the majority of angiosperms except for members of the families Orchidaceae, Podostemaceae and Trapaceae. The results of this process are two fusion products: (a) the zygote (fusion product of an egg and one of the male gametes) and (b) the triploid primary endosperm nuclei (arising from the fusion of the second male gamete with two polar nuclei). Multiplication of the former leads to the development of an *embryo* (diploid tissue) and the latter forms the *endosperm* (triploid tissue). The endosperm, which lacks any organogenic or vascular differentiation, may be regarded as the 'second embryo' which is modified to serve the nutrient requirements of the developing zygotic embryo. Absence of endosperm or its disfunction usually causes embryo abortion. The complete consumption of endosperm by the developing embryo results in *non-endospermous* seeds (legumes, cucurbits) and partial consumption *endospermous* seeds (cereals, castor bean, coconut, coffee). In endospermous seeds, the endosperm tissue stores reserve food substances which are ultimately digested during seed germination and utilised for initial growth of the seedling.

Initially, interest in endosperm culture centred around testing the growth and regeneration potentialities of this unique tissue. During the last few decades the technique of endosperm culture has been applied to raise triploid plants which have a significant role in plant improvement.

Lamp and Mills (1933) made the first attempt to grow young maize endosperm 10-25 DAP (days after pollination) on a nutrient medium comprising the extract of potato or young corn, mineral salts and dextrose. The endosperm cells adjacent to the embryo showed only slight proliferation on this medium; however, the success achieved in obtaining callus from immature maize endosperm in 1949 by La Rue and his associates generated interest among many workers to examine maize endosperm cultures from different angles. In a few instances they were able to induce root differentiation but only in one instance were roots and shoots induced. Subsequently, efforts were made to achieve differentiation in the callus derived from the immature endosperm of *Cucumis sativus*, *Lolium perenne* and *Ricinus communis*.

Since 1963 the Department of Botany at the University of Delhi has actively contributed to the field of endosperm culture. The maiden report demonstrating cellular totipotency of endosperm tissue (Johri and Bhojwani 1965) appeared from this school. Further contributions by many workers have been significant and it is recognised that endosperm culture can be applied to raise triploids for crop improvement (Hoshino et al. 2011).

9.1 The Technique

9.1.1 Explant

(I) Source of Material

In most cereals, the mature endosperm does not respond to cultural conditions. It is, therefore, important to note that cereal endosperm can proliferate only if excised during a proper period of development. Normally the endosperm of cereals undergoes certain changes 12 DAP, making this tissue unable to respond. The proper period when the endosperm tissue can be excised and cultured is 8-11 DAP for maize (Fig. 9.1), 8 *DAP Triticum aestivum and Hordeum vulgare,* 9-10 DAP *Lolium perenne* and 4-7 DAP *Oryza sativa.* Even the endosperm of some dicots (*Cucumis sativus*) exhibits similar behaviour and the tissue proliferates only when excised 7-10 DAP.

Members of Euphorbiaceae, Santalaceae and Loranthaceae have proved most amenable to endosperm culture. Association of embryo tissue in the initial stages seems essential for inducing proliferation of mature endosperm tissue in respect of the first two families. Similar requirement has been observed for neem (Chaturvedi et al. 2003b) and phlox (Razdan Tiku et al. 2014) endosperm cultures. The embryo can be removed as soon as proliferation of the endosperm has started. Presumably the embryo releases some factor(s) required to activate the cells to divide. In some species, such as *Croton bonplandianum* (Fig. 9.1 A, B) and *Putranjiva roxburghii the* 'embryo factor(s)' may be replaced by GA_3. Entire seeds are used as explant for such species (Bhojwani and Johri 1971, Srivastava 1973). Sometimes soaking the seeds in water for various periods (before excising the endosperm pieces) also eliminates the need for association of the embryo during initial stages of endosperm culture (Kagan-Zur et al. 1990). Seeds of *Sapium Sebiferum,* a bioenergy plant belonging to Euphorbiaceae, after surface sterilisation and washings are reported to produce callus if placed directly on the callus induction medium with growth regulators (Tian et al. 2012). Immature seeds provide explants for non-endospermous seeds (e.g., apple, citrus) and the endosperm of such seeds exhibits no dependence on the embryo factor for proliferation.

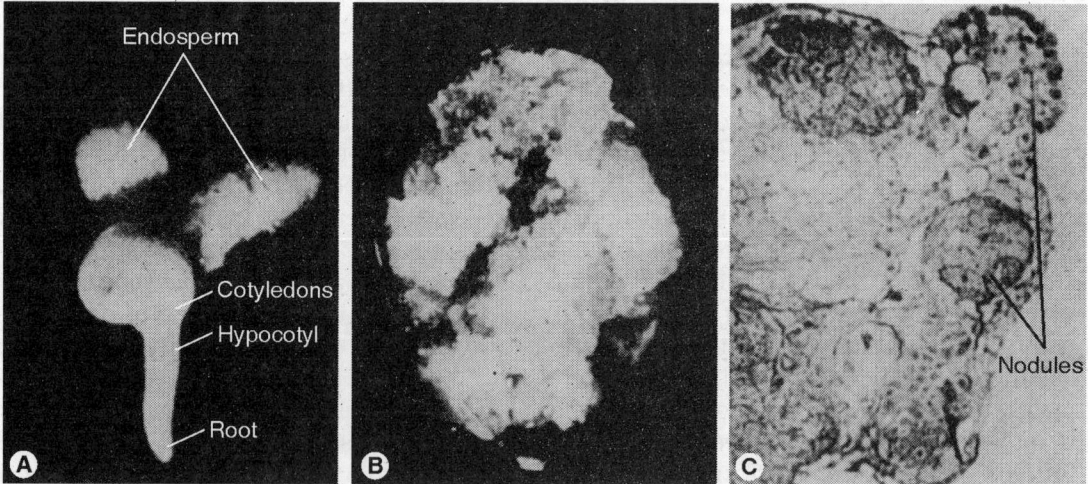

Fig. 9.1 A: 10-day-old culture of decoated seed *(Croton bonplandianum)* showing germinated embryo with enlarged cotyledons, elongated hypocotyl and root. B: Profusely growing endosperm callus. C: Section of a 4-week-old endosperm culture of Zea *mays* showing hypodermal nodules. (A and B after Bhojwani and Johri 1971; C after Sehgal 1969).

(II) Inoculation of Explant

Seeds having a massive endosperm are decoated, surface-sterilised with suitable disinfectant and, after two to three washings in sterile distilled water, planted on the nutrient medium. In families (Loranthaceae) in which the endosperm is surrounded by a sticky viscin layer, making it inconvenient to excise the endosperm tissue, the fruits are first surface-sterilised with 90% alcohol and then seeds (endosperm plus embryo) removed for dissection under aseptic conditions. For *in vitro* culture of an immature endosperm the entire seed or kernel is surface-sterilised and the endosperm tissue excised under aseptic conditions. Cereal endosperm explants are inoculated after they have been excised from the surface-sterilised immature ovaries obtained from spikelets of suitable DAP. The tops of kernels or immature ovaries (micropylar end) are cut off with a sterile knife and the exposed endosperm squeezed out and placed on the nutrient medium.

9.1.2 Nutrient Media

(I) Induction of Callus

The basal medium of White or MS (*see* Table 3.1) is generally used in endosperm culture. To induce callus from an immature endosperm the basal medium has been supplemented with tomato juice, yeast extract, grape juice and maize grain juice. For *Lolium* endosperm, only yeast extract was effective and in some cases the addition of asparagine or growth factors (1-2 mg l^{-1} 2,4-D or 1 mg l^{-1} IAA) was required. Generally, the establishment of dicotyledonous endosperm cultures requires the presence of an auxin (2,4-D or IAA), a cytokinin (kinetin or benzylamino purine) and a rich source of organic nitrogen (i.e., yeast extract or casein hydrolysate).

Sucrose (2-4%) has proved to be the best source of carbohydrate. *Asimina triloba* is an exception which shows response when starch is present in the medium as a carbon source (Lampton 1952). The optimum pH of the medium also seems to vary according to the species (*Asimina triloba*, 4.0; *Jatropha panduraefolia* and *Putranjiva roxburghii*, 5.0; *Ricinus communis* and *Zea mays*, 6.1).

(II) Organogenesis

In 1965, Johri and Bhojwani reported for the first time the organ-forming capacity of mature endosperms of *Exocarpus cupressiformis*. They observed the development of eight buds in a single explant ('seed') cultured on a medium supplemented with IAA, kinetin and casein hydrolysate (CH). These buds were triploid and, when cultured on a fresh medium of the same composition, formed a slow-growing callus that later differentiated shoot-buds. Thereafter, extensive studies were undertaken to raise full triploid plants from endosperm cultures and success was achieved in citrus, apple, actinidia, sandal-wood, ornamental and crop species which establishes totipotent nature of endosperm tissues (Table 9.1).

Shoot buds in loranthaceous and santalaceous parasites may differentiate directly from the peripheral cells of the endosperm although in most other plants it is preceded by a callusing phase. The supply of an exogenous cytokinin is necessary for the differentiation of shoot-buds from endosperm tissue. In 'seed' cultures of semi-parasites, such as *Scurrula pulverulenta* (Fig. 9.2) and *Taxillus vestitus*, a cytokinin alone induces bud formation. Both epidermal and hypodermal cells in *Scurrula* become meristematic, resulting in the formation of bud primordia, which is in contrast to the initiation of buds from the meristematic activity of epidermal cells in *Taxillus*. However, shoot-bud differentiation from the endosperm of *Dendrophthoe falcata* and *Leptomeria acida* required a combination of cytokinin and a low concentration of IAA or 1BA. Of the various cytokinins tested, 2-ip seems most effective although the addition of CH (2000 mg l^{-1}) to the medium is necessary for shoot-bud differentiation in *Dendrophthoe* and *Leptomeria*.

In autotrophs, organogenic differentiation from the endosperm is usually accomplished by callusing. On a medium supplemented with

Table 9.1 Shoots or triploid plants raised from culture of endosperm tissue

Family	Species
Actinidiaceae	*Actinidia chinensis*[a,b]
	A. chinensis × *A. melanandra*[k]
	A. arguata × *A. melanandra*
	A. arguata × *A. deliciosa*[k]
Annonaceae	*Annona squamosa*[c]
Caricaceae	*Carica papaya*[o]
Celastraceae	*Euonymus alatus*[q]
Euphorbiaceae	*Codiaeum variegatum*
	Emblica officinalis[d]
	Jatropha panduraefolia
	Putranjiva roxburghii
	Sapium sebiferum[p]
Gramineae	*Hordeum vulgare*[e]
	Oryza sativa
	Zea mays[f]
Juglandaceae	*Junglans regia*[g]
Liliaceae	*Asparagus officinalis*[h]
Loranthaceae	*Dendrophthoe falcata*
	Scurrula pulverulenta
	Taxillus vestitus
Meliaceae	*Azadirachta indica*[l]
Moraceae	*Morus alba*[m]
Polemoniaceae	*Phlox drummondii*[n]
Roasaceae	*Prunus persica*
	Pyrus malus
	P. communis[i]
Rutaceae	*Citrus grandis*
Santalaceae	*Exocarpus cupressiformis*
	Santalum album
Solanaceae	*Lycium chinensis*[j]
Umbelliferae	*Petrosetinum hortense*

[a] Mu et al. (1990)
[b] Gui et al. (1982)
[c] Nair et al. (1986)
[d] Sehgal and Khurana (1985)
[e] Sun and Chu (1981)
[f] Zhu et al. (1988)
[g] Tulecke et al, (1988)
[h] Liu et al. (1987)
[i] Zhao (1988)
[j] Gu et al. (1985)
[k] Kin et al. (1990)
[l] Chaturvedi et al. (2003b)
[m] Thomas et al. (2000)
[n] Razdan Tiku et al. (2014)
[o] Sun et al. (2011)
[p] Tian et al. (2012)
[q] Thammina et al. (2011)

For other species *see* Bhojwani and Razdan (1983).

Further reports have also appeared about induction of callus from endosperm cultures of coconut, parsley, tomato, *Ricinus* and *Vitis*. Root formation has also been observed in endosperm-derived calli of *Croton bonplandianum*, *Euphorbia geniculata* and *Leptomeria acida*.

2,4-D, kinetin and yeast extract the endosperm of *Croton* and *Jatropha* forms a profusely growing callus which remains unorganised. Differentiation is possible only when the callus tissue is transferred to the basal medium. Whereas endosperm calli of croton develop only roots, those of *Jatropha* and *Putranjiva* form roots as well as shoots. Both these organs in *Putranjiva* differentiate as a bipolar axis on a medium supplemented with IAA, kinetin and CH.

9.1.3 Development of Shoot Buds and Plantlets

Buds initiating directly from the endosperm have limited growth. The full growth can only be achieved if they are excised and cultured on a fresh medium. With manipulations in the media constituents calli proliferating from buds may differentiate to form shoot buds or haustoria. An increased concentration of kinetin may prevent haustoria formation (*Taxillus*) and favour initiation and growth of shoot buds. A low concentration of auxin (IAA or IBA) appears necessary for kinetin-induced shoot-bud differentiation in *Dendrophthoe* and *Leptomeria* (Nag and John 1971).

Plantlet formation from endosperm callus may follow the embryogenic mode of development. Embryo differentiation occurs when the proliferated tissue is transferred from a callusing medium to a basal medium with or without gibberellin. Sometimes an increase in the concentration of salts to twice their level and GA_3 to 5 mg l^{-1} in the callusing medium is essential for inducing embryogenesis, as in *Citrus* for example (Gmitter et al. 1990).

Interestingly, in *Taxillus* the position of the endosperm on the medium significantly affects the differentiation of buds and ultimate plantlet formation. Endosperm halves planted with their cut surfaces in contact with the medium are more responsive than cultures with cut ends away from the medium. Almost all cultures produce buds when the cut portion is placed in contact of the medium. If the cut end is turned away from the medium, only 30% buds are produced.

Injury to the endosperm is likely to enhance the frequency of shoot-bud formation. In *Taxillus,* the presence of an embryo adversely affects bud differentiation from the endosperm.

9.1.4 Maintenance of Cultures

To induce callus, the endosperm cultures (castor, bean, maize, phlox) are maintained in darkness or diffused light (*see* also Razdan Tiku et al. 2014). Differentiation, however, takes place when the calli are transferred to bright light (2000-4000 lx). Activated charcoal seems essential to initiate callus from the cellular endosperm of *Cocos nucifera* L. (Kumar et al. 1985). Further, success in obtaining a callus is attributed to the initial association of zygotic embryo with endosperm

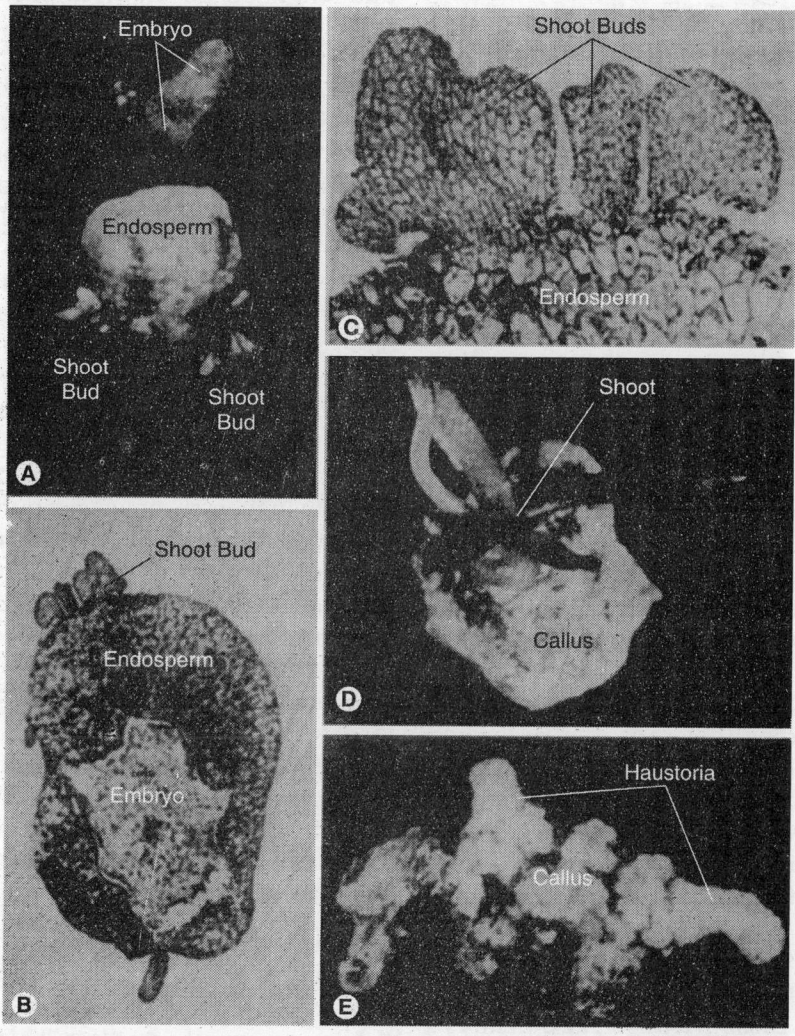

Fig. 9.2 *Scurrula pulverulenta* endosperm culture. A: old culture of 'seed' on White's medium supplemented with zeatin (10^{-5} M) showing differentiation of shoot-buds. B: Transection of cultured 'seed' in which buds are arising from the periphery of the endosperm. C: Portion from B confirming the peripheral origin of buds. D and E: Endosperm buds cultured on a medium enriched with IAA, kinetin and CH. Note the formation of shoot and haustoria from callused buds (Courtesy: Dr S.S. Bhojwani).

tissue in cultures for 2-weeks which after callus initiation is removed (Chaturvedi et al. 2003b, Razdan Tiku et al. 2014). The temperature necessary for the growth of an endosperm callus as well as differentiating cultures is reported to be around 25°C.

9.1.5 Biochemical Markers in Endosperm Cultures

Racchi and Manzocchi (1988) cultured immature endosperm of maize and identified a protein (26 kd peptides) characteristic of pigmented tissues. This showed that anthocyanin and proteins could be used as biochemical markers in maize endosperm cultures. Further, in long-term cell cultures from maize endosperm, waxy and amylase-extender type mutations were observed and cells in these cultures were not completely dedifferentiated since the expression of zein (storage protein in maize seeds) was highly reduced (see Saravitz and Boyer 1987, Manzochhi et al. 1989).

9.2 Histology and Cytology of Callus

Zea mays endosperm excised 12 DAP comprises *a* comparatively homogeneous tissue with peripheral layers of meristematic cells. These cells undergo anticlinal and periclinal divisions adding to the girth of the outermost layer until it becomes four cells thick. The medium is fortified with 2,4-D, kinetin and YE to induce the differential growth at localised areas leading to the formation of meristematic nodules. These nodules arise either directly from the peripheral layer or from cells located immediately below the outermost layer. The callus initiates from the inner side of the endosperm in *Croton* which becomes visible when the endosperm bursts open into two halves. Similar observation was found with neem endosperm cultures (Chaturvedi et al. 2003b). At this stage the embryo is removed and proliferation commences all over the endosperm tissue. *Suntalum* endosperm starts callusing with the differentiation of concentric layers of meristematic tissue about

five to six layers below the epidermis. This tissue in *Osyris wightiana* and *Putranjiva roxburghii* develops to nodular meristerns in the presence of kinetin alone or in combination with IAA. Tissues derived from the endosperm in the members of Euphorbiaceae, Loranthaceae and Santalaceae readily differentiate tracheidal elements which may appear scattered or in clusters. The endosperm undergoes organogenic differentiation generally in such plants which exhibit tracheidal differentiation.

Most of the plants regenerated from endosperm cultures are triploid although cytologically, the endosperm tissue exhibits a high degree of polyploidisation of its cells associated with abnormalities such as chromosome bridges and lagging chromosomes. Hypoploid and aneuploid cells are as common as those showing normal triploid chromosome number. Cells with ploidy higher than $3n$ chromosomes have been observed in *Croton*, *Jatropha* and *Lolium*. These aspects of endosperm cytology are reported both *in vitro* and *in vivo*. Endosperm cells of mistletoes (*Dendrophthoe falcata*, *Taxillus vestitus* and *T. cuneatus*) are exceptions for remarkable stability in their chromosome number. Similarly, the endosperm cells of rye-grass remain triploid even in a 10-year-old callus in spite of the fact that the length of the culture period and media composition are known to influence polyploid mitosis in long-term cultures. Flow cytometry is an alternative procedure that has been employed to determine ploidy level of *Euonymus alatus* 'campactus' (Burning bush) plants derived from endosperem (Thammina et al. 2011).

9.3 Application of Triploids in Plant Improvement

Advances made in the technique of endosperm culture have enabled the production of full triploid plants (Table 9.1). Triploid plants are self-sterile and usually seedless. This trait increases edibility of fruits and is desirable in such plants as apple, banana, mulberry, grape, mango, watermelon and others which are commercially important for their edible fruits. In cereals or crops where grains or seeds are used, triploids are undesirable. Seed sterility in timber or fuel-yielding plants is not a serious setback because their vegetative parts are important and these plants can be multiplied vegetatively. The same holds true for other species which are grown for their vegetative parts. In such plants triploids show better performance than their relative diploid or tetraploids. For example, the triploids of *Populus tremuloides* have better quality pulpwood which is a characteristic important to the forest industry. Various trisomics from among triploids may also be useful in gene mapping.

Theoretically, the most common method of triploid production is through crossing a tetraploid with a diploid. This method has not been successful in most cases and hence endosperm culture is an alternative and possibly the only feasible technique in these contingencies. The demonstration of totipotency in triploid cells of endosperm origin has generated much enthusiasm for its application in triploid production under *in vitro* conditions. Higher yield and harvest index, increase in dry matter, tastier fruits and resistance to pathogens in comparison to normal diploids are some of the superior traits exhibited in endosperm-derived triploid plants (Bhojwani and Dantu 2013, Razdan Tiku et al. 2014).

Triploid production through endosperm culture technique although successful only in a limited number of species has scope for plant breeders who should apply this technique to various other economically important species or cultivars where triploids are desirable and conventional methods of plant improvement have been difficult or impossible (Thomas and Chaturvedi 2008, Hoshino et al. 2011). Intensive research is needed to design suitable media for induction of calli and their regeneration to plants. Growth-promoting factors of the endosperm are not necessarily specific to the embryo of the same species. This feature may enable the cultured endosperm tissue to support the growth of young embryos from a different plant species and allow raising of full plants, particularly in such isolated young or hybrid embryos which abort prematurely (*see* Chapter 11).

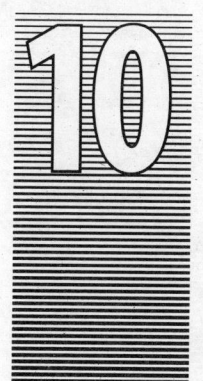

In Vitro Pollination and Fertilization

Pollination and fertilisation under *in vitro* conditions offer an opportunity for producing hybrid embryos among plants that cannot cross by conventional methods of plant breeding. In nature, intergeneric or interspecific hybridisation occurs less frequently. This is due to barriers hindering the growth of the pollen tube on the stigma or style. In such cases the style or part of it can be excised and pollen grains either placed on the cut surface of the ovary or transferred through a hole in the wall of the ovary (Leduc et al. 1992). This technique, called *intraovarian pollination*, has been successfully applied in such species as *Papaver somniferum, P. rhoeas Eschscholtzia californica, Argemone mexicana* and *A. ochroleuca.* Another approach to overcome the barriers to pollen tube growth is direct pollination of cultured ovules (*in vitro* ovular pollination) or excised ovules together with placenta (*in vitro* placental pollination). This technique of *in vitro* pollination was developed at the University of Delhi to produce hybrids among species of the Papaveraceae and Solanaceae (Maheshwari and Kanta 1964). The placental pollination of ovules has been successfully applied to overcome self-incompatibility in *Petunia axillaris* and to obtain haploids *Mimulus luteus* using the pollen from *Torenia fournieri*. Various other techniques developed to circumvent the prezygotic barriers to fertility include: (a) bud pollination, (b) stub pollination, (c) heat treatment of the style, (d) irradiation and (e) mixed pollination.

The development of seed through *in vitro* pollination of exposed ovules has been described as 'test-tube fertilisation', whereas the process of seed formation following stigmatic pollination of cultured whole pistils has been referred to as '*in vitro* pollination' (*see* Bhojwani and Razdan 1983, 1996). In these two processes, fertilisation of the egg occurs inside the ovule by gametes delivered by the pollen tube. Contrarily, the phenomenon of test-tube fertilisation in animal systems involves the *in vitro* fusion of excised eggs by free-floating sperms (male gametes). Considering the fact that male gametes in plants do not float freely and are delivered by the pollen tube, a general term *in vitro* pollination (IVP) has been used for *ovular pollination* (application of pollen to excised ovules), *ovarian pollination* (application of pollen to excised ovary*)*, *placental pollination* (application of pollen to ovules attached to the placenta) and *stigmatic pollination* (application of pollen to the stigma) under *in vitro* conditions (Fig. 10.1).

10.1 Methodology

10.1.1 Materials

Ovaries which are large and contain many ovules are the best experimental material. In members of the Solanaceae (*Nicotiana tabacum, N. alata, N.*

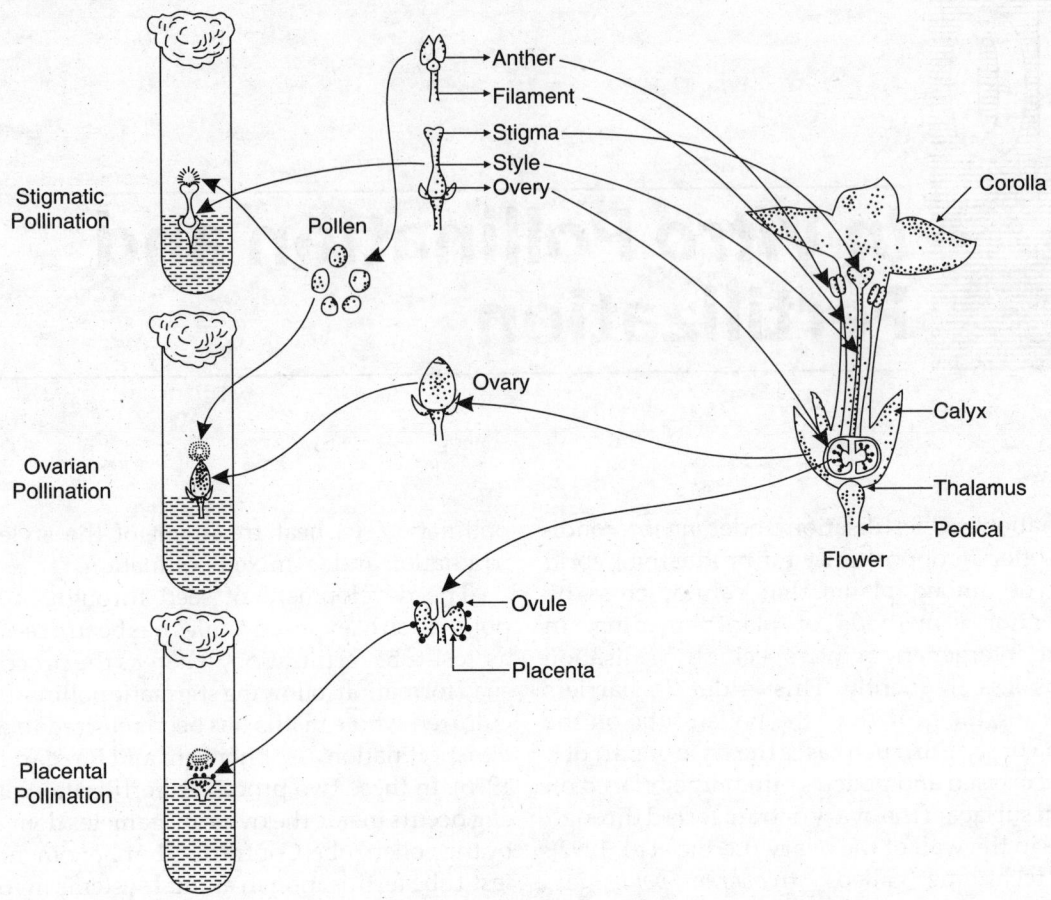

Fig. 10.1 Diagrammatic representation of in vitro pollination.

rustica, Petunia hybrida), Papaveraceae *(Papaver somniferum, Eschscholtzia californica, Argemone mexicana)* and Caryophyllaceae *(Melandrium album, M. rubrum, Agrostemma githago, Dianthus caryophyllus),* the placentae are covered with several hundred ovules. Large numbers of ovules remain undamaged during isolation of placentae, thereby contributing to success in their pollination *in vitro* and ultimate development of seeds. The other essential material is pollen, which should be viable and able to germinate. There must be abundant growth of pollen tubes all over the ovules and placentae in cultures. The germination of pollen *in vitro* may be difficult in

respect of some families but can be overcome by immersing ovules (example: *Brassica oleracea)* one day before pollination in 1% calcium chloride solution to favour the growth of pollen tubes (Zenkteler 1985).

To start an experiment on test-tube pollination it is essential to collect data on aspects such as time of anthesis, dehiscence of anthers, germination of pollen tube into ovules, viability of ovules and fertilisation inside the embryo sacs. Entry of the pollen tube through the micropyle of the ovules can be observed in stained squash preparations under the microscope.

10.1.2 Disinfection of Material

A reasonable disinfection of ovule and pollen is the principal requirement for *in vitro* pollination. Flower buds to be used as the female partner are emasculated before anthesis and bagged in order to prevent undesired pollination. The buds are brought to the laboratory for aseptic culture just before the anthers are at the stage of dehiscence. The whole pistils (after removing sepals and petals), or the ovaries alone, are sterilised by a quick rinse in 70% alcohol, surface-sterilised with a suitable agent (*see* Chapter 4) and finally washed thoroughly with distilled water. Ovaries of plants which grow in open air require a longer period of sterilisation than plants maintained in a clean and well-kept greenhouse. The ovary wall is then carefully peeled with sterile scalpels, forceps, or needles to expose the mass of bare ovules attached to the placenta. The entire placenta, or part of it bearing the ovules, is used in placental pollination. To perform *in vitro* stigmatic pollination the excised pistils are carefully surface-sterilised without wetting the stigma with the sterilant solution. Sometimes the entire pistil in which the placenta-bearing ovules have been exposed (by removing the ovary wall) are cultured to study the effect of simultaneous placental and stigmatic pollination in the same pistil.

To collect pollen under aseptic conditions, anthers removed from buds or open flowers are kept in sterile petri plates containing a presterilised filter paper until their dehiscence. The dehisced pollen is then aseptically deposited on the cultured ovules, placenta, or stigma depending on the nature of the experiment. Disinfection of anthers, particularly when plucked from open flowers, may be necessary using a sterilant. Generally, the pollen deposited directly on the cultured part of the pistil performs better than that spread on the medium around the ovules.

In graminaceous species (maize), ovaries are well protected by several layers of husks and surface sterilisation may not be necessary. Bagged spadices are collected 2-6 days after the emergence of silk from the husks, which are then severed with a scalpel to enable the outer and inner husks to be removed manually by means of sterilised instruments under aseptic conditions. Ovaries (4-10 in two rows) are removed from pieces of cob and transferred on the medium in petri plates in such a way that their silks remain hanging out of the plate after putting on the lid. Pollination is done directly on the silks or ovules *in vitro*. Twenty-four hours after pollination the silks are clipped off and the petri plates sealed. In some plants the bacterium *Erwinia stewartii* may be present as an endophytic contaminant which is not eliminated by surface sterilisation of the ear. Therefore, a brief rinse of the inner husks with 95% ethanol (Dhaliwal and King 1978) or a 30-min treatment with 1% Famosept (Sladky and Havel 1976) may be necessary.

10.1.3 Culture of Ovules and Ovary

(I) Ovules

The growth of pollen tubes attached to bare ovules is often inhibited by the presence of water on the surface of the ovules. This film of water should be dried with filter paper and, later, the dried ovules covered by the pollen grains. Raising seeds from ovules which contain globular or older embryos is comparatively easy, as reported on the basis of success achieved in *Gynandropsis gynandra, Impatiens balsamina, Nicotiana tabacum* and *Allium cepa*. However, 6 days after *in vitro* pollination ovules contain a single-celled zygote which requires more complex growth conditions (*see* Section 10.2.2). The technical procedure for self-pollinating and cross-pollinating ovules is the same. In both types of experimental materials the early growth of proembryo is more or less similar. For the development of subsequent embryonic stages, ovules which have been self-pollinated are usually kept on the placenta until seed formation while, contrarily, cross-pollinated ovules require placenta only during the initial 6-8 days of culture. Afterwards they can be transferred to a fresh medium without placenta. The beneficial effect of placental tissue associated with *Gynandropsis* ovules compensates for enrichment of the medium necessary for the growth and development of excised ovules in culture.

Ovule culture has proved to be a very useful technique for raising interspecific hybrids within the genus *Gossypium*, *Helianthus* and *Trifolium*; intergeneric hybrids between crosses of the genera belonging to members of Gramineae and Cruciferae (*see* Bhojwani and Razdan 1996). Normally these hybrids fail to develop *in situ* due to early embryo abortion and premature abscission of fruits. Ovule culture can be used to rescue them.

(II) Ovary

The technique of ovary culture was developed by Nitsch in 1951 who successfully reared ovaries (*Cucumis* and *Lycopersicon*) excised from pollinated flowers *in vitro* to develop into mature fruits. These fruits contained viable seeds but failed to attain a size comparable to fruits developed under natural conditions. Subsequently, there were reports on successful culture of excised ovaries from a number of species (*Linaria macroccana, Tropaeolum majus, Iberis amara, Hyoscyamus niger*) on a medium containing mineral salts and sucrose. The addition of vitamin B to the medium resulted in the development of fruits of normal size with viable seeds. Further enrichment by IAA or coconut milk induced even larger fruits than the *in vivo* formed fruits of *Anethum graveolens* (*see* Kanta and Maheshwari 1963a).

The floral envelopes (lemna and palea) play an important role in the development of the fruit and the embryo of monocots. Ovaries excised soon after pollination of *Triticum aestivum* and *T. spelta* develop in cultures only when the floret envelopes remain intact. This requirement of the floret envelopes associating with excised monocot ovules *in vitro* is known as the 'hull factor'. In the absence of. this factor, DNA synthesis and cell elongation of barley embryo cells can take place but cell division does not occur. Similarly, the

association of perianth while attempting *in vitro* Stigmatic pollination has been found necessary in dicots. Several interspecific and intergeneric hybrids can be produced between sexually incompatible parents in the family Cruciferae with the aid of ovary culture (Batra et al. 1990).

10.2 Factors Affecting Seed-set after *in vitro* Pollination

10.2.1 *Physiological State of the Explant*

The physiological state of the pistil at the time of excising the ovules or ovary influences the seed-set after *in vitro* pollination. Wetting the surface of ovules or stigma (in stigmatic pollination) may lead to poor pollen germination or bursting of the pollen tubes and, consequently, poor seed-set. Pollen germination on the stigma and growth of the pollen tube through the style influence the synthesis of proteins, which may sometimes inhibit entry of the pollen tube into the ovary. Hence it is necessary to ascertain in which part of the pistil the barrier exists. To improve the chances of success of *in vitro* pollination the level of incompatibility can be reduced by exicising the part synthesising inhibitory proteins and directly pollinating the remaining portion of the pistil under experimental conditions.

The time of excising the ovules from pistils has a definite influence on seed-set after *in vitro* pollination. Ovules excised 1-2 days after anthesis show a higher seed-set than those excised on the day of anthesis. In maize, the spike pollinated *in vitro* 3-4 days after silking yields a better seed-set.

10.2.2 *Culture Medium*

The nutrient medium plays an important role in supporting the normal development of ovary and ovules in cultures until seed formation. In the first report (Maheshwari 1958) on successful culture of ovules, nutrient medium comprised Nitsch's mineral salts, White's vitamins (*see* Table 3.1 for both constituents), and 5% sucrose. Ovules from *Papaver rhoeas* and *P. somniferum* were excised 6 DAP (days after pollination) and

Table 10.1 Medium widely used to culture *in vitro* pollinated ovules[a]

Constituents	Amount (mg l^{-1})
$CaNO_3 \cdot 4H_2O$	500
KNO_3	125
KH_2PO_4	125
$MgSO_4 \cdot 7H_2O$	125
$CuSO_4 \cdot 5H_2O$	0.025
$Na_2 \cdot MoO_4$	0.025
$ZnSO_4 \cdot 7H_2O$	0.5
$MnSO_4 \cdot 4H_2O$	3.0
H_3BO_3	0.5
$FeC_6O_5H_7 \cdot 5H_2O$	10.00
Glycine	7.5
Ca-pantothenate	0.25
Pyridoxine HCl	0.25
Thiamine HCl	0.25
Niacin	1.25
Sucrose	50000
Agar	7000

[a] See Kanta and Maheshwari (1963b).

Table 10.2 Medium used for raising intra-and interspecific hybrids from young fertilised ovules[a]

Constituents	Amount (mg l^{-1})
KNO_3	5055
NH_4NO_3	1200
$MgSO_4 \cdot 7H_2O$	493
$CaCl_2 \cdot 2H_2O$	441
KH_2PO_4	272
$FeSO_4 \cdot 7H_2O$	8.3
$Na_2 \cdot EDTA$	11
H_3BO_3	6.18
$MnSO_4 \cdot 4H_2O$	16.9
$ZnSO_4 \cdot 7H_2O$	8.6
KI	0.83
$Na_2MoO_4 \cdot 2H_2O$	0.24
$CuSO_4 \cdot 5H_2O$	0.025
$CoCl_2 \cdot 6H_2O$	0.024
Nicotinic acid	0.49
Pyridoxine HCl	0.82
Thiamine HCl	1.35
Inositol	180
D-Fructose	3600
Sucrose	40000
IAA	5×10^{-5} mol l^{-1}

[a] After Steward and Hsu (1978).

possessed a zygote or two-celled proembryo with a few endosperm nuclei. The growth of the embryo in the initial stages was slower but soon after the globular stage became rapid, measuring 0.93 mm in size compared to 0.65 mm embryos *in vivo*. Addition of kinetin or CH was essential to promote the initial growth of the embryo. Several orchid ovules isolated from pollinated ovaries can grow successfully on simple 10% sucrose solution, but the ovules of *Zephyranthes* (containing a zygote and primary endosperm nuclei) require coconut milk or casamino acids in Nitsch's medium. The promotive effect of casamino acids may be substituted by histidine, arginine, or leucine. In *Trifolium repens*, ovules (1-2 DAP) developed into mature seeds only when the culture medium was supplemented with juice prepared from young fruits of cucumber or watermelon. GA$_3$ (10 mg l^{-1}), in addition to fruit juice, further promoted seed development in these very young ovules.

For *in vitro* culture of pollinated ovules of most species a modified Nitsch's medium (Table 10.1) is widely used. However, the medium developed by Steward and Hsu (Table 10.2) has been recommended for raising intraspecific or interspecific hybrids from young fertilised ovules. Information regarding the effect of various growth regulators and other supplements to the basal medium on seed development is elementary although the presence of 10 μg l^{-1} IAA or 0.1 μg l^{-1} kinetin significantly improve the number of seeds per ovary. Higher levels of kinetin are generally inhibitory. According to Gengenbach (1985), the nitrogen source does not affect fertilisation frequency in the excised *in vitro* pollinated maize ovaries, but a source of reduced nitrogen as a complete amino acid

Table 10.3 Culture medium containing a complete amino acid mixture as the nitrogen source required for optimal development and growth of maize kernels[a]

Constituents	μ moles l^{-1}	Amino acid	mg l^{-1}
KH_2PO_4	3750	Aspartate	500
$MgCl \cdot 6H_2O$	1500	Threonine	198
$CaCl_2 \cdot 2H_2O$	1200	Serine	347
H_3BO_3	100	Glutamate	1761
$MnSO_4 \cdot H_2O$	100	Proline	621
$ZnSO_4 \cdot 7H_2O$	10	Glycine	168
KI	5	Alanine	476
$Na_2MoO_4 \cdot 2H_2O$	1	Cysteine	160
$CuSO_4 \cdot 5H_2O$	0.1	Valine	330
$CoCl_2 \cdot 6H_2O$	0.1	Methionine	117
Fe-EDTA	50	Isoleucine	237
Thiamine-HCl	0.4 mg l^{-1}	Leucine	1009
Folic acid	0.044 mg l^{-1}	Tyrosine	262
Niacin	1.2 mg l^{-1}	Phenylalanine	387
Sucrose	150 g l^{-1}	Lysine	88
Agar	5.5 g l^{-2}	Histidine	164
		Arginine	150
		Asparagine	300
		Glutamine	292
		Tryptophan	204

[a] After Gengenbach (1985).

mixture (Table 10.3) is required for optimal kernel development and growth.

Osmolarity of the culture medium also affects the development of excised ovules. Generally, ovules containing a globular embryo develop into mature seeds on a medium containing sucrose anywhere in the range of 4-10%, but young fertilised ovules having a zygote and a few endosperm nuclei, or ovules just after fertilisation require 6% and 8% respectively.

10.2.3 Storage Conditions

There is no precise information on the effects of temperature and light on test-tube fertilisation. Usually the first steps of this process occur at room temperature and without special lighting (Zenkteler 1980). Only later, the ovary cultures are maintained at 22-26°C and other suitable conditions favouring embryogenesis. Several days after test-tube pollination some ovules usually enlarge, indicating that pollen tubes have entered the embryo sacs and both embryo and endosperm may be developing. This can be confirmed by cytoembryological examination.

10.2.4 Genotype

The response of *in vitro* ovaries in relation to the seed-set depends on the species.

of crucifers (shed at three-celled stage) are difficult to germinate in cultures and a modified technique is required to obtain germinable seeds. Kameya et al. (1966) dipped *Brassica oleracea* ovules in 1% solution of $CaCl_2$, planted them on a slide precoated with 10% gelatin of 10 μm thickness and then pollinated with the pollen from freshly opened flowers. The slide was stored in a covered petri plate with a moist filter paper stuck to the lid. After 24 hr incubation the fertilised ovules were transferred to Nitsch's agar medium until seed-set.

10.3 Applications of *in vitro* Pollination

In plant breeding programmes the technique of *in vitro* pollination has potential applications in at least three different areas, viz. (a) overcoming self-incompatibility, (b) overcoming cross-incompatibility, (c) haploid production through parthenogenesis, and (d) production of stress-tolerant plants.

Petunia axillaris and *P. hybrida* are self-incompatible species. Germination of pollen is good on self-pollinated pistils but a barrier exists in the zone of the ovary and, as a result, the pollen tube cannot fertilise the ovule. The barrier in these taxa can be overcome by *in vitro* pollination as through this approach fertilisation and seed-set have been reported to occur normally. In *P. axillaris* incompatibility can also be overcome through *in vivo* bud pollination.

Successful culture of *in vitro* pollinated ovules has raised the possibility of producing hybrids which are unknown because of prefertilisation incompatibility barriers. Intraspecific, inters-pecific, intergeneric and interfamilial crosses attempted through *in vitro* ovular and placental pollination are summarised in Table 10.4, and Appendices 10.1 and 10.2. Another application of *in vitro* pollination reported is the production of haploids of *Mimulus luteus* cv. *Tigrinus grandiflorus* by pollinating its exposed ovules with *Torenia fournieri*. The haploids of *M. luteus* developed parthenogenetically, which otherwise could not have been obtained through anther culture. Such parthenogenetic development of haploids in cultures from *in vitro* pollinated but unfertilised ovules, has also been reported in respect of *Hordeum vulgare*, *Nicotiana tabacum* and *Triticum aestivum*.

Maize plants tolerant to heat stress have been produced through *in vitro* pollination at high temperature. Petolino et al. (1990) observed that at elevated temperature (38°C) only heat tolerant pollen were able to effect fertilization and the resulting maize plants expressed the trait for heat tolerance. Additionally these plants exhibited increased vigour and grain yield. Such an approach demonstrated the potential of applying *in vitro* pollination for producing plants resistant to environmental stress.

Although development of young hybrid embryos can be achieved in extremely wider crosses through *in vitro* pollination the efficiency of this technique needs much improvement. Inadequate knowledge of the suitable culture requirements of very young ovules has contributed to most of the failures of *in vitro* pollination. Moreover, other areas of plant tissue culture such as haploid production through anther, pollen and protoplast culture have overshadowed the potential of this technique.

10.4 *In vitro* Fertilization Using Isolated Single Gametes

A landmark technique developed recently in the field of plant biotechnology has been the successful fusion of male and female gametes isolated from higher plants *in vitro* and subsequent regeneration of the fusion product (single zygote) into embryo and finally a plant (Kranz and Lörz 1993). This process, known as *in vitro* fertilization (IVF), requires isolation of male gametes (sperms) from germinating pollen grain or tube and of the female gamete (egg) from embryo sac (which also contains two synergids, central cells and antipodal cells). *In vivo*, all flowering plants undergo double fertilization which involves two fusion events: (a) fusion of one male gamete with egg cell resulting in the development of the fertilized egg (zygote) into embryo, (b) fusion of the other male gamete with central cells leading to the formation, of endosperm (nurse tissue for the growing embryo). Both fusion events occur deep within embryo sac which is encased by ovular tissues inside the ovary. The zygote therefore remains surrounded by endosperm and tissues of the ovule, whereas *in vitro*, on the contrary, fusion of isolated gametes, development of zygote, embryo and

Table 10.4 Summarised data on *in vitro* ovular and placental pollination[a]

Species	Medium			Results	
	Basal medium[b]	Sucrose concen. (%)	Supplements (mg l^{-1})	Germinable seeds	Non-germin-able seeds
Self-pollination					
Agrostemma githago	N	5	CH (500)	+	
Antirrhinum majus	N	5	CH (500)		+
Argemone mexicana	N	5	CH (500)	+	
Brassica campestris	MS				
B. napus	MS				
B. oleracea	N				+
Clivia miniata	N	5	CH (500)		
Cucubalus baccifer	N	2	—		+
Dianthus caryophyllus	W	4	—	+	
Dicranostigma franchetianum	N	4	—	+	
Digitaria purpurea					
D. lutea					
Eschscholtzia californica	N	5	CH (500)	+	
Galanthus nivalis	N	5	CH (500)		+
Melandrium album	W,N	2,5	— , CH (500)	+	
M. rubrum	W	2	—	+	
Narcissus pseudonarcissus	N	5	CH (500)		+
Nicotiana alata	L.S.			+	
N. rustica	N	5	CH (500)	+	
N. tabacum	N	5	CH (500)	+	
Paeonia officinalis	N	5	CH (500)		+
Papaver somniferum	N	5	CH (500)	+	
Petunia axillaris	R.S.	4	—	+	
P. hybrida	N	4	CH (500) +	+	
P. parodii	N	4	CH (500) +	+	
Primula pubescens			IAA (0.1)	+	
Scopia carniolica					
Sisymbrium loescli					
Torenia fournieri					
Trifolium repens	MS	3	CH (100)	+	
Tulipa gesneriana	N	5	CH (500)		+
Zea mays	W,N,M.S.	17	Yolk of hen's egg (100 drops l^{-1})	+	
Z. mays	L.S.	15		+	
Z. mays	G.P.	5	GA$_3$ (10)	+	

Contd...

Table 10.4 Contd.

Species	Medium			Results	
	Basal medium[b]	Sucrose concen. (%)	Supplements (mg 1^{-1})	Germinable seeds	Non-germinable seeds
Z. mays	M.S.	7	CH (500) + IAA (1) + Kn (0.5) 5	+	
Z. mays	M.S.	5	GA$_3$ (10.4)	+	
Cross-pollination					
Brassica napus × B. campestris	M.S.	2			
B. oleracea × B. cretica					
B. oleracea × Diplotaxis tenuifolia					
Melandrium album × M. rubrum	W	2	—	+	
M. album × Viscaria vulgaris	W	2	—	+	
M. album × Silene schafta	W	2	—	+	
M. album × S. friwaldskyana	N	2	—		+
M. album × S. alpina	N	2	—		+
M. album × S. tatarica	W	2	—		+
M. album × Minuartia laricifolia	N	2	—		+
M. album × Viscaria pyramidata	N	2	—		+
M. album × Dianthus carthusianorum	W	2	—		+
M. album × D. serotinus	N	2	—		+
M. album × Campanula persicifolia	N	2		+	
M. album × Cucubalus baccifer	N,W	2	—		+
M. rubrum × M. album	N,W	2	—	+	
Nicotiana tabacum × Petunia hybrida	N	5			+
N. tabacum × Hyoscyamus niger	N	5			+
N. tabacum × Physosclaina praealta	N,M.S.	2	—		+
N. tabacum × Withania somnifera	N,M.S.	2	—		+
N. tabacum × N. debneyi	N,M.S.	2	—	+	
Petunia parodii × P. inflata	W	4	—	+	
Zea mays × A. mexicana	G.P.	5	GA$_3$ (10)	+	

[a] After Bhojwani and Razdan (1983, 1996).
[b] G.P. = Green and Phillips (1975); L.S. = Linsmaier and Skoog (1965); M.S. = Murashige and Skoog (1962); N = Nitsch (1951); R.S. = Rangaswamy and Shivanna (1971); W = White's medium as modified by Rangaswamy (1961).

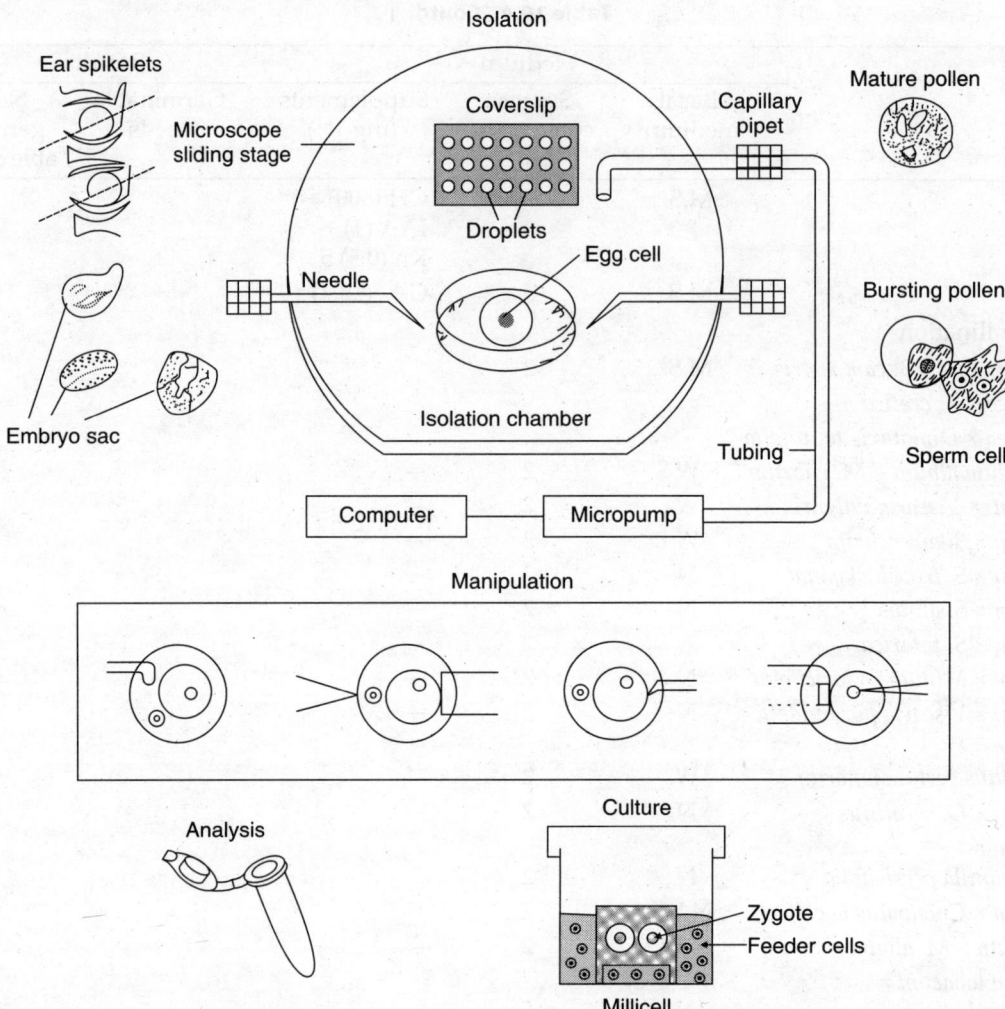

Fig. 10.2 Set-up for isolation, individual selection, and transfer of maize gametes. Ear spikelets are cut as indicated (dotted lines). After removing integuments, cells of the embryo sac are isolated from nucellar tissue pieces in the isolation chamber (plastic dish) manually with needles following transfer with a capillary into microdroplets on a cover slip. Subsequently, sperm cells are selected in the isolation chamber after release from the pollen grains by osmotic shock and transferred into microdroplets. Manipulations are performed in these microdroplets under microscopic observation: physiological and biochemical studies, cell fusion, and microinjection for further analysis and culture (after Kranz 1999).

plant take place in the absence of surrounding tissues. Thus, *in vitro* fertilization, using isolated higher plant gametes is different from that *in vivo*, and offers new opportunities for monitoring molecular events occurring during early stages of embryogenesis in flowering plants as well as in genetic engineering for crop improvement (Kranz and Schotten 2008, Okamoto 2011).

The methodology of gamete isolation, their *in vitro* fusion, and culture has been standardised for maize plants (*see* Kranz 1999). Isolation of sperm (male gamete) from pollen grains is executed by osmotic shock in mannitol solution (600 mosM kg^{-1} H_2O with mannitol), and viable egg (female gamete) isolated by microdissection of ovules incubated for 40-60 min, at $24 \pm 0.5°C$, in enzyme solution (0.75% pectinase, 0.25% pectolyase Y 23, 0.5% hemicellulase, 0.5% cellulase Onozuka with osmolarity adjusted to 600 mosM kg^{-1} H_2O with mannitol) set at pH 5. Microdissection is carried in 2µl microdroplet mannitol solution placed on UV-sterilised cover slip with the aid of microcapillaries connected to a computer-controlled dispenser to isolate and select the egg cell. The cover slip is then overlayed with 300 µl autoclaved mineral oil. This is followed by transfer of isolated sperm into microdroplet containing egg cell by use of micropump. After fixing two gametes on one of the electrodes by dielectrophoresis ($1MH_2$, 70 V cm^{-1}), egg is fertilized with sperm by electrofusion by supplying a single or maximum of three

negative DC-pulses (50 µg; 0.9-1.0 kV cm^{-1}). The *in vitro* fertilized eggs are subsequently removed from electrode and cultured on a semipermeable transparent membrane of a "Millicell-CM" dish (12 mm) filled with 0.1 ml of nutrient solution (modified MS medium supplemented with 1 mg l^{-1} 2,4-D, 0.2 mg l^{-1} kinetin, 600 mos M kg^{-1} H_2O with glucose, pH 5.5). The dish is then inserted in the middle of 3 cm petri plate containing 1.5 ml of a feeder cell suspension (embryogenic) layer of another maize line and maintained at $26 \pm 1.0°C$ in light/dark cycle of 16/8 hr with an approximate light intensity 50 $\mu Em^{-2}s^{-1}$.

A high frequency (90%) of sperm-egg fusion was reported (Kranz and Lörz 1993) and only fertilized eggs divided to form embryos from which regenerated full plants within 86 days after fusion (Kranz 1999). Set-up for isolation, individual selection and transfer of maize gametes for *in vitro* fertilization are summarized in Fig. 10.2. For applications and more descriptive account of *in vitro* fertilization, refer to Zenkteler and Bagniewska-Zadworna (2001).

IVF system has also been established for rice. Fusion of gametes can be performed using electrofusion and resultant zygote forms a cell wall with additional nucleolus. The zygote subsequently divides into an asymmetric two-celled embryo which undergoes further divisions to develop into early globular embryo. From such embryos plants can be regenerated (For detailed procedure, refer to Okamoto 2011).

APPENDIX 10.1
Distant *in vitro* Placental Pollination among Angiosperms[a]

Female parent	Male parent	Pollen germination on placenta
Caryophyllaceae	Solanaceae	
Melandrium album	*Datura stramonium*	++[1,2]
Melandrium album	*Nicotiana tabacum*	++[1,2]
Melandrium album	*Datura candida*	++
Melandrium album	*Petunia violacea*	+
Melandrium album	*Atropa belladonna*	+
Melandrium rubrum	*Nicotiana tabacum*	++
Silene vulgaris	*Nicotiana tabacum*	+
	Fabaceae	
Melandrium album	*Vicia faba*	++
Melandrium album	*Cytisus albus*	++
Melandrium album	*Lathyrus latifolius*	++
Melandrium rubrum	*Ulex europeus*	++
Lychnis coronaria	*Pisum sativum*	++
	Araceae	
Melandrium album	*Spathiphyllum grandifolium*	++
Silene vulgaris	*Spathiphyllum grandifolium*	++
Lychnis coronaria	*Spathiphyllum grandifolium*	++
	Asclepiadaceae (pollinia)	
Melandrium album	*Asclepias purpurescens*	++
Melandrium album	*Vincetoxicum hirsutum*	++
Melandrium album	*Asclepias syriaca*	++
Brassicaceae	Brassicaceae	
Brassica napus	*Arabis caucasica*	++[1,2]
Brassica napus	*Sinapis album*	++[1,2,3]
Brassica napus	*Sinapis arvensis*	++[1,2,3]
Brassica napus	*Bunias orientalis*	++[1,2]
Brassica napus	*Hesperis matronalis*	++[1,2]
Brassica oleracea	*Alliaria officinalis*	+
Brassica oleracea	*Barbarea vulgaris*	++[1,2]
Sinapis arvensis	*Brassica napus*	++[1,2,3]
Brassica oleracea	*Diplotaxis tenuifolia*	++[1,2,3]
Brassica napus	*Moricandia arvensis*	++[1,2,3]

Contd...

Appendix 10.1 Contd.

Female parent	Male parent	Pollen germination on placenta
	Fabaceae	
Brassica oleracea	*Lathyrus latifolius*	++[1]
Brassica oleracea	*Lotus corniculatus*	++
Fabaceae	Asclepiadaceae	
Lathyrus odoratus	*Asclepias purpurescens*	++
Robinia pseudoacacia	*Vincetoxicum hirsutum*	++
Laburnum anagyroides	*Vincetoxicum hirundinaria*	++
	Solanaceae	
Vicia faba	*Atropa belladonna*	+
Vicia faba	*Nicotiana tabacum*	+
Pisum sativum	*Nicotiana tabacum*	+
Solanaceae	Asclepiadaceae	
Datura Candida	*Asclepias syriaca*	++
Atropa belladonna	*Asclepias purpurescens*	++
Nicotiana tabacum	*Asclepias syriaca*	++
	Araceae	
Atropa belladonna	*Spathiphyllum grandifolium*	++
Nicotiana tabacum	*Spathiphyllum grandifolium*	++
Petunia hybrida	*Spathiphyllum grandifolium*	++
Liliaceae	Asclepiadaceae	
Ornithogalum nutans	*Vincetoxicum hirsutum*	++
Allium molly	*Asclepias purpurescens*	++
Hemerocallis middendorfii	*Asclepias purpurescens*	++[1]

[a] After Zenkteler and Bangniewska-Zadworna (2001)
+ sporadic germination
++ abundamt germination
[1] pollen tubes entered the micropyles
[2] pollen tubes inside the embryo sacs
[3] globular embryos formed

APPENDIX 10.2

Gymnosperm Pollen Germination on Cultured Stigmas and Placentae of Angiosperms[a]

Female parent	Male parent	Pollen germination status	
		On stigma	On placenta
Caryophyllaceae	Pinaceae		
Melandrium album	Pinus wallichiana		++[1]
Melandrium album	Pinus mugo	nil	++
Melandrium album	Ephedra distachya*		++[1]
Melandrium rubrum	Pinus mugo	nil	++
Melandrium rubrum	Pinus banksiana		++
Silene vulgaris	Pinus ponderosa		+
Lychnis coronaria	Pinus wallichiana	nil	++
Solanaceae			
Nicotiana tabacum	Pinus wallichiana		+
Atropa belladonna	Pinus wallichiana	nil	+
Petunia hybrida	Pinus wallichiana		+
Nicandra sp.	Pinus wallichiana	?	?
Datura stramonium	Pinus wallichiana	?	?
Nicotiana tabacum	Ephedra distachya*	nil	++
Liliaceae			
Allium molly	Pinus wallichiana		++
Allium odorum	Pinus wallichiana		++
Allium molly	Ephedra distachya*		++
Allium molly	Pinus mugo	nil	++
Paradisea liliastrum	Pinus wallichiana		incipient
Polygonatum multiflorum	Pinus wallichiana		++
Ornithogalum nutans	Pinus mugo		++
Hemerocallis middendorfii	Pinus mugo		++
Lilium martagon	Pinus wallichiana	incipient	++
Galtonia candida	Pinus wallichiana	nil	++
Fabaceae			
Lupinus polyphyllus	Pinus wallichiana		+
Lathyrus latifolius	Pinus wallichiana	nil	++
Pisum sativum	Pinus wallichiana		+
Vicia faba	Pinus mugo		+

Appendix 10.2 Contd.

Female parent	Male parent	Pollen germination status	
		On stigma	On placenta
Papaveraceae	Pinaceae		
Papaver rhoeas	*Pinus mugo*		+
Papaver rhoeas	*Pinus wallichiana*	nil	+
Papaver rhoeas	*Ephedra distachya**		++
Brassicaceae			
Brassica napus	*Pinus mugo*		+
Brassica napus	*Pinus ponderosa*	nil	++
Brassica napus	*Pinus wallichiana*		++

[1] some pollen tubes entered the micropyle
+ sporadic germination
++ abundant germination
[a] After Zenkteler and Bagniewska-Zasworna (2001).
* Ephedraceae

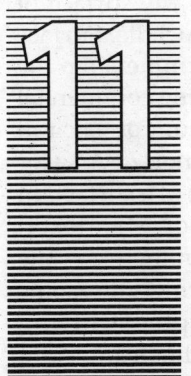

Zygotic Embryo Culture

Zygotic embryo culture is the aseptic isolation and growth of sexually produced embryos *in vitro* with the objective of obtaining viable plants. In 1904, Hannig published a paper describing the first systematic attempt to culture isolated mature embryos of angiosperms *Cochleria* and *Raphanus* (family: Cruciferae) aseptically. He successfully raised transplantable seedlings from the embryos cultured on a semi-solid medium containing mineral salts and sugar. Subsequently, many workers obtained plants by culturing excised embryos from mature seeds especially of species which failed to germinate due to embryo abortion. Laibach (1925, 1929) cultured excised embryos from seeds of an inter-specific cross *Linum perenne* × *L. austrianum* and succeeded in raising hybrid plants. The hybrids of this cross did not occur in nature because the seeds formed after crossing shrivelled, became very light and were incapable of germination. This provided a stimulus for further progress in the field of zygotic embryo culture. Through this technique it has been possible to rescue hybrid embryos which normally abort prematurely.

The technique of zygotic embryo culture is also useful for exploring the nutritional and physical conditions required for embryonic development, skipping seed dormancy in order to shorten the breeding cycle, determination of seed viability and microcloning of the source

material. Conventionally, the term 'embryo' has been used for zygotic embryos; hence in this book 'embryo culture' refers to the culture of zygotic embryos.

11.1 Types of Embryo Culture

According to Pierik (1989), there are in principle two types of embryo culture:

(I) Culture of Immature Embryos

This type of embryo culture is mainly used to grow immature embryos originating from unripe or hybrid seeds which fail to germinate. Excising such embryos is arduous and generally a complex nutrient medium is required to raise them to produce plants. The chances of success in this type of culture depend largely on the developmental stage of the excised embryo.

(II) Culture of Mature Embryos

Mature embryos are excised from ripe seeds and cultured mainly to avoid inhibition in the seed for germination. This type of culture is relatively easy as the embryos require a simple nutrient medium containing mineral salts, sugar and agar for growth and development.

Monnier (1980) compared the development of immature embryos (starting from globular stage) *in vivo* and *in vitro*. It appears from his studies that the embryos *in vitro* have a bulkier growth, retarded morphogenic expression (demonstrated by a pear-shape and longer globular stage), development of one cotyledon initially in dicots against two developing *in vivo* simultaneously, and a tendency towards polycotyledonous development. As far as possible, physiologically uniform embryos of the same size must be cultured, which can be achieved if the plants are raised and maintained under controlled conditions. For excising embryos at a specific stage of development, the selected plants should bear flowers and fruits after regular intervals in order to ensure a sufficient supply of the material needed. Seed legumes and crucifers are good starting materials for embryo culture.

To obtain embryos of a specific age, artificial pollination of freshly opened flowers is necessary, and it may be desirable to prepare a calendar showing the relationship between different stages of embryo development with DAP (day after pollination). However, if the goal is to obtain plants from abortive seeds, the embryos should be excised prior to the onset of abortion.

11.2 Technique

11.2.1 Surface Disinfection

Embryos of seed plants normally develop inside the ovules which, in turn, are covered by ovaries. Since they already exist in a sterile environment, disinfection of the embryo surface is unnecessary unless the seed coats are injured or a systemic infection is present. Instead, mature seeds, entire ovules, or fruits are surface-sterilised and the embryos removed aseptically from the surrounding tissues. In orchids, in which the seeds are minute with highly reduced seed coat and a functional endosperm is absent, whole ovules are cultured and treated as embryo cultures. Entire capsules of orchids are surface-sterilised and seeds removed under aseptic conditions. These are then spread in a single layer on the surface of an agar medium using a sterile needle. Surface sterilisation is carried out by immersing the material in hypochlorite-containing commercial bleach (5-10% Clorox, 0.45% sodium or calcium hypochlorite) for 10-5 min or ethanol (70-75%) for 5 min. A small amount (0.01-0.1%) of a surfactant (Tween-20, Tween-80, Teepol, or Mannoxol) added to the disinfection solution increases the tissue wettability. Magnetic stirring, ultrasonic vibrations, or a vacuum applied during soaking of the plant material in the disinfectant reduces the possibility of trapping air bubbles on the surface of the plant materials. In the case of seeds harbouring microbes internally, or embryos becoming contaminated during excision, resterilisation becomes necessary. For maize seeds and excised embryos the treatment may include a dip in 70% alcohol plus 5-10 min exposure to 2.6% sodium hypochlorite solution (*see* Hu and Wang 1986).

11.2.2 Excision of Embryo

The embryo-excision operation is performed aseptically in a laminar airflow hood (*see* Chapter 4). A stereomicroscope (90X) equipped with a cool-ray fluorescent lamp or fibreglass illuminator is required for excision of small embryos. The commonly used dissecting tools are forceps, dissecting needles, scalpels, razor blades and Pasteur pipettes. Mature embryos can be isolated with relative ease by splitting open the seeds. Soaking a hard-coat seed (*Iris, Cyclamen*) in water for a few hours to a few days before sterilisation makes its dissection easier. The excision of smaller or immature embryos requires careful dissection particularly if they are embedded in a liquid endosperm. In such a case the incision is made at the micropylar end of the young ovule and pressure applied at the opposite end to force the embryo out through the incision. The procedure adopted while isolating embryos at various stages of development is given in Appendices 11.1 and 11.2.

11.2.3 *Embryo-Endosperm Transplant*

Finding nutritional requirements for the growth of embryos which in particular abort at very early stages of development is a problem often confronted in embryo culture studies. Ziebur and Brink (1951) demonstrated that *in vitro* growth of excised immature embryos (300-1100 μm long) of *Hordeum* was considerably enhanced on surrounding them with endosperm tissues excised from another seed of the same species. Kruse (1974) implanted immature embryos arising in a cross *Hordeum × Secale* on cultured barley endosperm and recorded significant improvement in the frequency of hybrid plant development vis-à-vis the traditional method of embryo culture. Generally, an endosperm older than the embryo by 5 days was more efficient as a nurse tissue than one of the same age as the embryo.

The endosperm transplant technique for culturing young (immature) embryos is explained in Fig. 11.1. Williams and De Lautour (1980) inserted an excised hybrid embryo into a cellular endosperm dissected from a normally developing ovule of one of the parents, or a third species, and cultured the nurse endosperm with the transplanted embryo on the nutrient medium. Using this technique they raised interspecific and intergeneric hybrid plants of forage legumes which could not be reared by growing embryos directly on the medium.

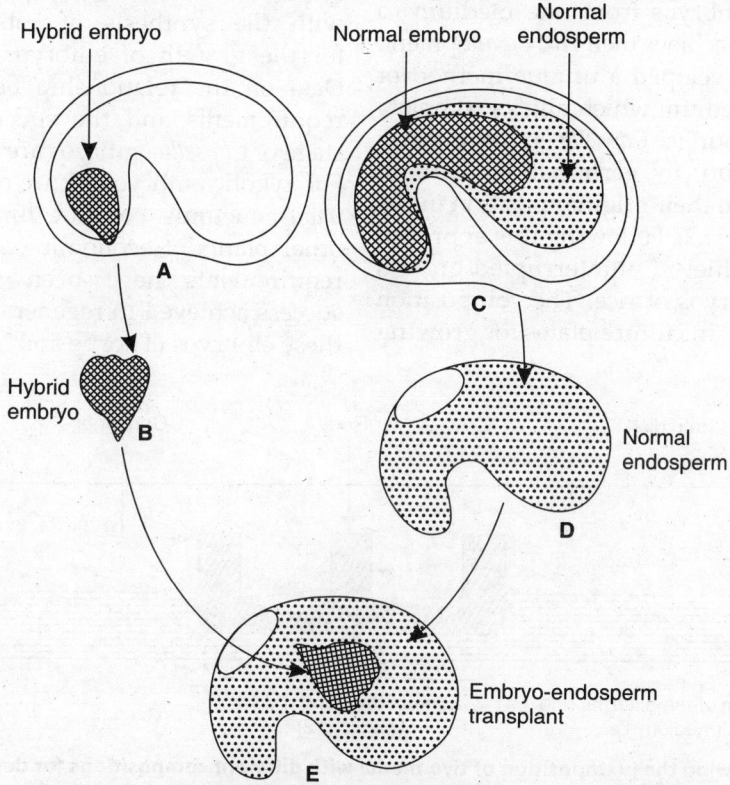

Fig. 11.1 Endosperm transplant technique developed by Williams and De Lautour (1980) to culture hybrid embryos isolated from crosses of various species of *Trifolium*, *Lotus* and *Ornithus*. A: The hybrid embryo in the ovule in which endosperm development has failed; B: Excised hybrid embryo; C: Normally developed intraspecific or self-pollinated ovule with cellular endosperm enclosing a torpedo to heart-shaped embryo; D: To provide transplant endosperm, the ovule is dissected and the normal embryo pressed out leaving an exit hole; E: The hybrid embryo is inserted into the normal endosperm through the exit hole.

11.3 Nutritional Requirements

The nutritional requirements of an embryo during its development *in vivo* constitute two phases: (a) *heterotrophic phase*—an early phase wherein the embryo is dependent and draws upon the endosperm and maternal tissues, and (b) the *autotrophic phase*—a later phase in which the embryo is metabolically capable of synthesising substances required for its growth, thus becoming fairly independent for nutrition. The critical stage at which the embryo passes from the heterotrophic phase into the autotrophic varies with the species.

Media constituents for *in vitro* growth of young or immature embryos also differ from those of mature embryos. This often necessitates the transfer of embryos from one medium to another in order to achieve their full development. Monnier (1976) developed a unique method of double-layered medium which allows complete development of younger *Capsella* embryos (early globular stage) up to germination without moving them from their original position in the culture plate (Fig. 11.2). Following this approach Ko et al. (1983) achieved uninterrupted growth of immature embryos of rice. The composition of two media used in culture plates for growing

Capsella and rice embryos respectively, are summarised in Table 11.1.

Embryos excised at or near the mature stage are completely autotrophic and grow on a simple medium comprising mineral salts with a carbohydrate energy source. When immature or heterotrophic embryos are cultured, the medium requirement becomes progressively complex to permit the expression of the total developmental potential of an embryo. This requires testing of a number of standard tissue culture media (*see* Chapter 3) as well as making modifications in their constituents empirically. The shift from the heterotrophic to autotrophic phase of embryogenesis is dependent upon the progressive activation of critical enzyme systems or biochemical pathways concerned with the synthesis of substances necessary for the growth of embryos at a specific age. Data on the relationship between nutritional requirements and the specific developmental stage of *Capsella* embryos are given in Table 11.2. For zygotic embryo culture of *Brassica juncea, B. napus*, cucumis, coconut, *Arabidopsis* and many other plants of economic value the nutritional requirements have been standardized and success achieved in regeneration of plants from these embryos (Thorpe and Yeung 2011).

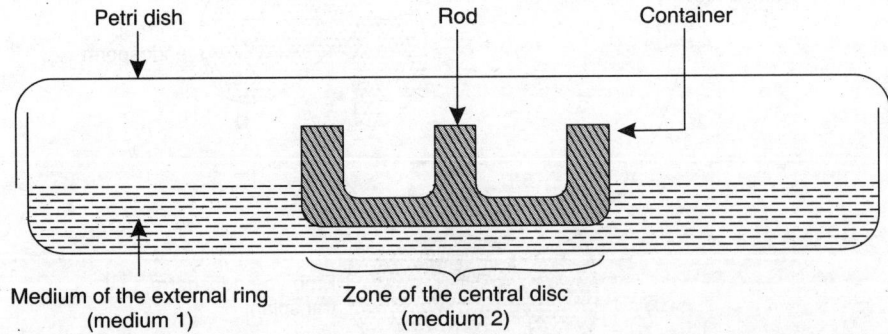

Fig. 11.2 Device allowing the juxtaposition of two media with different compositions for development of younger embryos up to germination without moving them in the culture dish. The first agar medium is liquefied by heating and then poured around the central glass container. After cooling and solidification of the medium, the container is removed. This medium will give the external ring. A second medium of a different composition is then poured into the central ring represented by the empty space formed by removing the glass container. Embryos are cultivated on the second medium in the central ring. As a result of diffusion the embryos are subjected to the action of variable medium with time (after Monnier 1976).

Table 11.1 Composition of media (Ml and M2) for uninterrupted growth of *Capsella* globular embryos[a] and media (SR and A) for culturing immature rice embryos[b]

Constituents	M1 external ring	M2 central zone	SR embryo growth	A callusing
Macronutrients				
KNO_3	1,900	1,900	2,730	2,275
NH_4NO_3	990	825	720	600
$CaCl_2 \cdot 2H_2O$	484	1,320	352.8	211.1
KH_2PO_4	187	170	179.5	136
$MgSO_4 \cdot 7H_2O$	407	370	296	185
KCl	420	350	—	—
Micronutrients				
Na_2EDTA	37.3	—	37.3	37.3
$FeSO_4 \cdot 7H_2O$	27.8	—	27.8	27.8
$MnSO_4 \cdot H_2O$	33.6	33.6	10	10
$ZnSO_4 \cdot 7H_2O$	21	21	2	2
H_3BO_3	12.4	12.4	3	3
KI	1.66	1.66	0.75	0.75
$Na_2MoO_4 \cdot 2H_2O$	0.5	0.5	0.25	0.25
$CuSO_4 \cdot 5H_2O$	0.05	0.05	0.025	0.025
$CoCl_2 \cdot 6H_2O$	0.05	0.05	0.025	0.025
Organic				
Myoinositol	—	—	100	100
B_1 (Thiamine)	0.1	0.1	—	—
Thiamine HCl	—	—	10	10
Nicotinic acid	—	—	1	1
Pyridoxine HCl	—	—	1	1
Glutamine	—	600	—	—
CH	—	—	250	—
Hormone				
KIN	—	—	0.2	0.2
NAA	—	—	0.2	—
2,4-D	—	—	—	2–4
Sucrose	—	180,000	60,000	30,000
Agar	7,000	7,000	9,000	9,000

[a] After Monnier (1976)
[b] After Ko et al. (1983)

Table 11.2 Progressive embryogenesis in *Capsella bursa-pastoris*[a]

Developmental stage	Length of embryo (μm)	Nutritional requirements
Early globular	20-60	Unknown for embryos smaller than 40 μm
Late globular	61-80	Basal medium (macronutrient salts[b] + micronutrient saltsc[c] + vitamins[d] + 2% sucrose) + IAA (0.1 mg l^{-1}) + kinetin (0.001 mg l^{-1}) + adenine sulphate (0,001 mg l^{-1})
Heart-shaped	81-450	Basal medium alone
Torpedo-shaped	451-700	Macronutrient salts[b] + vitamins[b] + 2% sucrose
Walking-stick-shaped and mature embryos	700 and larger	Macronutrient salts[b] +2% sucrose

[a] After Raghavan (1966).
[b] Macronutrient salts (mg l^{-1}): 480 Ca(NO$_3$)$_2$ · 4H$_2$O, 63 MgSO$_4$ · 7H$_2$O, 63 KNO$_3$, 42 KCl, and 60 KH$_2$PO$_4$.
[c] Micronutrient salts (mg l^{-1}): 0.56 H$_3$BO$_3$, 0.36 MnCl$_2$ · 4 H$_2$O, 0.42 ZnCl$_2$, 0.27 CuCl$_2$ 2H$_2$O, 1.55 (NH$_4$)$_6$Mo$_7$O$_{24}$ · 4H$_2$O and 3.08 ferric tartrate.
[d] Vitamins (mg l^{-1}): 0.1 thiamine hydrochloride, 0.1 pyridoxine hydrochloride, 0.5 niacin.

11.3.1 Mineral Salts

Inorganic nutrients of MS, B5, and White's media (Table 3.1) with certain degrees of modifications are the most widely used basal media for embryo culture work. Monnier (1978) modified the MS medium and formulated a new mineral solution which promoted higher survival rate of cultured immature embryos of *Capsella*. This medium (Table 11.3) contained higher levels of potassium (by adding 350 mg l^{-1}KCl) and calcium (double concentration of CaCl$_2$) reduced levels (approximately half) of ammonium (NH$_4$NO$_3$) and FeEDTA, and double concentration of MS micro-nutrients. Proembryos (8-cell globular embryos) of *Brassica juncea* successfully matured to normal embryos using double-layer culture system of Monnier involving two complex semisolid culture media differing only in osmolarity. Even zygotes of barley cultured in another medium Kao 90* divided to form microcalli which after co-cultivation with embryogenic microspores of the some plant in

liquid Kao 90 medium further grew to form mature calli and fertile plants. Similar success was achieved in cultures of isolated zygotes of wheat, rice *japonica* cv Taipei 309 and maize. Immobilised wheat zygotes (in 100 μl droplets of 0.75% agarose) co-cultured with appropriate population of embryogenic microspores in liquid medium also developed into normal embryo, which subsequently germinated to form fertile plants (c.f. Bhojwani and Dantu 2013).

11.3.2 Carbohydrates

Sucrose is the most commonly used source of energy for embryo culture. Addition of maltose, lactose, raffinose, or mannitol may be required in the embryo culture of some species (*Zea mays*). There are reports suggesting superiority of glucose over sucrose in the culture medium in supporting embryo growth of *Carex lucida*, several species of the family Rosaceae and Lilium hybrids. *Brassica juncea and Heracleum spondylum* embryos grow effectively in a medium

* Kao 90 medium is modified MS medium with reduced NH$_4$NO$_3$ (165 mg l^{-1}), vitamins and organic acids of KM medium (see Chapter 12), coconut milk (20 mg l^{-1}), vitamin-free casamino acids (250 mg l^{-1}), 9% maltose and 2,4-D (1 mg l^{-1}).

Table 11.3 Nutrient media for *Capsella* and barley embryo cultures

Constituents	Amount (mg l^{-1})	
	Capsella[a]	*Barley*[b*]
Macronutrients		
KNO_3	1900	—
$CaCl_2 \cdot 2H_2O$	880	740
NH_4NO_3	825	—
$MgSO_4 \cdot 7H_2O$	370	740
KCl	350	750
KH_2PO_4	170	910
Micronutrients		
KI	1.66	—
H_3BO_3	12.4	0.5
$MnSO_4 \cdot H_2O$	33.6	3
$ZnSO_4 \cdot 7H_2O$	21	0.5
$Na_2MoO_4 \cdot 2H_2O$	0.5	0.025
$CuSO_4 \cdot 5H_2O$	0.05	0.025
$CoCl_2 \cdot 6H_2O$	0.05	0.025
Ferric citrate	—	10
Na_2EDTA	14.9[c]	—
$FeSO_4 \cdot 7H_2O$	11.1[c]	—
Vitamins		
Vitamin B_1	0.1	—
Vitamin B_6	0.1	—
Thiamine HCl	—	0.25
Pyridoxine HCl	—	0.25
Inositol	—	50
Ca-pantothenate	—	0.25
Amino acids		
Glutamine	400	400
Alanine	—	50
Cysteine	—	20
Arginine	—	10
Leucine	—	10
Phenylalanine	—	10
Tyrosine	—	--
Sucrose	120000	34000
Agar (Difco)	7000	6000
pH	5	5

[a] After Monnier (1978).
[b] After Norstog (1973).
[c] 2 ml of a stock solution containing 5.57 g $FeSO_4 \cdot 7H_2O$ and 7.45 g Na_2EDTA l^{-1}
[*] Malic acid (1 g dissolved in 50 ml of water and pH set to 5 with NH_4OH) is further added to the constituents in the table for barley embryo culture.

containing glucose, fructose, ribose and xylose (in Table 11.4), galactose, mannose, or mannitol. Glucose has also been found essential for the root growth of mature *Ginkgo* embryos.

Glucose and sucrose, besides nutrition, also maintain osmolarity of the culture medium which is critical with respect to the age of embryo. Mature embryos grow fairly well at low sucrose concentrations but younger embryos require higher levels of carbohydrate. However, immature embryos of *Capsella* can also grow on low sucrose concentration (2%) provided specific growth factors are added to the basal medium. Various concentrations of sucrose used in embryo cultures depend on the species and size/age of the embryo (Table 11.5).

Table 11.4 Composition of the medium used for the culture of globular embryos of *Brassica juncea*[a,b]

Constituents	Amount (mg l^{-1})	Constituents	Amount (mg l^{-1})
Macroelements			
NH_4NO_3	200	$KH_2PO_4 \cdot 2H_2O$	100
KNO_3	1500	$Na_2EDTA \cdot 2H_2O$	37
$CaCl_2 \cdot 5H_2O$	850	$FeSO_4 \cdot 7H_2O$	28
$MgSO_4 \cdot 7H_2O$	400		
Microelements			
KI	0.75	$Na_2MoO_4 \cdot 2H_2O$	0.25
H_3BO_3	3	$CuSO_4 \cdot 5H_2O$	0.025
$MnSO_4 \cdot H_2O$	10	$CoCl_2 \cdot 6H_2O$	0.025
$ZnSO_4 \cdot 7H_2O$	2		
Sugar			
Sucrose	4000	Mannose	100
Glucose	2000	Rhamnose	100
Fructose	100	Cellobiose	100
Ribose	100	Sorbitol	100
Xylose	100	Mannitol	100
Organic acids[c]			
Sodium pyruvate	20	Malic acid	40
Citric acid	40	Fumaric acid	40
Vitamins and amino acids			
Inositol	500	Nicotinic acid	0.1
Glutamine	200	Pyridoxine-HCl	0.1
Thiamine-HCl	1	D-Biotin	0.01
Other supplements			
Cocount water[d]	300 (ml l^{-1})	Casein hydrolysate	100
Agarose (low gelling temperature Sea Plaque)	6000		

[a] After Liu et al. (1993).
[b] Medium was filter-sterilised.
[c] pH adjusted to 5.5 with NH_4OH.
[d] Obtained from green coconuts.

Table 11.5   Some examples of sucrose concentrations used in embryo cultures[a]

Plant name	Size/age of embryo	Sucrose concentration
Barley	0.2–0.8 mm	9%
	60 µm long	12%
Flax	globular and heart-stage	12%
Soybean	0.2–0.4 mm heart-stage	8%
Maize	0.3–1 mm	5%
Ilex	heart-stage	4%
Rice	immature (5–6–day–old)	4%
Cucumis	heart-shaped (0.1–0.8 mm)	3.5%
Capsella	globular embryo	12–18%
	(smaller than 80 µm) proembryo	2% + IAA (0.1 mg l^{-1}), kinetin (0.001 mg l^{-1}), and adenine sulphate (0.001 mg l^{-1})

[a] Compiled from Hu and Wang (1986).

11.3.3   Nitrogen and Vitamins

Embryos which have an excellent enzyme system can reduce nitrates to ammonium. Ammonium nitrate is significantly superior to KNO_3, $NaNO_3$ and $(NH_4)_2HPO_4$. Especially the presence of NH^+_4 in the medium has been found essential for proper growth and differentiation of embryos.

Various amino acids and their amides have been tested for embryo culture. Hannig (1904) reported that asparagine enhanced embryo growth but some workers found glutamine to be a superior source of nitrogen for the embryos of some species (*Capsella bursa-pastoris, Arabidopsis thaliana, Reseda odorata*) since asparagine drastically inhibited their growth. Other amino acids are either marginally promotive or inhibitory.

Casein hydrolysate (CH), an amino acid complex, has been widely used as an additive to embryo culture media. The sensitivity of CH may vary with the species. The optimum level of CH for *Hordeum vulgare* may be 500 mg l^{-1}, whereas *Datura tatula* embryos grow at CH concentration of 50 mg l^{-1}. Amino acids, CH and amides are autoclavable and can be sterilised together with inorganic nutrients of the media.

11.3.4   Natural Plant Extracts

Van Overbeek et al. (1941) observed that *Datura* globular embryos (0.2-0.5 mm long) either failed to grow or developed feebly, turning into callus in a culture medium containing minerals and sugar. However, if the medium was supplemented with non-autoclaved coconut milk (CM) these embryos increased in length without signs of precocious germination. This was the turning point in the culture of immature embryos. The authors suggested the presence of some 'embryo factor' in the liquid endosperm of coconut milk (CM) which presumably makes up for deficiencies of certain sugars, amino acids, growth hormones and other critical metabolites of the culture medium. CM effectively stimulates the growth of excised young embryos of sugarcane, barley, tomato, carrot, interspecific hybrids of *Vigna* and fern species.

Natural extracts from other parts or tissues of plant species have also been tested for the presence of the 'embryo factor'. Kent and Brink (1947) reported that promotion of embryonic growth and inhibition of precocious germination of immature barley embryos could be achieved by water extracts from dates, bananas, hydrolysates of wheat-gluten and tomato juice. These extracts were as effective as CM. Matsubara (1962) studied the activity of alcohol diffusates from young seeds of *Iris, Lupinus, Datura, Sechium, Zea, Thea, Camellia, Prunus, Quamoclit* and *Ginkgo* megagametophytes on the growth of very small embryos in cultures. Extracts from *Datura* and *Sechium* had comparative effects as CM, but *Lupinus* seed extracts proved to be twice as effective. Supplementing the nutrient medium with peat extract is reported to support the best growth of *Cyclamen* embryos (club-shaped).

Given the desire to develop purely synthetic media, attempts have been made to Substitute the 'embryo factor' of CM with chemically defined substances. To promote the growth of barley proembryos, CM could be substituted by phosphate-enriched White's medium fortified with glutamine and alanine as major amino acids and five other amino acids as the minor sources of N_2, at pH 4.5. The survival rate of the embryos was considerably increased when the concentration of KC1, KNO_3 and certain organic components was increased 5-10 times. This modified Norstog's medium (Table 11.3) proved the best of the various recipes tested for the culture of 250 μm-long embryos of five varieties of *Hordeum distichum*.

11.3.5   Growth Regulators

Auxin and cytokinins are not generally used in embryo cultures since they induce callus formation. At very low levels (0.01 mg l^{-1}) GA promotes embryogenesis of young barley embryos without inducing precocious germination (Norstog 1979) and effectively stimulates growth in excised heart-shaped embryos of *Phaseolus* even if the suspensor has been excluded (Yeung and Sussex 1979). There are reports that ABA also has a similar effect on barley *and Phaseolus* embryos. (Also see Fischer-Iglesias and Neuhaus 2001.)

11.3.6   pH of Medium

Excised embryos grow well in a medium with a pH 5 to 7.5. This is within range of the pH

(6.0) of ovular sap. Generally the medium pH is adjusted 0.5 unit higher than the desired pH in order to compensate for uncontrollable change in its value during the autoclaving process.

11.3.7   Incubation Conditions

The majority of workers arbitrarily adopt an incubation temperature of 25 ± 2°C to support embryonic growth and germination. This is not necessarily optimum although it may appear satisfactory. Sometimes the optimum temperature for an embryo culture may vary with the genotype in the same species. In species (*Zamia, Phaseolus,* cotton) adapted to warm temperature the embryo culture shows a favourable response at elevated temperatures (27-30°C), whereas incubation temperature to culture embryos of species (*Brassica* hybrids, rice, barley) occurring in cold regions or seasons range from 17-22°C (*see* Hu and Wang 1986).

Some workers believe light is not critical for embryo growth *in vitro* but many recent studies have revealed that it is advisable to incubate immature embryos of barley, flax, *Aegilops* × *Hordeum* hybrid, and interspecific hybrids of *Allium* in darkness before transferring them to light for germination. The rudimentary heart-shaped embryos of *Ilex* species are light sensitive. A type of recalcitrant secondary dormancy is induced when these embryos are excised and exposed to 4000 1x or more for a 4 hr-period during the initial four days of incubation. Such dormancy is irreversible and cannot be removed even on subsequent dark incubations or by a wide range of GA treatments. This suggests an initial dark incubation (4 days) of embryos in cultures is essential, following which they can grow to a mature stage even under continuous light regimes.

11.4   Role of Suspensor in Embryo Culture

Detailed structural, cytological and biochemical studies reveal an active involvement of the suspensor in embryo development. The suspensor is an ephemeral structure found at the radicular end of the proembryo and attains maximum development by the time the embryo reaches the globular stage. Generally, it is difficult to excise the suspensor along with the embryo because of its small size and delicate structure. In most of the studies embryos have been cultured without a suspensor. Older embryos (500 µm or more in length) appear to grow well with or without a suspensor. However, some painstaking studies have shown that in cultures the presence of a suspensor is critical, particularly for the survival of young embryos. Yeung and Sussex (1979) and Yeung and Meinke (1993) observed that attachment of the suspensor with young embryos of *Phaseolus coccineus,* or its placement in close proximity (when detached) of an embryo, on the culture medium strongly stimulated the development of embryos to maturity compared to those cultured without the suspensor. The requirement of the suspensor may be substituted by the addition of GA or ABA to the culture medium (see Section 11.3.5).

Feeder cell cultures have been found to provide necessary nutrients for the growth of zygotes or early proembryos of wheat, rice and tobacco thereby substituting the functions of suspensor and endosperm (Haslam and Yeung 2011). Selected feeder cells such as microspore suspension cells, living ovules, mesophyll protoplasts or cells from suspension cultures are placed in culture vessel in which is inserted a sterile mini-dish having bottom covered with membrane. Zygotes or proembryos are placed inside on the membrane and entire system sealed in order to maintain in the culture room. Feeder cells provide nutrients necessary for the zygote or proembryo for onward growth. One interesting finding of feeder system is that zygote or proembryo of a particular plant will effectively grow only when feeder cells are from different source-plant.

11.5 Precocious Germination

Plant physiologists and biochemists conceive embryo development as a linear progression from zygote formation to germination. According to Walbot (1978), embryo development can be classified into five stages (Table 11.6). Initially, the cell divides meristematically to form a group of cells called the proembryo. This is followed by tissue differentiation during which the meriste-matic activities become localised. After its full development the embryo enters a phase of dormancy characterised by metabolic quiescence and developmental arrest of cells. Dormancy may last a few days or several months or even years depending on the plant species and environmental conditions.

Table 11.6   Major stages in the development of *Phaseolus* **embryos[a]**

Stage	Characteristics
1. Cleavage and differentiation	Cell division with little growth; differentiation of all major tissues
2. Growth	Rapid cell expansion and division
3. Maturation	Little or no cell division or expansion, synthesis and storage of reserve materials
4. Dormancy	Developmental arrest
5. Germination	Renewed cell expansion and division; embryo growth

[a] After Walbot (1978).

Excised immature plant embryos on a nutrient medium tend to bypass the stage of dormancy and cease to undergo the linear embryogenic mode of development. The embryos develop instead into weak seedlings. The phenomenon of seedling formation without completing normal embryogenic development is called precocious germination. Dietrich (1924) described the precocious germination of excised immature embryos as 'Kunstliche Fruhgeburt'. The objective of culturing immature embryos, therefore, has been to understand the factor(s) that regulate the orderly *in ovulo* development of embryos under natural conditions, among its other applications. As described under Section 11.3, CH in the culture medium promotes further embryogenic development and delays germination in the cultures of excised immature embryos of barley. A high sucrose level (12-18%) in the medium also reportedly inhibits the precocious germination of immature embryos. Other exogenous factors that seem to suppress precocious germination are reduced O_2 tension, elevated temperature and high light intensity. Walbot (1978) has proposed that an ABA-like substance may prevent precocious germination by suppressing water uptake by the embryo in nature which is essential for germination. An ABA-related signal transduction pathway is implicated in the expression of both LEA (Late Embryogenesis Abundant) genes and storage proteins (Cuming et al. 1996). There is clear evidence in *Nicotiana plumbaginifolia* about involvement of ABA in the maintenance of dormancy (Grappin et al. 2000). Apparently the leaching of these germination inhibitors from young embryonic cells under cultural conditions results in precocious germination of embryos.

11.6 Morphogenesis in the Cultures of Seeds with Partially Differentiated Embryos

In some flowering plants (belonging to 19 families) fully developed seeds contain embryos that lack differentiation into radicle, plumule and cotyledons. Orchid seeds and those of root-parasitic members of Orobanchaceae harbour an unorganised (globose) embryo from which seedlings arise without undergoing further embryogenic differentiation. In such cases the embryonal end proximal to the micropyle is regarded as the 'radicular pole' and distal to the micropyle as the 'plumular pole'. Interestingly, only one of the poles (monopolar pattern) is involved in the development of the seedling. The plumular pole of the orchid embryo enlarges to form a spherule-like structure called *protocorm* which, after turning green and having attained a certain size, differentiates roots and shoots.

Unlike orchids, seedlings of the root-parasitic members of Orobanchaceae are derived from the radicular pole of the embryo. This pole gives rise to a 'radicular cylinder' whose tip penetrates the root of the host while that portion of the cylinder that remains outside proliferates into an irregular mass of tissue called the *tubercle*. It is from the tubercle that the shoot eventually differentiates. The monopolar pattern of seedling development in *Orobanche aegyptica* can be modified to a bipolar pattern by media manipulations (Kumar and Rangaswamy 1977). On a TB medium (salts after Tepfer et al. 1963, glycine, niacin, thiamine HC1, calcium pentothenate and sucrose) supplemented with CM or yeast extract, the germination was monopolar through the radicular cylinder, but the addition of IAA (0.1 mg l^{-1}), kinetin (0.5-10 mg l^{-1}), GA_3 (5-30 mg l^{-1}) or strigol (0.01 μg l^{-1}) to this medium induced a bipolar pattern of seedling development. In a bipolar pattern the plumular pole differentiates shoot buds and radicular pole roots. Substitution of sucrose in the TB medium by glucose, mannose, or raffinose, also favoured bipolar germination.

Root parasites are dependent on host stimulus for shoot development since in nature most of them require a host contact for normal seedling morphogenesis. In cultures, the addition of an extract or exudate from the host root to the nutrient medium stimulates seed germination of these parasites independent of the host. This stimulus may also be provided by cytokinins (*Striga euphrasioides*), gibberellins (*Orobanche ludoviciana; O. aegyptica*), sacopotelin (*Striga asiatica*) and strigol (*O. lutea*). High light intensity (3000 lx) favoured shoot development without host stimulus in seed cultures of *Sopubia delphinifolia*.

Germination of the *Cuscuta* embryo is also monopolar. It has a distinct plumule but lacks a radicle. During embryogenesis or in the mature embryo there is no indication of the existence of a protoderm, root meristem, or hypocotyl. During *in vivo* seed germination the plumule forms a shoot but the radicular end of the root system is altogether absent. Attempts to induce artificial rooting on plants raised from embryo cultures yielded negative results. Loranthaceous stem parasites and leafy mistletoes form distinct haustorial structures from embryos comparable to those formed in nature without host or plant extract in the medium.

11.7 Organogenic Potential of Embryo Callus

An embryo callus is reported to possess a high regenerative capacity compared to those derived from mature organs, such as the leaf, stem and root. This is especially true of crucifers, cereals and millets. The age of the embryo considerably influences the regenerative ability of maize calluses. Green and Phillips (1975) obtained a differentiating callus from maize embryos excised 18 DAP. No differentiation occurred if the callus originated from a mature embryo. Immature embryos of oats, barley, sorghum, wheat, rice, rye, rye-grass and triticale are good explants for initiating callus capable of plant regeneration. The suspension cultures of *Pennisetum americanum*, known to differentiate whole plants from isolated protoplasts, were derived from immature embryos. To obtain a callus with morphogenetic potential excised immature embryos are placed on an agar medium with the scutellum facing up in the presence of 2,4-D alone or in combination with cytokinin. An embryo callus of *Picea abies* regenerated numerous shoot buds but only in one instance did the root differentiate (*see* Hu and Wang 1986). In tree species, such as neem (*Azadirachta indica A. Juss*) immature zygotic embryos showed high regenerative potential as these embryos differentiated into various regenerants viz. somatic embryos, shoot buds and neomorphs in cultures. Plantlets, however, were obtained from neomorphs and adventitious shoot buds (Chaturvedi et al. 2004a).

11.8 Practical Applications

11.8.1 Rescuing Embryos from Incompatible Crosses

In interspecific and intergeneric hybridisation programmes, incompatibility barriers often prevent normal seed development and production of hybrids. Although there may be normal fertilisation in some incompatible crosses, embryo abortion results in the formation of shrivelled seeds. Of several cases studied, poor or abnormal development of the endosperm caused embryo starvation and eventual abortion. Isolation of hybrid embryos before abortion and their *in vitro* culture may circumvent these strong post-zygotic barriers. The most useful and popular application of embryo culture, therefore, has been to raise rare hybrids by rescuing embryos of incompatible crosses. With improvement in techniques of embryo culture, and exploration of its media requirements, full plants have been raised in respect of a large number of interspecific and intergeneric incompatible crosses (Table 11.7).

11.8.2 Overcoming Dormancy and Shortening Breeding Cycle

Long periods of dormancy in seeds delay breeding work, especially on horticultural and crop plants. Using embryo culture technique the breeding cycle can be shortened in these plants. For example, the life cycle of *Iris* was reduced from 2-3 years to less than one year. Similarly, it was possible to obtain two generations of flowering against one in *Rosa* sps. Overcoming seed dormancy and shortening of their breeding cycles have also been reported in weeping crab apple (*Malus* sps.), hollies (*Ilex*), American basewood (*Telia americana*) and other plants.

Germination of excised embryos is regarded as a more reliable test for rapid testing of viability in seeds, especially during the dormancy period.

11.8.3 Overcoming Seed Sterility

In early-ripening fruit cultivars, seeds do not germinate because their embryos are still immature. Using the embryo culture method it is possible to raise seedlings from sterile seeds of early ripening stone fruits (sweet cherry), peach, apricot, and plum.

"Makapuno" coconuts are very expensive and most relished for their characteristic soft

Table 11.7 Embryo culture of some interspecific and intergeneric crosses[a, b]

Plant species	Purpose of embryo culture
Abelmoschus esculentus × *A. manihot* hybrid	Overcome non-viability
A. esculentus × *A. moschatus*, *A. tulerculates* × *A. moschatus* hybrids	Overcome non-viability
Agropyron tsukushiense × *H. bulbosum* hybrid	Attempted polyhaploid production
Allium cepa × *A. fistolusum* hybrid	Study of embryo development
Avena fatua	Overcome seed dormancy
Brassica campestris × *B. oleracea* hybrid	Following ovule culture to overcome non-viability Study of media supplements
Brassica pekinensis × *B. oleracea* hybrid	Overcome non-viability
Carthamus tinctorius × *C. lanatus* hybrid	Overcome non-viability
Cattleya, Laelia and other orchids	Induce embryo growth in the absence of symbiotic fungus
Cerasus vulgaris × *C. tomentosa* hybrid	Overcome non-viability
Chrysanthemum boreale × *C. pacificum* hybrid	Overcome non-viability
Cicer arietinum × *C. pinnatifidum*	Overcome non-viability[b]
C. arietinum × *C. bijugum*	Overcome non-viability[b]
C. arietinum × *C. judiacum*	Overcome non-viability[b]
Colocasia esculentum × *C. antiquorum*	Overcome self-sterility of seeds
Corchorus capsularis × *C. olitorius* hybrid	Overcome non-viability
Cucumis spp. (*C. metuliferus* and *C. melo*)	Study of self- and cross-compatibility relationships
Cucumis spp. interspecific hybrids	Overcome non-viability
Cucurbita pepo × *C. moschata* and reciprocal hybrids	Overcome non-viability
Datura discolor × *D. stramonium* and other interspecific hybrids	Overcome non-viability
D. innoxia × Tree *Datura* (Brugmansia?)	Overcome non-viability
D. stramonium × *D. ceratocaula* and other interspecific hybrids	Overcome non-viability
Drosophyllum lusdanicum	Overcome natural seed dormancy
Elaeis guineensis	Overcome natural seed dormancy
Gossypium arboreum × *G. herbaceum* hybrid	Overcome low seed-set
G. hirsutum × *G. barbadense* hybrid	Overcome non-viability
G. davidsonii × *G. sturtii* hybrid	Overcome non-viability
G. hirsutum × *G. arboreum* and other interspecific hybrids	Overcome non-viability
Hordeum brachyantherum × *H. vulgare* and other interspecific hybrids; *H. californicum* × *Secale cereale* and other intergeneric hybrids	Overcome non-viability

Contd...

Table 11.7 Contd.

Plant species	Purpose of embryo culture
H. jubatum × *Secale cereale* hybrid	Overcome non-viability
H. sativum × *H. bulbosum* hybrid	Overcome non-viability
H. vulgare × *H. bulbosum* hybrid	Overcome non-viability
H. vulgare × (*H. compressum* × *H. pusillum*), *H. vulgare* × *H. hexopodium* hybrids	
Hordeum × *Tritcum*	Overcome non-viability
Hordeum × *Agropyron*	
Hordeum × *Secale* intergeneric hybrids	
Impatiens hookeriana × *I. campanulata*	Overcome non-viability
Iris (tall bearded) × *I. tectorum* (crested *Iris*) hybrid	Overcome non-viability
I. munzii × *I. sibirica*	Overcome non-viability
"Caesar's brother" hybrid,	Overcome non-viability
I. pallida × *I. macrantha*	Overcome non-viability
I. pallida × *I. chamaeris* hybrids	
I. pseudocorus × *I. versicolor* hybrid	Overcome non-viability
Lathyrus clymenum × *L. articulatus* hybrid	Overcome non-viability
Lens culinaris subsp. *culinaris* × *L. ervoides*[b]	Overcome non-viability
L. culinaris subsp. *inclinaris* × *L. nigricans*[b]	Overcome non-viability
L. culinaris subsp. *inclinaris* × *L. odemensis*[b]	Overcome non-viability
Lilium auratum platyphyllum × *L. henryi*, *L. longiflorum* × *L. candidum* hybrids	Overcome non-viability
L. henryi × *L. regale* hybrid	Overcome non-viability
L. longiflorum interspecific hybrids	Overcome non-viability
Lilium speciosum "album" × *L. auratum*, *L. speciosum* "Rubrum" × *L. auratum* hybrids	Overcome non-viability
Linum perenne × *L. austriacum* hybrid	Overcome non-viability
Lotus corniculatus × *L. filicaulis* and other interspecific hybrids	Overcome non-viability
L. pedunculatus × *L. tenuis* hybrid	Overcome non-viability
L. tenuis × *L. corniculatus* hybrid	Overcome non-viability
Lycopersicon esculentum × *L. peruvianum* hybrid	Overcome non-viability
Medicago sativa and unnamed interspecific hybrids	Overcome embryo abortion due to self-sterility and interspecific non-viability
Melilotus officinalis × *M. alba* hybrid	Overcome non-viability
Ornithopus sativus × *O. compressus* Pitman	Overcome non-viability
O. pinnatus × *O. sativus* hybrids	
Oryza paraguaiensis × *O. brachyantha* and other interspecific hybrids	Overcome non-viability

Contd...

Table 11.7   Contd.

Plant species	Purpose of embryo culture
O. sativa × *O. minuta,* *O. sativa* × *O. sp.* (Paraguay) hybrids	Overcome non-viability
O. sativa × *O. officinalis* and other interspecific hybrids	Overcome non-viability
O. sativa × *O. schweinfurthiana* and other interspecific hybrids	Overcome non-viability
Phaseolus coccineus × *P. acutifolius,* *P. coccineus* × *P. vulgaris, P. vulgaris* × *P. acutifolius (P. vulgaris* × *P. coccineus)* × *P. acutifolius* hybrids	Overcome non-viability
P. vulgaris × *P. acutifolius* hybrids	Overcome non-viability
P. vulgaris × *P. lunatus* hybrid	Overcome non-viability
P. vulgaris × *P. polyanthes*	Rescue of hybrid embryo using pod culture technique[b]
Pinus lambertiana × *P. armandi,* *P. lambertiana* × *P. koraiensis* hybrids	Facilitate germination and overcome hybrid non-viability
Prunus avium (sweet cherry), *P. cerasus* (sour cherry), *P. persica* (plum), *Pyrus communis* (pear), *Malus domestica* (apple) (intervarietal hybrids)	Overcome low seed viability
Ribes nigrum × *Grossulaeia reclinata* hybrid	Overcome non-viability
Solanum melongena cv. *sonepat* *local* × *S. khasianum* hybrid	Overcome non-viability
S. nigrum × *S. luteum* hybrid	Overcome non-viability
Trifolium ambiguum × *T. hybridum* hybrid	Overcome non-viability[a,b]
T. ambiguum × *T. occidentale,* *T. ambiguum* × *T. montanum,* *T. repens* × *T. isthmocarpum* hybrids	Overcome non-viability[a,b]
T. repens × *T. nigrescens* and other interspecific hybrids	Overcome non-viability[a,b]
T. ambiguum × *T. repens* hybrid	Overcome non-viability[a,b]
T. repens × *T. uniflorum* hybrid	Overcome non-viability[a,b]
T. pratense × *T. sarosiense* hybrid	Overcome non-viability[a,b]
Tripsacum dactyloides × *Zea mays* hybrid	Overcome non-viability
Triticum aestivum × *Aegilops* spp. hybrids	Overcome non-viability
T. durum × *Elymus arenarius* and other intergeneric hybrids	Overcome non-viability
T. durum abyssinicum × *Secale cereale* hybrid	Overcome non-viability
Vigna umbellata × *V. angularis* hybrid	Overcome non-viability
V. radiata × *V. angularis* hybrid	Overcome non-viability

[a] For details see Collins and Grosser (1984).
[b] For detailed reference see Thorpe and Yeung (2011).

fatty endosperm in place of a liquid endosperm. Under normal conditions the coconut seeds fail to germinate. In an attempt to achieve seed germination and complete transmission of makapuno trait, De Guzman et al. (1971) obtained 85% success in raising field-grown makapuno trees with the aid of embryo culture. The same procedure can be applied for propagation of *Maranta leuconeura, Colocasia esculenta, Musa balbisima,* and *Pinus armandii* × *P. koriensis.*

11.8.4 Production of Monoploids

Embryo culture has been used in the production of monoploids (haploids) of barley. With the cross *Hordeum vulgare* (♀) × *H. bulbosum* (♂), fertilisation occurs normally but thereafter chromosomes of *H. bulbosum* are eliminated, resulting in the formation of monoploid *H. vulgare* embryo which can be rescued by embryo culture (*see* Chapter 8). Following the bulbosum technique, haploids were obtained from a cross *Agropyron tsukushiense* × *Hordeum vulgare* and other cereals. For details, see the review published by Kott and Kasha (1985).

11.8.5 Clonal Micropropagation

The regenerative potential is an essential prerequisite in non-conventional methods of plant genetic manipulations. Because of their juvenile nature, embryos have a high potential for regeneration and hence may be used for in vitro clonal propagation. This is essentially true of conifers and graminaceous members.

Complete plantlets *in vitro* from conifers were achieved in long-leaved pine (*Pinus palustris*) through embryo culture (Sommer et al. 1975). This was followed by plantlet formation and their clonal micropropagation from embryos of *P. elliottii, P. radiata, P. regida, P. monticola, P. taeda, P. virginiana* and *P. sabiniana.* Cotyledonary buds were also induced in *P. coulleri* (Patel and Berlyn 1983) and *P. strobus* (Minocha 1980). Both organogenesis and somatic embryogenesis have been induced in major cereals and forage grasses from embryonic tissues. Generally, callus derived from immature embryos of cereals has the desired morphogenetic potential for regeneration and clonal propagation. As an exception, plant regeneration of orchid grass can be achieved in subcultured tissues derived from mature embryos. Probably these embryos are very small and the degree of development as well as organisation may be similar to that of immature embryos of large cereal grains.

Plant embryo culture is an invaluable breeding technique as far as it is possible to synthesise hybrids from incompatible crosses. However, the number of hybrid seedlings rescued in many instances is extremely low due to the difficulty in growing very young embryos. Further viability decreases with age of the embryo in most of the incompatible crosses. Efforts are needed to identify the requirements for embryos of progressively younger stages in major crop species.

11.8.6 Biotechnological Applications

Plasma membranes of desiccated zygotic embryos are generally perforated with large pores. Taking advantage of this characteristic, Topfer et al. (1989) transformed some cereals and legumes by imbibing their mechanically isolated zygotic embryos in a solution containing plasmid DNA with chimeric gene NPT-II. Rajasekaran (2013) reported biolistic transformation of cotton zygote embryos isolated from mature seeds. This procedure offers quick evaluation of gene constructs in cotton cells, organs and tissues arising from germination of transformed embryos, particularly in fibres which originate as single cells from maternal epidermal layer of cotton ovules. This approach should encourage scientific workers to explore the potential application of zygotic embryo culture in genetic transformation of agronomic, horticultural and forest species together with investigation of molecular aspects of embryonal growth and differentiation. For an overview of genetic transformation protocols using zygote embryos as explants of angiosperms and gymnosperms refer to Tahir et al. (2011).

APPENDIX 11.1

Procedure for Isolating Dicot Embryos at Various Stages of Development from *Capsella bursa-pastoris*[a]

1. Keep disinfected capsules in a few drops of the sterile culture medium (Fig. 11.3A).
2. Remove outer walls by an incision in the region of the placenta and pull two halves apart with forceps to expose ovules (Fig. 11.3B).
3. Detach a single ovule from the placenta and place in the depression (cavity) of a new sterilised slide containing a drop of medium.
4. Split the ovule longitudinally using a sharp mounted blade (Fig. 11.3C) to isolate the half containing the small or immature embryo (Fig. 11.3C).
5. Carefully tease apart the ovular tissues in order to free the entire embryo along with the attached suspensor.
6. To excise older embryos, make a small incision in the ovule on the side lacking the embryo (Fig. 11.3D) and apply pressure with a blunt needle to set the embryo free.

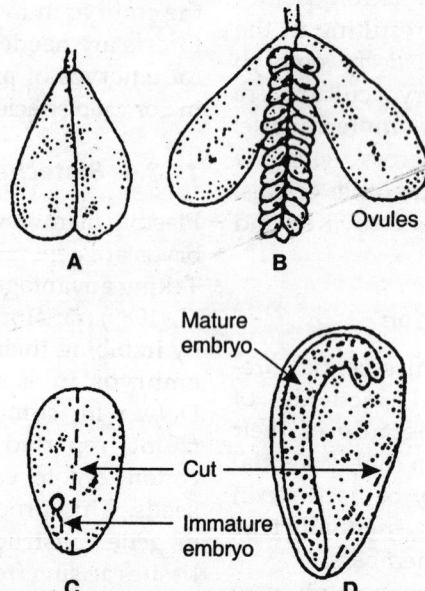

Fig. 11.3 Isolation of embryos from *Capsella bursa-pastoris* (procedure and explanation of figure given in Appendix 11.1).

[a] After Raghavan (1996).

APPENDIX 11.2

Procedure for the Isolation of Immature Monocot (Barley) Embryos[a]

1. Rinse the disinfected caryopsis with sterile water (Fig. 11.4A).
2. Place it in a sterile petri dish with the rachilla underneath and the lemma above (Fig. 11.4B).
3. Remove lemma, fruit wall and seed-coat in order to expose the embryo (Fig. 11.4C).
4. Carefully isolate the entire embryo with plumule, scutellum and radicle with the help of a knife and inoculate on the medium inside the culture tube (Fig. 11.4D).

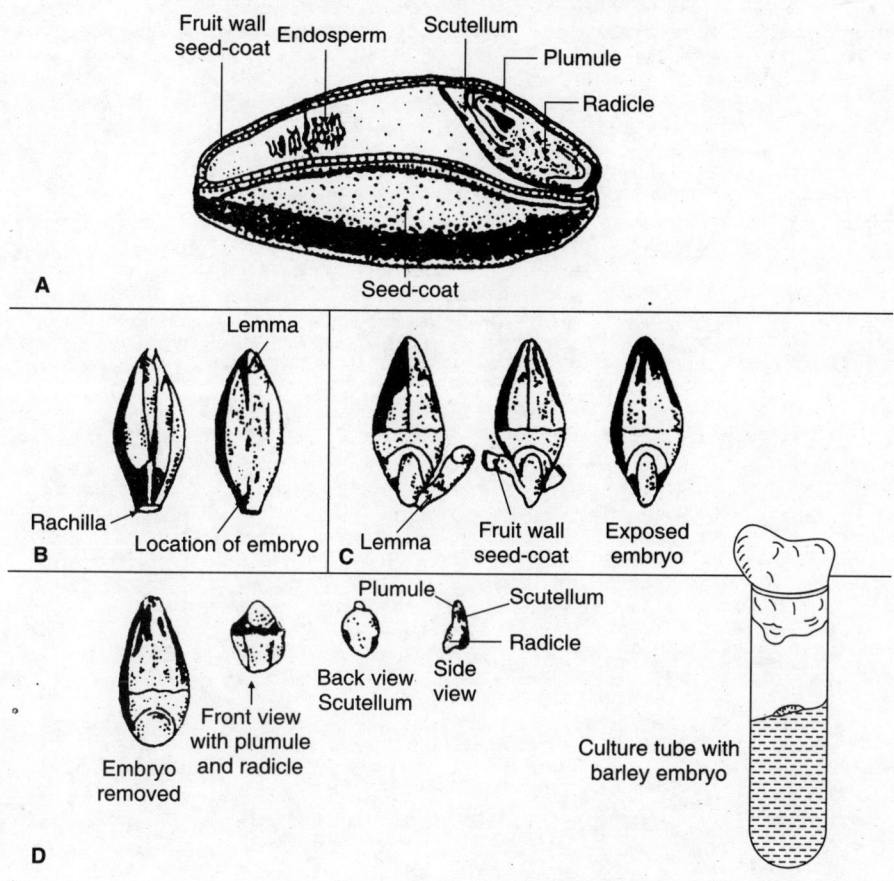

LONGITUDINAL SECTION OF CEREAL GRAIN

Fig. 11.4 Embryo culture of barley.

---

[a] Adopted from Pierik (1989).

Somatic Hybridisation and Cybridisation

In plant breeding programmes many desirable combinations of characters cannot be transmitted through conventional methods of genetic manipulations. Only recently has "another process, other than the sexual cycle, become available for higher plants that can lead to genetic recombination" (Cocking 1979). This non-conventional genetic procedure involving fusion between isolated somatic protoplasts (wall-less naked cells) under *in vitro* conditions and subsequent development of their product (heterokaryon) to a hybrid plant is known as *somatic* hybridisation. Since the first report on protoplast fusion-derived somatic hybrid plants of *Nicotiana glauca* (+) *N. langsdorfii* by Carlson et al. (1972), somatic hybridisation has opened up several possibilities for the parasexual manipulation of plants. Protoplast culture provides excellent opportunities for research on plant improvement: first, by exploring genetic variations among the existing crops and then attempting transfer of the available genetic information from one species to another through fusion of protoplasts isolated from somatic tissues of these crops (Razdan and Cocking 1981).

Plastids and mitochondrial genomes (cytoplasmically encoded traits) are inherited maternally in sexual crossings. Through the protoplast fusion process the nucleus and cytoplasm of both parents are mixed in the hybrid cell (heterokaryon). This results in various nucleo-cytoplasmic combinations. Sometimes interactions in the plastome and genome contribute to the formation *of cybrids* (cytoplasmic hybrids). Cybrids, in contrast to conventional hybrids, possess a nuclear genome from only one parent but cytoplasmic genes from both parents. The process of protoplast fusion resulting in the development of cybrids is known as *cybridisation.* In cybridisation, heterozygosity of extrachromosomal material can be obtained, which has direct application in plant breeding. Although there have been limitations in obtaining novel plants with desirable agronomic traits of many economically important species by somatic hybridisation, or cybridisation, studies during last two decades have revealed that the process of protoplast fusion may be a useful tool for the induction of genetic variability and combination of traits which do not exist in nature (*see* Razdan 1984-85; Bajaj 1989a,b; Gleba and Shlumukov 1990, Bhojwani and Razdan (1996); Xu and Wei (1997), Hall 1999, Sangthong et al. 2009, Skálová et al. 2012).

12.1 Isolation of Protoplasts

Methods of protoplast isolation can be classified into three main groups: (a) mechanical (*non-*

enzymatic), (b) sequential enzymatic (*two-step*) and (c) mixed enzymatic (*simultaneous*) procedures.

Mechanical isolation is done by cutting plasmolysed tissue with a sharp-edged knife and releasing the protoplasts by deplasmolysis. The principal deficiency of this approach is that the protoplasts released are few in number; mechanical isolation is thus only of historical importance now. Isolation of protoplasts mechanically from higher plants was pioneered by Klercker in 1892. Generally, protoplasts were isolated from highly vacuolated cells of storage tissues (onion bulbs, scales, radish root, mesocarp of cucumber and beet root).

Cocking (1960) used a concentrated solution of cellulase enzyme (prepared from cultures of the fungus *Myrothecium verrucaria* to degrade cell walls and demonstrated the possibility of large-scale protoplast isolation from higher plants. Further progress in enzymatic isolation of protoplasts was achieved as soon as cellulase and macerozyme enzymes became available commercially in 1968. Takebe et al. (1968) employed sequential or two-step procedure for isolating mesophyll protoplasts using commercial preparations of enzymes. The sequential approach involves initial incubation of macerated plant tissues with pectinases which, in turn, are then converted into protoplasts by a cellulase treatment. However, Power and Cocking (1968) mixed two enzymes together (simultaneous procedure) and isolated protoplasts in one-step. In this mixed enzymatic approach, plant tissues are plasmolysed in the presence of a mixture of pectinases and cellulases, thus inducing concomitant separation of cells and degradation of their walls to release the protoplasts directly. Most workers use the one-step method because it is less time-consuming and reduces the chances of microbial contamination by eliminating some steps.

The use of commercially available enzymes (Table 12.1) has enabled the isolation of protoplasts from practically every plant tissue in which cells have not acquired lignification. However, the reproducible potential for regeneration has been observed mostly in protoplasts isolated from leaf mesophyll tissues or cell suspension cultures. Addition of potassium dextran sulphate (0.5% w/v) or antioxidant PVP-10 (MW 10,000) to enzyme solution is recommended for isolation of viable protoplasts of recalcitrant deciduous tree species (Ochatt and Power 1992). Isolation of viable and culturable protoplasts in large quantities is affected by several factors and it is necessary to establish optimum conditions for a system empirically.

12.1.1 Source of Protoplasts

(I) Leaves

The leaf is the most convenient and popular source of plant protoplasts because it allows isolation of a large number of relatively uniform cells. Protoplast isolation from leaves involves five basic steps: (a) sterilisation of leaves, (b) removal of epidermal cell layer, (c) pre-enzyme treatment, (d) incubation in enzyme and (e) isolation by filtration and centrifugation (Fig. 12.1). The procedure is detailed in Appendix 12.1.

Sometimes it is difficult to peel the epidermis from leaves of some species (monocots). In such cases the leaf material is cut into small pieces (ca 1 mm^2) and then combined with vacuum infiltration. This procedure allows adequate infiltration of enzymes into the leaf cells. As soon as the vacuum is removed the leaf pieces will sink and eventually release the mesophyll protoplasts. Through this approach a large yield of protoplasts can be obtained in comparatively short periods.

(II) Callus Cultures

Young callus cultures are also ideal material for obtaining large quantities of protoplasts. Older callus cultures tend to form giant cells with thick cell walls which are usually difficult to digest; therefore, young actively growing callus is subcultured and used after two weeks for protoplast isolation.

Table 12.1 Commercially available enzymes more often used for protoplast isolation

Enzyme	Source	Supplier
Cellulases		
Cellulase Onozuka R-10	*Trichoderma viride*	Kinki Yokult Mfg. Co. Ltd. 8-12, Shingikancho Nishinomiya, Japan
Cellulase Onozuka RS	*T. viride*	Yokult Honsha Co., Tokyo, Japan
Cellulase YC	*T. viride*	Seishin Pharma Co. Ltd. 9-500-1, Nagareyama Nagareyama-shi, Chiba-ken, Japan
Cellulase CEL	*T. viride*	Cooper Biomedical Inc. Malvern, PA, USA
Cellulysin	*T. viride*	Calbiochem, San Diego, CA, USA
Driselase	*Irpex lacteus*	Kyowa Hakko Kogyo Co. Ltd., Tokyo, Japan
Meicelase P-l	*T. viride*	Meiji Seiki Kaisha Ltd., No. 8, 2-Chome Kyobashi, Chou-Ku, Japan
Hemicellulases		
Helicase	*Helix pomatia*	Industrie Biologique Francaise, Gennevilliers, France
Hemicellulase	*Aspergillus niger*	Sigma Chemical Co., St. Louis, MO, USA
Hemicellulase H-2125	*Rhizopus* sp.	Sigma, Munchen
Rhozyme HP 150	*Aspergillus niger*	Genencor Inc., South San Francisco, CA, USA
Pectinases		
Macerase	*Rhizopus arrhizus*	Calbiochem, San Diego, CA, USA
Macerozyme R-10	*R. arrhizus*	Yakult Honsha Co., Tokyo, Japan
PATE	*Bacillus polymyxa*	Farbwerke-Hoechst AG, Frankfurt, FRG
Pectinol	*Aspergillus* sp.	Rohm and Haas Co. Independence Hall West, Philadelphia, PA 19105, USA
Pectolyase Y-23	*Aspergillus japonicus*	Seishin Pharma Co. Ltd., Japan
Zymolyase	*Arthrobacter luteus*	Sigma Chemical Co., USA

Plant regeneration from cells isolated from cultured tissues could have substantial application in agriculture. Plants regenerated from populations of single cells may retain all the essential characters of a cultivar or clone but selectively alter undesirable genes. For example, sugarcane plants derived from callus cells have been demonstrated to provide clones having all the characteristics of the parent cultivar but for some improved traits such as disease resistance, enhanced yield potential and higher sugar content (Liu and Chen 1978). Isolation of protoplasts from callus tissues is one of the most refined procedures of single-cell regeneration efforts which have resulted in obtaining variant *protoclones* (protoplast propagated clones) in agriculturally important species (*see* details in Chapter 14).

(III) Cell Suspension Cultures

Cell suspension cultures also provide excellent source materials for isolating protoplasts. A high-density cell suspension is centrifuged. After removing the supernatant, cells are incubated in an enzyme mixture (cellulase + pectinase) in a culture flask placed on a platform shaker for 6 hr to overnight depending on the concentration of enzymes and protoplasts isolated (Fig. 12.2) following the protocol mentioned in Appendix

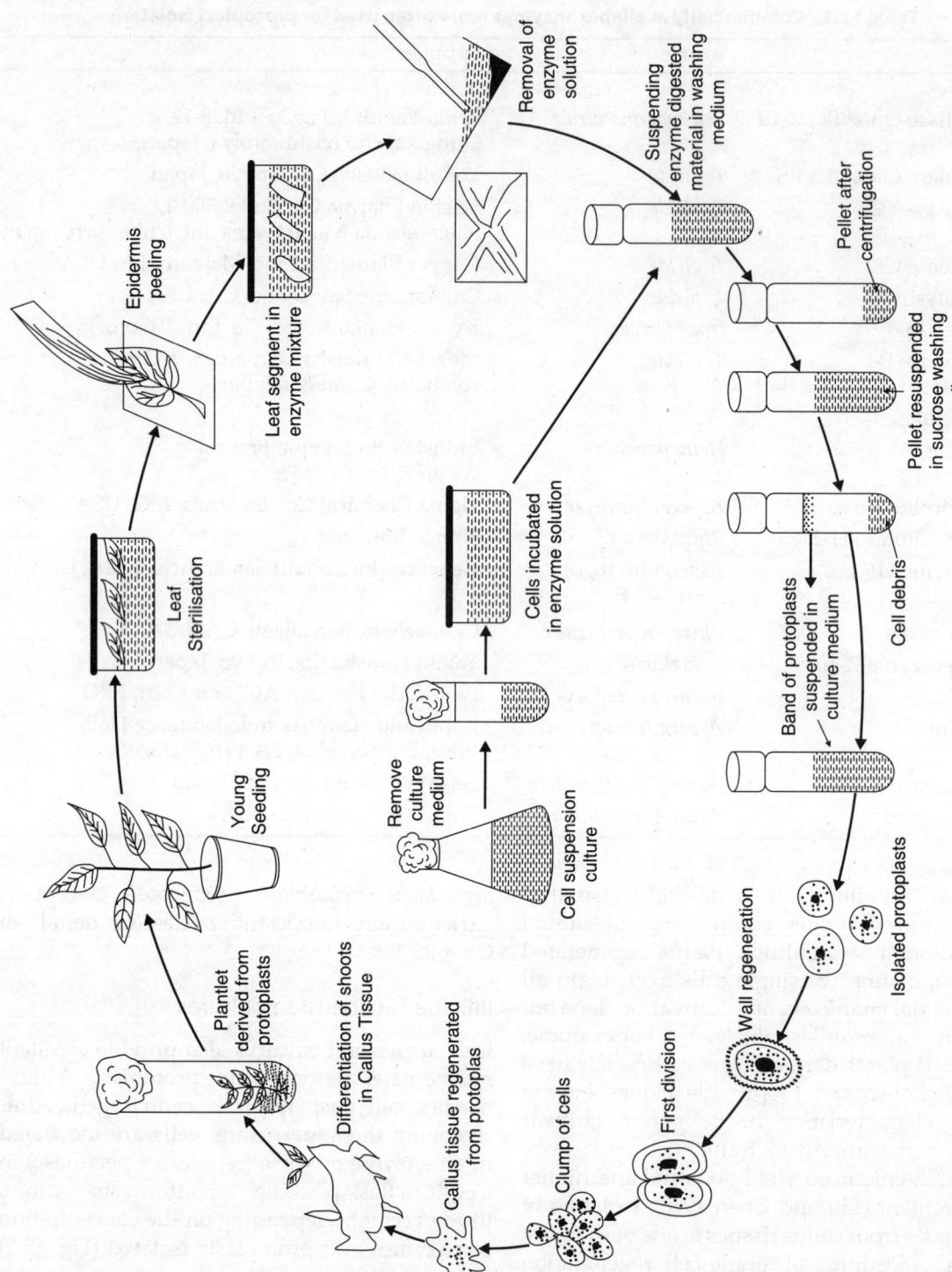

Fig. 12.1 Steps involved in isolation, culture and regeneration of protoplasts.

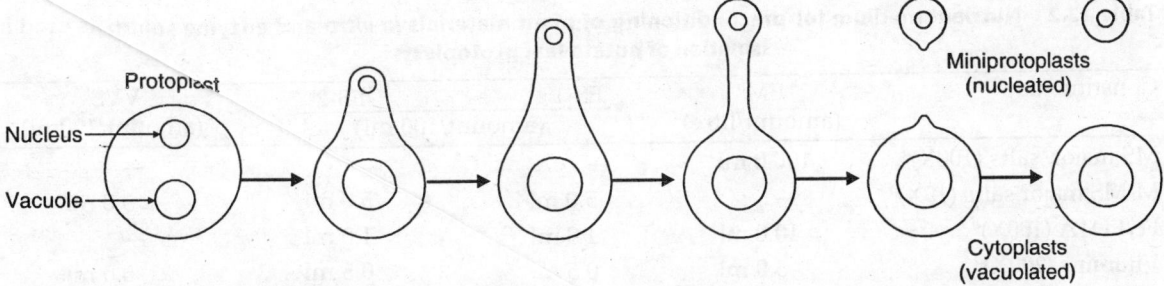

Fig. 12.2 Scheme showing fragmentation of a protoplast during centrifugation in iso-osmotic density gradients (observations based on Lörz 1984).

12.1. A lower concentration of enzymes helps to prevent the formation of aggregates in the cell suspension in order to obtain better yields.

Protoplasts originating from cell suspension cultures have demonstrated the potentiality to regenerate plants in species in which attempts to induce sustained division of mesophyll protoplasts have failed. The success in formation of plantlets from protoplasts of morphogenetic cell culture origin in cereals (pearl millet, sorghum and barley cv. Golden Promise) proved cell suspension cultures are a viable source of totipotent protoplasts.

(IV) Preconditioned Plant Materials

Mesophyll protoplasts of some crop plants have a low morphogenetic response. This is because of the fact that the physiological state of growth of a donor plant under natural conditions largely affects the regeneration potential of protoplasts in these systems. On the contrary, tissue-culture regenerated plants (axenic cultures) are maintained under uniform physiological conditions and, therefore, provide materials (leaf) preconditioned for protoplast isolation and regeneration. This approach is particularly essential for regeneration of potato protoplasts. Fully expanded leaves excised from 3- to 4-week-old plants raised from virus-free meristem-tip cultures provided a source of high-yielding, viable and regenerating protoplasts of tetraploid ($2n = 4x = 48$) and dihaploid ($2n = 2x = 24$) potato (*Solanum tuberosum*) cultivars and related wild species (*S. chacoense*). In other crops

(amphidiploid *Brassica napus* cv. Victor tomato) also the *in vitro* propagated shoot cultures provide protoplasts that regenerated several hundred colonies used for studies in morphogenesis and DNA virus investigation (*see* Thomas et al. 1982; Shahin 1984). Components of the medium used for preconditioning of plant material *in vitro* and enzyme solutions for isolation of potato mesophyll protoplasts are given in Tables 12.2 and 12.3.

Axenic shoot cultures are also valuable for providing juvenile material suitable for protoplast isolation in woody perennial species (Ochatt and Power 1992). For reproducible high plating efficiency of protoplasts isolated from field-grown plant material, it is desirable that source plants are preconditioned by growing under controlled conditions of light, temperature and humidity such as in greenhouse.

12.1.2 Preparation of Cytoplasts and Miniprotoplasts

The immediate product of protoplast fusion is a binucleated cell (heterokaryon) with mixed cytoplasm of both parents. During fusion experiments, particularly at interspecific and intergeneric levels, some of the problems associated with the combination of both nuclear genomes and mixed cytoplasmic materials can be overcome by replacing one or even both fusion partners by *subprotoplasts* (protoplast fragments). Another approach to achieve a more defined combination of nuclear and cytoplasmic materials is to inactivate the nuclear genome of

Table 12.2 **Nutrient medium for preconditioning of plant materials *in vitro* and enzyme solutions used in isolation of potato leaf protoplasts**[a]

Constituents	PM[b] (amount/litre)	ES-1[c] (amount/100 ml)	ES-2[c]	VVS[b] (amount/100ml)
MS major, salts (100X)[d]	100.0 ml	—	—	—
M-MS major salts (10X)[d]	—	5.0 ml	5.0 ml	5.0 ml
Fe/EDTA (100X)[d]	10.0 ml	1.0 ml	1.0 ml	—
Vitamins (200X)[d]	5.0 ml	0.5 ml	0.5 ml	5.0 ml
Micronutrients (100X)[d]	10.0 ml	1.0 ml	1.0 ml	10.0 ml
Casein hydrolysate	—	5.0 ml	5.0 mg	10.0 mg
MES	97.0 mg	9.7 mg	9.7 mg	97.0 mg
Cellulase R-10	—	0.5%	—	—
Macerozyme R-10	—	0.1%	—	—
Cellulysin	—	—	2.0%	—
Macerase	—	—	0.25%	—
PVP-10	—	1.0 g	1.0 g	—
Sucrose	30.0 g	10.27 g	10.27 g	10.27 g
Kinetin	0.2 mg	—	—	—
Agar	7.0 g	—	—	—
pH	5.8	5.65	5.65	5.8

[a] after Shahin (1984).
[b] Medium autoclaved for 15 min.
[c] Solution filter-sterilised.
[d] Stock solutions presented in Table 12.3.
Abbreviations: PM—Medium for obtaining preconditioned plant material; ES-1—Enzyme solution designed for isolation of protoplasts from tetraploid materials; ES-2—Enzyme solution to obtain protoplasts from dihaploid and wild species; WS—Washing medium.

one of the fusion partners by irradiation in order to produce "physiological" subprotoplasts. The ripening pericarp of some solanaceous species produce subprotoplasts naturally whereas *in vitro* they are formed by 'budding' of protoplasts or by elongation of the protoplasts during plasmolysis. Subprotoplasts can be collected as a distinct fraction from protoplast preparations.

Isolated protoplasts can be experimentally induced to fragment into types of subprotoplasts called *miniprotoplasts* or *cytoplasts*. The term miniprotoplast was coined by Wallin et al. (1978) for subprotoplasts having nuclear material. Miniprotoplasts can divide and may be able to regenerate into plants as in *Solanum niger* (Lesney et al. 1986). Equivalent terms given to miniprotoplasts are *karyoplast* (evacuolated subprotoplast) or *nucleoplast*. Cytoplasts are

nuclear-free subprotoplasts which do not divide but are useful in the process of cybridisation. The fragmentation of protoplasts is achieved through application of different centrifugal forces created by discontinuous gradients during the process of centrifugation. Additional exposure of isolated protoplasts to cytochalasin B in combination with centrifugation has also been found beneficial for enucleation. Protoplasts are dramatically deformed and 'drawn out' (Fig. 12.2) due to centrifugal forces during centrifugation. Different densities of the cellular components then allow the enucleation of protoplasts in iso-osmotic density gradients. Components of high density (nucleus) are oriented toward the bottom and those having less density (vacuoles) move to the top of the centrifuge tube. After prolonged centrifugation either the nuclei surrounded

Table 12.3   Stock solutions used to prepare enzyme solutions and media for isolation and culture of potato leaf protoplasts[a]

Constituents	Amount/litre	Constituents	Amount/litre
MS major salts (100X)		H_3BO_3	620.0 mg
KNO$_3$	19.0 g	$MnSO_4 \cdot 4H_2O$	2.23 mg
NH$_4$NO$_3$	16.50 g	$Na_2MoO_4 \cdot 2H_2O$	25.0 mg
CaCl$_2 \cdot$ 2H$_2$O	4.40 g	$CuSO_4 \cdot 5H_2O$	2.50 mg
MgSO$_4 \cdot$ 7H$_2$O	3.70 g	$CoCl_2 \cdot 6H_2O$	2.50 mg
KH$_2$PO$_4$	1.70 g	Vitamins (200X)	
Modified MS major salts		Myoinositol	20.0 g
(M-MS 100X)		Thiamine HCl	100.00 mg
KNO$_3$	19.0 g	Pyridoxine HCl	100.0 mg
CaCl$_2 \cdot$ 2H$_2$O	4.40 g	Nicotinic acid	1.0 g
MgSO$_4 \cdot$ 7H$_2$O	3.70 g	Glycine	500.0 mg
KH$_2$PO$_4$	1.70 g	Biotin	10.0 mg
Micronutrients		Folic acid	100.0 mg
(100X)		Fe/EDTA (20X)	
KI	83.0 mg	Na$_2$EDTA	373.0 mg
ZNSO$_4 \cdot$ 7H$_2$O	920.0 mg	FeS0$_4 \cdot$ 7H$_2$0	278.0 mg

[a] After Shahin (1984).

by some cytoplasmic material are pinched off, giving rise to nucleated miniprotoplasts or cytoplasmic materials without nuclei form enucleated (vacoulated) cytoplasts. Both types of subprotoplasts eventually become spherical after removal from the gradient.

The composition of the gradient, speed of centrifugation and treatment with cytochalasin B depend on the specific type of protoplast used for enucleation. A broad framework within which a procedure can be developed for a particular protoplast system has been outlined by Lörz (1984) and other workers. Suitable components for establishing gradients for protoplast centrifugation are inorganic salts, sugars and modified silica gels (Percoll). The composition of iso-osmotic density gradient with cytochalasin (Gradient A) and without cytochalasin (Gradient B) is given in Table 12.4. Different solutions (starting from I to IV) are gently layered without mixing in the centrifuge tube to form a discontinuous gradient. Centrifugation of Gradient A is performed in a

swinging-bucket rotor containing 10 ml tubes (each loaded with $2\text{-}5 \times 10^5$ protoplasts), using a speed between 20,000 and 40,000 g for 15-20 min at 37°C; or 45-90 min at 12°C for Gradient B. After centrifugation, enucleated cytoplasts are located in the top fraction of the gradient and nucleated miniprotoplasts form a band between layers I and II. Both subprotoplast fractions are collected by Pasteur pipettes, resuspended in the medium and used in the fusion process.

Table 12.4   Gradients for enucleation of protoplasts[a]

Gradient A		Gradient B
	IV	Protoplasts in 0.22 M CaCl$_2$
Protoplasts in culture medium	III	0.5 M mannitol, 5% Percoll
1.5 M sorbitol, 50 μg/ml cytochalasin B	II	0.48 M mannitol, 20% Percoll
Saturated sucrose	I	0.45 M mannitol, 50% Percoll

[a] After Lörz (1984).

12.1.3 *Viability of Protoplasts*

Isolated protoplasts should be healthy and viable in order to undergo sustained divisions and regeneration. Methods used to check cell/protoplast viability are described in Section 5.8. Further tests include: (a) observation of cyclosis or cytoplasmic streaming, (b) oxygen uptake by protoplasts measured by an oxygen electrode and (c) photosynthetic activity. These tests indicate active metabolism in protoplasts. Exclusion of Evan's blue dye by intact membranes and staining with FDA are some of the common methods used to determine the viability of freshly isolated protoplasts.

12.2 Culture of Protoplasts

The first step in the protoplast culture is the development of a cell wall around the membrane of isolated protoplasts. This is followed by induction of divisions in the protoplast-derived new cell giving rise to a small cell colony. By manipulation of the nutritional and physiological conditions in the nutrient media, cell colonies may be induced to grow a callus continuously or to regenerate whole plants. The culture methods and requirements of isolated protoplasts are often similar to those of single cells, as described in Chapter 5.

Isolated protoplasts or their hybrid cells are cultured either in a liquid or agar medium. The common practice of using a liquid culture medium includes either incubating protoplasts/heterokaryons in a thin layer or as small drops of nutrient medium inside a petri dish which, in turn, is covered by another petri plate and finally sealed with parafilm. The culture dish is then maintained at low-light or dark conditions at 25°-28°C.

For culturing protoplasts in the nutrient medium containing agar the procedure given for Bergmann's cell plating technique (*see* Chapter 5) may be followed. About 2 ml aliquots of isolated protoplasts of suitable density (ca 10^3-10^5 cells ml^{-1}) are mixed with an equal volume of agar nutrient medium, the temperature of which should not exceed 45°C. On solidification of agar, the culture plates are sealed and maintained in an inverted position at 25°-28°C. With this method, individual protoplasts or heterokaryons can be conveniently observed under a microscope and plating efficiency readily determined. Let it be noted, however, that some workers have found the plating protoplasts in a liquid nutrient medium on the top of a layer of solidified agar medium suitable for regeneration and further development of protoplasts.

12.2.1 *Culture Media*

(I) Nutritional Components

Protoplast culture media generally comprise nutrients similar to those required for callus and suspension cultures. However, the concentration of iron, zinc and ammonium used in plant tissue culture media may be too high for protoplast culture. Mostly the salts of B5 and MS media (*see* Chapter 5) with some modifications have been found suitable. Increasing the calcium concentration in a protoplast culture medium two to four times over the normal concentration of a cell culture medium may be beneficial in preserving membrane integrity (Torres 1989). Generally, protoplast culture media contain 3-5% sucrose but in some systems (tobacco protoplast cultures) a lower sugar (1.5%) content is required. Organic nitrogen in the form of CH is generally included in the medium as reduced nitrogen, and NH_4NO_3 (20 mmol l^{-1}) as reduced inorganic nitrogen.

The vitamins used for protoplast culture are the same as used in standard tissue culture media. Both types of growth substances (auxins and cytokinins) are used in the protoplast culture media in various combinations in order to induce cell wall formation and divisions in isolated protoplasts. Cereal protoplasts require 2,4-D which may be either sufficient (if used alone) or better in combination with cytokinin (Vasil and Vasil 1980). However, 2,4-D as the sole auxin source generally leads to loss of morphogenetic potential in the protoplast-derived callus. Other auxin sources are NAA or IAA. The commonly used cytokinins are BAP, kinetin, 2iP, or zeatin. Although the exact combination of the two types of growth hormones in the medium varies according to the species, it has been observed that protoplasts from actively growing cell cultures may find a high auxin/kinetin ratio suitable to induce divisions, while those derived from highly differentiated cells (leaf tissue) require a high kinetin/auxin ratio for regeneration.

Supplementing the culture medium with antioxidants glycine and/or PVP-10, n-propylgallate, glutathione, etc. improves the regenerative response in protoplasts of *Prunus avium* (sweet cherry), *Beta vulgaris* and *Lolium perenne*. The mesophyll protoplasts of *Brassica napus* show divisional frequency increasing two-fold on addition of 2% Ficoll to the culture medium (*see* Bhojwani and Razdan 1996).

(II)   Osmoticum

During isolation and culture, protoplasts require osmotic protection until they regenerate a strong wall. Inclusion of an osmoticum in both isolation and culture media prevents rupture of protoplasts. A variety of ionic as well as nonionic solutes have been tested for adjusting the osmotic potential of various solutions used in protoplast isolation and culture. The most widely used osmotica in a protoplast culture medium as well as in an enzyme mixture are sorbitol, mannitol, glucose, or sucrose. Protoplasts are more stable in a slightly hypertonic solution. For mesophyll protoplasts (cereals and pea) sorbitol or mannitol may be a suitable osmotic stabiliser, whereas sucrose has been preferred over glucose or mannitol to culture protoplasts of potato, sweet pea, brome grass and cassava (Michayluk and Kao 1975). For tobacco suspension cultures galactose and fructose have been used as osmotica. The osmolarity of the medium is gradually reduced by periodic addition of a few drops of fresh medium as soon as the protoplasts have regenerated walls and undergone divisions.

Ionic substances (335 mmol l^{-1} KCl and 40 mmol l^{-1} MgSO$_4$ · 7H$_2$O) improve the viability of protoplasts and yield cleaner preparations. Usually enzyme solutions are supplemented with certain salts (5-100 mmol^{-1} CaCl$_2$) along with non-ionic osmotic stabilisers. Cocking and Peberdy (1974, *see* Appendix 12.1) developed a cell-protoplast washing (CPW) solution containing salts and a suitable osmoticum to provide stable and cleaner preparations. The CPW solution can be used during enzyme incubation and washing of protoplasts. The enzyme incubation period depends on its concentration in the solution and the type of material used.

12.2.2   *Multiple Drop Array (MDA) Screening*

In 1977, Potrykus and co-workers developed the MDA technique for systematically screening a large number of multiple combinations of media constituents for protoplast culture. This technique was principally designed for cereal (maize) leaf protoplast cultures since they did not respond to culture regimes that regenerate protoplasts like of solanaceous members. The MDA screening technique uses hanging droplets of 40 µl as the experimental unit. Each droplet represents one combination of factors to be tested. The droplets are arranged in a regular array of 7 × 7 drops on the lid of a 9-cm petri dish. To test seven different auxins (factors) in combination with four different cytokinins in the medium, each of these factors is used in seven different concentrations. The whole experiment, therefore, includes 4 × 7 petri dishes. Since each petri dish contains 49 droplets this results in a total of 4 × 7 × 49 = 1372 two-factor combinations. This experiment can be performed by one person 5-6 hr in addition to the time required for media preparation, protoplast isolation and culture evaluation (for details of this procedure, *see* Harms et al. 1979; Harms 1984).

12.2.3   *Plating Density*

A density of 1×10^4 to 1×10^5 protoplasts ml^{-1} is optimal since at such high densities the cell colonies arising from individual protoplasts tend to grow at a fairly early stage in culture. Experiments on somatic hybridisation and muta-genesis, however, require cloning of individual cells; this can only be achieved by plating protoplasts at low densities (100-500 protoplasts ml^{-1}). At low density the development process of individual protoplasts, or heterokaryons, can be easily monitored, thereby enabling isolation and identification of hybrid colonies in the absence of a stringent selection system.

The nutritional components of the most commonly used culture media are not sufficient to induce divisions in protoplasts plated at low density. Kao and Michayluk (1975) developed a complex protoplast culture medium (KM 8p) in which individually cultured protoplasts (*Vicia hajastana*) are capable of dividing until callus formation. This medium (Table 12.5) also induced faster divisions in mesophyll proto-plasts of alfalfa, pea, potato, and potato + tomato fusion

Table 12.5   KM 8p medium for culturing protoplasts at low density[a]

Constituents	Amount (mg l^{-1})	Constituents	Amount (mg l^{-1})
Mineral salt		Vitamin-free casamino acid[c]	125
NH$_4$NO$_3$	600	Coconut water	10 ml l^{-1}
KNO$_3$	1900	(from mature fruits,	
CaCl$_2 \cdot$ 2H$_2$O	600	heated to 60°C for	
MgSO$_4 \cdot$ 7H$_2$O	300	30 min and filtered	
KH$_2$PO$_4$	170	Organic acids	
KCl	300	(adjusted to pH 5.5 with	
Sequestrene 330 Fe[b]	28	NH$_4$OH) Sodium pyruvate	5
KI	0.75	Citric acid	10
H$_3$BO$_3$	3.00	Malic acid	10
MnSO$_4 \cdot$ H$_2$O	10.00	Fumaric acid	10
ZnSO$_4 \cdot$ 7H$_2$O	2.00	Vitamins	
Na$_2$MoO$_4 \cdot$ 2H$_2$O	0.25	Inositol	100
CuSO$_4 \cdot$ 5H$_2$O	0.025	Nicotinamide	1
CoCl$_2 \cdot$ 6H$_2$O	0.025	Pyridoxine-HCl	1
Sugars		Thiamine-HCl	10
Glucose	68400	D-Calcium pantothenate	0.5
Sucrose	125	Folic acid	0.2
Fructose	125	p-Aminobenzoic acid	0.01
Ribose	125	Biotin	0.005
Xylose	125	Choline chloride	0.5
Mannose	125	Riboflavin	0.1
Rhamnose	125	Ascorbic acid	1
Cellobiose	125	Vitamin A	0.005
Sorbitol	125	Vitamin D$_3$	0.005
Mannitol	125	Vitamin B$_{12}$	0.01
Hormones	Soybean × barley		Soybean × pea or N. glauca
2,4-D	1		0.2
Zeatin	0.1		0.5
NAA	—		1

[a] Sterilised by filtration.
[b] Geigy Chemical Corp., Ardsley, N.Y.
[c] Difco Laboratories, Detroit, USA

products plated at low densities. The cultures in this medium are placed in darkness since 8p medium turns phytotoxic under strong light.

(I) Feeder Layer Technique

Another approach to culture protoplasts at low-density is the *feeder layer technique*. Raveh et al. (1973) prepared a feeder cell layer by exposing tobacco cell suspension protoplasts (10^6 cells ml^{-1}) to an X-ray dose of 2×10^3 R, which inhibited division of cells but allowed them to remain metabolically active. Irradiated protoplasts were washed three times to remove toxic substances due to irradiation and then plated in soft agar medium at a density 2.4×10^4 ml^{-1}. Non-irradiated protoplasts of low density (10-100 protoplasts ml l^{-1}) were plated over this feeder layer. Protoplasts of the same species or different species can be used as a feeder layer (Jain et al. 1998).

(II) Co-culture of Protoplasts

Protoplasts from two different species have also been co-cultured to promote their growth or that of the hybrid cells. Metabolically active and dividing protoplasts of two types are mixed in a liquid medium and plated together so that there is cross-feeding between the two types. This enables the protoplasts or cells at low density to undergo sustained divisions. The co-culture method is generally used where calli arising from two types of protoplasts can be morphologically distinguished. For example, mechanically isolated hybrid cells co-cultured with protoplasts isolated from an albino strain will develop green colonies that are readily distinguishable from non-green colonies of albino types (Menczel et al. 1978).

(III) Microdrop Culture

Microdrop culture has been successfully used to culture hybrid cells of *Nicotiana glauca* (+) *Glycine max* (Kao 1977) and *Arabidopsis thaliana* (+) *Brassica campestris* (Gleba and Hoffmann 1979). The technique requires specially designed Cuprak

dishes which have a smaller outer chamber and a large inner chamber. The inner chamber carries numerous numbered wells, each with a capacity for 0.25-25 µl droplet of nutrient medium. Individual protoplasts or heterokaryons per 0.25-25 µl droplet of nutrient medium are transferred by Drummond pipette to each well of the inner chamber of the Cuprak dish. The outer chamber is filled with sterile water to maintain humidity inside the dish. After covering it with a lid, the dish is sealed with parafilm and maintained at optimal light and temperature conditions. The size of the droplet is critical for division of a single protoplast or heterokaryon, as it gives a ratio of cell/volume of culture medium equal to a cell density $2-4 \times 10^3$ ml^{-1}. An increase in the size of the droplet would decrease the effective plating density.

(IV) Other Treatments

Electroporation treatment of protoplasts stimulate their divisions and regeneration (Ochatt and Power 1992). Protoplasts are suspended in buffer solution four times their plating density and exposed to high voltage (250-2000 V) DC pulses (10-50 us) at intervals of 10 s. Electroporation could enhance DNA synthesis in isolated protoplasts which in turn promotes early gene expression for differentiation and regeneration. Low voltage electric treatment is also reported to enhance division in protoplasts of *Medicago sativa* (Dijak et al. 1986) and *Trifolium subterraneum* (Zhongyi et al. 1990). Heat-shock treatment (45°C for 5 min, followed by 10 s on ice) has a stimulatory effect similar to electroporation treatment on protoplast division and regeneration of *Pennisetum squamulatum* (Gupta et al. 1988).

12.3 Protoplast Regeneration

12.3.1 Formation of Cell Wall

The process of cell-wall formation may be completed in two to several days although protoplasts in culture generally start regenerating a cell wall within a few hours after isolation. Protoplasts lose their characteristic spherical shape once the wall formation is complete. The regeneration of cell wall can be demonstrated using Calcalfluor White ST (American Cyanamide Co., USA) fluorescent stain.[1] The freshly formed cell wall is composed of loosely arranged microfibrils, the process requiring an exogenous supply of a readily metabolised carbon source (sucrose) in the nutrient medium. Ionic osmotic stabilisers in the medium are reported to suppress the development of a proper wall.

There is a direct relationship between wall formation and cell divisions. Protoplasts with a poorly developed wall often show budding and those which are not able to regenerate a proper wall fail to undergo normal mitosis.

12.3.2 Development of Callus/ Whole Plant

Soon after the formation of a wall around the protoplasts, the reconstituted cells show considerable increase in size and first divisions generally occur within a week. Subsequent divisions give rise to small cell colonies. After 2-3 weeks macroscopic colonies are formed which can be transferred to an osmotic-free medium to develop a callus. The callus may be induced to undergo organogenic differentiation, or whole plant regeneration, following the procedures described in Chapter 6.

The steps involved during the isolation, culture and regeneration of protoplasts are represented in Fig. 12.1. Since the first report (Takebe et al. 1971) of plant regeneration from isolated protoplasts, the list of species exhibiting this potentiality has increased. Initially, the majority of taxa which yielded protoplasts capable of regenerating callus or whole plants belonged to Solanaceae but subsequent advances in protoplast technology resulted in differentiation of shoots, or plants from isolated protoplasts of several other families, including forage legumes, fibre or pulp crops, various tree species, and even cereals, such as pennisetum, rice and wheat (Table 12.6). The number has increased to the extent that 368 species or subspecies of higher plants covering 161 genera of 48 families yield protoplasts capable of regenerating into plants (Xu and Xue 1999).

12.3.3 Regeneration via Embryogenesis

Plant regeneration of protoplasts via embryogenesis mostly occurs in species belonging to Gramineae, Umbelliferae, Rutaceae and leguminosae (forage legumes), among angiosperms, and Pinaceae of Gymnosperms (Xu and Xue 1999). More importantly, it should be noted that potential for embryogenesis is eventually lost due to repeated subculture of embryogenic cell culture and this problem can be circumvented by cryopreservation of embryogenic calli or cell suspensions.

12.4 Protoplast Fusion

The fact that isolated protoplasts are devoid of walls makes them easy tools for undergoing fusions in vitro. An important aspect has been that incompatibility barriers do not exist during the cell fusion process at interspecific, intergeneric, or even interkingdom levels (Ahkong et al. 1975). Thus plant protoplasts represent the finest single cell system that could offer exciting possibilities in the fields of somatic cell genetics and crop improvement.

12.4.1 Spontaneous Fusion

During the enzymatic degradation of cell walls some of the adjacent protoplasts may fuse together to form homokaryons (homokaryocytes). These plurinucleate cells sometimes contain 2-40 nuclei, a phenomenon attributed to

---

[1] The presence of a wall can be tested by incubating protoplasts in 0.1 or 0.01% Calcafluor solution, in an appropriate osmotic stabiliser, for 5 min. The protoplasts are then washed to remove excess dye and mounted on a slide in an osmotically suitable medium. Calcafluor binds to the wall material and fluoresces under a mercury vapour lamp with an excitation filter BG12 and suppression filter K510. Tinapol solution (Geigy, U.K. Ltd.) behaves in a manner similar to Calcafluor.

Table 12.6 Some examples where shoot differentiation or plant regeneration has been achieved from cultured protoplasts[a,b,c]

Family	Species	Family	Species
Compositae	*Cichorium intybus*		*M. sativa* cultivars
	Lactuca sativa cultivars		*M. varia*
	L. serriola		*Psophocarpus tetragonolobus*
	L. saligna		*Trifolium hybridum*
	Petasites japonicus		*T. repens*
	Senecio vulgaris		*T. rubens*
Cruciferae	*Arabidopsis thaliana*	Liliaceae	*Asparagus officinalis*
	Brassica campestris		*Hemerocallis* sp.
	B. carinata	Linaceae	*Linum usitatissimum*
	B. juncea		*L. strictum*
	B. napus		*L. lewissii*
	B. nigra	Magnoliaceae	*Liriodendron tulipfera*
	B. oleraceae var. *capitata*	Ranunculaceae	*Ranunculus sceleratus*
	Sinapis alba	Rutaceae	*Citrus aurantifolia*
Cucurbitaceae	*Cucumis sativus*		*C. grandis*
Euphorbiaceae	*Manihot esculenta*		*C. limon*
Gramineae	*Bromus inermis*		*C. medica*
	Oryza sativa		*C. paradisi*
	Pennisetum americanum		*C. reticulata*
	Saccharum spp.		*C. sinensis*
	Triticum aestivum	Salicaceae	*Populus tremula*
Leguminosae	*Glycine argyrea*		*P. alba* × *P. grandidentata*
	G. canescens		*P. nigra* var. 'Butulifolia' ×
	G. clandestina		*P. trichocarpa*
	G. max	Santalaceae	*Santalum album*
	Medicago arborea	Solanaceae	*Atropa belladonna*
	M. coerulea		*Capsicum annuum*
	M. difalcata		*Datura metel*
	M. falcata		*D. meteloides*
	M. glutiniana		*D. innoxia*
	M. hemicyla		*Hyoscyamus muticus*

Contd...

Table 12.6 Contd.

Family	Species	Family	Species
	Lycopersicon esculentum		S. dulcamara
	Nicotiana acuminata		S. etuberosum
	N. alata		S. fernandezianum
	N. debneyi		S. gilo
	N. glauca		S. khasianum
	N. langsdorffii		S. luteum
	N. longiflora		S. lycopersicoides
	N. otophora		S. melongena
	N. paniculata		S. nigrum
	N. plumbaginifolia		S. pennellii
	N. suaveolens		S. phureja
	N. sylvestris		S. phureja × S. chacoense
	N. tabacum		S. pinnatisectum
	Petunia hybrida		S. torvum
	P. inflata		S. tuberosum cultivars
	P. parodii		S. tuberosum tetraploid
	P. parviflora		clones,
	P. violacea		diploid clones
	Salpiglossis sinuata		S. uporo
	Solanum aculeatissimum		S. viarum
	S. aviculare		S. xanthocarpum
	S. brevidens	Ulmaceae	Ulmus species
	S. chacoense	Umbelliferae	Daucus carota
			Foeniculum vulgare

[a] For detailed references *see* Bhojwani and Razdan (1983, 1996) and Bajaj (1989a,b).
[b] Several other species in which callus formation without shoots from isolated protoplasts have been achieved are not listed.
[c] About 368 species belonging to 48 families of Angiosperms and Gymnosperms yield protoplasts regenerating plants, shoots or embryos (*see* Xu and Xue 1999).

expansion and subsequent coalescence of the plasmodesmatal connections between the cells. More frequent homokaryon formation has been observed in protoplasts isolated from dividing cultured cells. However, the sequential method of protoplast isolation or exposure of the cells to a strong plasmolyticum would sever the plasmodesmatal connections and, consequently, reduce the frequency of spontaneous fusion.

12.4.2 Mechanical Fusion

The giant protoplasts of *Acetabularia* have been fused mechanically by pushing together two protoplasts. This fusion does not depend upon the presence of fusion-inducing agents. However, in this procedure protoplasts are likely to get injury. Protoplasts released from meiocytes (*Lilium* and *Trillium*) in enzyme solutions readily fuse by gentle tapping in a depression slide and some of the di- and trinucleate cells reportedly complete the second division, forming tetrad configurations in culture media (Ito and Maeda 1974).

12.4.3 Induced Fusion

Freshly isolated protoplasts can be induced to undergo fusion, irrespective of their origin, with the help of a range of *fusogens* (fusion-inducing agents) e.g., $NaNO_3$, artificial sea-water, lysozyme, high pH/Ca^{++}, polyethylene glycol (PEG), antibodies, Concavalin A, polyvinyl alcohol, electrofusion, dextran and dextran sulphate, fatty acids and esters. Of these, the following treatments have yielded success in producing somatic hybrid plants.

(I) $NaNO_3$ Treatment

Induced fusion by $NaNO_3$ was first demonstrated by Power et al (1970). Isolated protoplasts were cleaned by floating in sucrose osmoticum. Transfer of the protoplasts in 0.25 M $NaNO_3$ solution and subsequent centrifugation promoted the fusion process. Carlson et al. (1972) used this method for producing the first somatic hybrid plant by fusing protoplasts of *Nicotiana glauca* and *N. langsdorffii*. This procedure results in a low frequency of heterokaryon formation and protoplasts are markedly altered in their uptake capabilities.

(II) HIGH pH/Ca^{++} Treatment

This method was developed by Keller and Melchers (1973) for fusing two different lines of tobacco protoplasts and is now commonly used. Isolated protoplasts are incubated in a solution of 0.4 M mannitol containing 0.05 M $CaCl_2$, with pH at 10.5 (0.05 M glycine-NaOH buffer) and temperature 37°C. Aggregation of protoplasts generally takes place at once and fusion occurs within 10 min. Many intraspecific and interspecific somatic hybrids have been produced using this procedure.

(III) PEG Treatment

PEG has been used as a fusogen in a number of plant species because of the reproducible high frequency of heterokaryon formation. About 0.6 ml of PEG solution (dissolve 1 g of PEG, mol. wt. 1500, in 2 ml of 0.1 M glucose, 10 mM $CaCl_2$, and 0.7 mM KH_2PO_4) is added in drops to a pellet of protoplasts in the tube. After having capped the tube, protoplasts in PEG are incubated at room temperature for 40 min. Occasional rocking of tubes helps to bring the protoplasts in contact. This is followed by elution of PEG by the addition of 0.5-1 ml of protoplast culture medium in the tube after every 10 min. Occasional rocking of tubes after every 10 min. Preparations are now washed free of fusogen by centrifugation and the protoplasts resuspended in the culture medium.

Both the mol. wt. and the concentration of PEG are critical in inducing successful fusions. PEG less than 100 mol. wt. is not able to produce tight adhesions while that ranging up to 6000 mol wt. can be more effective per mole in inducing fusions. At higher mol. wt., however, PEG produces too viscous a solution for easy handling. Treatment with PEG in the presence of/or by eluting with high pH/Ca^{++} is reported most effective in enhancing the fusion frequency and survivability of protoplasts.

After treatment with fusogen, protoplasts are cultured following the standard procedures (*see* Section 12.2). The nature of the change at the membrane surface that leads to fusion is not clear. Some electrophoretic evidence suggests the elimination of a charge (by $NaNO_3$) on the protoplast surface. PEG either provides a bridge by which Ca^{++} can bind membrane surfaces together or leads to a disturbance in the surface charge during the elution process. Details regarding the procedure for using various other fusogens are given by Kao and Michayluk (1989).

(IV) Electrofusion

Studies in the past few years have shown that electric fields can be used for protoplast fusion. This procedure, called *electrofusion*, has been found to be simpler, quicker and more efficient than chemically-induced fusions. More importantly, cells after electrofusion do not show cytotoxic responses as generally found in protoplasts or heterokaryons subjected to PEG treatment.

These aspects and the demonstrations using electric pulses to introduce foreign DNA into plant cells (electroporation, *see* Chapter 13), have further heightened interest in the application of electrofusion in somatic cell genetics.

Senda et al. (1979) first attempted fusion by positioning two microelectrodes with the help of a micromanipulator at the ends of adjoining *Rauwolfia* protoplasts. Success in inducing fusion was achieved with brief 5-12 amp DC pulses restricted to single protoplast pairs only. Subsequently, Zimmermann and Scheurich (1981) demonstrated that batches of protoplasts could be fused by electric fields by devising a protocol which is now widely used. This protocol involves a two-step process. First, the protoplasts are introduced into a small fusion chamber (Fig. 12.3) containing parallel wires or plates which serve as electrodes. Second, a low-voltage and rapidly oscillating AC field is applied, which causes protoplasts to become aligned into chains of cells (pearl chains) between the electrodes (Fig. 12.4a, b). This creates complete cell-to-cell contact within a few minutes. Once alignment is complete, the fusion is induced by application of a brief spell of high-voltage DC pulses (0.125-1 kV cm^{-1}). A high voltage DC pulse induces a reversible breakdown of the plasma membrane at the site of cell contact, leading to fusion and consequent membrane reorganisation. The entire process, starting from the introduction of the protoplasts inside the chamber and their transfer to culture media, can be completed in 5 min or less.

Heterokaryons produced by electrical fusion divide in the culture medium and have demonstrated the capability of regenerating somatic hybrid plants in some cases. Shoot

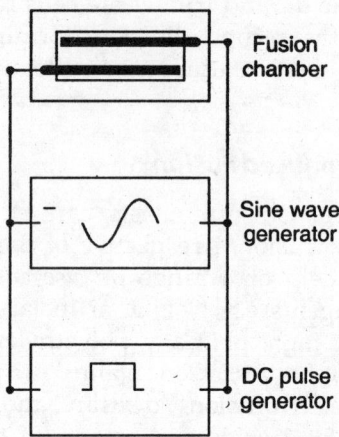

Fig. 12.3 Schematic representation of electrofusion equipment. A fusion chamber containing two parallel electrodes is connected to a high-frequency oscillator (sine-wave or AC-field generator) and a DC-pulse generator.

Note: The equipment also includes an electrical gate to disconnect the AC field from the circuit during delivery of DC pulse (after Bates 1989).

or somatic hybrid plant regeneration after electrofusion of protoplasts has been reported in combinations: *Nicotiana tabacum* (+) *N. tabacum*, *N. plumbaginifolia* (+) *N. tabacum*, *N. glauca* (+) *N. langsdorffii*, and *Solanum tuberosum* (+) *S. phureja*. In addition to these, callus regeneration has been achieved from protoplast combinations of *Brassica napus* (+) *B. napus* and *Solatium brevidens* (+) *N. rustica* (*see* detailed references in review by Bates 1989, Motomura et al. 1995).

12.5 Selection of Somatic Hybrids and Cybrids

12.5.1 Somatic Hybrids

Generally, 20-25% protoplasts may be involved in a fusion event although heterokaryon formation as high as 50-100% has been reported. Thus there is a basic need for selection of the hybrid cells or fusion products since after fusion treatments the protoplast populations consist of a heterogenous mixture of unfused parental types, homokaryons, and heterokaryons. The various selection procedures employed are:

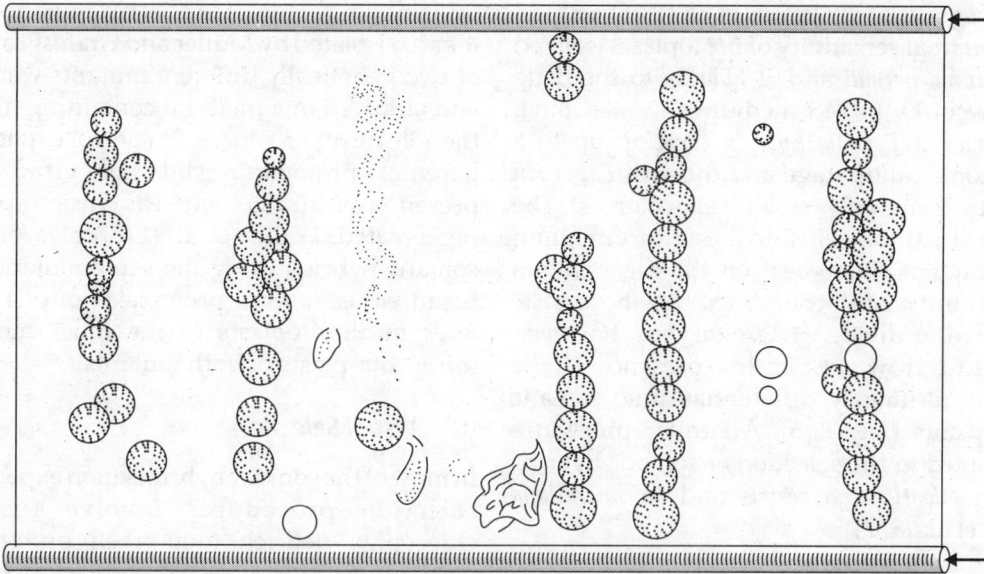

Fig. 12.4a Mesophyll protoplasts of tobacco aligned into pearl chains under the influence of an AC field (100 V/cm, 0.6 MHz). Arrow shows the electrodes at the top and bottom of the field of view (adopted from Bates 1989).

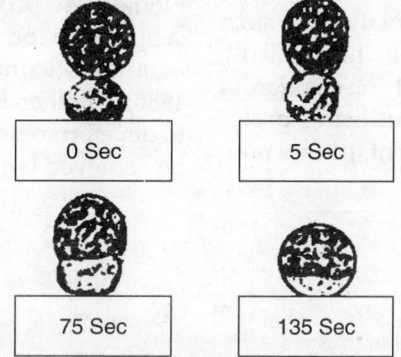

0 Sec	5 Sec
75 Sec	135 Sec

Fig. 12.4b Fusion of dielectrophoretically aligned oat mesophyll protoplasts following application of a single DC pluse (800 V/cm, 15 µS duration) (adopted from Bates 1989).

(A) Biochemical Basis for Complementation and Selection

Heterokaryons in fusions involving mesophyll protoplasts from the two parental types cannot be identified and biochemical markers are required allowing only their growth in cultures to form somatic hybrid plants. Carlson et al. (1972) demonstrated the value of a biochemically based selection procedure for somatic hybridisation of *Nicotiana* species. This selection procedure was based upon a prior knowledge of the nutritional requirements of mesophyll protoplasts isolated from the genetically tumorous *Nicotiana glauca* and *N. langsdorffii*. Protoplasts of the hybrid were able to grow in culture to form calli, whereas parental types failed to develop into calli. A truly useful selection system, however, would be one which does not rely upon prior knowledge of the hybrid plants. Other parameters of biochemical complementation in somatic hybrids need to be applied.

(i) *Drug sensitivity:* Power et al. (1976) utilised the differential sensitivity of protoplasts isolated from *Petunia parodii* and *P. hybrida* to the drug actinomycin D. In MS medium the mesophyll protoplasts of *Petunia hybrida* develop up to a macroscopic callus stage and those of *P. parodii* divide to form only small cell colonies. The addition of actinomycin D to the culture medium apparently has little effect on the regeneration potential of *parodii* protoplasts, but those of *P. hybrida* fail to divide. Heterokaryons, however, are able to grow despite the presence of the drug and ultimately differentiate into somatic hybrid plants (Fig. 12.5). A similar procedure was adopted in the selection of somatic hybrids between *Nicotiana sylvestris* and *N. knightiana* (Maliga et al. 1977).

(ii) *Auxotrophic mutants:* The selection of somatic hybrids as a result of complementation by auxotrophic mutants may be useful as only the hybrid lines are expected to survive in the minimal medium. Although isolation of such mutants of higher plants is somewhat difficult, Glimelius et al. (1978) succeeded in selection of numerous somatic hybrids by utilising proto-plasts of nitrate reductase-deficient (nitrate non-utilising) and chlorate-resistant mutant lines of tobacco isolated by Müller and Grafe. Protoplasts of two genetically different mutants were fused and cultured in a medium containing nitrate as the sole nitrogen source. In control experiments, parental protoplasts did not grow in the presence of nitrate whereas fusion products regenerated. Wallin et al. (1979) also produced somatic hybrids using the same mutants. They fused either normal protoplasts of one mutant with miniprotoplasts of the other mutant or miniproto-plasts of both mutants.

(B) *Visual Selection*

In most of the somatic hybridisation experiments, selection procedures involve fusion of chlorophyll-deficient (non-green) protoplasts of one parent with the green protoplasts of the other parent since this facilitates visual identification of heterokaryons at the light microscope level. Non-green protoplasts are isolated from cell suspension or cultured cells, epidermal cells, or antibiotic-induced albino plantlets (Razdan 1980). Further selection of these heterokaryons to develop somatic hybrid plants in cultures may be achieved by:

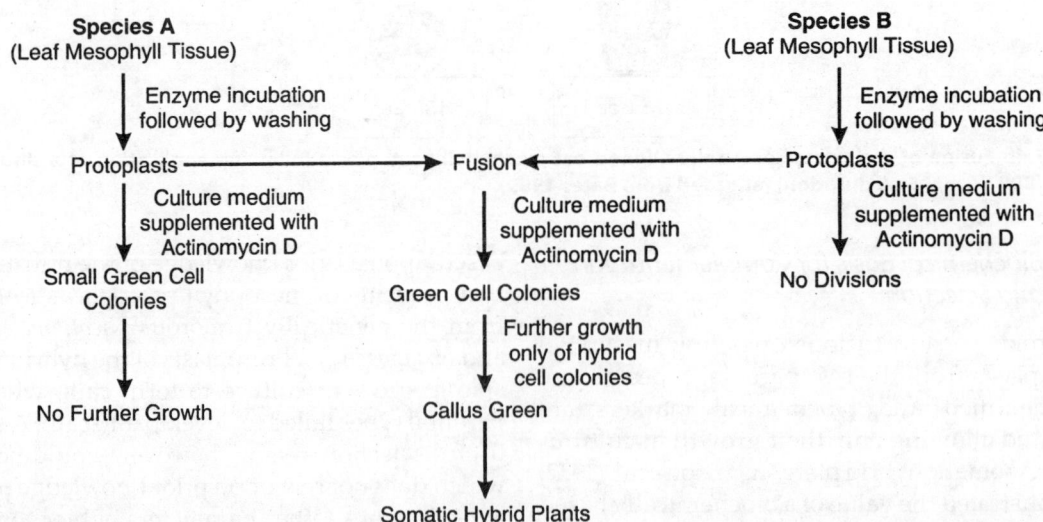

Fig. 12.5 Scheme illustrating the selection of somatic hybrids utilising the differential sensitivity of the mesophyll protoplasts to drug actinomycin D (example: interspecific somatic hybrid between *Petunia parodii* and *P. hybrida*).

(i) *Complementation selection coupled with differential media growth*: Visual selection procedure is coupled with complementary natural differences in the sensitivity of parental protoplasts to media constituents which enable only the hybrid cells to develop in cultures and regenerate plants. For example, wild type (mesophyll) protoplasts of *Petunia parodii* were fused with albino protoplasts isolated from cell suspension cultures of *P. hybrida, P. inflata* and *P. parviflora* in separate experiments performed by Cocking and co-workers at University of Nottingham, UK. In all these combinations green *parodii* protoplasts got eliminated at the small colony stage, while the protoplasts of the other parent developed colourless colonies (Fig. 12.6a). Hybrid components, contrarily, proliferated into green calli and, subsequently, somatic hybrid plants. Similar procedures were followed by other scientists in the selection of interspecific somatic hybrids in *Daucus, Datura* and other genera.

In experiments on intergeneric somatic hybridisation, however, Krumbiegel and Schieder (1979) used the scheme in which the parental protoplasts and heterokaryons were allowed to develop calli in cultures. The morphological differences in the resultant three types of calli permitted the identification of the hybrid tissue, which could then be selected out to regenerate somatic hybrid plants (Fig. 12.6b).

(ii) *Mechanical isolation:* Individual heterokaryons can be identified visually under a light microscope, isolated mechanically by means of a Drummond pipette and cloned in microdrop cultures (*see* Section 12.2.3). Gleba and Hoffmann (1979) used this technique for producing somatic hybrid plants: *Arabidopsis thaliana* (+) *Brassica campestris*. This approach suffers from the fact that it requires special culture media for each particular hybrid-cell type to divide and form clusters. Hence an alternative has been suggested/namely microdrop culture of single cells using feeder layers or nursing of heterokaryons by co-culture with phenotypically different protoplasts.

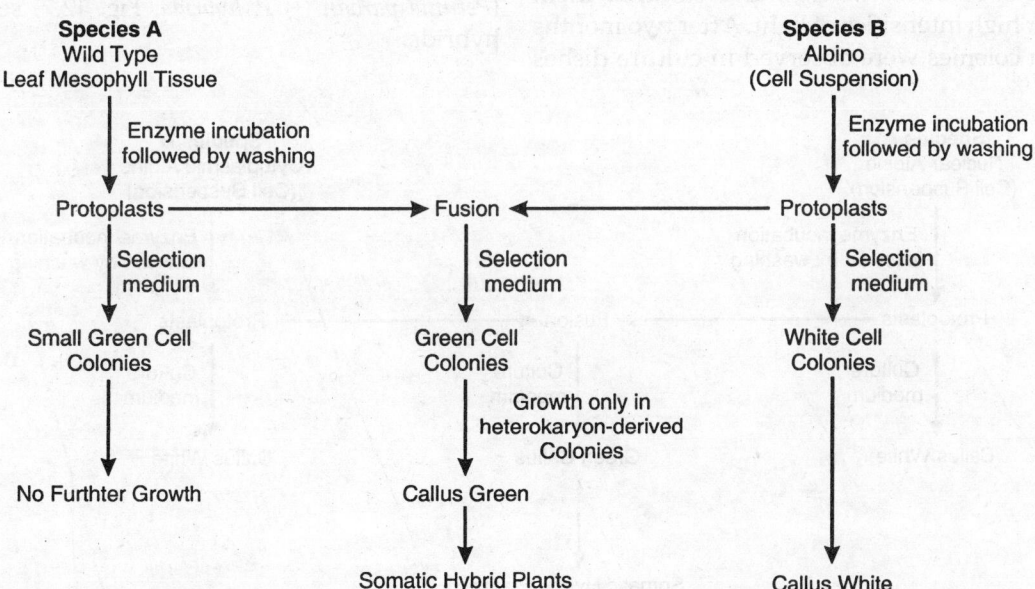

Fig. 12.6a Visual selection procedure coupled with differential growth of parental protoplasts to media constituents enabling only hybrid callus to regenerate plants (examples: *see* text).

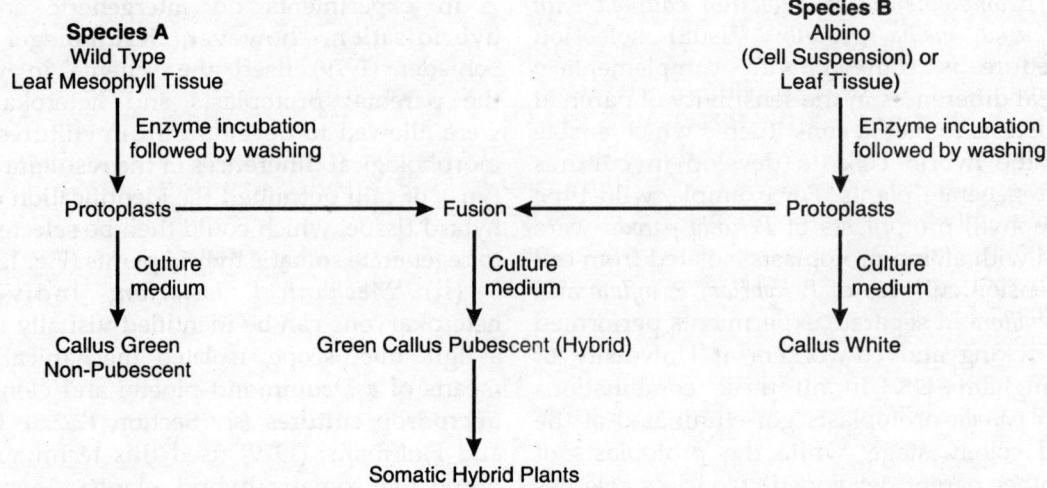

Fig. 12.6b Selection scheme used in the intergeneric somatic hybridisation of *Atropa belladonna* (wild type) with *Datura innoxia* (albino).

(C) *Use of Non-Allelic Albino Mutants for Complementation Selection*

This selection system was developed by Melchers and Labib in 1974. They fused haploid chlorophyll-deficient and light-sensitive protoplasts of *Nicotiana tabacum* and cultured them under high intensities of light. After two months green colonies were observed in culture dishes

as a consequence of complementation between the two albino mutants. On further culturing, these green colonies regenerated somatic hybrid plants. Non-allelic albino mutants have been used successfully to produce intraspecific (*Datura innoxia* (+) *D. innoxia*) and interspecific (*Petunia parodii* (+) *P. hybrida*, Fig. 12.7) somatic hybrids.

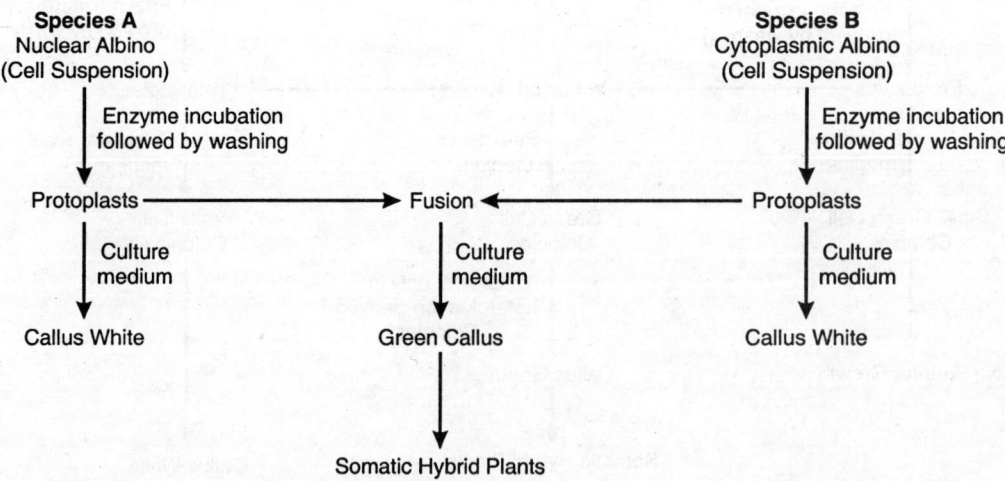

Fig. 12.7 Use of non-allelic albino mutants for complementation and selection of somatic hybrids (example: *Petunia parodii* (+) *P. hybrida*).

(D) Flow Cytometric Analysis

Various laboratories used techniques of flow cytometry and fluorescent-activated cell sorting for the analysis of plant protoplasts whilst maintaining their viability. These techniques were applied for the sorting and selective enrichment of heterokaryons. The hybrid calli derived from this sorted material are reported to regenerate hybrid plants in *Nicotiana* (*see* Afonso et al. 1985, Glimelius 1987). The procedures established for screening of somatic hybrid plants through fluorescent-activated sorting of fused protoplasts have been comprehensively described by Galbraith (1989).

12.5.2 Cybrids

Normally, cybrids are formed during fusion of protoplasts from two phylogenetically distant species. In such a situation there is disharmony between nuclear and cytoplasmic genomes within the hybrid cell. Regenerants from such a cell will have plastomes from both parental species but the functional genome of only one species due to chromosome elimination. The plants derived from such regenerants will, therefore, genetically be hybrids only for cytoplasmic traits. Procedures followed for cybrid selection include:

(A) Inactivation of Donor Cells by Irradiation

To ensure an effective cytoplasmic gene transfer, the cells of cytoplasm-donor species are irradiated. Irradiation with X-rays or gamma rays, in doses of 50 to 300 Gy, is effective in partial or complete inactivation of donor cells. This treatment completely arrests division of the non-fused cells and serves as a selection factor for screening cybrids. Gamma fusion has been successfully employed in bringing about interspecific and, occasionally, intergeneric combinations both at the nuclear and organelle level (Negrutiu et al. 1989). These nucleo-cytoplasmic interactions can occur even in such products of gamma fusion as are derived from protoplasts belonging to sexually incompatible species. The division in these fusion products, and subsequent cells, will give rise to a cell colony in which non-directional elimination of chromosomes from the donor partner occurs. The extent of elimination varies, however, depending on the donor genome and defined culture conditions. Plants arising from such fusions prove to be asymmetric hybrids with introgression of a limited amount of genetic information with novel chloroplast/mitochondrial segregations as well as alloplasmic substitutions (cytoplasm of one parent substituted by another) (Fig. 12.8).

(B) Fusion of Cytoplasts with Protoplasts

An alternative method for effecting the transfer of cytoplasmic traits is by fusion of cytoplasts with protoplasts. Maliga et al. (1982) demonstrated that streptomycin resistance encoded by chloroplasts could be transferred via cytoplasts. Similarly, Tan (1987) claimed to have obtained cybrids by fusion of cytoplasts from *Petunia hybrida* and protoplasts of *Lycopersicon peruvianum*.

12.6 Assessment of Somatic Hybrid and Cybrid Nature of Plants

The nature of somatic hybrids and cybrids is assessed on the basis of morphological and karyological features of the plants derived from protoplast fusion. These observations are further confirmed by an analysis of the electrophoretic pattern of isoenzymes of hybrids and the two parents. The subunit polypeptide pattern of ribulose-1,5-biphosphate carboxylase (RUBP case) also confirms the presumed plant as somatic hybrid or cybrid. The fraction-1 protein has two subunit polypeptides: the small subunit polypeptides are coded by nuclear DNA and the large subunit by chloroplast DNA (cp DNA). These subunits serve as markers in assessing the nature of hybridity. ECO R1 restriction nuclease fragmentation of cp DNA, followed by the separation of the fragments on gel electrophoresis, and tentoxin sensitivity of the chloroplasts, have proved useful in

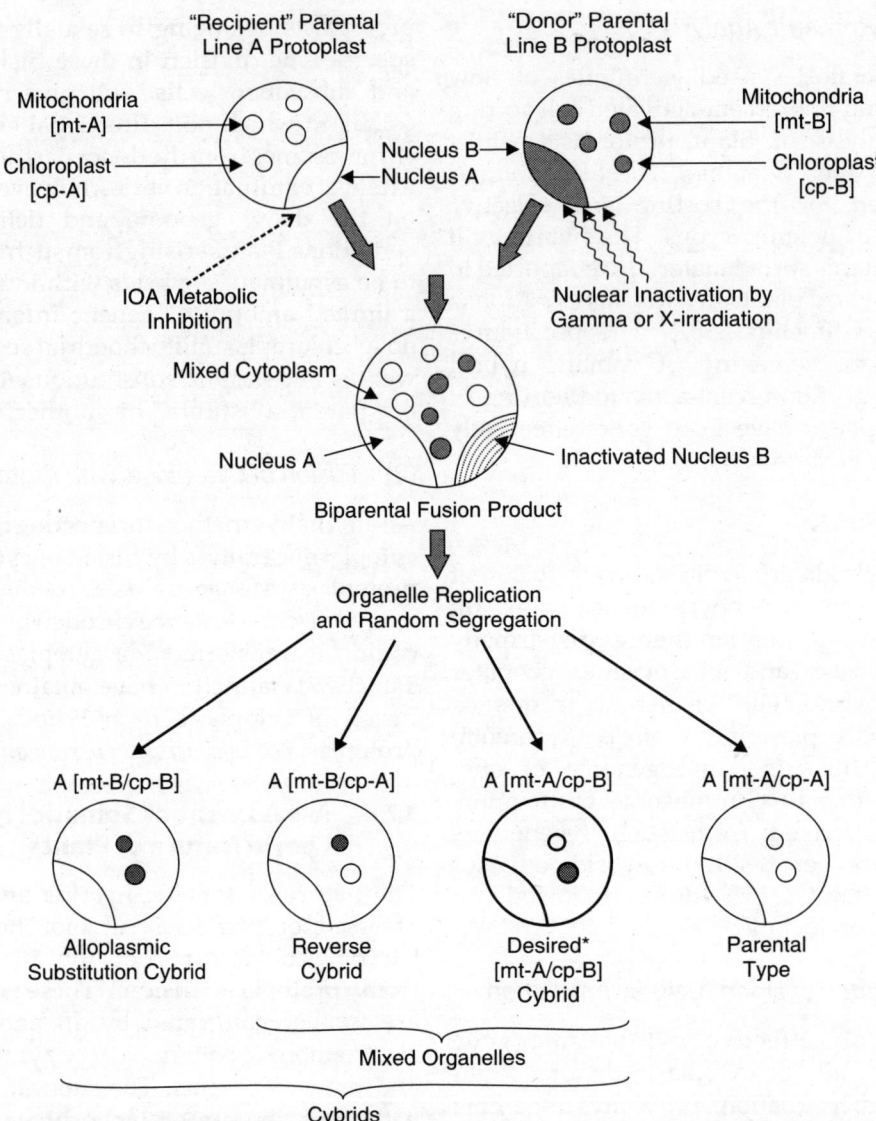

Fig. 12.8 Protoplast fusion and possible organelle segregation (after Yarrow 1999).

characterisation of the chloroplast genome in plants obtained from protoplast fusion.

Molecular techniques such as Restriction Fragment Length Polymorphism (RFLP) and Random Amplified Polymorphic DNA (RAPD) also serve to characterise somatic hybrids and cybrids. Southern blot analysis using species-specific repetitive DNA (rDNA) radioactive and non-radioactive probes can be used to analyse nuclear genomes in somatic hybrids (*see* Pehu 1991, Xu et al. 1993) or cybrids. Molecular markers in plant genotyping appears to have both advantages and disadvantages, however, some of the molecular markers used in recent past are still indispensable which include SSRs (single sequence repeats), SNPs (single nucleotide polymorphism), microarray-based markers-SFPs (single feature polymorphism), DArt (diversity array technology) and NGS (next generation sequencing) which are considered high throughput markers (*see* for details Mir and Varshney 2013).

12.7 Practical Applications of Somatic Hybridisation and Cybridisation

12.7.1 Means of Genetic Recombination in Asexual or Sterile Plants

Somatic cell fusion appears to be the only approach through which two different parental genomes can be recombined among plants that cannot reproduce sexually. Further protoplasts of sexually sterile (haploid, triploid and aneuploid) plants can be fused to produce fertile diploids and polyploids. There are several reports describing the amphidiploid and hexaploid plants produced from fusion of haploid protoplasts of tobacco. Protoplasts isolated from dihaploid potato clones have been fused with isolated protoplasts of *Solanum brevidens* to produce hybrids of practical breeding value (Fish et al. 1988). Haploid protoplasts from an anther-derived callus of rice cultivars, upon fusion also produce fertile diploid and triploid hybrids (Toryama and Hinata 1988).

12.7.2 Overcoming Barriers of Sexual Incompatibility

In plant breeding programmes, sexual crossings at interspecific or intergeneric levels often fail to produce hybrids due to incompatibility barriers. The bottlenecks in sexual hybridisation may, therefore, be overcome by somatic cell fusion. In some cases somatic hybrids between two incompatible plants have also found application in industry or agriculture.

Schieder (1978) obtained amphidiploid *Datura innoxia* (+) *D. discolor*, and *D. innoxia* (+) *D. stramonium*, by fusing their diploid mesophyll protoplasts. These hybrids did not exist in nature as conventional breeding procedures

proved unsuccessful. Somatically produced amphidiploids of these combinations *of Datura* species are propagated for industrial uses as they demonstrate heterosis and higher (20-25%) scopolamine content than in the parental forms.

Nicotiana repanda, N. nesophila and *N. stockonii* are resistant to a number of diseases but are not sexually crossable with tobacco (*N. tabacum*). However, fertile hybrids have been reported in combination *N. tabacum* (+) *N. nesophila* and *N. tabacum* (+) *N. stocktonii* by protoplast fusion (Appendix 12.2). Somatic hybridisation of dihaploid and tetraploid potato protoplasts with isolated protoplasts *of Solanum brevidens, S. phureja* and *S. pennellii* resulted in the synthesis of fertile, partially amphieuploid plants possessing important agricultural traits, e.g., resistance to potato leaf virus, potato virus Y and Erwinia soft rot (*see* Collins and Edwards 1998). Using this approach, tomato (*Lycopersicon esculentum*) hybridised somatically with a number of wild species has resulted in the synthesis of hybrids which are fertile and used in breeding programmes. Interspecific somatic hybridisation involving species that are sexually incompatible with eggplant (*Solanum melongena*) has also resulted in the production of amphidiploids with traits resistant to verticillium wilt (Guri and Sink 1988).

Rapeseed (*Brassica napus*) is a natural amphidiploid of *B. oleracea* and *B. campestris*. Schenk (1982) was the first to resynthesise rapeseed *in vitro* using protoplast fusion. Somatic hybridisation between *B. napus* and *B. nigra* cultivar, possessing the gene for resistance to *Phoma lingam*, yielded amphidiploid plants carrying this gene. These hybrids possess all the three *Brassica* genomes (A, B and C) and are now incorporated in breeding programmes (Sjodin and Glimelius 1989a,b). Somatic hybrids have been produced parasexually by protoplast fusion between *Brassica juncea* (a major oilseed crop of the tropical world) and the sexually incompatible species *Diplotaxis muralis* (Chatterjee et al. 1988) and *Erica sativa* (Sikdar et al. 1990). Fertile somatic hybrids by protoplast fusion *Gossypium hirsutum* + *G. klotzschianum*, *G. hirsutum* + *G. bickii*, *G. hirsutum* + *G. stockii* and *G. hirsutum* (4x) + *G. davidsonii* (2x), *G. hirsutum* L. + *G. triloba*, facilitate gene transfer from its wild relatives (Sun et al. 2006, Yu et al. 2012).

The potential of somatic hybridisation in perennial tree breeding is best illustrated by interspecific and intergeneric somatic hybridisation among citrus species. Somatic hybrids produced through these experiments are amphidiploids featuring characteristics for scion improvement and increased rootstock potential. Other combinations resulting in the synthesis of somatic hybrids and cybrids are summarised in Appendix 12.2. In addition to examples in this Appendix interspecific somatic hybrids have been produced in more genera (e.g. *Helianthus*, *Lupinus*, etc.) although much emphasis is now on transformation of single protoplasts (Chapter 13).

12.7.3 *Cytoplasm Transfer*

Power et al. (1975) fused mesophyll protoplasts of *Petunia* with cultured cell protoplasts of the crown gall of *Parthenocissus* and selected a line which contained the chromosomes of only *Parthenocissus* but exhibited some of the cytoplasmic properties of *Petunia* for sometime. This was followed by direct application of cybridisation in agricultural biotechnology by transfer of cytoplasmic male sterility from *Nicotiana techne* to *N. tabacum* (Belliard et al. 1978), *N. tabacum* to *N. sylvestris* (Zelcer et al. 1978, Aviv et al. 1980) and *Petunia hybrida* to *P. axillaris* (Izhar and Power 1979). Besides cytoplasmic male sterility, the genophore of the cytoplasm codes for a number of practically important traits, such as the rate of photosynthesis, low or high temperature tolerance and resistance to diseases or herbicides. Recent experiments on cybridisation have resulted in plants with reconstructed cytoplasm combining mitochondrial DNA (mt DNA) and cp DNA encoded traits from both parents.

The best example illustrating the potential for protoplast fusion in reconstructing cytoplasm for practical purposes is the genus *Brassica*. Two desirable traits coded by cytoplasmic genes have been genetically manipulated through interspecific cybridisation between different species of *Brassica*. These traits include cytoplasmic male sterility (cms) and resistance to atrazine herbicides. The cms gene in *Brassica* plants, *Diplotaxis muralis* and *Raphanus sativus* is of *alloplasmic* (the nucleus of one species into a foreign cytoplasm) origin. *Raphanus sativus* is of interest because it leads to complete male sterility. Cms restorer genes have been introduced into rapeseed (*Brassica napus*) from this plant. Mutants resistant to atrazine herbicide have also been discovered both in *Brassica napus* and *B. campestris*. Protoplast fusion experiments (conducted in various laboratories) have resulted in the synthesis of cybrid plants with reconstructed cytoplasm combining both cms (coded by *Raphanus* mt DNA) and low temperature tolerance or atrazine resistance (coded by *Brassica* cp DNA). Similarly, cytoplasmic genes coding for atrazine resistance and cms have been transferred into cabbage, rice and potato (*see* Bajaj 1989b, Gleba and Shlumukov 1990). Cybrid technology has been used to transfer antibiotic resistance in potato and in studies on organellic interactions in citrus, tomato, tobacco and petunia (*see* Yarrow 1999).

12.8 Present Perspective

In somatic hybridisation and cybridisation, the essential prerequisite is that parental protoplasts and their fusion products regenerate to whole plants. Research in the past decade has shown that plants can be raised *in vitro* from isolated protoplasts of species belonging to a range of angiosperm families (Table 12.6). Somatic hybrids have been produced between sexually compatible as well as incompatible species. It could be possible to overcome prezygotic embryo/ endosperm (*Petunia parodii* + *P. inflata*, Power et al. 1979) and postzygotic (*Datura innoxia* + *D. stramonium*, Schieder 1978; *Petunia parodii* + *P. parviflora*, Power et al. 1980) incompatibility barriers by protoplast fusion. Experiments on intergeneric somatic hybridisation have also been successful in some cases such as potato + tomato somatic hybrids (Melchers et al. 1978) and synthesis of 'Arabidobrassica' (Gleba and Hoffmann 1979).

Symmetric somatic hybridisation results in fertile interspecific or intergeneric somatic hybrids as evident by resynthesis of *Brassica napus* through fusion of *B. oleracea* and *B. campestris* protoplasts (Sunderberg et al. 1987). Intergeneric somatic hybrids *B. napus* + *Arabidopsis thaliana* (Forsberg et al. 1994), and *B. napus* + *Thlaspi perfoliatum* (Fahleson et al. 1994), are also symmetric and fertile. Somatic hybrids that comprise amphiploid chromosome numbers are symmetric, whereas in most cases there is selective elimination of one of the parental chromosomes during the process of regeneration and morphogenesis in fusion product of protoplasts. The latter result either in the formation of asymmetric somatic hybrids or cybrids. The nuclear chromosomes of *Solanum phureja* were eliminated preferentially in somatic hybrids *Solanum tuberosum* + *S. phureja* (Pijnacker et al. 1987) and these hybrids were phenotypically closer to the other parent (*S. tuberosum*) and therefore asymmetric. The significance of asymmetric somatic hybridisation in cytoplasmic gene transfer nevertheless seems to have potential for improvement of agro-horticultural traits (Section 12.7.3). With successes and subsequent advancements (Appendix 12.2) in protoplast technology it is desirable that efforts be concentrated on important plant species which have potential in industry, food production or for disease resistance. Crops which have not yielded satisfactory results through conventional methods of genetic manipulation need to be aided by non-conventional *in vitro* techniques such as somatic hybridisation/cybridisation, embryo culture, etc. to manifest their full potential. This calls for a 'broad spectrum' approach for the genetic improvement of crops. Even somatic hybrids of sexually compatible plants may exhibit new variations as a result of interactions between plastomes donated by parental species during protoplast fusion. The technique of cybridisation, besides transfer of male sterility, can be adopted for the introduction of genes for resistance into the new species. The modification of plants with respect to nitrogen fixation has also been contemplated though with little success through transformation of protoplasts by uptake of exogenous DNA, or organelles, carrying this trait. Further, genetically heterogenous clones can be derived from protoplast culture and fusion which display a high frequency of variations for several agronomic traits. These aspects have been described in subsequent chapters.

The above developments suggest possibility of an immense potential for somatic cell genetics in crop improvement. However, the genetic diversity that can be generated via somatic cell fusion is still poorly understood. This is because only a very limited number of the synthesised somatic hybrids or cybrids have been fertile or amphiploids. Induction and control over the degree of species-specific chromosome elimination in wide or distant somatic hybridisation requires to be mastered in order to understand the mechanism of producing desirable asymmetic nuclear hybrids (*see* Gleba and Shlumukov 1990, Pelletier and Romagosa 1993). Gametosomatic hybridization involving fusion between male gamete (microspore tetrad or young-stage pollen) or female gamete (egg cell) protoplast and isolated single cell somatic protoplasts could facilitate combination of both nuclear and organelle genomic combination in hybrid protoplasts of cucumis, petunia and other plants resulting in cybrids or somatic hybrids (Sangthong et al. 2009, Skálová et al. 2012).

APPENDIX 12.1

Procedure for Isolation of Protoplasts

From leaf cells by simultaneous method

1. Select fully expanded leaves from 7- to 8-week-old plants growing in a greenhouse or growth cabinet with controlled conditions of light, temperature and humidity.
2. Surface-sterilise the leaves by first immersing in 70% ethanol for 30 s followed by rinsing in 0.4-0.5% sodium hypochlorite solution for about 30 min.
3. Peel the lower epidermis with fine forceps and cut out the peeled areas with a fine scalpel.
4. Place the peeled leaf pieces on a thin layer of 600 mmol l^{-1} mannitol-CPW[a] solution (autoclaved) in such a way that the peeled surface is in contact with the solution.
5. After about 30 min replace the mannitol-CPW solution by a filter-sterilised solution of enzyme containing 4% cellulase, 0.4% macerozyme, 600 mmol l^{-1}[*] mannitol and CPW salts.
6. Incubate the petri plate with enzyme solution and leaf pieces in darkness at 24-26°C for 16-18 hr.[*]
7. Gently squeeze the leaf pieces with a Pasteur pipette to liberate the protoplasts. Remove the large debris by passing through a 60-80 μm mesh sieve.
8. Transfer the sieved protoplast solution to a screw-cap centrifuge tube and spin at 100 g for 10 min.
9. Remove the supernatant and resuspend the sediment in 860 mmol l^{-1} sucrose (prepared in CPW solution) in a screw-cap centrifuge tube and spin at 100 g for 10 min.
10. Green protoplasts form a band at the top of the sucrose solution; transfer them with a pipette to another tube. Add the protoplast culture medium to the protoplasts and centrifuge at 100 g for 3 min. Repeat such washing three times.
11. After the final wash add enough culture medium to achieve a protoplast density of 0.5 x 10^5 to 1x10^5 ml^{-1}.
12. Plate the protoplasts following any of the procedures described in the text.

From mesophyll cells of cereals[b]

1. Select leaves from 5-6-day-old seedlings by cutting at the base of leaf. Discard the apical (0.5 cm) region.
2. Surface-sterilise the leaves in 0.1% Zephiran-10% ethanol for 5 min.
3. Wash twice in a solution containing 600 mmol l^{-1} sorbitol and 10 mmol l^{-1} CaCl$_2$.
4. Cut leaves into transverse (1-2 mm wide) strips and transfer them to conical flasks containing enzyme solution (0.5% macerozyme, 1% hemicellulase, 2% cellulysin, 600 mmol l^{-1} sorbitol, pH 5.4). The proportion of enzyme solution for each gram of leaf tissue is 10:1.
5. Infiltrate the leaves under partial vacuum for 3-5 min.
6. After 2 hr filter the leaf digest first through a 0.7 mm mesh nylon sieve and then through a 0.05-mm sieve.
7. Transfer the material that has passed through the sieves to centrifuge tubes and spin at 50 g for 90s.
8. Remove the supernatant and wash the pellet thrice with washing medium.
9. Suspend the protoplasts in the nutrient medium and plate them in the culture medium.
10. Seal the culture plates with parafilm. Place them in large glass plates lined with filter paper moistened with 0.001% CuSO$_4$. Finally store in darkness at 23°C.

From root nodules (example: Trifolium sp.)[c]

1. Wash the root nodules (1-5 mm in length) excised from aseptically grown plants inoculated with *Rhizobium trifolii* in protoplast dilution buffer (PDB).[d]
2. Cut the nodules into four pieces and again wash in PDB.
3. Incubate the nodule pieces in the enzyme solution (4% cellulysin, 2% macerase, 1% driselase in PDB, pH 5.8) in darkness at 23°C.
4. After 3-4 hr dissociate the partially digested nodules by passing through the orifice of a Pasteur pipette and continue incubation for a further 90 min.
5. Sieve through a 50 urn nylon mesh and wash the digested tissue twice in PDB by spinning at 200 g for 10 min.
6. Suspend the washed protoplasts in 30% sucrose solution and spin at 100 g for 10 min.
7. Collect the intact protoplasts from the top of the sucrose pad and plate them in a nutrient medium following usual procedures.

From suspension cultures

1. Transfer the high-density cell suspension to the centrifuge tube and spin at 100 g for 10 min.
2. Remove the supernatant, transfer the cells in the culture dish containing the enzyme solution (2% Rhozyme, 4% Meicelase, 0.3% Macerozyme, 13% Mannitol, CPW salts, pH 5.6) onto a platform shaker for 6 hr to overnight.[*]
3. Transfer the enzyme digested cells to a screw-cap centrifuge tube and spin at 100 g for 10 min.
4. Follow steps 10-12 given for isolation of leaf protoplasts. Protoplasts isolated from cell suspension cultures are chlorophyll deficient.

---

[a] Cell-protoplast washing (CPW) medium contains (mg l^{-1}): KH_2PO_4 (27.2), KNO_3 (101), $CaCl_2 \cdot 2H_2O$ (1480), $MgSO_4 \cdot 7H_2O$ (246), KI (0.16), $CuSO_4 \cdot 5H_2O$ (0.025), pH 5.8 (Cocking and Peberdy 1974).
[b] *See* Scott et al. (1978).
[c] *See* Gresshoff et al. (1977).
[d] PDB buffer comprises: 250 mmol l^{-1} sorbitol, 250 mmol l^{-1} mannitol, 10 mmol l^{-1} K_2HPO_4, 2 mmol l^{-1} $CaCl_2$, pH 5.8.
[*] Enzyme incubation period depends on the concentration and type of enzymes used as well as on the nature of source material.

APPENDIX 12.2

Somatic Hybrid and Cybrid Plants Obtained through Protoplast Fusion[a]*

Interspecific Hybridisation[b]

Brassica napus + B. campestris
B. napus + B. carinata
B. napus + B. nigra
B. napus + B. oleracea
B. oleracea + B. campestris
B. spinescens + B. juncea
Citrus sinensis + C. lima
C. sinensis + C. pardisi
C. sinensis + C. unshiu
Datura innoxia + D. candida[c]
D. innoxia + D. discolor[c]
D. innoxia + D. sanguinea[c]
D. innoxia + D. stramonium[c]
Daucus carota + D. capillifolius
D. capillifolius + D. carota
Dianthus chinensis + D. barbatus
D. chinensis + D. caryophyllus
Helianthus annuus + H. giganteus
Lotus corniculatus + L. coimbrensis
Lycopersicon peruvianum + L. esculentum
L. peruvianum + L. pennellii
Medicago sativa + M. falcata.
Nicotiana glauca + N. langsdorffii
N. debneyi + N. tabacum
N. gossei + N. plumbaginifolia
N. rustica + N. sylvestris
N. sylvestris + N. knightiana[c]
N. tabacum + N. alata
N. tabacum + N. glauca
N. tabacum + N. glutinosa
N. tabacum + N. knightiana
N. tabacum + N. nesophila[c]
N. tabacum + N. otophora
N. tabacum + N. rustica[c]

N. tabacum + N. stocktonii[c]
N. undulata + N. bigelovii
Oryza sativa + O. brachyantha
O. sativa + O. echingeri
O. sativa + O. officinalis
O. sativa + O. perrieri
Petunia hybrida + P. parodii
P. hybrida + P. inflata
P. parodii + P. hybrida
P. parodii + P. inflata
P. parodii + P. parviflora[c]
Solanum melongena + S. ethiopicum
S. melongena + S. integrifolium
S. melongena + S. khasianum
S. melongena + S. nigrum
S. melongena + S. sisymbrifolium
S. melongena + S. torvum
S. tuberosum + S. brevidens
S. tuberosum + S. chacoense
S. tuberosum + S. circaeifolium
S. tuberosum + S. pennellii
S. tuberosum + S. phureja

Intergeneric Hybridisation[b]

Arabidopsis thaliana + Brassica campestris
Barbarea vulgaris + B. napus
Brassica campestris + Barbarea vulgaris
B. carinata + Camelina sativa
B. juncea + Diplotaxis muralis
B. juncea + Erica sativa[c]
B. juncea + Trachystoma ballii
Brassica oleracea + Moricandia arvensis
B. napus + Erica sativa
B. napus + Thlaspi perfoliatum
Citrus aurantifolia + Feronella lucida
C. aurantifolia + Swinglea glutinosa

Contd...

C. reticulata + Citropus gabunensis

C. sinensis + Atlantia ceylanica

C. sinensis + Murraya paniculata

C. sinensis + Poncirus trifoliata

C. sinensis + Severinia disticha

Datura innoxia + Atropa belladonna[c]

Daucus carota + Aegipodium podagaria

D. carota + D. capillifolius

D. carota + Petroselinium hortense

Diplotaxis hara + B. napus

D. ibicensis + B. napus

D. muralis + B. napus

D. muralis + B. juncea[c]

Lycopersicon esculentum + Solanum rickii

L. esculentum + Solanum lycopersicoides

Moricanda arvensis + Brassica oleracea

Nicotiana tabacum + Lycopersicon sp.

Oryza sativa + Echinochloa oryzicola

Populus (hybrid) + Hibiscus sabdariffa

Sinapis turgida + Brassica oleracea

S. turgida + B. nigra

Solanum lycopersicoides + Lycopersicon esculentum

S. tuberosum + Lycopersicon esculentum[c]

S. tuberosum + L. pimpinellifolium

---

* Besides these combinations, several hybrid cell lines are reported in which plant regeneration has not been achieved.

[a] *See* Bhojwani and Razdan (1983, 1996), and Bajaj (1989b), for individual references of each combinations.

[b] In some combinations sexual hybrids can be raised.

[c] Known to be sexually incompatible combinations.

Genetic Transformation

One of the most current areas of plant cell and tissue culture developed for exploring potential benefits in agriculture has been the transfer and expression of foreign genes into plant cells. The transformation of bacterial cells by uptake of DNA from another bacterium and subsequent integration of this foreign DNA into the genetic material of the host is well established. However, genetic modification of higher plants by introducing DNA into their cells is a highly complex process. For a cell to be transformed permanently, a foreign nucleotide sequence has to be introduced in such a way that alien genetic information should be able to express in the transformed cells. Availability *of restriction endonucleases* (which cut duplex DNA molecules into discrete fragments), development of the DNA-DNA hybridisation technique and marker genes for selection of transformed cells have enabled incorporation of exogenous DNA into cells of *Monera*, fungi, animals, and higher plants (*see* Lurquin 1989). Subsequent studies have shown that new genetic information added as a result of these processes of recombinant DNA technology in eukaryotic plants expresses not only at the cellular and later, whole organism level, but also can be transmitted in successive generations of transformed individuals. Such transgenic plants, called *GM* (genetically modified) plants, are not known to exist in

nature and could have some advantages over established methods of gene transfer in plants either through sexual process or by somatic hybridisation (*see* reviews by Bajaj 1989b, Cocking and Davey 1987, Schell 1987, Schell and Vasil 1989, Simpson and Herrera-Estrella 1989, Gorden-Kamm et al. 1990, Logemann and Schell 1993, Smith 2000, 2013, Trigiano and Gray 2000, 2010). Another area explored now is *molecular forming*, i.e., the use of crops for the large-scale production of valuable recombinant proteins (pharmaceutical proteins) in transgenics (Fasella and Hussain 2014).

In this chapter the systems and methodology used for genetic transformation of higher plants are briefly summarised. An attempt is made to assess the fate of foreign genes in transgenic plants and their application to agricultural biotechnology.

13.1 Transformation through Uptake of Foreign DNA by Seeds and Seedlings

The pioneering work on genetic modification of plants by introducing isolated DNA was done with whole plant systems such as seeds and seedlings. These studies seemed to reveal that exogenously applied DNA is not only taken up by plants, but also integrate into the host genome. The evidence for integration and replication of the exogenous DNA was based mainly on iso-pycnic centrifugation in cesium chloride (CsCI) gradients. Two major types of experiments, referred to as *integration and replication,* were conducted. In the integration experiment, radioactive labelled DNA of density different from the host DNA was administered to the seedlings. This was followed by isolating DNA from seedlings, or their excised parts, after a period of metabolism for analysis on CsCI gradients. Evidence for integration of the radioactive donor DNA was ascertained by the occurrence of a radioactive band approximately intermediate in density between the host and donor DNA. The intermediate density band was further analysed by sonication and denaturation. Experiments of this type demonstrated the 'integration' *of Micrococcus lysodeikticus* DNA into barley root, *E. coli* and *Streptomyces coelicolor* DNA into *Arabidopsis* and T_4 DNA into *Matthiola incana* seedlings. The replication experiments were also conducted in a similar manner and differed only in respect of the donor DNA, which was non-radioactive. These experiments also showed evidence of bacterial-plant DNA complexes in tomato, barley and *Arabidopsis*.

Several workers have tried transformation experiments using a similar donor DNA with the same host or different plant systems but were unable to reproduce the same results. They concluded that there was neither integration nor replication of the exogenous DNA inside the host cells. In fact, the donor DNA ultimately degraded in the host and the degraded products were used for endogenous DNA synthesis. Regarding the claim that exogenous DNA could also be taken up by seeds was probably accounted for by adsorption of DNA on the cell walls and cell membrane (*see* details and references in the review by Kleinhofs and Behki 1977).

A new thrust, however, on genetic transformation of whole plant systems seems to have developed as a result of some success achieved with cereals. The reported rhizobial infection of non-legumes to form an effective nodular

nitrogen-fixing symbiosis within the dicot plants belonging to the genus *Parasponia* (Trinick 1982) generated such interest as to extend nodular symbiotic association to monocots, especially the cereals. Cocking and co-workers (Al-Mallah et al. 1989) reported induction of nodular structures on the roots of rice seedlings when these roots were treated with a cellulase-pectolyase enzyme mixture followed by inoculation with either *Rhizobium,* or *Bradyrhizobium,* in the presence of PEG. This first report of inducing nodular structure on a rice plant by *Rhizobia in vitro* was thought to open new vistas in the genetic engineering of plants as it may be possible to extend this technique for effective nodulation in respect of other cereals as well as non-legumes. Further investigations are required to determine whether nodular structures on transformed rice are effective in nitrogen fixation. Otherwise, other strains of genetically engineered *Rhizobium,* or its wild type, may be used for this purpose.

13.2 Uptake of DNA by Pollen

In 1976, Hess and co-workers incubated *Nicotiana glauca* pollen with DNA isolated from *Nicotiana langsdorffii* and then used this pollen to pollinate emasculated *N. glauca* flowers. They reported that this experiment led to the production of transformed plants which developed tumours at the wound site on the stem. This is a characteristic feature exhibited by the sexual hybrids arising between these two species. Ohta (1986) mixed pollen from maize plant homozygous for aleurone and various endosperm marker colours with DNA extracted from a strain homozygous dominant for all these traits. They achieved a high pollen transformation efficiency. These observations suggested that pollen may act as vehicles for the transfer of foreign genes into plants.

In another approach, de la Peña et al. (1987) manually injected young floral tillers of rye plant with a solution of plasmid DNA carrying a chimeric gene consisting of T-DNA nopaline synthase gene flanking coding sequence of the Tn 5 *neo* gene (expressing for antibiotic aminoglycoside resistance). Injected flowers were pollinated with normal pollen and progeny seeds plated on kanamycin medium. Of the 3000 seeds, only two seedlings were isolated showing traits of kanamycin resistance as well as the *nos-neo* gene. At the time of injection the male germ cells lacked a surrounding callose wall and hence the authors concluded that the injected DNA was taken up by male gametes. However, this method needs wider applicability as it would bypass all the problems associated with plant cell or protoplast cultures by enabling the direct recovery of transgenic plants from gametes treated *in situ*. Mention may be made in this context about pollen-tube pathway transformation described in section 13.3.4 (H).

13.3 Transformation of Protoplasts

The isolated protoplast system initially appeared most promising for genetic modification of cells. In the absence of a wall around the plasma membrane, protoplasts are not only able to fuse but can also take up chloroplasts, nuclei, microorganisms and isolated foreign DNA. This feature prompted workers to investigate various modes of genetic modifications in cells using the protoplast system.

13.3.1 *Uptake of Organelles*

(I) Chloroplast Transplantation

Several workers reported the uptake of chloroplasts by albino protoplasts isolated from *Petunia*, tobacco and carrot plants during the years 1973-1974. This required a simple treatment—exposing a mixture of albino protoplasts and chloroplasts to PEG in a manner similar to protoplast fusion (*see* Chapter 12). EM studies revealed that PEG caused fusion of the two external membranes of the plastids and while entering the protoplasts both the plastid and protoplast membranes fused, causing disruption of these organelles (Davey et al. 1976). In view of this there was a shift in emphasis from direct uptake to transfer of chloroplasts through the fusion of protoplasts. Using this approach it could be possible to restore photo-autotrophy in albino protoplasts by fusion with wild type (green) protoplasts in a number of species (*see* Giles 1989). Medgyesy et al. (1985) used a gamma-irradiated chloroplast mutant of *Nicotiana* species for interspecific chloroplast transfer by protoplast fusion and in the process produced cybrids, whereas Cseplo et al. (1986) demonstrated that protoplast fusion was suitable to rescue the chloroplast mutants of *N. plumbaginifolia* from undesirable nuclear backgrounds.

The practical value of chloroplast transformation, via protoplast fusion, in plant improvement could be demonstrated by transfer of genes for herbicide resistance between related species. For example, isolation of mutants by cell selection (*see* Chapter 14) in populations of protoplast-derived colonies yielded triazine-resistant *Nicotiana* plants. The resistant trait was observed to arise as a result of the mutation in the chloroplast gene *psb-A* (*see* Maliga 1985). Similarly, inhibitors of bacterial protein synthesis were used to induce chloroplast mutations in protoplast cultures of *N. plumbaginifolia*. In the process it could be possible to isolate mutants resistant to antibiotics and photosynthesis-inhibiting herbicides (Cseplo et al. 1986). The bacterial APR (3) II gene from transposon Tn 5 (which confers resistance to the related amino glycoside antibiotics, neomycin, kanamycin, and G418) and the *cat* gene from transposon Tn 9 (which confers resistance to chloramphenicol, an inhibitor of protein synthesis on 70S ribosomes) are selectable marker or reporter genes (*see* Section 13.4) that have been found suitable for transfer of mutant chloroplast genes (Gray 1989).

Fluhr (1989) described how the interaction between chloroplast genes from weeds to sexually incompatible crop plants results in the control of light-mediated proteins expressed by *rbcs* and *cab* gene families. The transfer of chloroplast genes from weeds to sexually incompatible crop plants is another area that may have potential in future. Thus, studies on chloroplast uptake and exchange have played a useful role in understanding the nuclear-chloroplast interaction, physiology of isolated plastids, and transfer of herbicide resistance.

(II) Mitochondrial Transplantation

There is little evidence regarding the direct uptake of isolated mitochondria by protoplasts. However, there are reports on incorporation of mito-chondrial DNA (mt DNA) through protoplast fusion leading to the formation of cybrids (*see* Chapter 12). Mitochondrial genomes were first analysed in *Nicotiana* cybrids (Belliard et al. 1979) and, subsequently, in other cybrids produced by somatic hybridisation. It may also be possible to transfer traits such as toxin sensitivity of Texas male sterile cytoplasm encoded by mt DNA in maize (Lonsdale 1987).

(III) Nuclear Transplantation

Potrykus and Hoffmann (1973) reported uptake of isolated nuclei by mesophyll protoplasts of *Petunia,* tobacco and maize. Transplantation success improved from 0.5% to 5% in the presence of PEG and Ca^{++}. Since then the methods for isolation and uptake of nuclei by protoplasts have been refined to facilitate expression of the transferred genome in the cells. Saxena et al. (1986, 1987) transplanted *Vicia hajastana* nuclei in protoplasts of an auxotrophic cell line of *Datura innoxia* and isolated prototrophic cell colonies expressing for *Vicia* genomic DNA in cultures. However, these colonies on an appropriate regeneration medium developed shoots typical of wild type *Datura*.

The uptake of isolated chromosomes, instead of complete nuclei, may be another area to explore for transfer of foreign genes into plant cells. Development of the flow cytometry technique enabled the introduction of isolated chromosomes into protoplasts of *Haplopappus* (De Laat and Blaas 1984), *Nicotiana* (Verhoeven et al. 1987) and *Petunia* (Conia et al. 1987). Only metaphase chromosomes were found suitable. Isolated chromosomes can also be introduced into protoplasts by microinjection (De Laat and Blaas 1987). In most of these experiments

transplanted chromosomes were eliminated due to incompatibility.

Transfer of micronuclei into protoplasts has also been attempted. Amiprophos-methyl (APM) treatment of potato and carrot cells promoted accumulation of a large number of metaphases. The scattered single or groups of chromosomes in these cells developed a nuclear membrane and formed micronuclei which could be sorted out by flow cytometry and then used for transplantation (Sree Ramulu et al. 1987). This may open prospects for application of micronuclei for transferring specific intact chromosomes for the purpose of gene mapping.

13.3.2 Uptake of Micro-organisms

Initial attempts to transfer whole bacteria *(Rhizobium)* into protoplasts with a view to utilising the transformed protoplasts for nitrogen fixation yielded little success. Therefore, the emphasis shifted to the use of viruses as vectors for genetic engineering studies. Studies on aspects such as mechanism of infection, nuclear-protein integration and expression of viral DNA in isolated protoplasts should be possible since in nature viruses do infect plants and multiply within their tissues. In early studies transducing phages were used to transfer bacterial genes into plant cell lines or callus tissues. Although bacterial genes appeared to express briefly in the plant cells or callus cultures treated with phages, there was no evidence of "transgenosis" (asexual transfer, expression, and inheritance of genetic information between donor and recipient organisms separated by evolution) in long term cultures (Kleinhofs and Behki 1977).

Experiments on uptake of TMV by isolated protoplasts were initiated in 1969 (Cocking and Pojnar 1969, Takebe and Otsuki 1969). Since then numerous reports have appeared on the incorporation of viral nucleic acids into protoplasts and using them as vectors in genetic transformation. Most plant viruses have encapsulated genomes and it is assumed that no specific virus receptor sites exist on the exterior of the plasma membrane. Viral based transforming agents may be of the integrating or non-integrating type. In selecting the progenitor (virus) for the transforming agent the factors generally considered are: (a) amenability to genetic engineering techniques, (b) packaging constraints in the particle and (c) development of a system in which expression can be studied. The cauliflower mosaic virus (CaMV) is the most amenable and widely chosen progenitor for genetic engineering studies. Moreover, the sDNA genome in CaMV has appropriate restriction sites compatible with the methodology developed for *Agrobacterium tumefaciens* plasmid (Ti plasmid). Brisron et al. (1984) were successful in introducing a chimeral CaMV carrying methotrexate resistance into turnip cells in such a way that the virus replicated normally in the transformed plants. Similarly, barley protoplasts infected with the ssRNA bromovirus (BMV), on having the chloramphenicol acetyltransferase (CAT) bacterial gene inserted into the RNA3 genome, transcribed for CAT activity (French et al. 1986). Geminiviruses may also be used for genetic transformations but their usefulness as vectors are still under study. Another novel approach under consideration in genetic engineering is the *Agroinfection* which involves a combination of viral and bacterial vectors. Various strategies developed for uptake of viruses by plant protoplasts and their use as transforming agents have been reviewed by Cassells (1989).

Use of fungal protoplasts as eukaryotic gene vectors for plant cell transformation has also been attempted. The uptake of fungal protoplasts requires a suitable experimental system since fungi and higher plants establish a range of nutritional relationships (heterotrophic, saptrotrophic, symbiotic etc.) in nature. Constabel et al. (1982) fused *Catharanthus roseus* protoplasts with fungal protoplasts in an attempt to transfer the secondary metabolite synthesising ability of fungus into a higher plant protoplast system in cultures. Later, Lynch et al. (1989) successfully mediated uptake of celery (*Apium graveolens*) protoplasts by *Aspergillus nidulans* and *Fusarium*

oxysporium protoplasts using a PEG solution, which was confirmed by ultra-structural studies and diamidino-2-phenylindole staining. Protoplasts of celery showed improvement in the level of viability after uptake by *A. nidulans,* whereas with *F. oxysporium* the level decreased. However, it remains to be seen whether uptake of fungal protoplasts can be exploited for secondary metabolite synthesis in higher plant cell or protoplast system.

13.3.3 *Transformation Using Agrobacterium System*

The effort expended in plant genetic transformation appeared to be in vain since studies using various systems for transfer of exogenous DNA into the genome of higher plants failed to give reproducible results. In the past few decades, however, molecular studies of crown gall disease *by Agrobacterium* plasmids and the use of recombinant DNA technology, renewed interest in the subject and further investigations were carried out using *Agrobacterium* for plant genetic manipulations.

Agrobacterium tumefaciens, a soil bacterium, has the ability to infect most dicotyledonous plants, usually at a wound site. The tissue around the infected wound develops a neoplastic growth known as a crown gall tumour. The tumour tissue excised from the plant is capable of growing on hormone-free medium independent of the bacterium. In cultures, the crown gall tissue produces a set of metabolites termed opines (amino acids or sugar derivatives which plant cells do not metabolise normally but which can be utilised as a carbon and nitrogen source by the *A. tumefaciens* strain responsible for induction of the tumour). *Agrobacterium* harbours the Ti plasmid, which is directly responsible for tumour induction. The transfer of small DNA segments from this plasmid and their integration into the genome of host cells induces tumour formation in the plants. The property of the Ti plasmid DNA segment (T-DNA) integrating with the genome of higher plants has attracted world-wide attention for using *Agrobacterium* in genetic transformation techniques.

Agrobacterium rhizogenes is another bacterium used in plant genetic manipulations. The infection by this bacterium causes hairy roots which are of clonal origin. Unlike *A. tumefaciens* tumours, the tumour produced by *A. rhizogenes* has the capability for regenerating mature fertile plants. The tumour-inducing properties of *rhizogenes* strains are also carried on a large Ri plasmid DNA (Ri T-DNA). Two segments of this DNA (TL-DNA and TR-DNA) are transferred to plant cells during infection by *A. rhizogenes*.

(A) Tumour-Inducing Regions of Plasmid Genome

Ti plasmids produce six types of opines: octopine, nopaline, mannopine, agropine, agrocinopine and leucopine. Most of the transformation studies have concentrated on octopine and nopaline type plasmids. Genome analysis of these two plasmids by recombinant DNA technology (i.e., transposon insertions and deletion mutagenesis in plasmid DNA) has revealed that the T-DNA

region and the *Vir* region of the Ti plasmid are essential for *Agrobacterium-mediated* transformation of plants (Fig 13.1).

The T-DNA region is a defined segment bordered by specific sequences and can be transferred into the plant cell nucleus or organelle genome. A single DNA segment of 23 kb is transferred from the nopaline type while two DNA segments (13.6 kb and 7 kb) from the octopine type may be transferred either separately or together. The T-DNA genes responsible for tumour formation express only after integration into plant DNA sequences. Mutations in T-DNA produce tumours of altered morphology or attenuated in oncogenicity. For example, mutations in genes 1 and 2 are reported to induce tumours with abundant shoots, thereby suggesting that the shoot formation in the wild type tumors is suppressed by proteins encoded by these two genes. Contrarily, mutations in gene 4 induce tumours producing abundant roots. This suggests that the root formation is normally suppressed by proteins synthesised by this gene in the wild type. One analogy that can

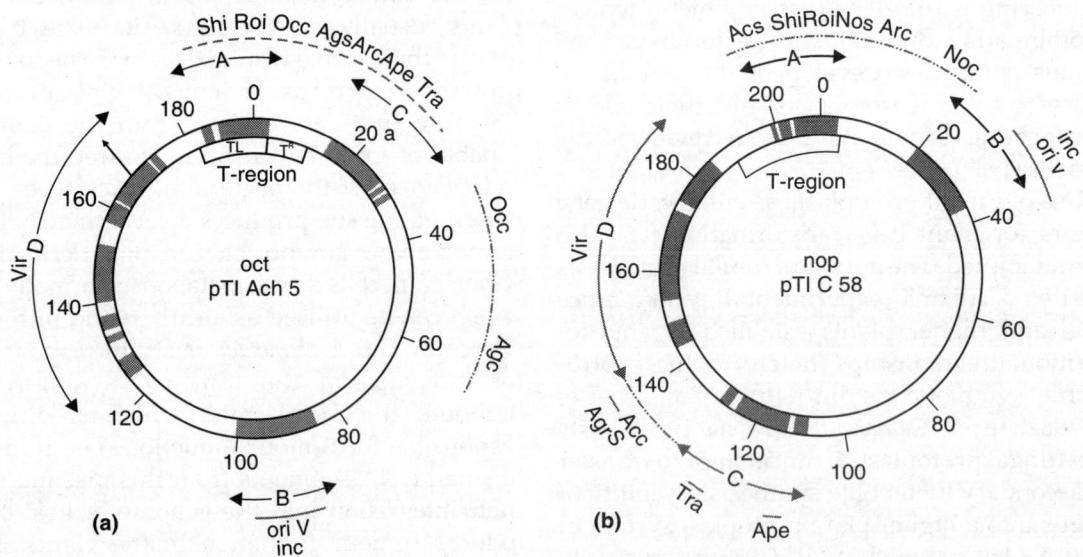

Fig. 13.1 Organisation of Ti plasmids: (a) octopine-type pTiAch5; (b) nopaline-type pTiC58. Abbrev: A—T-DNA; B—replication origin; C—conjugative functions; *D—Vir* region; *ocs*—octopine; *agroo*—agrocinopine; *nos*—nopaline; *agr*— agropine (adopted from Simpson and Herrera-Estrella 1989).

be drawn from these observations is that gene 4 has a cytokinin-like effect and genes 1 and 2 an auxin-like effect. The combination of these effects (suppressing root and shoot formation) in the wild type T-DNA leads to the formation of a characteristic crown gall.

The *Vir* region is closely linked to T-DNA in the Ti plasmid. It encodes functions necessary for transfer and integration of T-DNA in the plant genome.

Two auxin biosynthesis genes have been identified in the TR-DNA of the Ri plasmid. Four loci affecting virulence and tumour morphology are known to exist on the TL-DNA (White et al. 1985). Plants regenerated from hairy root tumours show the presence of TL-DNA. These plants exhibit an abnormal phenotype characterised by decreased apical dominance, adventitious rooting and wrinkled leaves. Since plants regenerated from *A. rhizogenes* are fertile, further investigations are required to a better understanding of developing Ri T-DNA as an efficient plant transformation system.

(B) *Construction of Plasmid Vectors*

(i) *Oncogenic vectors:* These are the simplest vectors and involve the insertion of foreign DNA sequences of interest into T-DNA with the help of intermediate vectors. By conjugating the intermediate vector plasmid carrying the desired foreign DNA sequences for expression in plant cells with *Agrobacterium* and then allowing recombination to occur with homologous T-DNA sequences inside the bacterium, the sequence of interest becomes inserted in-between T-DNA borders. The easiest and the most useful oncogenic vector has been pGV3851. This vector was constructed by deletion of the internal region of the nopaline T-DNA and replacing it with the pBR322 sequence. However, the right T-DNA region was retained along with transcript 4, which could induce oncogenic functions such as tumors with abnormal shoot development (*see* Zambryski et al. 1984). The plasmid pBR322 can be transferred by conjugation from E. *coli* to *Agrobacterium* and cointegrated into the left T-DNA of pGV3851 by

a single recombination event between the homologous sequences present in both plasmids. The fact to note is that pBR322 lacks the mobility (MOB) and transfer (TRA) functions which are necessary for conjugative transfer from *E. coli* to *Agrobacterium*. This necessitates the introduction of a helper plasmid into *E. coli* to enable pBR322 to undergo conjugation. Thus, newly constructed strains of oncogenic vectors harbouring foreign sequences of interest can be used to inoculate plants *in vivo* to induce transformed crown gall tissue with shoot tetratomas.

Oncogenic vectors have limited use because the normal mature plants cannot be regenerated from transformed tumor tissues. This system has been utilised, however, to study the light-regulated expression of foreign genes in plant cells.

(ii) *Non-oncogenic vectors:* These are the most widely used vectors since they allow regeneration of transformed-phenotypically normal plants. Non-oncogenic vectors also require intermediate vectors for insertion of the desired foreign genes into T-DNA. They are of the *co-integrative* or *binary* type.

1. *Co-integrative type:* Co-integrating transformation vectors are plasmids that cannot replicate in *Agrobacterium* and, therefore, need to be cointegrated into an endogenous Ti plasmid. The co-integration event is mediated through a segment of DNA common to both plasmids. This requirement for a homologous segment of DNA limits the vector to one or a few specific Ti plasmids. Two co-integrating plasmid systems developed are pGV3850 and SEV.

Vectors such as pGV3850 are deleted for all oncogenic functions encoded by T-DNA. Only border sequences of T-DNA and opine synthase genes are retained. The deleted sequences are replaced in a single recombination event by desired sequences of pBR322 strains in which the helper plasmids carrying MOB and TRA functions have already been introduced for allowing conjugation from *E. coli to Agrobacterium* (Zambryski et al. 1983). In this way, the desired foreign coding sequences become integrated between the T-DNA border regions. This co-integrated T-DNA non-oncogenic vector (Fig. 13.2) can now be used for transformation of plant cells. No crown gall tissue develops from transformed cells but wound callus tissue may be excised and cultured. Using a suitable combination of hormones, shoots or callus may be induced from this wound tissue. Transformed shoots may further be selected for antibiotic resistance, or screened for enzymatic activity, and eventually regenerated to form plants.

Variation in the co-integrate system results in the 'split-end' vector or SEV system (*see* Fraley et al. 1985). The main difference in this system is that the octopine Ti plasmid derivative happens to contain a left TL-DNA border with other genes (e.g., an antibiotic resistance gene from Tn903, *Vir* genes etc.), while the *E. coli* cloning vehicle contains a functional right T-DNA border. The two T-DNA border sequences are recombined on the Ti plasmid pGV3111 in the correct orientation using homologous sequences of these split-end

vectors. Thus, a functional T-DNA structure is reconstructed that allows transfer of the desired sequences to the plant cells.

Co-integrate vectors retain the T-DNA and *Vir* functions in a as configuration within the same plasmid.

2. *Binary type:* Represented by the 'Bin' series of plasmids (Bevan 1984), this class of vectors possesses the *Vir* and T-DNA (which is disarmed) functions on separate Ti plasmids. The plasmid carrying disarmed T-DNA is called miniplasmid and this bears alien gene along with selectable marker so that both genes get inserted as a unit into the plant genome. The *Vir* functions are supplied in *trans* on the second plasmid in order to drive T-DNA transfer into plant cells (Fig. 13.3). Binary vector is able to replicate both in *Agrobacterium* and *E. coli*. Gene construct (binary vector with inserted foreign gene, in sense or antisense orientation, between a promoter and a terminator along with selected marker) is cloned in *E.coli* and finally mobilised

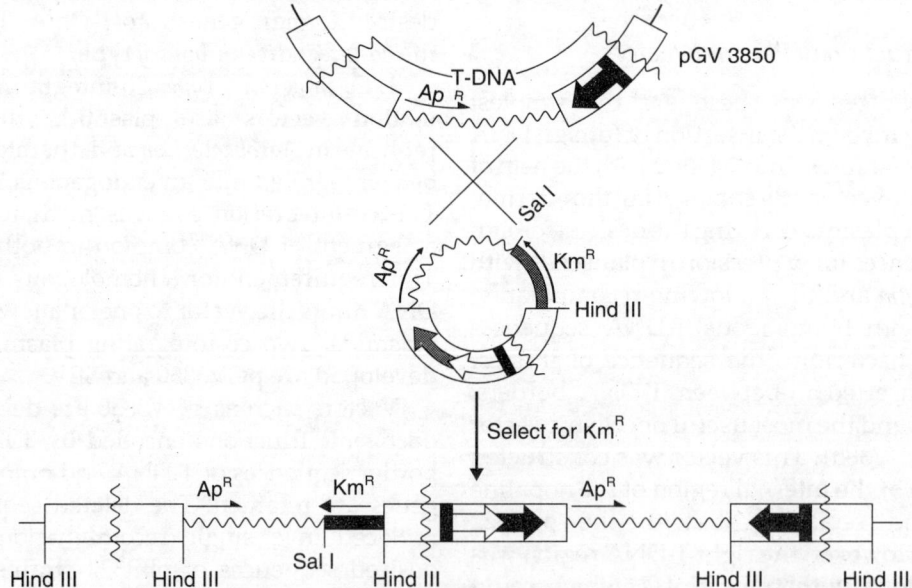

Fig. 13.2 Schematic representation of co-integrated non-oncogenic vector construction by homologous recombination event between pBR322 sequences of pGV3850 and the intermediate cloning vector. Jagged lines show T-DNA border sequences; black arrow—nopaline synthase sequences; white arrow—foreign coding sequences to be expressed in plant cells (after Simpson and Herrera-Estrella 1989).

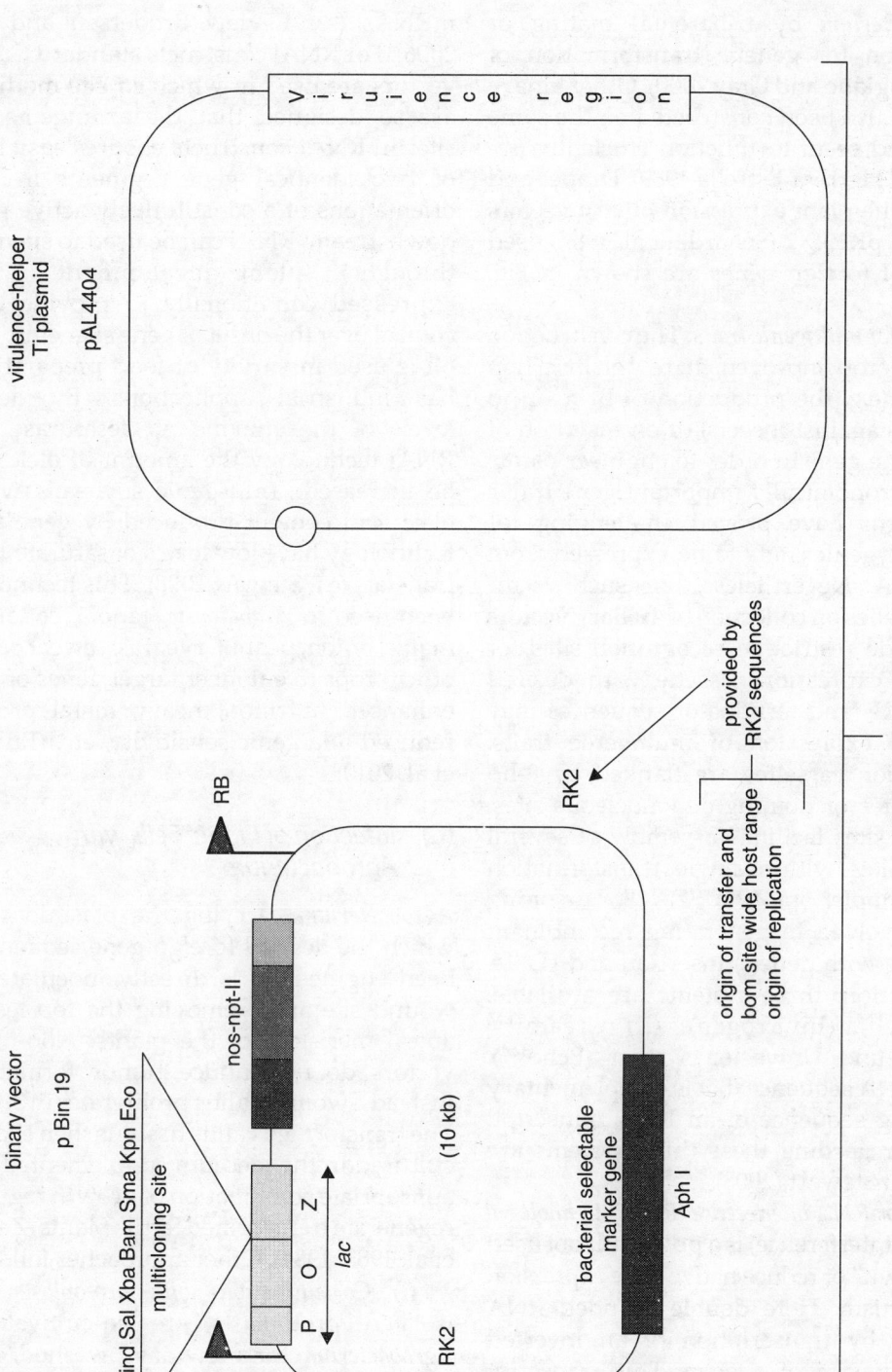

Fig. 13.3 Strategy for construction of a binary vector (adopted from Draper and Scott 1991).

into *Agrobacterium* by triparental mating or electroporation for genetic transformation of plants (*see* Trigiano and Gray 2000). Other binary vectors have also been constructed on the same principles and seem to function efficiently (*see* Simpson and Herrera-Estrella 1989), Draper and Scott 1991). The plant expression binary vectors pKYLX5 and pKYLX7 (Schardl et al. 1987) used in transfer of foreign genes are shown in Fig. 13.4.

3. *Vectors for multigenic traits*: The construction of vectors for crop improvement rely on insertion of one gene (e.g. the production of Bt toxin to protect crops against insects) or on insertion of more than one gene in order to engineer plants with may agronomically important gene traits. Vector designs have proved challenging for allowing polygenic traits to be expressed from single T-DNA. Nevertheless, one such vector design that relies on collection of axillary vectors which provide restriction recognition sites for insertion of expression cassette with desired promotes ORF, and terminator sequences may be used for expression of multigenic traits. The expression cassettes are flanked by 8-bp restriction sites or homing endonucleous sites. The homing sites facilitate assembly of several expression sites within single transformation vector (Example: pPZPRCS2). For cloning strategies involved in generating recombinant DNA vectors with gene(s) insertion and cDNA library insertion, three systems are available: (i) Gateway[TM] (Invitrogen), (ii) Creator[TM] (Clonetech), and Univector system (Echo[TM]). cDNA is a DNA sequence that is complementary to the coding sequence of an RNA transcript. The details regarding these three systems are mentioned in Stewart (2008).

4. *Vector for RNA interference (RNAi) technology*: RNAi (RNA interference) is a powerful tool used to "knockdown" or reduce native gene expression in the organism. Here double-stranded RNA is produced by transcription of an inverted repeated sequence of a gene. This transcript forms hairpin-loop structure that triggers RNAi pathway leading to degradation of homologous mRNAs (*see* Review Brodersen and Vionnet 2006). For RNAi constructs standard Gateway[TM] vectors are used in which *att* site modifications are so designed that the arrangement of *att* sites in RNAi constructs ensures easy insertion of two identical gene segments in opposite orientations of a constitutively active promoter downstream. RNAi can be used to silence genes throughout plant development or can be expressed conditionally to provide temporal control over the onset of gene silencing. Soybean oil is used in variety of food preparations and has industrial applications. By decreasing levels of the enzyme Δ_{12}-desaturase through RNAi technology the amount of oleic acid can be increased. Transgenic soybeans with high oleic acid content produced by gene silencing technology have low levels of saturated fats and transfats (c.f. Stewart 2008). This technology has been used in *Arabidopsis*, canola, cotton, coffee, maize, onion, peanut, ryegrass, sweat potato and other crops to engineer target genes of traits for enhanced nutrition, heavy metal production, reduced allergenic sensitivity, etc. (Rukavtsova et al. 2010).

(C) *Infection of Plants Cells with Agrobacterium*

Agrobacterium carrying the plasmid vector in which the desired foreign gene sequences have been engineered, is directly inoculated at the wound site after removing the top leaves and apical meristem of the plants. Non-oncogenic vectors do not induce tumor formation and instead a wound callus proliferates in 3–4 weeks. The transformed callus tissue is then excised and cultured in the medium supplemented .with an appropriate combination of growth regulators to regenerate transformed fertile plants (Zambryski et al. 1983, 1984). Other approaches followed are:

(i) *Co-cultivation with protoplasts:* Freshly isolated protoplasts are co-cultivated with *Agrobacterium* for a few days, washed, and then cultured in an antibiotic-containing medium (to eliminate bacteria). This is followed by

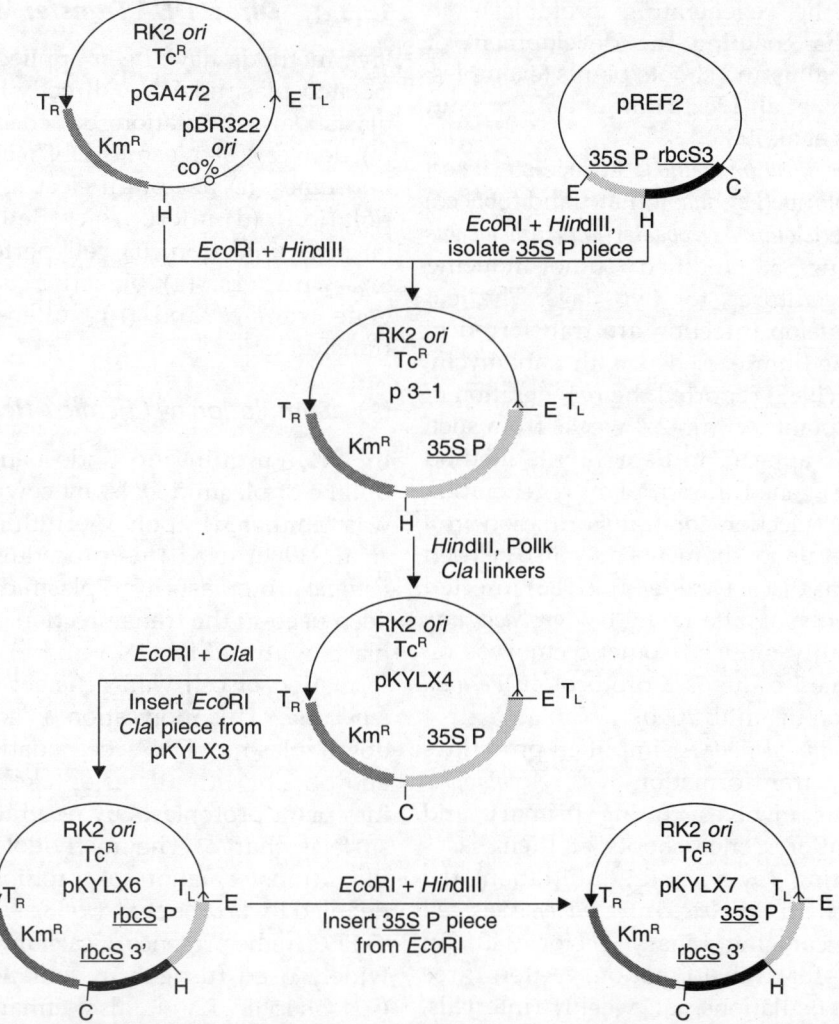

Fig. 13.4 Strategy for the construction of the plant expression shuttle vectors pKYLX5 and pKYLX7. The parent plasmid, pGA472 (An et al. 1985) is designed for *A. tumefaciens*-mediated transformation of plant tissues. The synthetic T-region, which is transferred and integrated in the plant genome, is comprised of a right border (T$_R$) and left border (T$_L$), flanking unique *Hin*dIII and *Eco*RI sites. The chimeric *nos-Aph(3′)II-nos* gene (Km^R) provides a selectable marker for plant transformation. The 35S promoter sequence in pREF2 was excised as an *Hin*dIII-*Eco*RI fragment, and cloned into pGA472, to generate p3-1. A unique *Cla*I site was introduced into the *Hin*dIII site, to generate pKYLX4. The *Hin*dIII site was not reconstituted in the process. To produce pKYLX5, the 0.9-kb *Cla*I-*Eco*RI fragment containing the *rbcS* P-MCS-*rbcS3′* expression cassette from pKYLX3 was introduced into pKYLX4. The *rbcS* P sequence on an *Eco*RI-*Hin*dIII fragment was replaced with the 35S prometer sequence contained on an *Eco*RI-*Hin*dIII of pREF2, generating the 35S P-MCS-*rbcS3′* expression cassette of pKYLX7. Plasmids are not drawn to scale. Both pKYLX5 and pKYLX7 are approximately 12-kb. Arrows: T-DNA borders (labelled appropriately as T$_L$ or T$_R$) and orientation (*see* details in Schardl et al. 1987).

transferring the regenerating protoplasts to suitable media enabling the development of transformed callus and shoots/plants (examples: tobacco, Krens et al. 1982; Hain et al. 1985; and potato, Ooms et al. 1986).

(ii) *Leaf disc infection method:* Surface-sterilised leaf discs of plants (*Petunia*, tomato, and tobacco) are inoculated *with Agrobacterium tumefaciens* strains having a modified tumor-inducing plasmid and cultured for two days. The leaf discs that develop infection are transferred to a selection medium enriched with kanamycin. Horsch et al. (1985) reported the regeneration of transformed plants within 2-4 weeks from such cultures. This appears to be a simple method that combines gene transfer, plant regeneration and effective selection for transformation in a single process. Shoot segments may also be tried in place of leaf discs. Deak et al. (1986) infected shoot segments of alfalfa with *Agrobacterium* and successfully induced somatic embryos on the transformed callus (*see* protocol *on Petunia* transformation in Smith 2000).

(iii) *Floral dip method:* A simplified procedure for genetic transformation of *Arabidopsis thaliana* plants involves cutting primary and secondary inflorescence shoots at their bases and inoculating (by vacuum infiltration) the wound sites with *Agrobacterium tumefaciens* cell suspensions carrying binary vector with an insert (gene for transformation). After three successive inoculations, at weekly intervals, treated plants are grown to maturity and seeds harvested. Progeny raised from these seeds is screened for transformants on a selective medium. This *in planta* transformation procedure (Bechtold et al. 1993, Katavic et al. 1994) was subsequently modified by an extremely simple 'floral dip' protocol (Clough and Bent 1998). In this protocol (Appendix 13.1), the flowering shoots are not cut but directly dipped in *Agrobacterium* cell suspension without vacuum infiltration and is followed as standard transformation procedure for *Arabidopsis*.

13.3.4   *Direct DNA Transfer Methods*

Five methods have been applied for uptake of isolated plasmid DNA directly by plant protoplasts: (a) stimulation by chemical 'helpers', (b) delivery of plasmid DNA encapsulated in liposomes, (c) use of an electric pulse (electropolation), (d) microprojectile bombardment, (e) microinjection, (f) cell perforations using SiC whiskers, (g) bioactive beads-mediated gene transfer and (h) pollen-tabe pathway transformation.

(A)   *Stimulation by Chemical Helpers*

In 1977, Lurquin and Kado demonstrated that uptake of plasmid DNA by cowpea protoplasts was stimulated by poly-L-ornithine (PLO). Davey et al. (1980) used this procedure to transform *Petunia* protoplasts by Ti plasmid DNA. PLO has been used in the transinfection of turnip protoplasts with CaMV DNA and to promote transformation of cell wall-deficient *Chlamydomonas reinhardii*. This polycation is known to slow down plasmid DNA degradation by DNases and possibly stimulates uptake of nucleic acids into plant protoplasts by neutralisation of their surface charge. The methodology describing PLO transformation of protoplasts has been detailed by Power et al. (1986).

PEG in the presence of calcium ions is the most widely used fusogen in somatic hybridisation (*see* Chapter 12). It also enhances uptake of spheroplasts by protoplasts (Hasezawa et al. 1981). Krens et al. (1982) first demonstrated the technique of direct DNA transfer into the plant protoplasts using this fusogen. The technique involves incubating protoplasts with naked DNA in the presence of PEG and $CaCl_2$. After gradual dilution of the mixture, the protoplasts are collected, washed and finally plated in a selective medium. An average transformation frequency achieved with tobacco protoplasts was in the range 10^{-2} and 10^{-5}. This technique has been applied for transformation of other dicot and cereal protoplasts with modifications.

The modifications include: (a) adding PEG after the DNA and (b) heat-shock treatment of protoplasts at 46°C followed by chilling on ice. Some examples of PEG/CaCl$_2$-stimulated DNA uptake by protoplasts are expression of chimeric genes in three monocotyledonous species (*see* Uchimiya et al. 1986, Paszkowski and Saul 1986) and transient expression of a chimeric *nos-cat* gene in mesophyll protoplasts of tobacco, wheat and maize (Paszty and Lurquin; cited by Lurquin 1989).

Isolated DNA co-precipitates with calcium phosphate microcrystals. This insoluble DNA complex is engulfed by animal cells through endocytosis. Within the cytoplasm of an animal cell, the microcrystals dissociate from the co-precipitate, allowing the DNA to reach the nucleus freely. This approach has also been tried with plant protoplasts (*see* Hain et al. 1985).

(B) Delivery of Plasmids Encapsulated in Liposomes

Liposomes are microscopic lipid molecules which are produced when phospholipids are dispersed in an aqueous phase. It is possible to generate unilamellar, multilamellar, neutral, positively charged or negatively charged liposomes (for details, *see* Caboche and Lurquin 1987). Lipsomes large enough to allow entrapment of DNA can interact with a variety of cells in such a way that plasmids will eventually be transferred to their nucleus. Negatively charged unilamellar liposomes are best suited to achieve transfer of RNA and DNA into the plant genome. For example, unilamellar liposomes composed of phosphatidylserine and cholesterol have been used for transfer and expression of a cloned cDNA copy of potato spindle tuber viroid, and *cos-neo* chimeric gene of *E. coli* plasmid in plant protoplasts (Deshayes et al. 1985, Faustmann et al. 1986). PEG promotes the efficient transfer of viral nucleic acids encapsulated in liposomes to intracellular compartments. Efforts were in progress then to achieve high DNA transformation frequency through the use of pH-sensitive liposomes targeting with lectins.

(C) Electroporation

The process by which macromolecules present in the extracellular medium are internalised by living cells on exposure to a brief electric pulse is termed electroporation. The technique allows transfer of chimeric genes into plant cells (protoplasts) by passing a short pulse of electricity which has a voltage peak value of 250-350 V with an RC constant in the millisecond range or a voltage peak value up to 2000 V with an RC constant in the microsecond range. Electroporation equipment can be easily built at economic rates. Minimum manipulation of protoplasts ensures their high survival rates. Using the electroporation technique, transient expression of foreign DNA has been demonstrated in maize (Fromm et al. 1985, D'Hallulin et al. 1999), wheat and rice (Lörz et al. 1985, Potrykus et al. 1985). However, in wheat gene transfer into intact scutellum cells could be achieved by electroporating zygotic embryos without any special treatment (Tahir et al. 2011). Bioactive-beads-mediated gene transfer was also considerably improved by application of electroporation (Murakawa et al. 2008a,b).

(D) Microprojectile Bombardment

A novel method using high-velocity micro-projectiles was developed for the delivery of foreign genes into plant cells or tissues. Exogenous DNA or RNA molecules can be coated (absorbed) to the surface of microscopic tungsten microprojectiles (particles: 4 μm dia) which are then accelerated to high velocities by a particle gun (Fig. 13.5), enabling them to pierce cell walls and membranes. Several cells can be bombarded by such microprojectiles carrying alien genes simultaneously. The particle bombardment, biolistics, particle gun acceleration, gene gun technology are terms interchangeably used for gene transfer using microprojectile bombardment.

Foreign DNA molecules could be introduced in about 90% cells in 1 cm^2 of onion epidermal tissue following bombardment

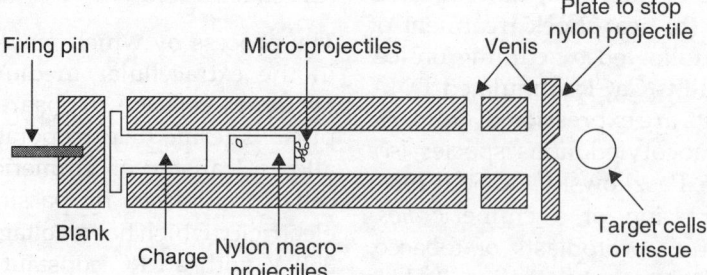

Fig. 13.5 Schematic diagram of particle gun for acceleration of microprojectiles (for further details *see* Klein et al. 1987).

with microprojectiles (Klein et al. 1987). The advantage in this approach is that naked nucleic acid molecules can be transferred directly into the intact plant cells or tissues and the steps required in the removal of cell walls to isolate protoplasts for transformation studies can be bypassed. Microprojectiles have also been used successfully to introduce foreign genes into maize (Klein et al. 1988, Gordon-Kamm et al. 1990) and soybean (McCabe et al. 1988) cells in suspension culture. In all these cells/tissues the transferred molecules expressed genetically. This approach may prove more useful over other methods followed in studies related to transient expression of foreign genes in plants (*see* Vasil and Thorpe 1994). Biolistic transformation of zygotic embryos offers a quick method of evaluation regarding gene constructs in cotton cells, organs, and even fibres originating from transformed embryos (Rajasekaran 2013). This method of transformation of wheat and pearl millet embryos provides valid proof of concept and fucntional genomics for production of transgenic plants (*see* Thorpe and Yeung 2011).

Another technique, *laser micropuncture*, was developed for direct transfer of a plasmid gene into animal cells since human culture cells transform at higher frequency and the genes express stably in the transformed cells (Tao et al. 1987). This method may also be extended to plant cells but with limitations (Badr et al. 2005).

AC CELL Technology (Christou et al. 1989) was developed as alternative to biolistic system. The principle of this technology was based on using electrical discharge as a method of particle acceleration and tested for transformation of soybean cell lines. Heritability of transgene, however, remained to few transformed lines and this technology could not be marketed commercially (Altman and Hasegawa 2012).

(E) *Microinjection*

Crossway et al. (1986) used a novel approach for direct gene transfer into plant cells using the microinjection technique (Fig. 13.6). Animal cells can be transformed at a high frequency on application of this technique. About 14% transformation frequency of tobacco mesophyll protoplasts was achieved and transformed cells could be identified without selection by Southern blot hybridisation. Some other examples of successful cell transformation by intracellular microinjection are *Nicotiana debneyi, Medicago saliva* and *Brassica napus*. In addition to protoplasts, microinjection has been applied to single cells, pollen tubes, immature embryos, pollen embryos and tillers of rye. Even metaphase chromosomes, isolated from kanamycin-resistant cell suspension of *Nicotiana pumbaginifolia*, microinjected into wild type protoplasts of this plant transformed successfully (De Laat and Blaas 1987).

For microinjection, protoplasts are surface attached on a slide by embedding in agarose, or polylysine, using a holding pipette. The DNA is transferred by injecting it with microneedles into the surface-attached protoplasts. Microinjected protoplasts are then cultured following the standard procedures (*see* Chapter 12). This technique is not economically viable since it requires sophisticated equipment and well-trained personnel.

(F) *Cell Perforations Using SiC Whiskers*

One of the simplest procedures developed for the delivery of DNA to recipient cells is perforating them by vortexing in the presence of silicon carbide (SiC) whiskers and the plasmid carrying the transforming gene. Cell perforations thereby allow direct entry of exogenous DNA and was used initially for genetic transformation of maize suspension cells and later extended to transform other plants (tobacco, *Agrostis, Lolium,*

American chestnut and wheat). SiC is known to be one of the hardest of "man-made" products and is used commercially as an abrasive as well as a component of saw blades. It is produced by carbonizing, at high temperature (in carbon monoxide), the silicate found within the cells of rice husks. The cell perforation method using SiC for the production of transgenic maize has been granted US patent 5302523. Detailed protocols and references on transformation of maize and other plants are given by Dunwell (1999) and Asad et al. (2008).

(G) *Bioactive-Beads-Mediated Gene Transfer*

Sone et al. (2008) reported a novel approach of using bioactive calcium alginate microbeads

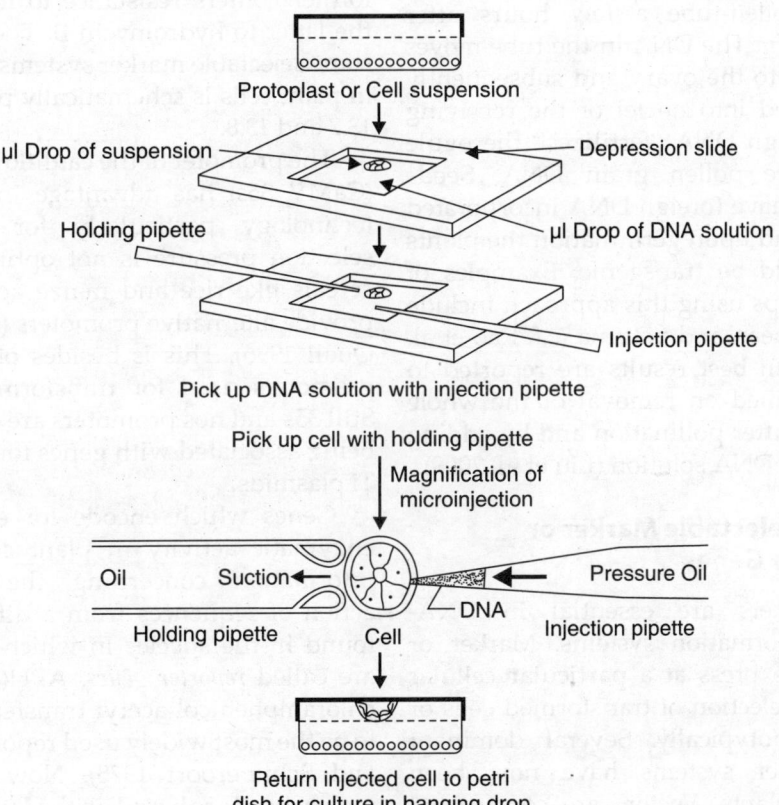

Fig. 13.6 Microinjection by the holding pipette method (developed by Crossway et al. 1986) followed by hanging drop culture.

(1.6μm) to entrap the foreign DNA which subsequently could be introduced into protoplasts via electroporation (Murakawa et al. 2008a). This procedure was later improved by linking DNA with liposomes. The resulting DNA-liposome complex after entrapment by the bioactive beads could be introduced into protoplasts using PEG fusion method. Using this approach four fold transformation efficiency of protoplasts was achieved but whole plants did not regenerate from transformed protoplasts thereby making its application questionable (Murakawa et al. 2008b).

(H) *Pollen-Tube Pathway Transformation*

This procedure involves the introduction of foreign DNA with desired trait into the germinating pollen-tube a few hours after pollination *in situ*. The DNA in the tube moves through style into the ovary and subsequently gets incorporated into nuclei of the receiving ovule. The foreign DNA "fertilizes" the ovule analogously like pollen grain DNA. Seeds harvested may have foreign DNA incorporated in its genome and upon germination the plants that grow would be transgenic. Examples of transformed crops using this approach include rice, maize, soybean and petunia (c.f. Rao et al. 2009). In soybean best results are reported to have been obtained on removal of the whole style 6-8 hours after pollination and by adding surfactant to the DNA solution (Liu et al. 2009).

13.4 Use of Selectable Marker or Reporter Genes

Selectable markers are essential in DNA-mediated transformation systems. Marker or reporter genes express at a particular cellular stage enabling selection of transformed cells or organisms phenotypically. Several dominant selectable marker systems have now been developed for plants. Vectors are constructed by placing the selectable marker gene that controls transcription signals in plant cells. The most widely used vector, pLGV1103 *neo,* has the aminoglycoside phosphotransferase APH(3)II gene fused to a *nos* promoter. APH(3)II is carried by transposon Tn5 and synthesises enzymes which inactivate the aminoglycoside antibiotics of the neomycin family (G418, neomycin and kanamycin). Many plant species are sensitive to aminoglycosides; the APH(3)II gene, therefore, confers a resistance phenotype. The enzyme, aminoglycoside acetyltransferase AAC(3), effectively modifies genta-mycin to an inactive form. When AAC(3) coding sequences are fused to a CaMV 35s promoter, they confer useful resistance to plant cells (Hayford et al. 1988). Other selectable marker genes that have proved useful are the dihydrofoliate reductase (DHFR) carried by transposon Tn7 and hygromycin phosphotransferase isolated from *E. coli.* The former confers resistance to methotrexate and the latter to hydromycin B. The construction of some selectable marker systems that can be used in plant cells is schematically presented in Figs. 13.7 and 13.8.

The promoter of the cauliflower mosaic virus (CaMV 35s) has advantage in transformation technology particularly for species where selection pressure is not optimal although in cereals like rice and maize actin or. ubiquitin provide alternative promoters (Christensen and Quail 1996). This is besides other constitutive promoters used for transformation purposes. Still 35s and nos promoters are extensively used being associated with genes found in T-DNA of Ti plasmids.

Genes which encode for easily detectable enzymatic activity in plant cells and supply information concerning the regulation or action of sequences from a different gene, not found in the species in which it will be used, are called *reporter genes*. APH(3)II, DHFR, and chloramphenicol acetyl transferase (CAT) used to be the most widely used reporter genes (Otten and Schilperoort 1978). Now reporter genes widely used are based on the Luciferase and Gus (beta-glucuronidase) systems (Ow et al. 1986, Jefferson 1987). The Pacific jelly fish (*Aequorea victoria*) green fluorescent protein (GFP) and

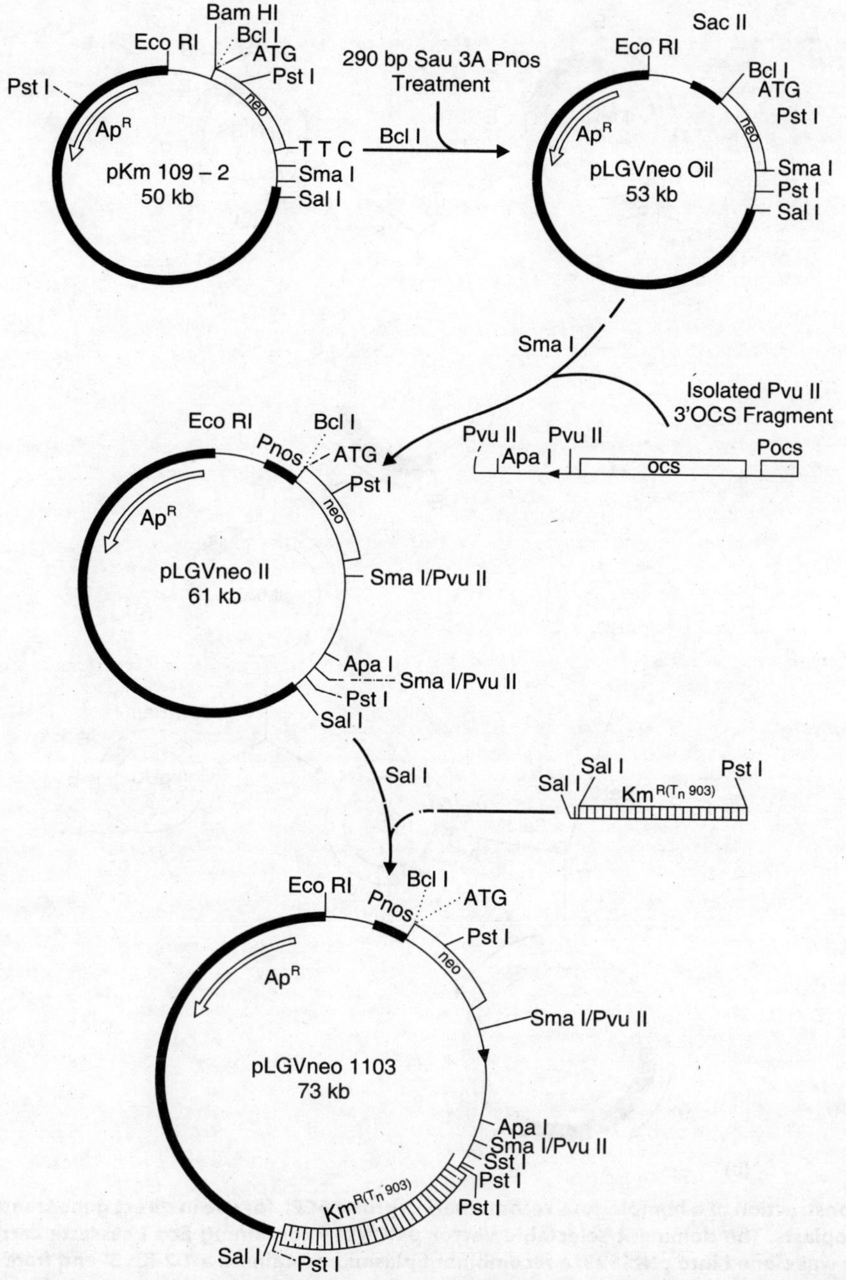

Fig. 13.7 Schematic representation of construction of a selectable marker system for use in plant cells. Small solid box—nopaline synthase promoter sequence. Small open box—NPT(II) coding sequence from Tn5. Large solid region—octopine. synthase polyadenylation signal sequence and hatched region—kanamycin resistance gene functional in bacteria and useful in selecting for Ti plasmid cointegrates (adopted from Simpson and Herrera-Estrella 1989).

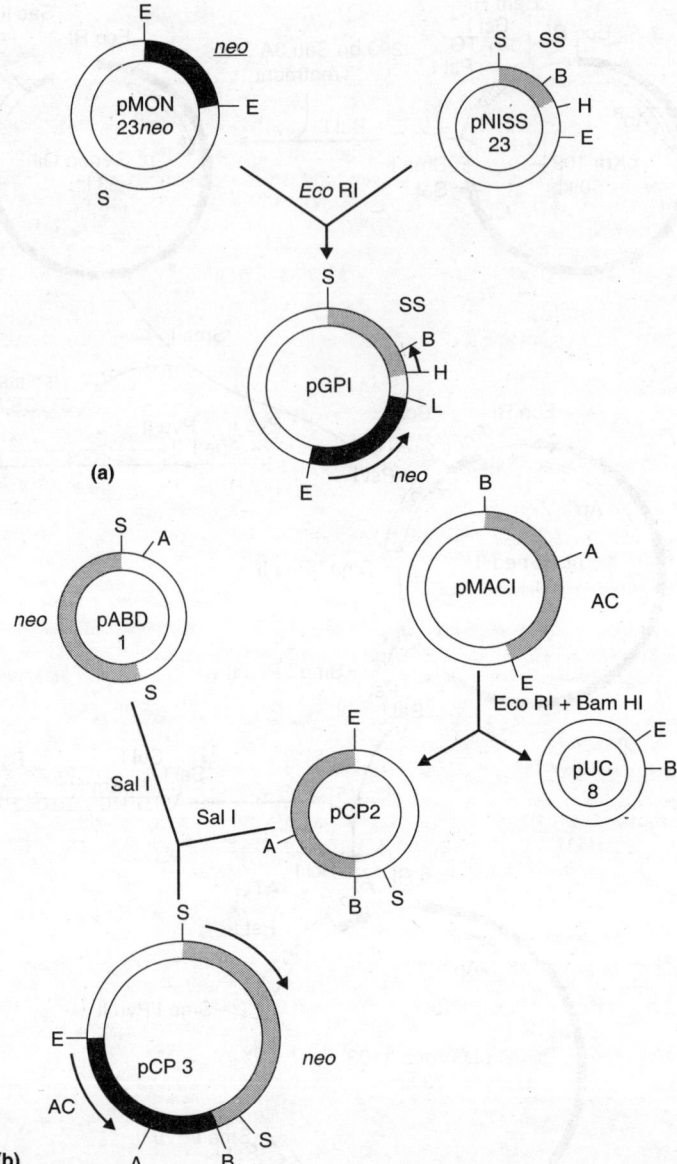

Fig. 13.8 (a) Construction of a homologous recombination probe pCPL for use in direct gene transfer into *Nicotiana tabacum* protoplasts. The dominant selectable vector pMON213 containing Eco I cassette carrying the *nos-neo* chimeric gene was cloned into pNtSS23, a recombinant plasmid containing a 1.2-Kb 5'-end fragment of a tobacco gene coding for small subunit (SS) of ribulose biphosphate carboxylase. Solid area—portion of the Rubisco SS gene; hatched area—chimeric *neo* gene; arrows—direction of transcription. The probe contains a single Bgl 11 site (designated B) convenient for linearising the plasmid inside its region of homology with the host chromosome.

(b) Homologous recombination probe pCP3 constructed for use in direct gene transfer into *Zea mays* protoplasts. This plasmid contains the *neo* gene flanked by the promoter and terminator regions of CaMV gene VI as well as a 3-Kb 5'-end fragment of a maize actin gene. Solid area—portion of the actin gene; hatched area—chimeric *neo* gene; arrows—direction of transcription. This vector contains a single Ava I site (designated A) useful for linearising the plasmid in its region of homology with the host chromosome (adopted from Lurquin 1989).

large variety of other such proteins are being developed for easy detection of transgenic plants or parts of plant and even pollen (Stewart 2008). Reporter genes have greatly facilitated analysis of the regulatory sequences which determine the qualitative and quantitative behaviour of tissue-specific or inducible genes. Applications of selection marker or reporter genes in plant improvement are discussed under Section 13.6 and for more details about these two categories of genes refer to Henry (2013). Methods are also being developed to produce marker-free transgenic plants. Strategies involved in the development of these methods, their applications and Limitations are reviewed by Darbani et al. (2007).

13.5 Integration and Expression of Foreign DNA in Plant Cells

After mediating the transfer of foreign DNA into plant cells there are two aspects that require further investigation: (a) whether foreign sequences inserted into T-DNA would be co-transferred, and (b) functional expression of foreign sequences in plant cells. Transposon insertion mutagenesis of the T-DNA demonstrated that the inserted Tn7 sequences carrying the foreign gene are co-transferred with the T-DNA. Calli regenerated from transformed cell lines have cells in which almost any DNA sequence inserted up to 50 Kb within T-DNA would be co-transferred.

Initial studies on the expression of foreign gene sequences transferred inside the plant cells did not bear fruitful results. Attempts to study the expression of antibiotic genes encoded by transposons (prokaryotic), yeast alcohol dehydrogenase gene, mammalian β-globin, and interferon genes in eukaryotic plant cells were unsuccessful (Barton et al. 1983, Shaw et al. 1983). The logical answer to this lack of expression was to construct a plasmid vector containing transcription initiation (promoter) and termination sequences (terminator) of genes known to be functional in plant cells. Between these two sequences there should be a cloning site to insert coding sequences of foreign genes of interest. The first reported expression of an engineered gene in plant cells based on this principle involved the promoter and 5'-flanking sequences of nopaline synthase gene recombined to the coding sequence, and 3'-end of the octopine synthase. The presence of octopine in transformed plant tissue using this construction confirmed that the chimeric *non-ocs* gene was expressed. Similarly, the construction involving the fusion of coding sequences of the CAT gene from the bacterial plasmid pBR 325 to nopaline synthase promoter 3' sequences also expressed for CAT enzymatic activity in plant cells. Other plasmid vectors constructed for the purpose are described under Section 13.3.3.

Studies have revealed that removal of the right 25-bp repeat region of T-DNA eliminates the capacity of the Ti plasmid to transfer DNA to plant cells, whereas deletion of the left 25-bp repeat region of T-DNA has little or no effect. Therefore, the right 25-bp repeat (a *cis* acting element) plays a crucial role in the directional mechanism of T-DNA transfer. Further, it is suggested that during T-DNA transfer single-stranded intermediates are formed which could represent the molecular form mobilised into the plant cells (*see* Stachel et al. 1987).

13.6 Applications in Plant Improvement

Agricultural crops suffer a great loss due to infection by plant pathogens or insect attacks. Application of herbicides for controlling these losses and destroying weeds in the fields is uneconomical. Using genetic transformation techniques it is possible to introduce genes for herbicide tolerance, bacterial, fungal or virus resistance in some important agricultural crops. Engineering plants for improvement in nutritional value, production of biofuels and pharmaceuticals, abiotic stress tolerance are other aspects are areas where gene transformation technology has proved successful. Advances made have resulted in the progress and growth of plant biotechnology.

13.6.1 Engineering for Herbicide Tolerance

Shah et al. (1986) selected *Petunia hybrida* cells for tolerance to increased levels of glyphosphate, a widely used broad spectrum herbicide. The tolerance to the herbicide was related by the authors to the overproduction of enolpyruvyl-shikimate-3-phosphate (EPSP). The EPSP synthase gene isolated from the herbicide tolerant cell lines was cloned into a plant expression vector utilising CaMV 35s promoter. Thus, the chimeric EPSP synthase gene with a high level of expression was constructed and finally transferred to *Petunia* cells. Only transformed cells proliferated to a callus in the medium supplemented with glyphosphate, whereas non-transformed cells and controls did not divide due to inhibition by this herbicide. Transgenic plants obtained from the callus derived from transformed cells showed tolerance to glyphosphate herbicide two to four times the normal dose required to kill the wild type *Petunia* plants. Mutant EPSP synthase genes from both bacterial and plant genomes reportedly express in genetically engineered tobacco plants. Similarly, plants have been engineered for selective tolerance to other herbicides such as Sulfonylurea, Imidazoline, Phosphinothricin, Bromoxyline and Triazine (*see* review by Padgette et al. 1989, Slater et al. 2008).

RoundUP Ready soybeans were one of the first herbicide tolerant transgenic crops widely accepted by farmers and are reported to utilise upto 87% of the total soybeans grown in the United States. Glufosinate is a herbicide naturally produced in *Steptomyces* bacteria; it kills plants by inhibiting enzyme glutamine synthetase (GS) responsible for glutamine synthesis in plants by utilizing excess plant nitrogen ammonium. Spray of glufosinate inhibits GS and results in the increase of ammonium concentration inside plants to toxic levels. Resistance to glufosinate could be achieved by engineering plants to express protein with ingredient product Libterty™. Herbicide glufosinate gets inactivated when sprayed onto plants engineered for this protein and such plants are known to carry trait "LibertyLink". Transgenic corn, canola and cotton varieties with LibertyLink trait have been produced in the United States. Similar strategy was used to develop buctril-cotton resistant to herbicide bromoxynil (*see* details in Stewart 2008). In China, scientists at CRI (Cotton Research Institute) of Shanxi Academy of Agricultural Sciences in collaboration with scientists at CSIRO in Australia succeeded in producing 1,4-B tolerant cotton using gene tfdA encoding 2,4-D monooxygenase isolated from the bacterium *Alcaligenes eutrophus* (*see* Fasella and Hussain 2014).

13.6.2 Engineering for Virus Resistance

In agriculture, the losses due to reduction in the yield of crops can be prevented to some extent by cross-protection (inoculation of plants with mild strains of virus to prevent more virulent strains from infecting the plant and causing severe disease symptoms). This method has some disadvantages: (a) the mild strain (protecting virus) may undergo mutation to a highly virulent form leading to crop losses, (b) the protecting strain may act in synergism with a non-related virus, (c) the protecting strain may be a severe pathogen of another crop and (d) the protecting strain is likely to cause a small but significant loss in the yield. Attempts have been made to achieve tolerance for virus in plants by genetic engineering methods. Abel et al. (1986) synthesised a chimeric gene containing cDNA of the coat protein (CP) gene of TMV inserted on Ti plasmid of *Agrobacterium tumefaciens* from which tumour-inducing genes had been removed. Fertile transgenic plants regenerated from *Petunia* cells, to which this chimeric gene had been transferred, also expressed resistance to virulent TMV. Similarly, transgenic tobacco and tomato plants expressing resistance against TMV infection have been obtained using TMV-CP gene. CP-mediated protection against cucumovirus (CMV), potexvirus (PVX), alfalfa mosaic virus (ALMV), tungro virus, etc. are other examples of genetically engineered virus resistance (*see* Hemenway et al. 1989).

The striking success of RNAi-mediated resistance to papaya ringspot (PRSV) was achieved by expressing the coat protein (CP) gene of a mild strain of PRSV in papaya (Gonsalves 1998). In Hawaii, the entire papaya plantation between 1992-1995 was virtually threatened of the ringspot virus because of which transgenic papaya crops expressing CP gene were produced and these crops remained virus resistant over the years. As a result, two transgenic papaya varieties 'SunUP' and 'Rainbow' were developed and released for commercial cultivation. Other examples using similar strategy include RNAi-mediated resistance to cucumovirus (CMV) in transgenic squash and feathery mottle virus resistance (FMV) in sweet pea.

13.6.3 Expression of Luciferase and GUS Gene in Plant Cells

As described in Section 13.4 reporter or indicator genes encode for enzymes that allow convenient and sensitive biochemical detection of gene expression and regulation in transformed cells. The Luciferase gene from firefly (*Photinus pyralis*) has been developed as a novel tool for studying gene expression in bacteria, mammalian and plant cells (*see* Ow and Howel 1989). The Luciferase gene encodes for an enzyme that catalyses the light emission through ATP-dependent oxidation of luciferin. Ow et al. (1986) fused luciferase (*luc*) cDNA with the CaMV 35s RNA promoter. Further, a 3'-end fragment from nopaline synthase (*nos*) gene was fused behind the *luc* cDNA. The entire CaMV-/uc-nos chimeric gene was inserted into the multicopy plasmid vector pHC19. This new construction having the luciferase gene and the *nos* fragment with a poly A signal (known to be functional in plant cells) was incorporated into carrot protoplasts by the electroporation method and into tobacco protoplasts using *Agrobacterium tumefaciens*. Extracts from electro-porated carrot cells produced light when mixed with substrates luciferin and ATP in cultures. Transgenic plants

regenerated from these cells also emitted light on 'watering' with luciferin. Luciferase, therefore, has value in the plant genetic engineering as a cell marker to identify transformed cells or as a tag to follow targeting of proteins to various destinations within and outside the cells of transformed plants. It can also be used as a reporter gene for studying gene expression in various genetic crosses as well as a probe for a variety of plant functions.

The wide acceptance of GUS as a reporter gene in genetic transformation studies is mainly due to the fact that it can be quickly assayed without using radioactive substance. Activity of this gene results in cleavage of β-glucuronide bond in a variety of useful substrates, some of which produce fluorescent or coloured products. GUS gene is fused to the promoter of the gene construct to introduce it into plant cells. Though plants generally lack appreciable GUS activity, however, in transformed tissues, GUS mimics the expression of the original gene and reports its activity. Thus GUS can be used as a sensitive background-free measure of gene expression (*see* Trigiano and Gray 2000, Henry 2013).

13.6.4 Engineering for Antibiotic and Non Viral Pathogen Resistance

Transgenic rice and maize plants have been regenerated by somatic embryogenesis from cell suspension-derived protoplasts electroporated with plasmid carrying the NPTII gene under control of the 35s promoter from CaMV. Heat-shock treatment of freshly isolated protoplasts prior to electroporation proved stimulatory for enhancing protoplast division and increasing the number of developed kanamycin-resistant colonies following exposure of protoplast-derived transformed cells to the antibiotic in the medium (Zhang et al. 1988, Rhodes et al. 1989). This could provide an opportunity for assessing the possibilities of introducing novel (non-selectable) genes into the economically important cereals. Similarly, transgenic plants of *Petunia*, tobacco, tomato, *Arabidopsis* and *Brassica napus* have been generated using a marker system

AAC(3) coding sequences fused to the CaMV 35s promoter that confers resistance to gentamicin (Klee and Rogers 1989). Other marker systems used for conferring resistance to antibiotics are described under Section 13.4.

Engineering resistance to ward off bacterial blight, a destructive disease caused by *Xanthomonas oryzae* pathovar. *Oryzae*, in domesticated rice *Oryza sativa* could be achieved by looking forward to alternative source of resistance to this disease. *Oryza longistaminata*, a native of Mali, is wild relative of rice which is resistant to bacterial blight but has low quality and yield of grain production. using genetic and molecular approaches an *R* gene called *Xa2l* was isloated from wild *O. longistaminata* and introduced into domesticated rice *O. sativa* by particle bombardment. The resultant transgenic rice carrying *AvZ* gene recognised by *Xa2l* demonstrated strong resistance to bacterial blight (Ronald 1997). The use of transgenic rice as food crop has generally remained low as compared to soybean and maize. Direct transfer of antifungal candidate genes could provide alternative approach to conventional breeding particularly for production of trees tolerant to *Phytophthora* spp. Corredoira et al. (2015b) used chestnut thaumatin-like protein gene (CsTL1) isolated from European chestnut cotyledons and transformed somatic embryogenic lines of this tree from which transgenic plants tolerant to fungus were raised.

13.6.5 Engineering for Resistance to Insect Attack

Bacillus thuringiensis, a gram-positive bacterium, has been used for many years as a safe biological insecticide because, while exhibiting high insect toxicity, it is completely non-toxic to vertebrates. The insecticidal activity of *B. thuringiensis* resides in crystalline inclusions produced during sporulation by the organism. These inclusions contain insecticidal crystal proteins (ICPs) which affect the midgut epithelium of sensitive lepidopteran, dipteran and coleopteran insects. Reports have recently appeared on cloning and characterization of an ICP gene (*bt* 2) from *B. thuringiensis* strain Berliner 1715. This gene encodes for protein BT 2 which is highly active towards several insects including some major agricultural pests.

The specificity of insecticidal activity of BT protein on a particular insect species is determined by the *cry* gene(s) carried by the bacterium. *Cry*-encoded BT proteins exist as protoxins and after activation through ingestion inside digestive tract of the insect these proteins are cleaved by proteases yielding toxins.

Experiments were conducted to design a modified *bt* 2 toxin gene better suited for efficient expression in plant cells. Two *Agrobacterium* expression vectors were constructed, one containing *bt: neo* fusion genes and the second a modified toxin gene *bt* 284. The latter encodes 610 aminoterminal amino acids of the BT 2 protein. To direct expression in plant cells, the two vectors were fused to the TR promoter of the *Agrobacterium* 2′ gene. Expression vectors fused to the TR promoter were then mobilised into *Agrobacterium* C58CI Rif^R pGV2260 in order to obtain a hybrid Ti plasmid through homologous recombination with the disarmed Ti plasmid of the *Agrobacterium* pGV2260 (Deblaere et al. 1987). This was followed by infection of tomato, potato and tobacco *leaf* discs with engineered strains of *Agrobacterium* pGV2260. Transformed shoots were selected on medium containing 50-200 μg/ml kanamycin and finally rooted on a non-selective medium. Plantlets that regenerated from the transformed shoots were not only resistant to kanamycin, but also expressed for the *bt* 2 toxin gene. Thus transformation of plants for all three crop species to insect tolerance has been possible using modified *B. thuringiensis* toxin genes (*see* details in review by Vaek et al. 1989).

Monsanto Company (USA) has developed insect resistance transgenic cotton plants, with the Bt gene as a dominant single trait. Bt cotton has been commercialised and widely grown in other parts of the world including China, but its production has become a subject of debate in some countries. Nevertheless, adoption of Bt crops has been quite extensive in many countries.

The presence of certain proteins in plants which inhibit the common mammalian digestive proteases has been identified as a natural mechanism of plant defence against insect attack. Trypsin inhibitor (CpTI), a 80 amino acid polypeptide, identified as an insecticidal component of cowpea seeds is active against a range of field and storage pests of economically important plants, including bruchid beetle (*Callosobrochus maculatus*). Insertion of the gene for expression of CpTII in tobacco plants created transgenic plants with enhanced resistance to *Helianthus virescens* (Hilder et al. 1987). Similarly, proteinase inhibitor gene from tomato in transgenic tobacco plants resulted in enhanced resistance to *Mendusa sexta* (Johnson et al. 1989). The prospects of genetic engineering of insect and pest resistance in forest trees is being increasingly explored considering impact of global climate change (Fasella and Hussain 2014).

13.6.6   *Engineering for Nuclear Male Sterility*

Introduction of male sterility is of advantage to a plant breeder since male sterile plants produce fertile female gametes but sterile male gametes. These plants can therefore be used as female parents without emasculation. Emasculation could be laborious, and at times tedious for plants having inflorescences with small and large number of florets, such as in cereals.

In hybrid seed production, cytoplasmic male sterile (CMS) lines have been used (Chapter 12), but these are found unstable under certain environmental conditions (Bhojwani and Razdan 1996). Goldberg (1988) identified nuclear gene TAZ9 from tobacco comprising the promoter PTA 29 and RNA coding gene 'barnase' from *Bacillus amyloliquefaciens*, which causes precocious degeneration of tapetum cells and arrest of microspore development. Transfer of this gene to other crops (oilseed, lettuce, tomato, cotton, maize, cauliflower, chicory, etc.) engineered male sterility in these

plants. Fertility restorer lines for barnase male sterile lines could be engineered by introducing a chimeric gene consisting of PTA 29 promoter fused to bacterial gene encoding barstar—an intracellular inhibitor of barnase (Mariani et al. 1992, Reynaerts et al. 1993).

13.6.7   *Improvement in Nutritional Value of Plants*

One of the notable examples of improvement in nutritional value of food crop is the development of golden rice, a transgenic plant producing grains in which endosperm has high level of β-carotene or provitamin A (Ye et al. 2000). Normal domesticated rice considered as good source of calories happens to be as staple diet in parts of southern Asia, sub-Saharan Africa and also widely consumed food crop all over the world. Yet this rice is not high in proteins and vitamins. Golden rice is advocated to provide provitamin A to millions of undernourished children who need it most, but there seems a concern that levels of β-carotene in this rice may not be sufficient to provide adequate nutritional benefit. Therefore another version of transgenic rice referred to as 'Golden Rice 2' was subsequently developed using phytoene synthase gene from maize instead of daffodil in original golden rice. This second version of golden rice is reported to accumulate over 20 times β-carotene higher than its predecessor (Paine et al. 2005). Since grains produced in these transgenic plants have bright yellow colour due to the presence of β-carotene, hence named "golden rice".

Oils are used in foods as a nutrient, the source of which are fatty acids produced by plants. They also have applications in manufacture of cosmetics, detergents, and plastics. Canola is the common name for the cultivated form of oil seed rape (*Brassica napus*) plant. Conventional methods of plant breeding contain the low levels of harmful glucosinolates and exucic acid in its oil for use in foods. Engineering canola plant with a thiosterase gene isolated from California bay tree (*Umbellularia californica*) resulted in accumulation of much higher levels of fatty acid and oil yield in transgenic plants making them even suitable replacement for palm and coconut oil as well as their products like margarine, shortenings and confectionaries (Altman and Hasegawa 2012). Improvement of soybean oil nutrient value has been discussed under section 13.3.3 (B).

13.6.8 Production of Pharmaceuticals and Biofuels

Production of oral vaccines in plants by gene transformation technology could be an alternative and cost effective approach in terms of storage, accumulation and administration of traditional vaccines. Introducing immunogenic protein in a food is likely to produce oral vaccine in transgenics that could be administered via consumption as food source. For example, production of surface antigen of the hepatitis B virus in transgenic potato demonstrated during clinical trials immune response to this virus in humans consuming these potatoes. in similar manner transgenic bananas act as source of oral vaccine.

Plants reportedly do not produce antibodies, but experiments on plants demonstrated expression of functional antibodies when encoding genes are introduced transgenically. Using this approach a product called *Caro RX* was produced in transgenic tobacco with planned application primarily in tooth pastes and mouth washing in order to prevent tooth decay caused by bacterium *Streptococcus mutans*. The vast majority of pharmaceuticals are transgenically produced in *E. coli*, yeast or mammalian plant cell cultures yet the transgenic plants offer economies of scale to grow and harvest large amounts of biomass expressing target product on relatively little land area (Stewart 2008). Glycosylation of proteins in plants and animals have potential differences and some patterns of glycosylation in plants, besides contamination of food products, are feared to cause allergy in humans on consumption of these drugs. Thereby this approach provides a hurdle in drug production. For more details on the production of vaccines and antibody in GM plants see Halford (2006).

Ever increasing demand for energy consumption world over has led to depletion of fossil fuels and need of the hour therefore is looking for alternative sources of energy. One suggested approach is use of transgenic plants to improve ethanol production (Himmel et al. 2007). Transgenic maize (corn) in the United States is dominant source of fermentable sugars which yeast degrade to ethanol. Compared to fossil fuels ethanol is biodegradable, renewable and cleaner to burn. In Brazil, however, sugarcane is the plant source of choice to produce ethanol because besides being perennial crop it has higher level of simple sugars superior for fermentation. Studies have also been made to explore the use of plant material high in cellulose as source for ethanol production. Corn stalks and leaves, wood chips, biomass crops such as switchgrass, fast growing willow, and poplar trees contain high cellulose energy that could potentially be converted to ethanol. Transgenic approaches are being explored to produce cellulose that would be more easily converted to simple sugars by microbes for alcohol production (Stephanopoulos 2007). Biodiesel can also represent alternative to petroleum-based diesel or fossil fuels. It is produced from oilseed crops through a process called transesterification. Transgenic application for biodiesel production on large scale is limited to soybean and canola as of now and used alone or in a blend with petroleum-based diesel. A matter of concern remains that genetically engineered biofuel crops will likely not be food or feed crops because of environmental, economic and political concerns. Companies will be opposed to altering oil seed or food crops for fuel purposes because of the viable chances of getting fuel and feed mixed (Stewart 2007).

13.6.9 Engineering for Abiotic Stress Tolerance

Abiotic stress caused by drought, salinity, high or low temperatures predominantly restrict agricultural productivity in regions where such environmental conditions exist. The continual rise in world population and decline in farmland owing to urbanisation/infrastructure development, etc impacted agriculturists to make all efforts in exploring methods that would lead to the utilization of non-arable land cultivable for crop production. Using transgenic approaches success has been achieved in the production of stress tolerant tobacco and *Arabidopsis* plants. Introduction of stress tolerant genes in crop plants is very much in need and there are reports of genetically engineered rice for salt tolerance and maize for drought tolerance (Grover et al. 2003, Mundre et al. 2006). Silicon carbide whisker-mediated embryogenic callus transformation of cotton (*Gossypium hirsutum L.*) and its regeneration into salt tolerant plants was reported by Asad et al. (2008).

Heat stress applied to many plants is known to induce heat shock protein (HSP) production which appear to counteract effects by stress caused by elevated temperatures by protecting or refolding denatured proteins. Of the five classes of HSPs four HSP 100, HSP 90, HSP 70 and HSP 60 are highly conserved in prokaryotes and eukaryotes. The fifth class comprises of small HSPs abundantly found in plants with no clear function. Attempts are being made to transform individual HSPs in *Arabidopsis*, carrot and tobacco to enhance heat tolerance (see details in Slater et al. 2008).

13.7 Perspective and Future of Transgenic Crops

The world population is projected to reach 9.4 billion by 2050 (Steward 2008). Immediate impact of this would be to double the crop production as it were in 1980s. The 10 May 2014 issue of Nature depicts that to feed all wheat consumers would require 400% more land over and above area occupied for growing this crop today under organic farming regimes. For growing various other crops would therefore require enormous land which appears challenging due to continual depletion of cultivable land resources as a result of urbanisation, industrialisation, and various forms of infrastructure development. Increase in crop productivity under the circumstances can occur by the use of best and latest that technology offers. In this context, the progress made in the field of DNA transfer technology has renewed interest for development of transgenic crops. Also, it has enabled a study of gene regulation at the molecular level (Grierson 1991). Through direct gene transfer, or *Agrobacterium*-mediated gene transfer, the expression of foreign genes has been observed in some important agricultural crops. Several of these can be stably transformed and mature plants regenerated (Table 13.1). Genetic transformation systems can be used to obtain transgenic plants of fibre (cotton), woody (*Populus, Pinus* and *Pseudotsuga*) and other crop secies (*see* Bajaj 1989b).

Bacterial chitinase gene introduced by DNA transfer technology is able to express in plant cells and its presence in transformed plants could improve their defence response to pathogens (Dunsmuir and Suslow 1989). Also the knowledge gained about regulation of seed-storage protein genes (Elliston and Messing 1989) maybe used in transformation of plants for this economic trait. Attempts on induction of nodular structures on rice roots (Section 13.1) were made with objective to brighten the scope for the possibility of transferring *nif* genes in non-leguminous plants.

Transgenic salt-tolerant tomato plants have been produced which accumulate salt in foliage and not fruit (Zhang and Blumwald 2001), but engineered polymine accumulation in tomato fruit enhances poly-nutrient (lycopene) content, juice quality and vine life (Mehta et-al. 2002). Some of the GM crops have been field-tested and released on commercial scale. In 1998, over 99% GM crop acreage was devoted to just two transgenic traits: herbicide resistance and insect resistance. Less than 1% GM acreage was used for growing 'quality traits'. Maize and soybean account for 75% GM crops, followed by cotton (12%), canola (11%) and potato (1%). USA is a

Table 13.1 Some examples of stable genetic transformation of plants after introduction of exogenous DMA[a]

Plant	Source	Mode of transfer	Vector used	Regeneration status
Alfalfa (*Medicago varia*)	Stem segments	*Agrobacterium* mediated	pTiB0542/ pGA471	Transformed callus and plants
Asparagus (*Asparagus officinalis*)	Stem segments	*Agrobacterium* mediated	C58pTiC58 wild type	Transformed callus
Carrot (*Daucus carota*)	Protoplasts	Direct DNA transfer	pNosCAT pGMVCAT	Transformed callus
Cotton (*Gossypium hirsutum*)	Hypocotyl segments	*Agrobacterium* mediated	pCMC204	Transformed plants
Potato (*Solanum tuberosum*)	Leaf or stem segments	Direct DNA transfer *Agrobacterium* mediated	TiB0542/ pGA472	Transformed callus and plants
Rice (*Oryza sativa*)	Protoplasts	Direct DNA transfer	pCN	Transformed plants
Rye (*Secale cereale*)	Floral tillers	Direct DNA transfer	pEV1103	Transformed plants
Soybean (*Glycine max*)	Hypocotyls, embryonic axes	*Agrobacterium* mediated	pNon9749 pTV1100	Transformed plants
Tobacco (*Nicotiana tabacum*)	Leaf discs	Direct DNA transfer *Agrobacterium* mediated	pTiB0542/ pGA472	Transformed callus and plants
	Protoplasts	Direct DNA transfer	pLGV1103 derivative	Transformed callus and plants
Tomato (*Lycopersicon esculentum*)	Leaf discs	*Agrobacterium* mediated	pMON200SE or derivatives	Transformed callus and plants
Walnut (*Juglans regia*)	Somatic embryos	*Agrobacterium* mediated Direct DNA transfer	pCEN594 pCEN200 pLGV1103	Transformed plants
Wheat (*Triticum monococcum*)	Protoplasts	Direct DNA transfer		Transformed callus

[a] For references *see* Simpson and Herrera-Estrella (1989).

largest GM crop producing country, whereas other countries like Argentina, Canada, Mexico, Spain, France and China have shown increased interest in the production of GM crops (Trigiano and Gray 2000, 2011). Transgenic plants of rice, wheat, brinjal, rose and *Arabidopsis* have been released on commercial scale (Dunwell and Wetten 2012). In China, national boisafety committee for genetic engineering granted approval for commercial cultivation of six GM crops namely cotton, soybean, maize, oilseed, rape, tomato and pepper (Halford 2006). GM technology is getting rapidly integrated into the food market. Introduction of such crops, together with cultivation of herbicide resistant GM soybean and Bt cotton, have caused concern in certain parts of the world. This concern is based on perceptions, such as (a) inadequate testing of GM crops, (b) unwanted spread of transgenes into wild species, by natural cross-pollination, with the fear that this results in evolution of herbicide resistant weeds, (c) overuse of Bt crops resulting in the development of resistant insects in nature, and (d) more importantly, health concerns, such as, transfer of antibiotic resistant genes from digested GM food to bacteria in gut, or possible unrecognized toxicity of transgenic proteins. Considering these concerns, and issues involved, GM crops are subjected to rigorous testing and their release on commercial scale is dependent on ascertaining their benign impact on the environment. Stringent GMO regulations have been formulated by the European Union (EU) which includes extensive case-by-case scientific evaluation of "new food" (GMOs) by European Food Safety Authority (EFSA). The report based on its evaluation is submitted to European Commission for granting or refusing

the adoption of new GMO for public use. As of August 2012, EU had authorised adoption of 48 GMOs adopted by EC within their territory (*see* for details Fasella and Hussain 2014). There is also safeguard clause that Member State(s) can invoke to temporarily prohibit use and/or sale of GMO if there are justified reasons to substantiate that approved GMO constitutes risk to human health and Environment. The Govt. of India on recommendation of Genetic Engineering Appraisal Committee, a regulatory body under Ministry of Environment and Forests, allowed field trials of 15 GM crops which also include rice, chickpea, brinjal and mustard for evaluation regarding safety of public health and environment (Hindustan Times 23 August 2014). In the United States, the policy for evaluation of a product of biotechnology is well regulated and in many cases three agencies namely FDA (Food and Drug Administration), EPA (Environment Protection Agency), and USDA (The United States Department of Agriculture) evaluate a product, whereas Canada assigns regulatory responsibility to three federal agencies CFIA (Canadian Food Inspection Agency), CEPA (Canadian Environment and Protection Agency) and Health Canada.

One of the most existing prospects suggested is the application of TILLING (Targeted Induced Local Lesions in Genomes) in selection of mutants without recourse to GM Technology. These mutants can be directly used in breeding programmes for genetic improvement of crops (*see* for details Comai and Henikoff 2006). Increased awareness, however, among the market forces and consumer recognition of resultant increased quality of GM crops can only help solve public concerns.

APPENDIX 13.1
Simplified Arabidopsis Transformation Protocol[a]

Present protocol (modified from Bechtold et al. 1993) is extremely simple. Authors have found that the MS salts, hormone, etc. make no difference, that OD of bacteria doesn't make much of a difference, that vacuum doesn't even make much of a difference as long as you have a decent amount of surfactant present. Plant health is still a major factor—healthy fecund plants make a big difference! With this method one should be able to achieve transformation rates above 1% (one transformant for every 100 seeds harvested from *Agrobacterium*-treated plants).

1. Grow healthy Arabidopsis plants until they are flowering. Grow under long days in pots in soil covered with bridal veil, window screen or cheesecloth.
2. (*optional*) Clip first bolts to encourage proliferation of many secondary bolts. Plants will be ready roughly 4-6 days after clipping. Clipping can be repeated to delay plants. Optimal plants have many immature flower clusters and not many fertilized siliques, although a range of plant stages can be successfully transformed.
3. Prepare *Agrobacterium tumefaciens* strain carrying gene of interest on a binary vector. Grow a large liquid culture @ 28°C in LB with antibiotics to select for the binary plasmid, or grow in other media. You can use mid-log cells or a recently stationary culture.
4. Spin down Agrobacterium, resuspend to OD_{600} = 0.8 (can be higher or lower) in 5% sucrose solution (if made fresh, no need to autoclave). You will need 100-200 ml for each two or three small pots to be dipped, or 400-500 ml for each two or three 3.5" (9 cm) pots.
5. Before dipping, add Silwet L-77 to a concentration of 0.05% (500 µl/L) and mix well. If there are problems with L-77 toxicity, use 0.02% or as low as 0.005%.
6. Dip above-ground parts of plant in Agrobacterium solution for 2 to 3 seconds, with gentle agitation, then see a film of liquid-coating plant. Some investigators dip inflorescence only, while others also dip rosette to hit the shorter axillary inflorescences.
7. Place dipped plants under a dome or cover for 16 to 24 hours to maintain high humidity (plants can be laid on their side if necessary). Do not expose to excessive sunlight (air under dome can get hot).
8. Water and grow plants normally, tying up loose bolts with wax paper, tape, stakes, twist-ties, or other means. Stop watering as seeds become mature.
9. Harvest dry seed. Transformants are usually all independent, but are guaranteed to be independent if they come off of separate plants.
10. Select for transformants using antibiotic or herbicide selectable marker. For example, *vapor-phase sterilize* seeds and plate 40 mg = 2000 seeds (resuspended in 4 ml 0.1% agarose) on 0.5X MS/ 0.8% tissue culture Agar plates with 50 µg/ml Kanamycin, cold treat for 2 days, and grow under continuous light (50-100 microEinsteins) for 7-10 days.
11. Transplant putative transformants to soil. Grow, test, and use!

For higher rates of transformation, plants may be dipped two or three times at seven-day intervals; one dip two days after clipping, and a second dip one week later. Do not dip less than 6 days apart.

---

[a] After Clough and Bent (1998).

Somaclonal and Gametoclonal Variant Selection

It is well known now that genetic variations occur in undifferentiated cells, isolated protoplasts, calli, tissues and morphological traits of regenerated plants. The cause of variation is mostly attributed to changes in the chromosome number and structure. Cytological heterogeneity in cultures arises mainly due to such factors as: (a) the expression of chromosomal mosaicism or genetic disorders in cells of the initial explants and (b) new irregularities brought about by culture conditions. In tissue cultures, such types of changes were generally discarded since the main objective was to raise genetically stable cultures. Recent investigations have revealed that cell or tissue cultures undergo frequent genetic changes (polyploidy, aneuploidy, chromosomal breakage, deletion, translocation, gene amplifications and mutations) and that these are also expressed at biochemical or molecular levels. Plant cell and tissue cultures, therefore, provide increased genetic variability relatively rapidly and without applying a sophisticated technology. The genetic variability in cultures expresses in the form of variant traits in regenerated plants which are then transmitted to the progeny through sexual (lettuce, tobacco) or vegetative (sugarcane, potato) propagation.

Variants selected in tissue cultures have been referred to as *'calliclones'* (from callus cultures; Skirvin 1978) or *'protoclones'* (from protoplast cultures; Shepard et al. 1980). In 1981, Larkin and Scowcroft coined a general term *'somaclonal variation'* for plant variants derived from any form of cell or tissue cultures, although Evans et al. (1984) prefer the term *'gametoclonal variation'* for variant clones specifically raised from gametic or gametophytic cells. Initial studies on sugarcane and potato generated much interest in somaclonal variation since some of the somaclones of these two crops had useful traits of agronomic interest. The diverse variation characteristic of somaclones highlights the fact that somaclonal variation may be an additional tool for crop improvement rather than an interesting scientific phenomenon (Evans and Sharp 1986a, b; Evans 1989; Bajaj 1990b, Jain et al. 1998, Pareek 2001, Gostimsky et al. 2005, Baer et al. 2007).

14.1 Source Material and Culture Conditions

Attempts directed towards the evolution of new crop varieties through somaclonal or gametoclonal variation are by and large influenced by genotype, explant source, duration of culture and growth hormone effects. These factors can be easily varied in respect of some crops, while in others the constraints of regeneration may limit flexibility in manipulating one or more of these critical variables.

14.1.1 Genotype and Explant

The genotype influences both the frequency of regeneration and the frequency of somaclonal variation. This can very well be exemplified by various potato cultivars from which the number of regenerated plants and somaclones varies under identical culture conditions. Sun et al. (1983) compared the frequency of polyploid regenerates in 18 varieties of rice and recovered multiploids only in the *indica* varieties but not in the *japonica*.

Explants are generally taken from any tissue, namely leaves, internodes, ovaries, roots and inflorescences. The source of the explant has often been considered a critical variable for somaclonal variation. Work with geraniums has shown that somaclonal variants could be recovered from *in vivo* root or petiole cuttings, but not from stem cuttings (Skirvin and Janick 1976). Pineapple (*Ananas cosmosus* (L.) Merr.) plants raised from calli of slip, crown and axillary buds showed alterations only in spine characters while those from calli of syncarp exhibited variations in leaf colour, spine, wax and foliage (Wakasa 1979). Van Harten et al. (1981) observed phenotypic alterations in 12.3% potato plants regenerated from leaf discs in contrast to 50.3% variant plants derived from callus of rachis and petiole. It is likely that different selective pressures would be exerted against different explant sources in cultures, resulting in a spectrum of somaclonal variation among regenerated plants.

14.1.2 Duration of Cell Culture

Many established cell cultures show chromosomal mosaicism. Several factors contribute to variation in the chromosome number such as (i) pre-existing chromosome mosaicism (polysomaty—the phenomenon of polyploidisation of body cells) in explants used for culture initiation, (ii) nuclear fragmentation associated with first cell division of callus initiation, (iii) endoreduplication or endomitosis occurring during cell culture initiation and (iv) abnormalities of the mitotic process leading to aneuploidy. While cell cultures may contain many abnormal cells, regeneration acts as a sieve mostly permitting the growth of stable diploid plants (*see* D'Amato 1977). In spite of this fact there are reports of polyploid and aneuploid plants regenerating *in vitro* from a large number of plant species. Polyploids have been recovered in tissue cultures of *Geranium*, ornamental *Nicotiana*, cultivated tobacco, tomato and alfalfa. Aneuploids generally arise from tissues of polyploid or hybrid plant origin (*Saccharum* hybrids, Heinz and Mee 1971) which can tolerate the loss or addition of a few chromosomes.

Barbier and Dulieu (1980), using a genetically marked explant source, demonstrated that genetic changes occurring in the first few mitoses in culture increase with the age of cultures. Similarly, Lörz and Scowcroft (1983) noticed an increase in frequency of genetic variability from 1.4 to 6% in protoplasts from a heterozygous tobacco by doubling the duration of cultures. Since longer periods (six months) are required to recover somatic hybrid plants in culture, it has been observed that such protoplast-derived hybrids are a richer source of variability than comparable sexual hybrids. This is evident from several unique somaclonal variants selected among somatic hybrids of some species (*see* Evans et al. 1983).

14.1.3 Growth Hormone Effects

Bayliss (1980) reported that high proportions of growth regulators effect karyotypic alterations in cultured cells. 2,4-D seems to induce chromosome variability in plants regenerated from tissue cultures of barley (Deambrogio and Dale 1980) and potato (Shepard 1981) when present in high proportions in the medium. Similarly, somaclonal variants of ornamental *Nicotiana* were obtained from leaf explants on a medium supplemented with 5-10 mmol^{-1} BAP (Bravo and Evans 1985). Growth hormones are essential for induction of organogenesis and shoot differentiation (*see* Chapter 5); however, high concentrations of these substances may not permit recovery of whole plants in tissue cultures and the proportion of hormones in the medium needs to be carefully monitored in establishing culture systems for *in vitro* propagation of somaclonal variants.

14.2 Molecular Basis of Variation

Variants may also arise as a result of more subtle changes due to single gene mutations in cultures which have cells apparently showing no karyological changes. Recessive mutations are not detected in RO plants (plants regenerated *in vitro* from any cell or tissue), but express in RI progeny (the progeny obtained after selfing of RO plants). The RI progeny segregates in Mendelian 3 : 1 ratio for the trait of interest. This further confirms the mutant nature of the variant. Somaclonal variants for single recessive gene mutations have been reported in respect of maize (Edallo et al. 1981), *Nicotiana sylvestris* (Prat 1983), rice (Fukui 1983, Sun et al. 1983) and wheat (Larkin et al. 1984). In some cases specific genetically marked strains (chlorophyll deficiency) have aided in evaluation of plants regenerated from cell cultures (Lörz and Scowcroft 1983).

Changes in the cytoplasmic genome have also been observed in somaclones. Gengenbach and his colleagues evaluated maize plants for two cytoplasmic traits: (a) sensitivity to the toxin secreted by *Drechslera maydis* race T.—causative agent of Southern maize leaf blight and (b) Texas male sterile (cms- T) cytoplasm. Both these traits are controlled by mt DNA. Gengenbach et al. (1977) observed that maize genotypes having male sterile cytoplasm were also sensitive to the toxin secreted by the micro-organism *D. maydis*. In an attempt to combine resistance to toxin with the cms-T trait in plants, they succeeded in selecting resistant plants associated with concomitant reversion to male fertility.

Another aspect of single gene mutation responsible for somaclonal variation relates to transposable elements. Chourey and Kemble (1982) detected variation as a result of insertion of plasmid-like DNA in the mitochondrial genome of cms-s maize cell cultures. Transposon-induced changes have been further reported in strains of tobacco (Lörz and Scowcroft 1983), alfalfa (Groose and Bingham 1984) and wheat (Ahloowalia and Sherington 1985). Increase or decrease in DNA methylation might account for transposable activity and single recessive gene mutations leading to variant selection in maize (Phillips et al. 1994).

Somaclonal variation may also be due to molecular changes caused by mitotic crossing over (MCO) in regenerated plants. This could include both symmetric and asymmetric variation. Single gene mutations by MCO may constitute a unique mechanism of inducing new genetic variations. Small changes in the structure of chromosomes could alter expression and genetic transmission of specific genes, such as deletion or duplication of a copy (or copies) of a gene, or gene conversion during repair processes. Further recombination, or chromosome breakage, at preferential region or 'hot spots' of particular chromosome(s) affects the genome in a disproportionate high frequency resulting in altered phenotypic expressions (Evans and Sharp 1986a).

Molecular studies have demonstrated that changes in the organelle DNA, isoenzyme and protein profiles correlate with the occurrence of somaclonal variation in plants (wheat, rice, potato, maize, barley and flax; Bajaj 1990b). Organelle DNA can be purified with relative ease and has a sequence complexity several orders of magnitude lower than the nuclear genome, making it an attractive subject for molecular investigations. Some restrictive enzymes can readily distinguish between internal and external cytosine-methylation patterns at the restriction site, which appear altered in somaclonal variants (Brown and Lörz 1986). Somaclones of wheat show alterations in gliadin profiles, Nor (nucleolar organiser) loci, and qualitative as well as quantitative differences in rDNA. Similarly, alcohol dehydrogenase zymograms of 645 maize plants obtained from embryo culture revealed one of them as a variant due to alterations at ADHI locus (Brettell et al. 1986). In variant-barley plants derived from callus cultures, variations with regard to rDNA spacer1 fragments and hordeins were observed. Comprehensive information on various aspects of the molecular basis of somaclonal variation has been provided by Ball (1990). The plausible pathways of DNA modification resulting in mutational events leading to somaclonal variation are summarised in Fig. 14.1. Molecular markers such as RFLP, RAPD, AFLP and IRAP (Inter-Retro-transposon Amplified Polymorphism) have found wide use in evaluation of somachonal variants (Gostimsky et al. 2005).

14.3 Isolation of Variants

14.3.1 Without Selection Pressure

Unorganised callus and cells, grown in cultures for various periods on a medium that contains no selective agent (toxic or inhibitory substance), are induced to differentiate whole plants. The regenerated plants are ultimately transferred to the field and screened for variation. Somaclonal variants of various crop species have been obtained using this approach.

(I) Sugarcane

The potential value of variant plants isolated from cell and tissue cultures was first recognised in respect of sugarcane. Work was initiated in the Fiji Islands to isolate subclones of sugarcane cv. Pindar for resistance to Fiji disease (caused by aphid-transmitted virus) and to downey mildew (caused by *Scelerospora sacchari*). The resistance was maintained by somaclones through several cane generations in the field. Of these, Pindar (70-31) performed better than the original cultivar (*see* Krishnamurthi and Tlaskal 1974, Heinz et al. 1997). In Australia, Larkin and Scowcroft (1981) explored the potential tissue culture variability for improvement of agronomically valuable sugarcane cv. Q101 to eye spot disease by *Helminthosporium sacchari*. Similarly, Liu (1981) identified a calliclone of sugarcane superior to the best local Taiwanese variety (F-160) in cane yield, sugar yield and resistance to smut disease caused by *Ustilago scitaminea*. Some other somaclones of

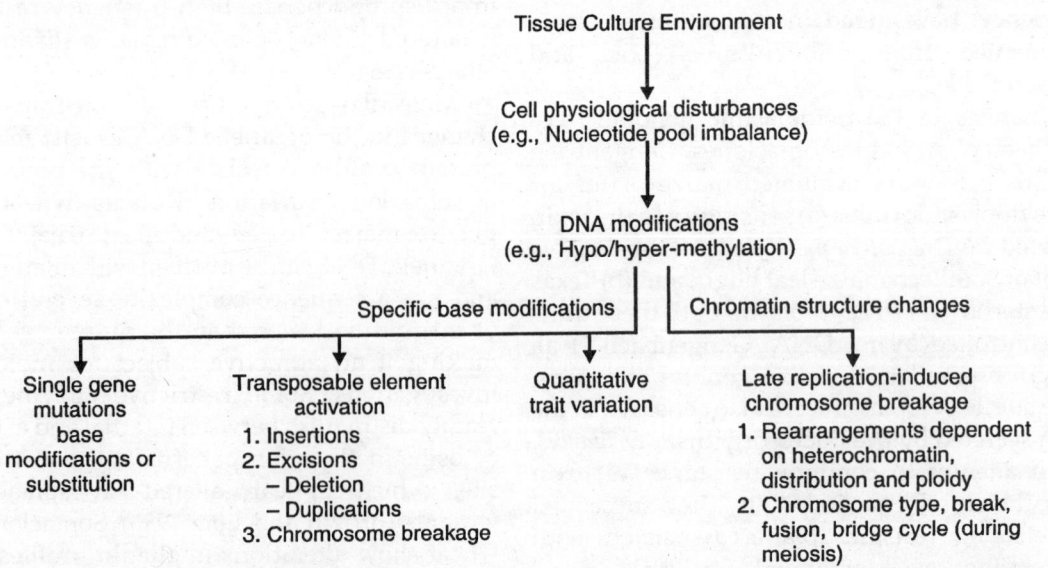

Fig. 14.1 Plausible pathways of DNA modification resulting in various mutational events leading to somaclonal variation (cited from: Ramawat 2000).

sugarcane for high cane yield, sugar content, etc. have been identified and released commercially (see Bhojwani and Dantu 2013). At Bangladesh Sugercane Research Institute somaclones of cane sugar produced *in vitro* under salinity stress were evaluated in the field and those showing better performance selected for release as salt tolerant lines of sugarcane (Kohinoor Begum and Islam 2015).

(II) Potato

High level heterozygosity of potato cultivars is essential for good plant vigour. This feature, coupled with the high sterility of plants, has complicated the application of conventional and mutation breeding methods to improve potato. Shepard et al. (1980) regenerated a large number of plants from mesophyll protoplasts of 'Russet Burbank' and recorded considerable variation in the protoclone population. Some of them were resistant to early blight (*Alternaria solani*) or to late blight (*Phytophthora infestans*). Protoclones resistance to virus Y and leaf-roll virus (PVY) have also been observed in the field (Thomson et al. 1986).

Genetic variability in potato arising from regeneration through tissue culture can be transmitted to the progeny through vegetative propagation (Sree Ramulu 1986). Variants isolated among regenerants of a callus from cv. Bintje produced, on vegetative multiplication, clones having traits for better yield and field resistance to late blight and common scab. The origin and potential of tissue culture variability in the improvement of potato varieties have been comprehensively covered in reviews by karp (1990) and Semal and Lepoivre (1990). Yet, the success achieved has been limited to a few cultivars only, thereby suggesting that the impact of somaclonal variation on the breeding or development of new commercial potato varieties has fallen short of its promise.

(III) Tomato

Evans and colleagues isolated tomato somaclones with variant phenotypes, such as recessive mutations for male sterility, resistance to *Fusarium oxysporium* jointless pedicel, tangerine virescent leaf, flower and fruit colour. Some of these variations in phenotypes were due to changes in the chromosome number (*see* Evans et al. 1987). These scientists at the DNA Plant Technology Co., U.S.A., developed somaclonal variants of tomato bearing fruits which had enhanced taste, a better texture, colour and high (20%) dry matter content. One such variant was patented and registered as a commercial variety in 1986 (*see* Marty l988).

(IV) Banana

Israeli et al. (1991) raised banana plants using *in vitro* techniques exhibiting a range of morphological variations: dwarfism, abnormal leaves, pseudostem pigmentation, persistence of flowers and split fingers. Some of these traits (dwarfism) were retained through successive vegetation cycles. All commercial banana cultivars are susceptible to *Fusarium* wilt disease and traditional breeding methods to select resistant lines have not been successful. Somaclones of Cavendish banana, on the contrary, showed resistance to the pathogen (Hwang 1991).

(V) Geranium

Following investigations on somaclonal variation among *Pelargonium* species, Skirvin and Janick (1976) developed an improved scented *Geranium* that they named 'Velvet Rose'. This is believed to be the first commercial cultivar originated through somaclonal variation. The new cultivar has symmetrical flowers with large fertile stamens, five-paired stigma, and sets seeds. The parental cultivar, on the contrary, has asymmetrical flowers with reduced-sterile anthers, a two-paired stigma, and never sets seeds.

(VI) Cereals and Grasses

The callus-derived plants of cereals and grasses can be a rich source of somaclones. Plants of these clones vary in height, size, shape of leaves, length of awns, fertility of tillers or spikes, or colour of seeds. Variants selected for their commercial potential are male sterility, gliadin protein regulation, frost resistance in wheat (Galiba and Sutka 1988), increased yield in rice and earliness in maize (Evans 1989), and fireblight resistance in pear (Viseur 1989). Among grasses, the species of *Lolium, Panicum, Paspalum* (C_4 forage grass) and *Pennisetum* hold some promise. Other examples of somaclonal variation based on single gene mutations in cereals are discussed under Section 14.2.

(VII) Lucerne

Protoclones of lucerne (*Medicago sativa*) regenerated from isolated mesophyll protoplasts reportedly showed variation for yield and disease resistance to *Fusarium solani, F. roseum* and *Rhizoctonia solani*. Some of these clones gave 60% higher yield than the parent lines (Johnson et al. 1984).

(VIII) Miscellaneous

Attempts have also been made to induce somaclonal variation in cucurbits, finger millet (*Eleusine coracana*), mustard, strawberry, peach, ornamental and forage legumes, *Fuchsia*, carnations, *Haworthia* and *Weigela*. For details of these studies, *see* Bajaj (1990b), Baer et al. (2007).

A celery somaclone (UC-T3) exhibiting significantly high resistance to *Fusarium* wilt disease (Heath-Pagliuso and Rappaport 1990) and pollen haploid somaclone (PHS) of rice developed by the Chinese are some other examples of commercial application of somaclonal variation in crop improvement. An improved variety of rice 'DAMA' was released through PHS method which combined anther culture with somatic tissue culture (*see* Heszky and Simon-Kiss 1992). Usefulness of cell cultures for development of 'Elite' germplasm and to enhance alien gene introgression in wide crosses have been ascertained. Some examples of heritable somaclonal variation in selected crops are given in Table 14.1.

Aeging of cell cultures may be one of the factors inducing somaclonal variation. Landley et al. (2015) observed cell culture aeging as highly mutagenic in coffee and raised somaclonal variant plants from long-term embryogenic cell lines in *Coffea Arabica* ($2x = 4x = 44$). Somaclonal variant plants had aneuploid (monosomic) chromosome number and were able to survive and remain viable due to allopolyploid nature of coffee.

14.3.2 *With Selection Pressure*

In this method variant cell lines are screened from cultures by their ability to survive in the presence of a substance in the medium that may be toxic/inhibitory, or under conditions of environmental stress. Variants may be obtained by direct selection, indirect selection, or by individually testing single cell-derived colonies. In the case of direct selection, the new variant cell types have a selective advantage over the rest of the population because of their tolerance to a specific toxic compound. As regards the indirect selection, the wild type cells are selectively killed and only the mutant recovered by transferring the surviving cells to a supplemented (enriched) medium. Isolation is carried out in most cases in suspension cultures or by plating single cells/protoplasts.

(I) Amino Acid Analogue and Amino Acid Resistance

Amino acid analogues either act as false-feedback inhibitors of amino acid biosynthesis or may be incorporated into proteins, resulting in a loss of enzyme function. Growth inhibition can usually be reversed by the addition of corresponding natural amino acid(s). Alteration in the feedback sensitivity results in the evolution of cell lines resistant to the analogue and consequent overproduction and accumulation of free amino acids. These variant cell lines maintain stable phenotypic expression in the absence of the analogue. The amino acid being overproduced may be metabolised or excreted into the culture

Table 14.1 **Some example of the veritable somaclonal variation in selected crops***[a]*

Crop	Trait(s)	Inheritance*	Reference
Tobacco	Resistance to toxin methionine sulfoximine	S	Carlson 1973
	Resistance to *Pseudomonas syringae*	S	Thanutong et al. 1983
	Leaf colour	S	Dulieu and Barbier 1982
	Leaf spots	S	Evans et al. 1984
	Tolerance to aluminium	S	Conner and Meredith 1985
	Resistance to herbicides chlorsulfuron and sulfometuron methyl	S	Chaleff and Ray 1984
Maize	Resistance to *Helminthosporium maydis* race T	Maternally inherited	Gengenbach et al. 1977
Tomato	Resistance to *Fusarium oxysporum*	S	Evans et al. 1984 Shahin and Spivey 1987
	Fruit colour	S	Evans and Sharp 1983
	Resistance to tobacco mosaic virus	S	Cassells et al. 1986
Wheat	Reduced wax, awning, glume colour, gliadin pattern	S	Larkin et al. 1984
	Adh isozyme	S	Brettelle et al. 1986
	Resistance to *Helminthosporium sativum*, BYDV (barley yellow dwarf virus)	S	Chawla and Wenzel 1987
Rice	Resistance to *Xanthomonas oryzae*	S	Ling et al. 1985
Brassica	Resistance to *Phoma lingam*	S	Sacristan 1982
	Seed colour	S	George and Rao 1983
Alfalfa	Resistance to *Fusarium oxysporum*	S	Hartman et al. 1984
Sugarcane	Resistance to Fiji disease	VP	Krishnamurthi and Tlaskal 1974
	Resistance to *Helminthosporium sacchari*	VP	Heinz et al. 1977, Larkin and Scowcroft 1983
Potato	Resistance to *Altenaria solani*	VP	Matern et al. 1978
	Resistance to *Phytophthora infestans*	VP	Shepard et al. 1980
	Mitochondrial DNA restriction pattern	Maternally inherited	Kemble and Shepard 1984
	Ribosomal DNA copy number	VP	Landsmann and Uhrig 1985

* S = Sexual transmission; VP = Transmission through vegetative propagation.
[a] For references in the Table see Brar and Jain (1998).

medium. Cell lines resistant to analogues of phenylalanine, proline, lysine, methionine and tryptophan have been isolated. In some 5-methyl-tryptophan-resistant carrot lines the production of tryptophan was accompanied by auxin independent growth and an increase in IAA levels (Widholm 1977, Sung 1979).

Some amino acids may inhibit growth if incorporated into the culture medium in specific combinations (lysine + threonine) or singly (valine, threonine). Selection against this synergistic growth-inhibition resulted in the development of resistant maize tissue cultures (Hibberd and Green 1982) and in plants derived from resistant M$_2$ embryos of barley (Bright et al. 1982, 1983). Similarly, tobacco cell lines resistant to glycine hydroxamate (overproducing free glycine) were selected (Lawyer et al. 1980).

Overproduction of an amino acid sometimes enhances the nutritional (protein) value of the variant cell lines. Rice lines overproducing lysine were shown to have higher protein-lysine content (Chaleff and Carlson 1975). However, no change was observed in the ratio of methionine to other amino acids in proteins of an alfalfa cell line overproducing methionine (Reish et al. 1981).

Numerous resistant cell lines have been isolated for amino acid and amino acid analogue traits. However, these traits are paralleled in plants regenerated from variant cell lines of only a few cases (Table 14.2).

(II)   Disease Resistance

A study of phytotoxins helps in understanding the mode and site of action at cellular and subcellular levels. Isolated single cells or protoplasts in cultures can be exposed to physical or chemical mutagens to induce mutations in cells for resistance to phytotoxin. Carlson (1973) reported that plants regenerated from ethyl methane sulphonate (EMS)-treated tobacco protoplasts, selected for resistance to methionine sulphoximide (MSO), showed enhanced tolerance to infection by *Pseudomonas tabaci*. The resistance was inherited as a single semi-dominant trait. Selection for resistance to *Helminthosporium* T-toxin (a host-specific toxin) in maize cultures was reported along similar lines by Gengenbach et al. (1977). In their study the embryogenic maize calli were exposed to the toxin in cultures and in the process they obtained resistant and fertile plants. The calli were derived from plants inheriting toxin-susceptible and male sterile traits maternally. Since cytoplasmic conversion to male fertility and T-toxin resistance did not transmit sexually in selfed progeny of T-cytoplasm maize, this example represents a typical case of somaclonal variation *(see* Maliga 1984).

(III)   Herbicide Rsistance

The growth of weeds in a population of agriculturally important crop is generally controlled by herbicides. As herbicides have a short residual life, they are applied repeatedly on the crops. Some crops become susceptible to a herbicide due to its repeated applications on them. Further this method of weed control is not economical since herbicides are quite expensive. Efforts have been made to follow *in vitro* approaches to obtain plants expressing tolerance to herbicides. Exposure of cell and protoplast cultures to various herbicides induces mutations that may yield tolerant cell lines which ultimately regenerate into plants. Examples of herbicide-resistant plants regenerated from cell cultures include *Nicotiana tabacum* (against amitrole, bentazone, chlorosulphon, isopropyl N-carbamate, phenme-difarm, picloram and paraquat), *Corydyllis* (against glyphosphate), and *Medicago saliva* (against glufpsinate). Some of the resistant traits are inherited as dominant or recessive mono-genie alleles (Table 14.3). Some more examples of herbicide resistance are given by Chaleff (1986), Weller et al. (1987), Ramawat (2000) and Pareek (2001). Herbicide tolerance can also be introduced into cells by somatic hybridisation or through gene transfer technology (*see* Chapters 12 and 13).

(IV)   Environmental Stress Tolerance

Environmental stress due to high salt levels in soil is a major constraint in the development of agriculture. Since the first report on *in vitro* regeneration of sodium chloride tolerant tobacco plants (Nabors et al. 1980), plant scientists have developed a number of cell lines and plants that can withstand the rigours of salinity (Table 14.3). Some plants retained the variant phenotype through two successive generations. Other examples and their details are given by Tal (1990).

Attempts to develop chilling tolerant cell lines were made by Dix in 1977 by exposing *Nicotiana sylvestris* cell lines to extreme low temperatures. However, the phenotype was not sexually trans-

Table 14.2 Amino acid and amino acid analogue resistant cell lines selected in cultures[a]

Selective agent	Species	Mutagen	Plant regen.	Resistance mechanism[b]	Genetics[b]
S-(aminoethyl)-L-cysteine	*Arabidopsis thaliana*	EMS	Yes	P,U	N.R.
	Daucus carota	None	No	N.R.	N.R.
	D. carota	None	No	A	N.R.
	Hordeum vulgare	Azide	Yes	N.R.	Single recessive allele
	Nicotiana tabacum	EMS	No	C	N.R.
	N. tabacum	UV or EMS	No	N.R.	N.R.
	N. sylvestris	None	No	N.R.	N.R.
Aminopterin	*Oryza sativa*	EMS	No	P	N.R.
	Datura innoxia (haploid)	None	Yes	N.R.	N.R.
Azaguanine	*Acer pseudoplatanus*	NTG	No	A	N.R.
	Glycine max	EMS	No	N.R.	N.R.
	Haplopappus gracilis	EMS	No	A	N.R.
	Medicago sativa	EMS	No	N.R.	N.R.
Azauracil	*H. gracilis*	EMS	No	C	N.R.
	Zea mays	EMS	No	F.A.	N.R.
Azetidine-2-carboxylic acid	*D. carota*	None	No		N.R.
	D. carota	None	No	P,U	N.R.
Bromodeoxy-uridine	*G. max*	NG[b]	No	P	N.R.
	M. sativa	EMS	No	N.R.	N.R.
	N. tabacum	None	Yes	N.R.	Single dominant nuclear allele
Ethionine	*D. carota*	EMS	No	U	N.R.
	D. carota	None	No	N.R.	N.R.
	M. sativa	EMS	No	P	N.R.
p-Fluorophenyl alanine	*A. pseudoplatanus*	None	No	A,S	N.R.
	A. pseudoplatanus	None	No	P,S	N.R.
	D. carota	None	No	P	N.R.
	D. carota	None	No	N.R.	N.R.
	D. carota	None	No	N.R.	N.R.
	D. innoxia	None	Yes	N.R.	N.R.
	N. tabacum	UV or EMS	No	N.R.	N.R.
	N. tabacum	None	No	A	N.R.
	N. tabacum	None	No	A	N.R.
	N. tabacum	None	No	S,U	N.R.
	T. tabacum	None	No	A,S	N.R.
	N. tabacum	None	Yes	N.R.	N.R.

Table 14.2 (Contd.)

Selective agent	Species	Mutagen regen.	Plant mechanism[b]	Resistance	Genetics[b]
6-Fluorotryptophan	*Petunia hybrida*	NG[b]	No	N.R.	N.R.
5-Fluorouracil	*D. carota*	None	Yes	N.R.	N.R.
Glycine hydroxamate	*N. tabacum*	None	Yes	P	Single dominant nuclear allele
Δ-Hydroxylysine	*N. tabacum*	EMS	No	N.R.	N.R.
	N. tabacum	UV or EMS	No	C	N.R.
Hydroxyproline	*D. carota*	EMS	No	U	N.R.
	H. vulgare	Azide	Yes	P	Single semi-dominant nuclear allele
Hydroxyurea	*N. tabacum*	None	Yes	N.R.	Single semi-dominant nuclear allele
Lysine plus threonine	*Z. mays*	Azide	Yes	P	Dominant nuclear allele
	Z. mays	None	Yes	F,P	N.R.
Methione sulfoximine	*N. tabacum*	None	Yes	P	Single semi-dominant allele or two recessive alleles
5-Methyltryptophan	*Catharanthus roseus*	None	No	A	N.R.
	D. carota	None	No	S,F	N.R.
	N. tabacum	UV or EMS	No	N.R.	N.R.
	N. tabacum	None	No	F	N.R.
	N. tabacum	None	Yes	N.R.	N.R.
	N. sylvestris	None	No	N.R.	N.R.
Seleno-amino acids	*N. tabacum*	None	No	P,U	N.R.
Thienylalanine	*N. sylvestris*	None	No	N.R.	N.R.
Threonine	*N. tabacum*	None	Yes	U	N.R.
Valine	*N. tabacum* (haploid)	UV	Yes	N.R.	Mendelian
	N. tabacum (haploid)	NG[b] or irradiation	No	N.R.	N.R.

[a] For detailed references of each cell line, *see* Flick (1983).
[b] Abbreviations for mechanisms: F—enzyme resistant to feedback inhibition; S—increase in synthesis of secondary metabolite; A—Increase in enzyme activity; P—increased pools of natural compound; U—altered uptake; C—cross-resistant to other anti-metabolites. N.R.—not reported, NG—N-methyl-N'-nitro-N-nitrosoguanidine.

mitted in plants regenerated from these lines. Similar results have been reported with regard to clonal variation for tolerance to PEG-induced water stress in cultured tomato cells.

(V) Auxotrophic Lines

Auxotrophs are continuously used for DNA transformation, somatic hybridisation and the study of metabolic processes. The use of haploid cell cultures and the application of effective

Table 14.3 Agriculturally useful mutant cell lines obtained *in vitro*[a]

Selective agent	Species	Mutagen	Plant regen.	Resistance mechanism[b]	Genetics[b]
Cold tolerance					
	Daucus carota	EMS	Yes[e]	N.R.	N.R
	N. sylvestris	None	Yes	N.R.	D
Herbicide resistance					
Asulam	*Apium gravaeolens*	None	No	N.R.	N.R.
Bentazone	*Nicotiana tabacum* (haploid)	γ-ray	Yes[c]	N.R.	Recessive nuclear allele
2,4-D	*Lotus corniculatus*	None	Yes	A	N.R.
2,4-D; 2,4,5-T; 2,4-DB	*Trifolium repens*	None	No·	N.R.	N.R.
Isopropyl-N-phenyl-carbamate	*N. tabacum*	EMS	Yes[c,d]	N.R.	N.R.
Paraquat	*N. tabacum*	X-ray	Yes[c]	N.R.	N.R.
Phenmedifarm	*N. tabacum* (haploid)	γ-ray	Yes[c]	N.R.	Recessive nuclear allele
Picloram	*N. tabacum*	None	Yes[c]	N.R.	Dominant nuclear allele; 3 linkage groups
Pathotoxin resistance					
Fusarium oxysporum	*Solanum tuberosum*	None	Yes	N.R.	N.R.
Helminthosporium maydis	*Zea mays* T-cms	None	Yes[c]	B	Cytoplasmic allele
Phytophthora infestans	*S. tuberosum*	None	Yes[c]	N.R.	N.R.
Salt tolerance					
	Capsicum annum	None	No	N.R.	N.R.
	Datura innoxia	None	Yes	N.R.	N.R.
	Kickxia ramosissima	None	Yes	N.R.	N.R.
	Medicago sativa	None	No	N.R.	N.R.
	Oryza sativa	None	No	N.R.	N.R.
	N. sylvestris	None	No	N.R.	N.R.
	N. sylvestris	None	No	N.R.	N.R.
	N. tabacum	EMS	Yes	N.R.	C

[a] *See* Flick (1983).
[b] Abbreviations for resistance mechanisms: A—Alteration in 2,4-D sterility conjugation; B—reversion to male sterility; C—transmitted sexually, but not at expected segregation ratios; D—not transmitted sexually. N.R.—not reported.
[c] Resistant plants.
[d] Sterile plants.
[e] Not expressed in plants.

mutagenic treatment have made feasible the recovery of a large number of auxotrophs. The first auxotrophic cell lines were reported by Carlson (1970) from dihaploid tobacco which grew slowly on a minimal medium and required specific metabolites to resume normal growth. Auxotrophic lines isolated in cell cultures of various plant species include requirement for

(a) pathotenate, adenine and isoleucine + valine in *Datura innoxia;* (b) histidine, tryptophan and nicotinic acid in *Hyoscyamus muticus;* and (c) isoleucine, leucine, and uracil in *Nicotiana plumbaginifolia.* The *Datura* lines were isolated in suspension cultures, the rest from leaf mesophyll protoplast cultures. Through cell fusion-complementation, auxotrophic lines of *N. plumbaginifolia* and *H. muticus* were identified to be recessive.

Chlorate, an analogue of nitrate, is converted by nitrate reductase enzyme into chlorite which is cytotoxic. Thus, in a chlorate-supplemented medium the wild type cells with nitrate reductase activity are killed while those deficient in the enzyme survive. Such nitrate reductase-deficient cell lines have a requirement for a reduced form of nitrogen such as a mino acids or ammonium succinate. Revertants to wild type could be tested by the ability of these cells to utilise NO_3 in the culture medium.

Nitrate reductase-deficient lines have been isolated in cultured cells of *N. tabacum, N. plumbaginifolia, H. muticus, D. innoxia, Rosa damascena* and *Petunia.* Two types of these cell lines have been identified: one mutant type *(nia)* is defective in the nitrate reductase apoprotein, while the other (cnx) is characterised by an impaired molybdenum co-factor (Mo-co). Such phenotypes have been used as complementing markers in the selection of somatic hybrids. Various other auxotrophic mutants and variant lines selected in cell cultures are given in Table 14.4.

(VI) Antibiotic Resistance

Cell lines resistant to the antibiotics streptomycin, linomycin, kanamycin, chloramphenicol and cycloheximide have been developed from various plant species. Streptomycin-resistant lines of *Nicotiana tabacum* are cytoplasmic, whereas those of *N. sylvestris* are controlled by recessive nuclear genes. Both the cell lines are isolated on the basis of their ability to green on selective medium.

Linomycin-resistant lines in *N. plumbaginifolia* and *N. sylvestris* were also selected on the basis

of greening. Resistance for linomycin has been identified as a chloroplast-controlled trait and inherits maternally. The kanamycin- and chloramphenicol-resistant lines of *N. sylvestris* and cyclohexamide resistance in tobacco and carrot lines are reported to be epigenetic phenotypes *(see* also Table 14.5).

For details and references of variant or mutant lines described in this chapter *see* reviews by Flick (1983), Maliga (1985), Evans and Sharp (1986a, Jain et al. 1998).

14.4 Nature of Gametoclonal Variation

The genetic material in somatic cells and tissues is normally equationally distributed through the process of mitosis. Contrarily, the gametes, which are products of meiosis, receive half of the genetic complement with alleles following Mendel's laws of segregation and independent assortment. To distinguish somatic-derived somaclones and gametic-derived gametoclones three parameters are considered. First, both dominant and recessive mutant genes induced by gametoclonal variation will express directly in haploid plants regenerated from microspores of diploid anthers (since they will have only one copy of each gene). This allows analysis of gametoclones (RO) directly to identify new variants. Second, recombinants recovered in gametoclones would be the result of meiotic crossing over. Third, the gametoclone can be used only after having stabilised by doubling its chromosome number.

The value of gametoclonal variation in crop improvement is evident from the development of double-haploid lines by anther culture of F_1 hybrid plants of wheat and rice (Zeng 1983). Anther culture has been used for the recovery of recombinant plants of an F_1 wheat hybrid between Xian nog 5675 (a variety with white glumes, top awn, clavate spike and short stalk) and Jili (a variety with red glumes, awn, fusiform spike and tall stalk). Several double-haploid plants were recovered that expressed mixed characters of the two parents.

Table 14.4 Auxotrophic mutants and various other variant cell lines selected in cultures[a]

Nutrient requirement/ phenotype	Species	Mutagen	Plant regen.	Resistance mechanism[b]	Genetics[b]
A. Auxotrophs					
Adenine	*Datura innoxia* (haploid)	EMS	No	N.R.	N.R.
p-aminobenzoic acid	*Nicotiana tabacum* (haploid)	EMS	Yes	N.R.	Single recessive nuclear allele
Arginine	*N. tabacum* (haploid)	EMS	Yes	N.R.	Single recessive nuclear allele
Biotin	*N. tabacum* (haploid)	EMS	Yes	N.R.	Single recessive nuclear allele
Histidine	*Hyoscyamus muticus* (haploid)	NG	Yes	N.R.	N.R.
Hypoxanthine	*N. tabacum* (haploid)	EMS	Yes	N.R.	Two recessive nuclear alleles
Isoleucine	*N. plumbaginifolia* (haploid)	γ-rays	Yes	A	N.R.
Lysine	*N. tabacum* (haploid)	EMS	Yes	N.R.	N.R.
Nicotinamide	*H. muticus* (haploid)	NG	Yes	N.R.	N.R.
Pantothenate	*D. innoxia* (haploid)	EMS	No	N.R.	N.R.
Proline	*N. tabacum* (haploid)	EMS	No	N.R.	N.R.
Undefined temperature sensitive	*N. tabacum* (haploid)	None[e]	Yes	B	N.R.
Uracil	*N. plumbaginifolia*	γ-rays	No	N.R.	N.R.
B. Miscellaneous Carbohydrate metabolism					
Glycerol utilisation	*N. tabacum*	None	Yes	N.R.	Single[c] dominant nuclear allele
Maltose utilisation Plant growth regulator	*Glycine max*	None	No	N.R.	N.R.[d]
Abscisic acid resistant	*N. tabacum*	EMS	No	N.R.	N.R.
Auxin heterotrophic and auxin resistant	*N. sylvestris* gall	γ-rays	No	C	N.R.
Carboxin resistant	*N. tabacum*	None	Yes	N.R.	N.R.
Cell wall	*D. innoxia*	Azide	No	D	N.R.
UV-resistant	*Rosa damascena*	None	No	E	N.R.

[a] Adopted from Flick (1983).
[b] Abbreviations for mechanism: A—threonine deaminase deficient; B—decreased ornithine decarboxylase, S—adenosyl methionine decarboxylase, nitrate-reductase activity; chlorophyll deficiency; increased PAL activity; C—regulation of auxin synthesis; D—alkaline phosphatase secretor/glucosidase secretor; E—increased polyphenolics. N.R.—not reported.
[c] Homozygous dominant lethal.
[d] Stable in absence of selection.
[e] BUdR selection.

Table 14.5 Selection of cells for antibiotic resistance

Selective agent	Species	Mutagen	Plant regen.	Resistance mechanism[b]	Genetics[a]
Amphotericin B	*Nicotiana tabacum*	EMS	No	O	N.R.
Chloramphenicol	*N. sytvestris* (haploid)	None	Yes[b]	N.R.	N.R.
Colchicine	*Acer pseudoplatanus*	None	No	C	N.R.[c]
Cyclohexamide	*Daucus carota*	None	Yes	D	N.R.[c]
	N. tabacum	EMS	Yes	D	N.R.[c]
Kanamycin	*N. tabacum*	None	Yes	C	Inconclusive[d]
	N. sylvestris (haploid)	None	Yes	R	N.R.
Nystasin	*N. tabacum*	EMS	No	O	N.R.
Streptomycin	*N. tabacum*	None	Yes	R,P	Single cytoplasmic allele
	N. sylvestris	None	Yes	N.R.	Single recessive nuclear allele
	N. sylvestris (haploid)	None	Yes	P	Single cytoplasmic allele

[a] Abbreviations: C—cross-resistant to streptomycin; D—detoxification; O—overproduction for natural compound; P—normal plastid development; R—alteration of chloroplast ribosomal protein. N.R.—Not reported.
[b] Possible chimera.
[c] Mutant not stable.
[d] Erratic segregation ratios implying undefined nuclear genetic control (Source: Flick 1983).

Variation in plants regenerated from gametophytic tissue has been reported in some cases due to uncovering 'residual heterozygosity' (Evans and Sharp 1986a). The approach followed to uncover residual heterozygosity in the plants is to culture anthers from double-haploid and examine the variation that appears in subsequent cycles of androgenesis. However, in the double-haploid plants raised from highly inbred donor plants the residual heterozygosity as a possible source of variation is precluded. Examination of plants derived from the second cycle of the tabacco double-haploid anther culture revealed variation in yield, height, days to flower and total alkaloidal content. Similar variation was recorded in plants from double-haploid *N. sylvestris* raised after several cycles of androgenesis. Based on these observations de Paepe et al. (1983) suggested that gametoclonal variation is the result of factors that cause meiotic recombination and mutation prior to initiating anther culture. Mutations may, however, occur even after or by the doubling process. Further applications of doubled-haploid systems are discussed in Section 8.8.1. The genetic changes associated with use of microspores for regeneration equally contribute to gametoclonal variation. Various examples of gametoclonal variation are given in Table 14.6.

Variation in the chromosome number of gametes or gametophytic tissue plays an important role in gametoclonal variation. This is evident from a range of aneuploids and mixoploids recovered from anther cultures of wheat, maize, and sexual hybrid of wheat and triticale (Hu 1983). Another aspect to be taken note of is that variants uncovered following androgenesis are never recovered from somatic protoplast cultures (de Paepe et al. 1983). Further, several reports suggest that gene

Table 14.6   Examples of gametoclonal variation among some plants raised from gametophytic tissue cultures[a]

Species	Altered characteristics observed	Genetic	Androgenetic (A) or callus mediated (CM)
Brassica napus	Time to flower, glucinolate content, leaf shape and colour, flower type, pod size and shape	Mutation	A
Hordeum vulgare	Plant height, yield, fertility, neck length, days to maturity	Not examined	CM
	Grain yield, culm length, heading date	Transgressive segregation, gametophytic selection	CM
Nicotiana sylvestris	Leaf colour, leaf shape, growth rate, DNA content, repeated sequences of DNA	Crumpled leaf is nuclear, germ amplification	A
N. sylvestris (several cycles)	Flower length, plant height, leaf shape, leaf colour, exerted stigma, capsule weight, foliar outgrowths	Nucleus genes	A
N. tabacum	Yield, grade, time to flower, plant height, number of leaves, leaf shape, total alkaloids, % reducing sugars	Not examined	A
	Yield, plant height, number of leaves, % nornicotine	Not residual heterozygosity	A
	TMV resistance, root knot resistance	Variation in expected segregation ratios	A
	Days to flower, total alkaloids, yield per plant, leaves per plant	Residual heterozygosity; no cytoplasmic effects	A
	Yield reduction, plant height, days to flower, total alkaloids	Nuclear mutations except plant height is maternal; not residual heterozygosity	A
	Yield reduction, alkaloid content, leaf size, time of flowering, leaf quality	Induced changes, not residual heterozygosity	A
	Heterochromatin content, nuclear DNA content	Gene amplification	A
	Time of flowering, plant size, leaf shape, alkaloid content, number of leaves	Residual heterozygosity	A
	Altered temperature sensitivity	Cytoplasmic mutant	A

Contd...

Table 14.6   (Contd.)

Species	Altered characteristics observed	Genetic	Androgenetic (A) or callus mediated (CM)
Oryza sativa	Chloroplast content, time to flower, plant height, fertility	Some single genes induced in culture	CM
	Plant height, seed and leaf size, seed weight, tillering, flowering, % open hulls	Transmitted to progeny	CM
	Seed size, seed protein level, plant height, level of tillering	Transmitted to progeny	CM
	Waxy mutant	Single gene change	CM
Saintpaulia ionantha	RuBCase activity	Not examined	A
Secale cereale × *S. vavilotrii*	Fertility	Not analyzed	CM

[a] For detailed references *see* Evans and Sharp (1986a).

amplification contributes to reduction in the yield of gametoclonal variant plants (Dhillon et al. 1983). Thus, the mutation spectrum obtained by gametoclonal variation may differ from that obtained by somaclonal variation. Sometimes a combination of gametoclonal variation (PHS Method) and somaclonal variation could result in release of improved varieties, such as 'DAMA' rice (*see* Section 14.3.1) Gametosomatic hybridisation is being explored as alternative tool for somaclonal variant selection. (Skálóva et al. 2012, also *see* Chapter 12).

14.5   Applications in Plant Breeding

Somaclonal variation and gametoclonal variation represent useful sources of introducing genetic variations that could be of value to plant breeders. Single gene mutation in the nuclear or organelle genome may give the best available variety *in vitro* that has a specific improved character. In this manner somaclonal variation could be used to uncover new variants retaining all the favourable characters along with an additional useful trait, such as resistance to

diseases or an herbicide. These variants can then be field-tested to ascertain their genetic stability. Reciprocal crosses between desirable R_1 progeny and seed-derived controls would further stabilise the variants and help in seed-set among the promising lines. Gametoclonal variation, induced mostly by meiotic recombination during the sexual cycle of the F_1 hybrid, results in transgressive segregation to uncover unique gene combinations.

Various cell lines selected *in vitro* may prove potentially applicable to agriculture and industry. Plants regenerated have expressed traits for resistance to herbicide, pathotoxin, salt or aluminium. Additionally, the variability in cell cultures has played a useful role in the synthesis of secondary metabolites on a commercial scale (*see* Chapter 17).

The techniques employed for induction of somaclonal and gametoclonal variation are relatively easier than recombinant DNA technology. In particular, the improvement of crops with polygenic traits by conventional and non-conventional breeding methods have proved very difficult. Somaclonal variation may be a

Table 14.7 Aspects of improvement desired in some crop plants through somaclonal variation[a]

Crop plants	Improvement desired
Cereals	
Rice	Tungro virus and leaf hopper resistance
Wheat	Quality, high temperature tolerance
Sorghum	Borer and shoot fly resistance
Maize	Delinking of phytohormone effect from reproductive development
Legumes	
Pigeon-pea	Resistance to pod borer, fusarium wilt, high protein content
Bengal gram	
Oilseeds	
Groundnut	Resistance to *Cercospora, Fusarium, Aspergillus flavus*
Rapeseed	Male sterility, low pungency combined with insect resistance
Sunflower	Self-compatibility, resistance to *Alternaria* leaf blight, *Rhizoctonia, Fusarium* complex
Cotton	Insect resistance to gossypol free lines
Vegetables	
Potato	Virus and *Phytophthora* resistance
Tomato	Resistance to disease
Onion	
Fruits	
Grapes	Resistance to disease, short duration
Citrus	Higher solid content, increased shelf life
Watermelons	
Other(s)	
Sugarcane	Resistance to disease, short duration
Castor	Mutants with high fatty acids; oleic acid, ricinoleic acid

[a] *See* also Bhaskaran (1987).

proper technology for genetic manipulation of these crops (Table 14.7).

The occurrence of uncontrolled somaclonal variation in tissue culture raised plants is at present limiting the use of *in vitro* methods for clonal propagation of selected genotypes. Despite these drawbacks, somaclonally derived mutants are finding their way to the market and in agriculture. For example, Fresh World Farms in USA has been marketing a somaclone of tomato with altered colour, taste, texture and shelf-life (Larkin 1998). Further, American Cyanimid was expected to release imidazolinone herbicide-resistant maize to the market. Newly acquired variants of bermudagrass (*Cynodon dactylon*) called *Brazos R-3*, with increased resistance to fall armyworm (Croughan et al. 1994) and sweet pea *(Lathyrus sativus)* somaclone, which no longer accumulates neurotoxin in the grain (Yadav and Mehta 1995), are now available to farmers. Somaclonal variation and induced mutations in crop improvement are comprehensively dealt by Jain et al. (1998) and Baer et al. (2007).

PART IV
Application to Horticulture and Forestry

Production of Disease-Free Plants

Most of the horticultural and forest crops are infected by systemic diseases caused by fungi, viruses, bacteria, mycoplasma, aphids (insects), seeds and pollen. Pathogen attack does not always lead to the death of the plant but very often the infection caused by pathogens considerably reduces the yield and quality of crops. While pathogens are nearly always transferred in plants through vegetative propagation, viral diseases occur in virtually all seed-propagated as well as vegetatively propagated crop species (Kartha 1984). Eradication of pathogens is highly desirable to optimise the yields and also to facilitate the movement of materials across the international boundaries. While plants infected with bacteria and fungi may respond to treatments with bactericidal and fungicidal compounds, there is no commercially available treatment to cure virus-infected plants. The therapeutic chemicals capable of eradicating virus from infected plants are either not available or considered environment unfriendly though the use of virazole and vidarabine (antimetabolites) in the culture medium have resulted in the production of virus-free lily and apple plants (*see* Pierik 1989, Wang et al. 2009).

To produce disease-free plants a healthy nucleus stock could be developed by selecting out one or more healthy plants and then multiplying them vegetatively, but where the entire population of a clone is infected the only way to obtain a pathogen-free plant is through tissue culture.

Apical meristems in the infected plants are generally either free or carry a very low concentration of the viruses. However, the titre of the viruses increases in the older tissues corresponding to the increase in the distance from the meristem tips. Various reasons attributed to the escape of the meristems by virus invasion are: (a) viruses move readily in a plant body through the vascular system which in meristems is absent, (b) a high metabolite activity in the actively dividing meristematic cells does not allow virus replication and (c) a high endogenous auxin level in shoot apices may inhibit virus multiplication.

A knowledge of the gradient of virus distribution in shoot tips encouraged Holmes (1948) to obtain virus-free plants from infected individuals of *Dahlia* through shoot-tip cuttings. Morel and Martin (1952) applied tissue culture techniques for elimination of viral infection in plants. They cultured meristem tips excised from infected *Dahlia* and obtained disease-free plants. Since then the advances made in elimination of virus by tissue culture have been such that this approach has now become a popular horticultural practice (Table 15.1). Meristem-tip culture has also enabled plants to

Table 15.1 Plants from which viruses have been eliminated by tissue culture technique[a]

Family	Species	Virus eliminated[b]
Amaryllidaceae	*Hippeastrum* sp.	Mosaic
	Nerine	Nerine latent, unidentified
	Narcissus tazetta	AMV, NDV
Araceae	*Caladium hortulanum*	Dasheen Mosaic
	Calocasia esculenta	Dasheen Mosaic
	Xanthomonas brasiliensis	Unidentified
Bromeliaceae	*Ananas sativus*	Unspecified
Caryophyllaceae	*Dianthus barbatus*	Latent, Mottle, Ringspot
	D. caryophyllus	Vein Mottle
Compositae	*Chrysanthemum* sp.	Chlorotic Mottle, Complex of viruses, Green Flower Stunt Vein Mottle, Virus B
	Dahlia sp.	Complex of viruses *Dahlia* Mosaic, Tomato Aspermy Vein Mottle, Virus B
Convolvulaceae	*Ipomoea batatas*	Feathery Mottle, Hanmon Mosaic Internal Cork, Rugosa Mosaic, Synkuyo Mosaic
Cruciferae	*Armoracia lapathifolia*	CLMV, TuMV
	A. rusticana	TuMV
	Brassica oleracea	CbBRSV
	Nasturtium officinale	CMV, CLMV, TuMV
Daphneceae	*Daphne* sp.	AMV, CMV, RbRSV
	D. odora	Daphne Virus S
Euphorbiaceae	*Manihot* sp.	African Cassava, Mosaic, Cassava Brown Streak, Mosaic
Geraniaceae	*Pelargonium* sp.	CMV, Tomato Black, Ringspot ,Tomato Ringspot
Gramineae	*Lolium multiflorum*	RgMV
	Saccharum officinarum	Mosaic
Hydrangeaceae	*Hydrangea macrophylla*	Hydrangea Ringspot
Iridaceae	*Freesia* sp.	FrMV
		Freesia Virus I
		Phaseolus Virus 2
	Iris sp.	IMV
	Gladiolus sp.	Unidentified virus
Labiatae	*Lavandula* sp.	Dieback
Leguminosae	*Glycine max*	SMV
	Trifolium pratense	WCMV
Liliaceae	*Allium sativum*	GMV, OYDV, GYSV
	Asparagus officinalis	Unspecified
	Hyacinthus sp.	HyMV
	Lilium sp.	CMV, HyMV, Lily Symptomless Latent, LMV, unidentified
Moraceae	*Humulus lupulus*	Hop Latent and Necrotic Ringspot

Contd...

Table 15.1 (Contd.)

Family	Species	Virus eliminated[b]
Musaceae	*Musa* sp.	CMV, unidentified
Orchidaceae	*Cymbidium* sp.	Cymbidium Mosaic
Polygonaceae	*Rheum rhaponticum*	Tobacco, Rattle, CMV, ChLRV, Strawberry Latent Ringspot, TuMV
Ranunculaceae	*Ranunculus asiaticus*	Unidentified
Rosaceae	*Fragaria* sp.	Complex of viruses Crinkle, Edge, Latent A, Latent C, Pallidosis, Strawberry Yellow Edge, Vein Banding, Vein Chlorosis, Mottle
	Malus sp.	Latent viruses
	Rubus ideaus	Mosaic
Saxifragaceae	*Ribes grassularia*	Vein Banding
Solanaceae	*Nicotiana rustica*	AlfMV, CMV, ChLRV, AMV, Tobacco ringspot
	N. tabacum	Dark-green islands of TMV
	Petunia sp.	TMV
	Solanum tuberosum	Leaf Roll, Paracrinkle, PVA, PVG, PVM, PVS, PVX, PVY
Zingiberaceae	*Zingiber officinale*	Mosaic

[a] For detailed references of each species *see* Bhojwani and Razdan (1983, 1996); Grout (1999).
[b] Abbreviations:

AlfMV	=	Alfalfa Mosaic Virus	OYDV =	Onion Yellow Dwarf Mosaic Virus
AMV	=	Arabis Mosaic Virus	PVA =	Potato Virus A
CbBRSV	=	Cabbage Black Ringspot Virus	PVG =	Potato Virus G
ChLRV	=	Cherry Leaf Roll Virus	PVM =	Potato Virus M
CLMV	=	Cauliflower Mosaic Virus	PVS =	Potato Virus S
CMV	=	Cucumber Mosaic Virus	PVX =	Potato Virus X
FrMV	=	Freesia Mosaic Virus	PVY =	Potato Virus Y
GMV	=	Garlic Mosaic Virus	RbRSV =	Raspberry Ringspot Virus
GYSV	=	Garlic Yellow Streak Virus	RgMV =	Ryegrass Mosaic Virus
HyMV	=	Hyacinth Mosaic Virus	SMV =	Soybean Mosaic Virus
IMV	=	Iris Mosaic Virus	TMV =	Tobacco Mosaic Virus
LMV	=	Lily Mosaic Virus	TuMV =	Turnip Mosaic Virus
NDV	=	Narcissus Degeneration Virus	WCMV =	Wild Clover Mosaic Virus

be freed from other pathogens including viroids, mycoplasmas, bacteria and fungi (*see* Walkey 1978, Bhojwani and Razdan 1983, 1996; Kartha 1984, Pierik 1989, Grout 1999, Kane 2000, Khayat 2012).

15.1 Methods of Virus Elimination

Plants are often infected with more than one type of virus, including some not even known. A plant can be claimed as free of only those viruses for which specific tests have given negative results and accordingly labelled as 'free of a specific virus or pathogen'. However, a general term 'virus-free' is used by commercial horticulturists for plants freed of any type of virus. It must also be borne in mind that virus-free material can be reinfected if proper precautionary measures are not adopted. These are listed towards the end of this chapter.

15.1.1 Heat Treatment

Before the advent of meristem cultures the *in vivo* eradication of viruses from plants was achieved by heat treatment (thermotherapy) of whole plants (Hollings 1965). At temperatures higher than optimum many viruses in plant tissues are partially or completely inactivated with little or no injury to the host tissues. Heat treatment is given through hot water or hot air. The hot-water treatment effectively eliminates virus in dormant buds, whereas hot-air treatment is recommended for elimination of virus from actively growing shoots. Moreover, the survival rate of the host tissue is better in hot-air treatment. For hot-air treatment, actively growing plants are placed in a thermotherapy chamber adjusted at a temperature of 35-40°C for a period varying from a few minutes to several months. An adequate supply of humidity and light should be maintained during this period. Further, plants must have ample carbohydrate reserves to withstand the heat treatment. This may be achieved by pruning of the plants (Wang and Hu 1980). The temperature of the air should be gradually raised during the first few days until the desired temperature is reached. Small cuttings are taken from the shoot tips immediately after heat treatment and grafted onto healthy rootstocks.

Baker and Kinnaman (1973) eradicated all viruses from carnation shoot tips by continuous treatment of plants at 38°C for 2 months. They also recommended that a relative humidity of 85-95% must be maintained during hot-air treatment of carnation. A major limitation of thermotherapy for virus eradication has been that not all viruses are sensitive to this treatment. In particular, this type of treatment has proved effective against isometric and thread-like viruses, and diseases caused by mycoplasmas (Quak 1977). Some plants, such as potato require several months of treatment to eradicate virus (Leaf Roll Virus) in them. Prolonged heat treatments reportedly inactivate the resistance factor(s) in the host-plant tissues and may prove counterproductive by increasing the sensitivity of cured plants to pathogens.

Generally the percentage of plants that survive after heat treatment is small. Moreover, many viruses which could not be eradicated by this treatment have been eliminated by tissue (meristem tip) culture *with* or *without*

thermotherapy. In view of this, meristem-tip culture has become the most widely applicable approach to virus elimination. It may, however, be noted that extended low temperature (2-4°C) of the donor material could also be effective as found in elimination of Hop Latent Viroid from hop plants (Adams et al. 1996).

15.1.2 Meristem-tip Culture

The size of the explant is critical for virus elimination since in most of the systemic viral diseases the pathogen often establishes a gradient in plant tissues. It has been generally observed that virus-free plants regenerated *in vitro* are inversely proportional to the size of the meristem cultured. About 60% cassava (*Manihot esculenta*) plantlets regenerated from 0.4 mm meristems, isolated from sprouted cuttings of infected plants, happened to be mosaic-disease free. In contrast, plants regenerated from large meristems were indexed as diseased (Kartha and Gamborg 1975). For *Dahlia* or gapes 400 μm-500 μm long meristem-tips seems optimum size in culture to obtain from them virus-free plants, whereas for rose meristems large range of the size 200 μm-800 μm appears optimum to regenerate virus-free plants (c.f. Bhojwani and Dantu 2013). However, studies on plants regenerated from meristem cultures of potato plants infected with different viruses indicate that in addition to the size of the explant (meristem), the role of certain host-virus combinations contributes to the success of virus elimination (Mellor and Stace-Smith 1977). Accordingly, viruses labelled in order of increasing difficulty of elimination are potato leaf roll virus (PLRV), potato virus A (PVA), PVY, PYM, PVS, aucuba mosaic virus and spindle tuber viroid (PSTV). Nevertheless, using meristem tip culture virus free plants have been obtained in important crops like banana, grapes, strawberries, citrus, potatoes, bamboo and some tree spices (Khayat 2012).

(A) Explant Terminology

In attempts to recover pathogen-free plants

through tissue culture techniques, horticulturists and pathologists have designated the explants used for initiating cultures as 'shoot tip', 'tip', 'meristem', and 'meristem tip'. The portion of the shoot lying distal to the youngest leaf primordium and measuring up to about 100 μm in diameter and 250 μm length is called the *apical meristem*. The apical meristem together with one to three young leaf primordia measuring 100-500 μm constitute the *shoot apex* (*see* Cutter 1965, Hollings 1965). In most published works explants of larger size (100-1000 μm long) have been cultured to raise virus-free plants. The explants of such a size should in fact be referred to as *shoot tips*. However, for purposes of virus or disease elimination the chances are better if cultures are initiated with shoot tips of smaller size comprising mostly meristematic cells (Baker and Phillips 1962). Therefore, the term 'meristem' or 'meristem-tip' culture is preferred for *in vitro* culture of small shoot tips.

(B)   *Excision of Meristem Tips*

The *in vitro* techniques used for culturing meristem tips are essentially the same as those used for aseptic culture of plant tissues (*see* Chapter 4). Before initiating meristem cultures it is essential to ensure that explants are free of surface pathogens. Meristem tips can be isolated from apices of stems, tuber sprouts, leaf axils, sprouted buds of cuttings, or germinated seeds. These explants are so well protected with overlapping leaf primordia that a careful dissection should permit excision of aseptic explants without surface sterilisation treatment. However, a thorough disinfection is required for meristems originating from underground plant organs such as tubers, rhizomes, bulbs and corms. As a precautionary measure the shoot buds may also be surface-sterilised by a quick dip in 75% ethanol before excising explants despite the high degree of sterility maintained around meristem tips.

Meristem-tip explants used for virus elimination are too small to be isolated with unaided eyes. A stereoscopic microscope (8-40

magnification) with a suitable (preferably cool fibre) light source is required to excise active meristems. Care must be taken to prevent desiccation of shoot tips due to constant flow of air through the laminar air-flow cabinet and the heat generated by the light source attached to the microscope used for dissection. Performing dissection in a petri plate lined with a sterile moist filter paper helps to retard desiccation of small explants.

A set of forceps, needles and blades (mounted on separate handles of suitable lengths) are used to perform dissections. These instruments need flame sterilisation (*see* Chapter 4) after each step in the dissection process to avoid contamination of the explant. To excise meristem tips the bud is held under the microscope in one hand with a fine pair of forceps. This is followed by removing leaves and leaf primordia with the help of fine needles to expose the apical meristem, which appears as a shiny dome (Fig. 15.1). The meristem is then excised by giving a clean cut with a sharp blade mounted on a large handle. The same instrument may be used to transfer the tip to the culture medium.

(C)   *Physiological Conditions of Explant*

Meristem tips should preferably be taken from actively growing buds. The position of the bud appears to be important for providing an explant source in carnations and chrysanthemums. Meristem tips taken from terminal buds were observed to give better results than those from axillary buds (Stone 1963), but such a difference was not discernible in strawberry (Boxus et al. 1977). Often several axillary buds develop per shoot and, therefore, can be used to increase the overall production of virus-free plants. The percentage of virus-free plants can also depend on the season, especially with crops which display periodic growth. In temperate trees the growth of the plant is limited to only a very short period in the spring, after which the shoot apex remains dormant for a long time. In such cases meristem tips should be cultured during spring

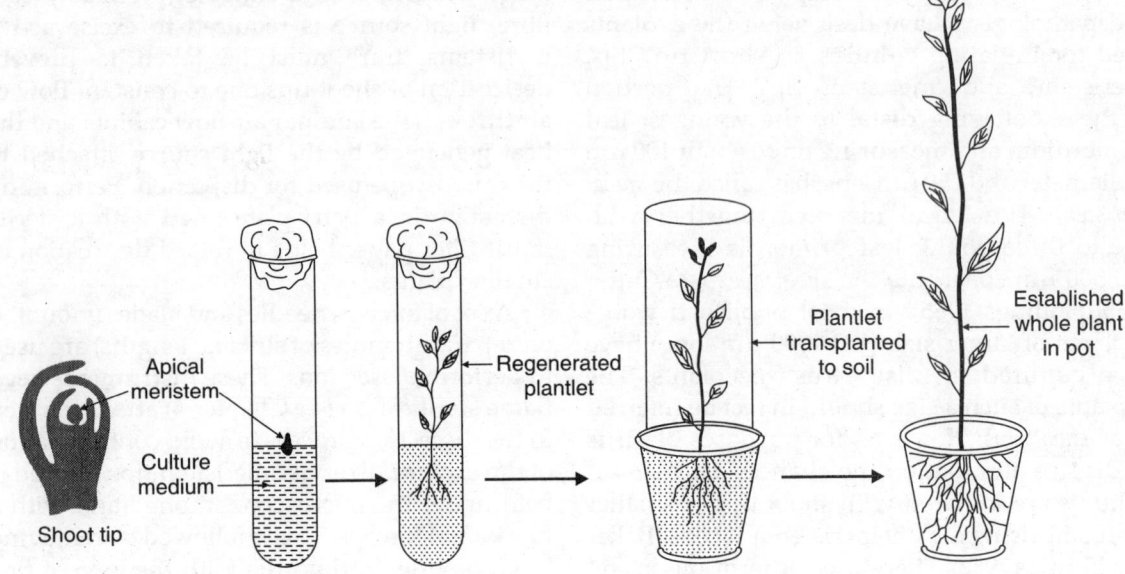

Fig. 15.1 Scheme showing steps in meristem-tip culture (adopted from Pierik 1989).

or only after breaking dormancy by suitable treatments (chilling or light effects).

The efficiency of meristem-tip culture will also depend on the genotype and rootability of shoots regenerated from them and their freedom from pathogens. To ensure prolific root growth in the shoots, the meristem-tip culture may be initiated in a particular season. For potato varieties, the meristem tips taken in spring and early summer root more readily than those taken later in the year (Quak 1977).

(D) *Thermotherapy and Meristem-Tip Culture*

Although the apical meristems are often virus-free, there may be exceptions. Some viruses are known to actually invade the meristemaric region of the growing tips in certain plants. These viruses have been identified as TMV, PVX and CMV (Walkey and Cooper 1976, Mori 1977). In such cases it has also been possible to obtain virus-free plants by combining meristem-tip culture with thermotherapy. This becomes necessary because neither of these viruses is readily eliminated by thermotherapy or meristem-tip culture alone. Heat treatment can be given to the mother plant before excising the meristem tips or, alternatively, tip cultures can be exposed to high temperatures. In most strawberry varieties heat treatment (36°C for 6 weeks) also improved the growth rate of plants regenerated from meristem-tips (Mullin et al. 1974). However, the duration of heat treatment has to be decided carefully since excessive exposure can damage the plant tissues. To avoid deterioration of meristem-tip cultures by continuous exposure to high temperature, the treatment with diurnal or daily cycles of high and low temperatures can be tried. For example, CMV in tissue cultures of *Nicotiana rustica* and *Stellaria indica* can be inactivated by following the diurnal cycles of 40°C (16 hr) + 22°C (8 hr) per day or the daily cycle of 2 days at 40°C and 2 days at 35°C. Addition of virazole in the growth medium seems to be effective for elimination of artichoke latent virus (ALV) in plants regenerated from shoot-tip culture combined with thermotherapy (Khayat 2012).

(E) *Culture Medium*

The composition of the culture media used for shoot-tip culture of some crop plants is given in Table 15.2. However, the suitability of the MS medium (*see* Chapter 3) for meristem-tip culture has been emphasised by several workers. Generally, a small amount (0.1-0.5 mg l^{-1}) of auxin or cytokinin, or both, is beneficial. Among auxins, IAA and NAA are widely used since 2,4-D promotes callusing of the explant. Of these, NAA is considered more effective due to its better stability. In angiosperms the meristematic dome in the shoot tip is not autonomous for auxin. According to Shabde and Murashige (1977), the auxin is probably synthesised by the second pair of youngest leaf primordia; therefore, the presence of exogenous auxin in the medium seems essential for the successful culture of meristem explants without leaf primordia. Those explants requiring only auxin, or cytokinin, probably have a high level of the other growth regulator not required by them endogenously in their meristems.

GA_3 (0.1 mg l^{-1}), in combination with BAP and NAA, is reported to be essential for plant regeneration from excised meristem tips (200-500 μm) of cassava (Kartha et al. 1974). In *Dahlia* meristem-tip cultures, the presence of GA_3 suppressed callusing and favoured better growth and differentiation. However, many other workers have found GA_3 has no appreciable effect on meristem cultures and may even prove inhibitory at higher concentrations.

Meristem-tip cultures are generally initiated on solidified (agar) medium although both liquid and semi-solid media have been tried. In cultures in which an agar medium induces callusing of the explants, a liquid medium is recommended. For liquid cultures a filter-paper bridge is prepared inside the culture vessel in such a way that the two arms of the filter paper dip into the medium and the platform on which the explant is placed remains above the medium (Fig. 15.2).

(F) *Storage Conditions*

Meristem-tip cultures require light incubation except those (e.g., *Pelaragonium*) which need dark incubation to minimise the effect of polyphenols. The optimum light intensity found for initiating potato tip cultures is 100 1x which should be further increased up to 200 1x after 4 weeks. As soon as the shoots attain a height of 1 cm the light intensity must be raised to 4000 1x (*see* Wang and Hu 1980).

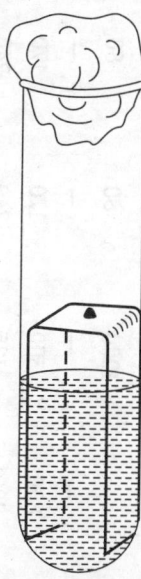

Fig. 15.2 Filter-paper bridge technique of culturing excised shoot tip in liquid medium (after Bhojwani and Razdan 1983).

15.1.3 *Chemical Treatment*

In the absence of effective therapeutic chemicals capable of eradicating virus from infected plants under field conditions, there are reports of some attempts to suppress viruses in plant tissues and protoplasts by the addition of some chemicals in the media. This is known as chemotherapy. However, these efforts produced conflicting results. In some instances virus multiplication was suppressed by the addition of cytokinins and other growth substances, while in others it was actually stimulated. Antimetabolites malachite green, thiouracil and acetylsalicylic acid had

Table 15.2 Culture media used for shoot-tip culture of some crop plants

Constituents	Media (amount in mg l^{-1})[a]								
	1	2	3	4	5	6	7	8	9[c]
NH_4NO_3	–	–	–	–	–	60	60	–	1650
KNO_3	125	125	125	200	125	–	–	125	1900
$(NH_4)_2SO_4$	–	–	–	–	1000	–	–	–	–
KCl	–	–	–	–	1000	80	80	–	–
$CaCl_2 \cdot 2H_2O$	–	500	500	–	–	–	–	500	440
$Ca(NO_3)_2 \cdot 4H_2O$	500	–	–	800	500	170	170	–	–
$MgSO_4 \cdot 7H_2O$	125	125	125	200	125	240	240	125	370
KH_2PO_4	125	125	125	200	125	40	40	125	170
$FeCl_3 \cdot 6H_2O$	–	–	–	–	1	–	–	–	–
Fe-citrate	–	–	–	–	5	5	–	–	–
$Fe(SO_4)_3$	–	25	–	–	–	–	–	25	–
$FeSO_4 \cdot 7H_2O$	–	–	–	–	–	–	27.8	–	27.85
$Na_2 \cdot EDTA$	–	–	–	–	0.1	–	37.3	1	37.25
$MnSO_4 \cdot 4H_2O$	–	0.8	–	–	1	–	22.3	0.05	22.30
$ZnSO_4 \cdot H_2O$	–	0.04	–	0.2	1	0.05	8.6	0.025	8.60
$NiCl_2 \cdot 6H_2O$	–	0.025	–	0.3	–	–	–	0.025	–
$MnCl_2 \cdot H_2O$	–	–	–	1.8	–	0.4	–	–	–
$COCl_2 \cdot 6H_2O$	(10 drops of Berthelot soln.)[b]	0.025	(10 drops of Berthelot soln.)[b]	–	0.03	0.05	0.025	0.025	0.025
$CuSO_4 \cdot 5H_2O$		0.025		0.08	0.03	–	0.025	0.025	0.025
$AlCl_3$				–	–	0.02	–	–	–
$H_2MoO \cdot H_2O$				0.02	–	–	–	–	–
$Na_2MoO_4 \cdot 2H_2O$	–	–	–	–	0.01	–	0.025	0.25	0.25
KI	–	0.25	–	–	1	0.6	0.83	0.25	0.83
H_3BO_3	–	0.025	–	2.8	1	0.1	6.2	0.025	6.2
Myoinositol	0.1	0.001	0.1	–	100	10	0.1	–	–
Ca-pantothenate	10	0.001	10	–	–	–	10	–	–
Inositol	–	–	–	–	–	–	–	–	100

Contd...

Table 15.2 (Contd.)

Constituents	Media (amount in mg l^{-1})[a]								
	1	2	3	4	5	6	7	8	9[c]
Nicotinic acid	1	—	1	5	1	1	1	—	0.5
Pyridoxine-HCl	1	0.001	1	1	1	1	1	—	0.5
Thiamine-HCl	—	—	—	1	1	1	1	1	0.1
Glycine	—	—	—	—	—	—	—	—	2.0
Biotin	0.1	0.001	0.01	—	0.01	0.01	0.01	—	—
Cystein	—	0.001	10	—	1	10	10	—	—
Adenine	—	—	—	—	0.1	5	5	—	—
AdSO$_4$	—	—	—	—	—	—	—	8	—
Casein hydrolysate	—	—	1	—	—	1	1	—	—
Sucrose	20000	—	20000	30000	20000	10000	30000	40000	—
Glucose	—	40000	—	—	—	—	—	—	—

[a] After: 1—Morel (1948); 2—Morel and Martin (1952); 3—Kassanis (1957); 4—Nielsen (1960); 5—Morel and Müller (1964); 6—Mori (1971); 7—Wang and Huang (1975); 8—Baker and Kinnaman (1973); 9—Mellor and Stace-Smith (1977).

[b] Berthelot solution contains (mg l^{-1}): MnSO$_4$ (2000), NiSO$_4$ (60), TiO$_2$ (40), CoSO$_4$ (60), ZnSO$_4$ (100), CuSO$_4$ (50), BeSO$_4$ (100), H$_3$BO$_3$ (50), Fe$_2$(SO$_4$)$_2$ (50000), KI (50), H$_2$SO$_4$ (sp. gr. 1.83) (1 ml).

[c] Medium used especially for potato shoot-tip culture contains growth hormones (mg l^{-1}); Kinetin (0.04), IAA (0.5), GA$_3$ (0.1). Addition of activated charcoal (2857) is optional. For media 1-8 the proportion of growth hormones varies according to the plant species.

little or a limited effect on virus elimination in meristem-tip regenerated plants, but the incorporation of nucleoside analogue ribavirin (syn. virazole) to the culture medium exerted a positive effect in eliminating PVX virus in potato plantlets regenerated from mesophyll protopasts (Shepard 1977), CMV and alfalfa mosaic virus in various types of cultured plant tissues (Pierik 1989). Another analogue, vidarbine, removed a virus from apple plants raised in cultures. Cyclo-hexamide and actinomycin-D are also known to inhibit virus replication in plants (Fasella and Hussain 2014).

15.1.4 Cryotherapy and Electrotherapy

Elimination of viruses, phytoplasmas and bacteria from shoot tips by exposing to liquid nitrogen is called cryotherapy. Thermotherapy followed by cryotherapy of shoot tips aids in elimination of nine viruses viz. banana streak virus A, cucumber mosaic virus, grape vine virus A, plum pox virus, potato leaf roll virus, potato virus Y, raspberry bushy dwarf virus, sweet potato feathery mottle virus and sweet potato chlorotic stunt virus. Sweet potato leaf phytoplasma and Hanglongbing bacterium (causing citrus "greening") are other pathogens eliminated using this approach. For a detailed methodology see review by Wang et al. (2009).

Electrotherapy involves exposure of shoots to electrical pulses (15-15mA) for 5-10 minutes. The shoots are then surface sterilized and cultured in suitable medium. This technique has been successfully used to eradicate viruses from potato, banana and tomato (Falah et al. 2009, Meybodi et al. 2011).

15.1.5 Other in vitro Methods

Callus cultures may also prove useful in raising plants free of virus. Based on the fact that distribution of viruses in plants is uneven, the callus derived from infected tissues does not carry the pathogen uniformly in all cells. This became evident when virus-free tobacco plantlets regenerated *in vitro* from TMV-infected callus

cultures (Hansen and Hildebrandt 1966). The possible reasons for the escape of some cells of a systematically infected callus from virus are: (a) virus replication is unable to keep pace with cell proliferation and (b) some cells acquire resistance to virus infection through mutagenesis. Virus-free plants have also been regenerated from shoot-tip calli of several other plant species.

Murakishi and Carlson (1976) utilised the uneven distribution of virus in tobacco leaf to obtain TMV-free plants. Explants (1 mm) isolated from the dark green islands (areas) of TMV-infected tobacco leaves were aseptically cultured and about 50% of the regenerated plants were virus-free. Similarly, White (1982) cultured TMV-induced lesions of *Nicotiana tomentosa* and successfully obtained 60% plants free from pathogens. Obviously, the dark green patches in the leaves of these plants were either virus-free or had a very low concentration of the virus. Calli derived from nucellus (*Citrus*), anther and floral meristems have also been used to raise virus-free plants. The drawbacks of virus elimination through calli cultures that need to be borne in mind are the genetic instability of cultured cells and low potential of plant regeneration in respect of some important crop species. In this context elimination of virus via somatic embryogenesis in anther callus has been employed to produce disease free plants of grape vine (Borroto-Fernandez et al. 2009).

Other *in vitro* methods that can be followed for regeneration of plants for disease resistance are somatic cell hybridisation, gene transformation and somaclonal variation. These have been discussed in previous chapters.

15.2 Virus Indexing

Testing plants for the presence or absence of viruses is called virus indexing. Every meristem tip, callus-derived or micrografted plant must be tested before using it as a mother plant to produce virus-free stock. Many viruses have a delayed resurgence period in cultured plants. This necessitates indexing of plants several times at periodic intervals and only those individuals which give consistently negative results should be labelled as 'virus-tested' for a specific virus. Methods followed for virus indexing are the sap transmission test, serology, EM examination, and Nucleic acid-based tests.

(A) Sap Transmission Test

This is the most sensitive of the three methods and can be easily performed on a commercial scale. Leaves are taken from the *test* plant (plant to be assayed for the presence or absence of virus) and ground in an equal volume (w/v) of buffer solution (0.1 mol l^{-1} sodium phosphate) using a mortar and pestle. The filtered-leaf extract (sap) is then smeared onto the leaf of an *indicator* plant (a plant susceptible to a specific virus or group of viruses) predusted with 600-grade carborandum powder. The indicator plant develops the characteristic symptoms of the virus should it be present in the sap. Some of the indicator plants used are *Chenopodium amaranticolor, C. quinoa, Gomphrena globosa, Nicotiana tabacum, N. benthamiana* and *N. clevelandii* (Klopmeyer 2000). The inoculated indicator plants are maintained in a greenhouse or aphid-proof cages separated from each other and from other plants.

(B) Serology

This test is performed by adding a drop of centrifuged sap from a test plant to a drop of antiserum taken from the blood of a rabbit. If the virus is present, the precipitation will take place due to the presence of specific antibodies in the blood. The enzyme-linked immunosorbent assay (ELISA) is one of the serological methods used to identify virus(es) based on antibody reaction (*see* Pierik 1989, Gamborg and Phillips 1995).

(C) EM Studies

EM observations are particularly useful for identifying latent viruses (viruses which exhibit no visible symptoms). This method is not usually implemented as specialised equipment and trained personnel are required to carry out EM studies. Immunosorbent Electron Microscopy (ISEM) procedure described by Barbara and Clark (1986) combines both serology and EM studies for detection of viruses.

(D) Nucleic-Acid-Based Tests

Several variations of RT-PCR (reverse transcriptase-polymerase chain reaction) and loop-mediated isothermal amplification (LAMP) methods detect even very low amount of viruses (RNA) in plants. For detection of viroids specific nucleic acid hybridizations are found useful. There are also reports of simultaneous assay of pathogens such as viruses, viroids and phytoplasmas using DNA microarray technology (Rowhani et al. 2005, Fasella and Hussain 2014).

15.3 Micrografting

Initiating meristem cultures of some crop species, particularly in woody perennials, is difficult. The shoots regenerated from meristem cultures of woody species quite often do not form roots. For such species micrografting is done, which enables them to become virus-free. The concept of micrografting evolved from the work of Morel and Martin (1952) on *Dahlias*. In this technique the meristems are grafted onto a virus-free rootstock (seedling) maintained and propagated *in vitro*. The first successful micrografting was carried out by Murashige et al. (1972) and, subsequently, Navarro et al. (1975) obtained *Citrus* plantlets in which two viruses had been eliminated. Micrografting has proved highly useful for elimination of virus in fruit crops, namely apricot, wine-grape, *Eucalyptus, Camellia japonica* and peach. An extensive appraisal of micrografting of fruit crops is given by Jonard (1986).

15.4 Eradication of Pathogens other than Virus

Mainly meristem-tip culture and callus cultures have been used to eliminate viruses in plants. There is evidence suggesting that these cultures may also contribute to the eradication of fungi, bacteria and mycoplasmas in the infected plants. Baker and Phillips (1962) successfully eliminated the fungus *Fusarium roseum f. cerialis* from carnation plants using meristem cultures. Another important bulbous ornamental, *Gladiolus*, is commonly attacked in France by *F. oxysporium f. gladioli.* Tramier (1965) obtained gladioli plants from meristem-tip cultures which were free from this fungus. Representatives of some important genera of fungi eliminated from various plants in cultures include *Verticillium, Phialphora, Phytophthora* and *Rhizoctonia.* In a similar manner, bacteria such as *Pseudomonas carophylli* and *Pectobacterium parthenii* have been excluded from carnation, *Xanthomonas pelargonii* from pelargonium and *X. begoniae* from begonia hybrids (*see* Pierik 1989).

Elimination of systemic bacterial and fungal pathogens from geraniums has been possible using *culture indexing.* This procedure, initially developed in 1940s to free *Chrysanthemum* of *Verticillium* wilt, was utilized to eliminate major bacterial (*Xanthomonas campestris* cv. *pelargonii,* Xcp) and fungal (*Verticillium* spp.) pathogen of geranium (Oglevee-O'Donovan 1993). Culture indexing is simplistic in its method and very effective in selection of plants free of systemic fungal and bacterial pathogens (Klopmeyer 2000). The steps entailed in culture indexing are mentioned in Appendix 15.1.

Some mycoplasma-like bodies cause aster yellow disease in carrot plants. Explants excised from these plants have been liberated of this pathogen by repeated transfers in cultures. Jacoli (1978) has shown that these mycoplasma bodies undergo gradual degeneration by employing this procedure and ultimately disappear within 80 days. (also *see* Section 15.1.4).

15.5 Applications and Limitations of Tissue Culture in Plant Disease Elimination

From the various examples given in this chapter, it is apparent that tissue culture techniques have a useful role in the eradication of systemic diseases in plants. Meristem-tip culture seems the most reliable method for pathogen elimination. Cryotherapy, electrotherapy and nucleic-acid-based assay are emerging potential tools of pathogen elimination in viruses (Rowhani et al. 2005, Falah et al. 2009, Meybodi et al. 2011). The overall operation involving the production, multiplication and maintenance of the *in vitro* regenerated plants requires a good knowledge of pathology. Viruses can be transmitted either mechanically, through an insect vector, or through other biological agents. The advantages of the tissue culture technique in raising disease-free plants can be offset, however, by increased susceptibility of these plants to attack from more severe viruses and fungi. One of the reasons for this increased susceptibility may be the altered nutritional and physiological state of the tissue culture-derived plants as a result of pathogen eradication. In the light of this the phenomenon of cross-protection and the application of gene-transfer technology may be more beneficial for incorporation of disease resistance in plants.

It is also essential to have a good knowledge of greenhouse maintenance to control the re-infection of disease-free plants. The procedures followed for virus elimination in plants and prevention of reinfection are given in Appendix 15.2 and Appendix 15.3, respectively.

APPENDIX 15.1

Elimination of Systemic Bacterial and Fungal Pathogens from Geranium by Culture Indexing[a]

1. Establish mother plants in an incubation greenhouse maintained at 25-30°C day/20-25°C night.
2. Remove top cuttings (5-8 cm long) from the mother plant and number them. Excise the bottom 2-cm portion of each cutting with sterile knife in the laminar airflow hood. Surface sterilise for 10 min in a commercial bleach solution.
3. Wash the disinfected-cutting portions with sterile distilled water three times, slice each cutting portion aseptically into thin (2 mm) sections.
4. Transfer sections onto a test tube containing sterile nutrient broth and incubate at 23-25°C (ambient temperature) for 10-14 days; presence of bacterial or fungal pathogen is confirmed in case the nutrient broth turns cloudy.
5. Discard original numbered cuttings as well as the mother plant from which sections tested pathogen-positive; retain only those cuttings (with mother plants), sections of which yielded no micro-organisms in the nutrient broth.
6. Grow all cuttings that tested pathogen (bacteria or fungi) free in the incubation greenhouse and after 3-4 months retest the plants obtained for detection of pathogen, following steps 1-5, to ensure negative pathogenicity.
7. At least 3-4 culture indexing should be done in a year to confirm that plants are pathogen-free.

APPENDIX 15.2

Procedure Followed for Virus Eradication in Potato Plants[b]

1. Plant a single-eyed tuber piece in soil under normal environmental conditions. As soon as the first sprout is 15 cm tall take a tip-cutting 6-8 cm long. Remove the lower leaves and apply rooting hormone to the cut surface and plant the cutting in a 10-cm peat pot of sterilised soil. Cover it with a glass beaker for 10 days.
2. After three to four weeks move the set-up to a growth chamber where the light is 3000-4000 Ix for 16 hr a day and the air temperature about 36°C during the day and about 33°C at night.
3. Two weeks later pinch out the tip of the young plant to promote the growth of axillary shoots.
4. Give heat treatment for 6 weeks and then remove one axillary shoot for bud excision, leaving at least two leaves on the plant to encourage further growth.
5. From the excised shoot remove the lamina of the leaves in such a way that the small basal portions of the petiole are left on the stem. Enfold the shoot in wet paper to prevent wilting.
6. Excise the meristem tip under a binocular by pulling back the petiole stump to expose the axillary bud and breaking of its rudimentary leaves with a dissecting needle. It is advisable to cut off the meristem tip when the two youngest leaf primordia are still left using a fine fragment of blade fixed in a suitable holder.
7. Transfer the excised meristem to a suitable culture medium.
8. Store the cultures at about 23°C in a 16-hr light regime.

---

[a] After Klopmeyer (2000).
[b] After Mellor and Stace-Smith (1977).

9. Transfer the regenerated rooted-shoots to the soil. Maintain the plants under high humidity initially. In varieties in which shoots fail to root, their grafting onto healthy rootstock may be tried.
10. Index the plants first while transplanting in soil and then at regular intervals. Indicator plants used for potato viruses are *Gomphrena globosa* and *Chenopodium amaranticolor*.

APPENDIX 15.3

Procedure for Prevention of Reinfection of Disease-free Plants Raised in Cultures

1. Plants freed from pathogens should be maintained in greenhouses designed in a manner that ensures minimum chances of reinfection by pathogens and the attack by aphids.
2. Natural carriers of diseases such as insects and nematodes should be controlled by continuous spraying in the greenhouses.
3. On entering a greenhouse strict hygiene is necessary. One should wear washed-clean clothes and shoes. Hands must be disinfected as well as grafting and inoculating knives.
4. The pots and substrates should be germ-free.
5. In case proper precautionary measures cannot be taken in the greenhouses, it is recommended that disease-free plant material be maintained *in vitro*.
6. Individual plants in the greenhouse should be so separated as to not allow them any contact. This applies to both types of test or indicator plants.
7. Indexing of each plant for diseases should be carried out continuously at periodic intervals.

16

Clonal Propagation

In nature, the methods of plant propagation may be either *asexual* (by multiplication of vegetative parts) or *sexual* (through generation of seeds). Sexually propagated plants demonstrate a high amount of heterogeneity since their seed progeny are not true-to-type unless they have been derived from inbred lines. Asexual reproduction, on the other hand, gives rise to plants which are genetically identical to the parent plant and thus permits perpetuation of the unique characters of the cultivars. Multiplication of genetically identical copies of a cultivar by asexual reproduction is called *clonal propagation*. A genetically identical population derived from a single individual by asexual reproduction constitutes a *clone*. Plants such as banana, grape, fig and chrysanthemum produce little or no viable seeds. Vegetative multiplication is the only method for the propagation of such crops. The most widely used *in vivo* methods of cloning agricultural crops include cuttings of vegetative parts, layering, grafting and budding. These processes have been successfully applied for clonal propagation of potato, apple, pear, arboricultural crops, ornamental bulbs, tuberous plants and so forth.

Clones also generate through a process called *apomixis* (seed development without meiosis or fertilisation). Since apomixis is restricted to only a few species, horticulturists have adopted the methods of asexual or vegetative reproduction for clonally multiplying the selected cultivars. The advantages of vegetative reproduction over sexual reproduction are: (a) faster asexual multiplication in contrast to seed propagation in plants with a long life cycle, (b) bypassing the undesirable juvenile phase associated with seed-raised plants of woody perennials or some cultivars by propagating them vegetatively directly from the adult material and (c) establishing gene banks by multiplying variants among clonally propagated plants.

The *in vivo* clonal propagation of plants is often difficult, expensive and even unsuccessful. Tissue culture methods offer an alternative means of plant vegetative propagation. Clonal propagation through tissue culture (popularly called *micropropagation*) can be achieved in a short time and space. Thus, it is possible to produce plants in large numbers starting from a single individual. Use of tissue culture for micropropagation was initiated by G. Morel (1960), who found this as the only commercially viable approach for orchid propagation. Since then several crop species have been micropropagated and recipes are now available which can be adopted by growers on industrial scale (*see* Bhojwani and Razdan 1983, 1996, Amoo et al. 2011, Paek et al. 2011).

As described in previous chapters, the techniques of single cell and protoplast culture also enable thousands of plants to be derived within a short space and time. Products of this rapid vegetative propagation can be regarded as clones only when it is established that the cells they comprise are genetically identical. However, in practice, isolated cells/protoplasts, callus and cell suspension cultures may consist of many abnormal and deviant idiotypes (Mantell et al. 1985), and their regenerants may not conform to a true-to-type clonal propagation situation. These regenerants may generate novel types of genetic and epigenetic variations collectively referred to as somaclonal variation (*see* Chapter 14). The most widely used methods of clonal propagation, therefore, rely on the proliferation and induction of axillary or apical shoot meristems *in vitro,* the aspects of which are discussed in this chapter.

16.1 Techniques

In vitro clonal propagation is a complicated process requiring many steps or stages. Murashige (1978a,b) proposed four distinct stages that can be adopted for overall production technology of clones commercially. Stages I-III are followed under *in vitro* conditions, whereas Stage IV is accomplished in a greenhouse environment. Debergh and Maene (1981) suggested an additional stage 0 for various micropropagating systems. The adoption of all these stages (Fig. 16.1) not only simplifies the daily operation, accounting, or product cost, but also allows for greater ease in communication with other laboratories (Chu and Kurtz 1990). Thus, a particular plant can be marketed or requested by specifying its stage. Establishment of a reproducible system with well-characterised stages is a prerequisite for promotion of projection targets as well as schedules in the commercial production of plants.

16.1.1   Stage 0

This is an initial step of micropropagation in which stock plants used for culture initiation

Stage 0   Selection and maintenance of stock plants for culture initiation
↓
Stage I   Initiation and establishment of aseptic culture (Main steps: explant isolation, surface sterilisation, washing, and establishment on appropriate culture medium)
↓
Stage II   Multiplication of shoots or rapid somatic embryo formation using a defined culture medium
↓
Stage III Germination of somatic embryos and/or rooting of regenerated shoots *in vitro*
↓
Stage IV   Transfer of plantlets to sterilised soil for hardening under greenhouse environment (in a few cases this stage may also include the *in vivo* rooting of regenerated shoots by skipping Stage III)

Fig. 16.1 Major stages of *in vitro* clonal propagation.

are grown for at least 3 months under carefully, monitored glasshouse conditions. Stock plants are grown at a relatively low humidity and watered either with irrigation tubes or by capillary sand beds or mats. This stock plant preconditioning stage also includes measures to be adopted for reduction of surface and systematic microbial contaminants (*see* Chapter 15, also Knauss 1976).

16.1.2   Stage I

Murashige (1974) defined this stage as the initiation and establishment of aseptic cultures. The main steps involved are preparation of the explant from stock plants (stage 0) followed by its establishment on a suitable culture medium. Cultures are initiated from explants of several organs but shoot tips and axillary buds are most often used for commercial micropropagation. Procedures to surface-sterilise the explant and induce a healthy growth in the culture medium defined for each species may be devised as discussed in Chapters 3 and 4. It may also be advisable to control microbial contamination within explant tissues in case such efforts at Stage 0 were not successful. Stage I lasts 3 months to 2 years and requires at least four passages of the subculture.

Usually explants carrying a preformed vegetative bud are suitable for enhanced axillary branching. When the objective is to produce virus-free plants from an infected individual, it becomes obligatory to use cultures derived from submillimetre shoot tips (*see* Chapter 15). If stock plants are tested virus-free, the most suitable explants are nodal cuttings. There are some disadvantages in using small-sized explants for micropropagation. Small shoot-tip explants have a low survival rate and show slow initial growth. Meristem-tip cultures may also result in the loss of certain horticultural traits exhibited by the presence of virus (example: clear-vein character of *Geranium* cv. Crocodile, Fig. 16.2). Therefore, subterminal or slightly older segments are desirable which can withstand the toxic effects of sterilisation agents much better than the terminal cuttings. In some ornamentals (*Cordyline terminalis* and *Dracaena godseffiene*), terminal buds are used to initiate cultures since lateral buds fail to produce reculturable shoots (Miller and Murashige 1976). For rhizomatic plant (strawberry, Boston fern), runner tips are commonly used.

For multiplication through adventitious bud formation, with or without callusing, explants are derived from root, stem, leaf or nucellus based on their natural capacity to form adventitious buds. Leaf-base and scale-base explants are highly meristematic in monocots and joined with basal plate. Cultures in these plants are initiated using explants along with a small piece of basal plate. Immature zygotic embryos are suitable explants for establishing embryogenic cultures for propagation of woody species although in citrus, and mango, nucellar cultures are highly embryogenic (*see* Bhojwani and Razdan 1996).

16.1.3   *Stage II*

This stage takes up the bulk of micropropagation activity using a defined culture medium that stimulates maximum proliferation of regenerated shoots. Various approaches followed for micropropagation include:

i) Multiplication through the growth and proliferations of meristems excised from apical and axillary shoots of the parent plant.

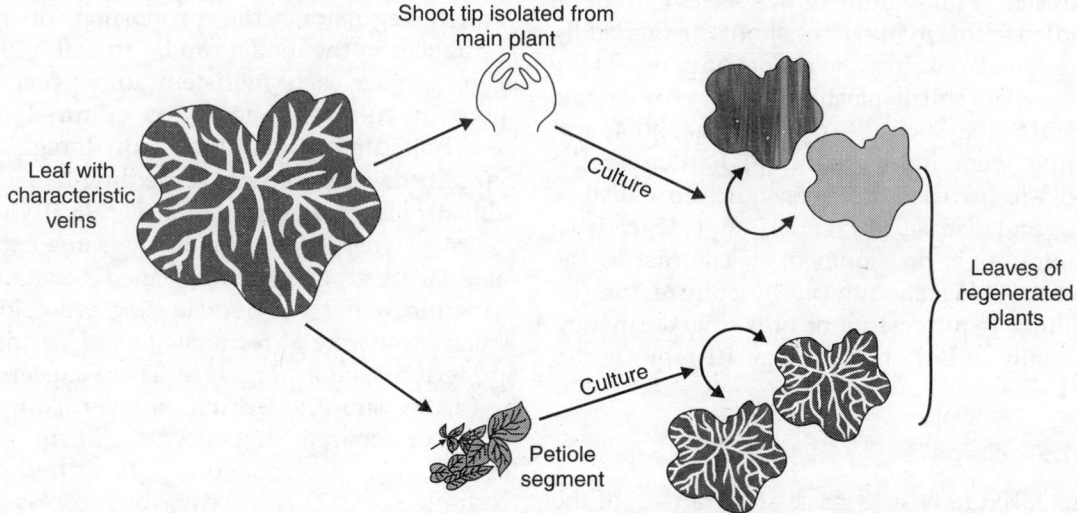

Fig. 16.2 Shoot-tip culture and petiole culture of *Geranium* cv. Crocodile, in which the clear-vein character of leaves has been attributed to infection by a virus or virus-like agent. Note that the clear-vein character is transmitted in leaves of plants regenerated from petiole-segment culture but not of shoot-tip culture (adopted from Cassells et al. 1980).

ii) Induction and multiplication of adventitious meristems through processes of organogenesis or somatic embryogenesis directly on explants.

iii) Multiplication of calli derived from organs, tissues, cells, or protoplasts and their subsequent expression of either organogenesis or somatic embryogenesis in serial subcultures. Shoots obtained from these calli can be further multiplied following procedures (a) and (b).

A passage or harvest cycle generally requires 4 weeks. Shoots are harvested from the multiplying culture to either be sold as a stage-II product or carried on to stage III. Generally, stage II lasts 10-36 months with a large number of subcultures of similar age.

16.1.4   Stage III

Shoots proliferated during Stage II are transferred to a rooting (storage) medium. Sometimes shoots are directly established in soil as microcuttings to develop roots. Since such a possibility depends on the particular species and, at present, a large number of species cannot be handled in this manner, the shoots are generally rooted *in vitro*. When the shoots or plantlets are prepared for soil, it may be necessary to evaluate several factors such as (i) dividing the shoots and rooting them individually, (ii) hardening the shoots to increase their resistance to moisture stress and disease, (iii) rendering plants capable of autotrophic development in contrast to the heterotrophic state induced by culture and (iv) fulfilling requirements of breaking dormancy, especially of bulb crops. Stage III requires 1-6 weeks.

16.1.5   Stage IV

Steps taken to ensure successful transfer of the plantlets of stage III from the aspetic environment of the laboratory to the environment of the greenhouse comprise Stage IV. Unrooted Stage II shoots are also acclimatised in suitable compost mixture or soil in pots under controlled conditions of light, temperature and humidity inside the greenhouse. In such cases Stage III is skipped. Supplying bottom heat-aids to pots with plantlets or cuttings and maintenance of a dense fine-particle fog system within the greenhouse enhances the rooting process. Complete plants can also be established in the artificial growing media such as soil-less mixes, rockwood plugs, or even sponges. It takes 4-16 weeks for the finished product (i.e., plant size in the range of 3-6 in) to be ready for sale or shipment.

16.2   Multiplication by Axillary and Apical Shoots

Axillary and apical shoots contain quiescent or active meristems depending on the physiological state of the plant. Vascular plants with an indeterminate mode of growth have in their leaf axils subsidiary meristems with the potential for growing into a shoot. However, only a limited number of axillary meristems have the capacity to develop *in vivo* if the type of branching of a particular species displays apical dominance. Since the mechanism of apical dominance has been demonstrated to be under control of various growth regulators, the proportion of these substances in the media can be so manipulated as to induce each meristem to regenerate a shoot in cultures. Shoot tips cultured on a basal medium containing no growth regulators typically develop into single seedling-like shoots with strong apical dominance (Fig. 16.3). On the contrary, when the shoots of the same explant material are grown on culture media containing cytokinin, axillary shoots develop precociously which proliferate to form clusters of secondary and tertiary shoots (Fig. 16.4). These clusters can be further sub-divided into smaller clumps of shoots or separate shoots which, in turn, will form similar clusters when subcultured on a fresh-medium. This subdivision process may continue indefinitely provided the basic nutrient formulations are adequate for normal growth. About 5-10 multiplication rates can be achieved on a regular 4-8-week micropropagation cycle, which may ultimately lead to extremely impressive rapid clonal propagation levels in the range of 0.1-3.0 x 10^6 within one year (*see* Mantell et al. 1985).

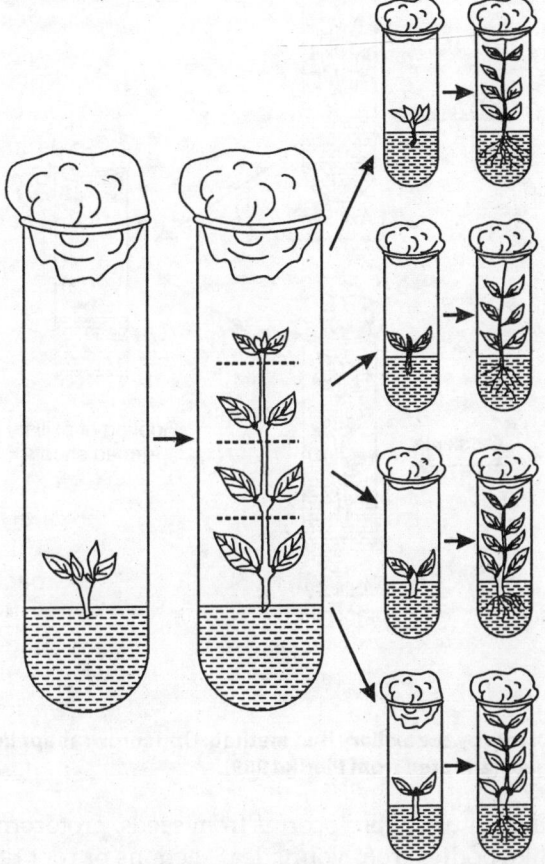

Fig. 16.3 Single-node method of clonal propagation. Note that the seedling-like shoot developing from each node shows strong apical dominance (adopted from Pierik 1989).

In general, the technique of proliferation by axillary shoots is applicable to any plant that produces regular axillary shoots and responds to cytokinins such as BAP, 2iP and zeatin. Many forest and orchid tree species are good candidate for *in vitro* clonal propagation using axillary shoots.

Apical shoots (1-5 mm) are normally cultured on media containing mixtures of auxin (0.01-0.1 mg l⁻¹) and cytokinin (0.05-0.5 mg l⁻¹). The level of cytokinin is raised in subsequent subcultures

to induce an acceptable rate or proliferation without yellowing or distortion of shoots. If the presence of cytokinin in the media inhibits root development, cultured material is often transferred in Stage III to a rooting medium which contains either no or reduced levels of cytokinin.

16.3 Multiplication by Adventitious Shoots

Adventitious shoots are stem and leaf structures that arise naturally on plant tissues located in sites other than at the normal leaf axil regions. Whereas axillary buds are 'preformed meristems', adventitious regeneration events occur at sites where meristems do not exist. Thus adventitious shoot regeneration is often dependent on the presence of organised explant tissue. These structures include stems, internodes, leaf blades, cotyledons, root elongation zone, bulbs, corms, tubers and rhizomes, amongst others. Almost every one of these organs can be used as a cutting in conventional clonal propagation (e.g., leaves of *Begonia* and some other ornamentals produce shoots on a large scale). A similar type of adventitious shoot development can be induced in cultures by using a suitable explant from perconditioned plant material and appropriate levels of growth regulators in the medium. Bulbs and corms grow from meristems at the base of leaves and scales where they join the basal plate. These meristematic regions regenerate multiple shoots on a suitable culture medium. Levels of fidelity in propagated material of these species are quite high because adventitious shoots arise only from single epidermal cells (example: adventitious shoots from stems of *Tulipa*). Compared to conventional propagation techniques the *in vitro* micropropagation methods result in overall multiplication level of × 5 to × 10 per annum (*see* Mantell et al. 1985).

Continuous propagation by adventitious shoot proliferation from bulbs and corms can be achieved by cultivating two vertically split pieces (ca 10 mm) of shoot bases. Clusters of shoots develop from around the abaxial surfaces

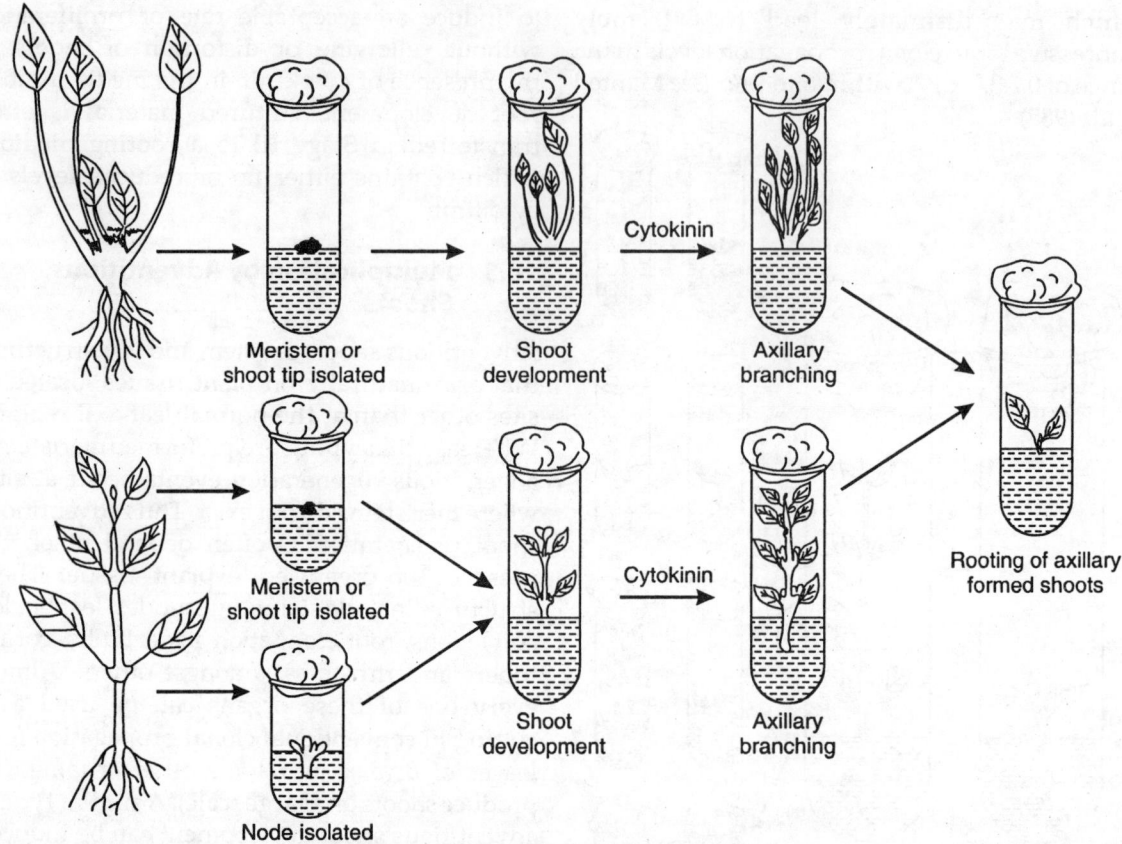

Fig. 16.4 Scheme showing the *in vitro* clonal propagation of plants by the axillary bud method. Upper row: as applied to rosette plants; bottom row: applied to plants which elongate (adopted from Pierik 1989).

of developing leaves and scales (Fig. 16.5a). Senescence and dormancy in such cultured materials can be prevented *in vitro* by trimming of shoots to within 2-3 mm of the basal plate (Fig. 16.5b). This approach has been found to ensure continuously productive cultures of *Iris, Lilium* and *Tulipa* hybrids for indefinite periods. For bulbs with a strong apical dominance *(Narcissus, Nerine, Allium* and *Hyacinthus)*, it is necessary to destroy the main apex. This can be achieved by making two shallow vertical cuts at right angles to each other up to the level of the basal plate (Fig. 16.5c). When large enough, the regenerating shoots are, in turn, cross-cut to obtain repeated cycles of regeneration (Fig. 16.5d). *Phalaenopsis* orchids have high economic value in floricultural industry. Micropropagation of these orchids via

the culture of protocorms from seeds, protocorm-like bodies from young leaf sections or root tips have been successfully achieved and the details of recipes developed are mentioned by Paek et al. (2011).

Clonal propagation by adventitious embryo formation is another useful approach followed for many important plant species. Adventitious embryos can arise directly from a group of cells within the original explants or from primary embryoids. Orchids produce a large number of embryoids at the tip of leaves *in vivo*, while cultivars *of Citrus* and *Mangifera* develop poly-embryos from the nucellar tissue. There are many other species which develop embryo *in vivo* from diverse organs and tissues. Adventitious embryoids and diploid and can be used as

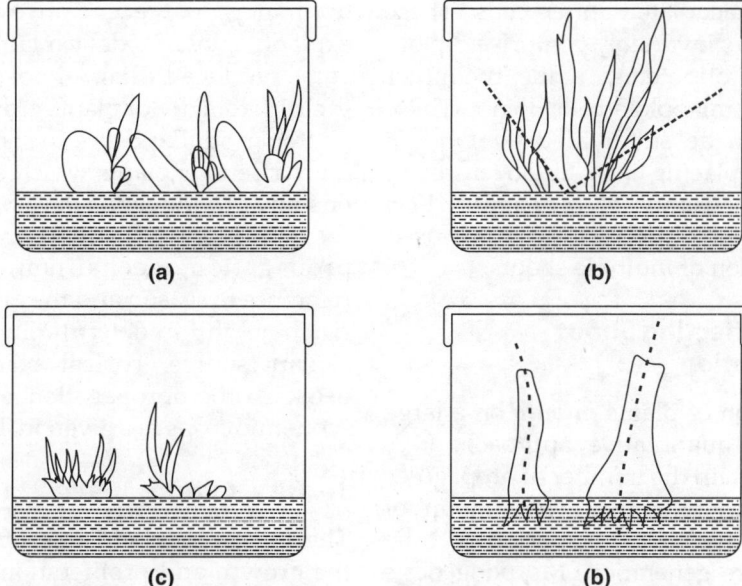

Fig. 16.5 Some approaches followed for micropropagation of bulbs and corms: (a) regeneration of shoot clusters; (b) trimming shoot to within 2-3 mm of basal plate; (c) vertical cuts at right angles to each other for removing dominance of main axis; (d) cross-cutting minicorms and bulbils to obtain repeated cycles of propagation *in vitro.* **Dotted lines indicate the orientation of cuts and divisions (after Mantell et al. 1985).**

clonal material. Similarly, adventitious embryos obtained *in vitro* by inducing embryogenesis (*see* Chapter 7) on explants are good materials for clonal propagation. For example, leaf pieces of coffee trees form embryos directly when cultured on a basal MS medium containing high levels of cytokinin. Teixeira da Silva and Tanaka (2010), however, developed thin cell layer (TLC) culture method for rapid and enhanced clonal multiplication of *Cymbidium* hybrid Twilight Moon, which by traditional methods of mircropropagation was difficult to propagate.

16.4 Multiplication through Callus Culture

Differentiation of plants from cultured cells via shoot-root formation (*de novo* origin) or somatic embryogenesis, where applicable, can be the fastest method of shoot multiplication and cloning of plant species. However, the cultures in which calli are produced tend to be of low value as a means of micropropagation. The most serious

drawback in the use of callus cultures for shoot multiplication is the genetic instability of their cells, due to which the initial plant regeneration capacity of the tissue may decline with the passage of time. Notwithstanding this, *in vitro* propagation *via* organogenic or embryogenic calli is unavoidable in some economically important species of cereals, forage legumes, freesias, citrus, coffee, forest and tropical palm trees. Even plant regeneration from protoplasts also requires passage through at least one callus stage. The production of many thousands of plantlets from calli either derived from cell suspensions or isolated protoplasts constitutes unique cases of cloning, such as 'calliclones' and 'protoclones'. Such clones commonly exhibit somaclonal variations (Chapter 14).

Genetically stable calli have also been derived from explants of various plants representing *Lilium, Chrysanthemum* and tomato (*see* Mantell et al. 1985). In these types of calli, slow-growing meristematic cells from which shoots and embryoids arise are derived from the peripheral layer of highly vacuolated inner cells. These meristematic layers invariably comprise diploid cells expressing totipotency while the inner layers made up of mixoploid cells do not. These types of calli can be subdivided by random dissection or by placing in a homogeniser to produce many thousands of propagules in a single operation. Each propagule may be used in mass propagation of multiple shoots.

16.5 Factors Affecting Shoot Multiplication

Clonal propagation of plants *in vitro* on a large scale requires a quantitative approach. It is necessary to ascertain the number of propagules which can be regenerated in a given amount of culture material over a given period of time or a single culture generation. Morphogenesis and proliferation rate of culture depend on the various factors influencing the relative incidence of organogenesis or embryogenesis. This necessitates optimisation of conditions associated with *in vitro* vegetative plant propagation.

16.5.1 Physiological Status of Plant Material

Successful cultures are rarely obtained from senescing tissues. Explants isolated from the more recently produced parts of a plant are more regenerative than those from older regions. The regenerative potential of tissue culture diminishes with each year of maturation, particularly in perennial woody species. Generally, somatic embryogenesis is associated with cultures established from embryo explants rather than mature or non-embryonic tissues. Reports about season-linked regenerative capacity of tissue cultures (Litz and Canover 1981) have also appeared in the literature. Papaya tissue cultures can be established in hot summer months, whereas flower-stem explants of *Tulipa* give rise to shoots only when excised during the dry storage (dormant) phase. Once elongation of stems has commenced, following

dormancy, the regenerative potential of tulip explants is lost. Nodal explants of *Dioscorea alata* yams produce axillary shoot growth only when excised from donor plants growing under a 16-hr photoperiod. Exposure to shorter photoperiods gives rise to explants which either do not grow or show prolific callus development.

Familiarity with the donor plant's natural propagation mechanism, especially with reference to season and the growth stage, is very much helpful in determining the more suitable explant source. Typical examples of explants used in micropropagation of some important economic crops are given in Table 16.1.

16.5.2 Culture Media

The importance and role of culture media in the growth and proliferation of tissue cultures have been described in Chapter 3. The standard tissue culture media are more suitable for achieving Stages I and II of micropropagation. Only Stage III requires some modifications. The relative proportion of mineral components and two types of growth-regulating substances in the culture medium largely govern the overall micropropagation performance of a culture system. Therefore, a general concept proposed by Skoog and Miller (1957) stating that organ differentiation in plants is regulated by an interplay of auxin and cytokinin (Chapters 5 and 6) should work as a guide when developing a new medium for a new plant for micropropagation. GA_3 has been used in the shoot proliferation medium to improve shoot elongation, rate of multiplication, growth and quality of shoots in woody species (Brand and Lineberger 1992). For micropropagation, the relative concentrations of NH_4^+ and K^+ in the medium influence the number and size of shoots produced from explants. These mineral components also affect the induction of somatic embryogenesis in cultures. To induce adventitious root formation, after axillary shoot propagation, cytokinin is usually omitted and auxin added. Depending on the species, root formation may occur on a medium without hormones. Gibberellins and

Table 16.1 Explants used in micro propagation of some economically important ornamental and horticultural plants[a]

Group	Plant	Explant
Bulbs and corms	*Allium, Lilium, Tulips, Narcissus*	Scale leaves, leaf bases, Stems, floral parts[b]
	Nerine	Scale leaves, stems
	Iris	Scale Leaves, stems, inflorescences
	Gladiolus	Corms, stems
	Freesia	Corms, leaf bases, stems, floral parts, inflorescences
House plants	*Saintpaulia ionantha*	2-mm thick petiole section
	Petunia hybrida	Small leaf pieces
	Kalanchoe blossfeldiana	Stem pieces
	Begonia × hiemallis	Petioles and leaf lamina
	Goxinia sp.	Petioles and leaf lamina
Orchard crops	*Malus domestica*	Root pieces
Conifers	*Picea abies*	Needles from upper half of buds

[a] After Mantell et al. (1985).
[b] Pedicels, ovaries, petals, or anther filaments.

abscisic acid in the medium are also reported to inhibit root formation. Activated charcoal may also induce the formation of adventitious roots in some species. Its presence in the medium reduces the light supply to *in vitro* regenerated shoots and helps in removing inhibition by the adsorption of all such compounds released in cultures.

16.5.3 Culture Environment

Although green, the *in vitro* regenerated shoots grow as heterotrophs since they derive all of their nourishment (organic as well as inorganic) from the culture medium. Several studies have indicated that light, apparently absorbed by photosynthetic pigments in cultured tissues, plays an important role in inducing the morphogenesis of these tissues. Kato (1978) observed that bud induction in excised leaf segments of lily *Heloniopsis orientalis* was to a certain extent controlled by the photosynthetic system in its cells. The optimal light intensity found for shoot multiplication in most of the species is 100 1x. The quality of light also controls the organogenic differentiation and growth of shoots in cultures. For example, blue

light (467 nm) induced bud formation in tobacco shoots (Seibert 1973) and even doubled the number of lettuce shoots (Kadkade and Seibert 1977) regenerated from callus cultures. Several micropropagation systems have now been described in which morphogenesis is induced by red/far-red light treatments. The red light stimulates induction of flower buds and far-red light root production (*see* Tran Thanh Van 1977).

Photoperiodic effects are dependent on the relative sensitivity of the individual species (*see* Section 16.5.1). A diurnal iliumination of 16 hr day and 8 hr night is generally found satisfactory for multiplication and proliferation of shoots although there are exceptions such as cauliflower which require 9 hr daylight regimes.

Temperature also forms an important part of the culture environment which, in most micropropagating cultures, is maintained constant around 25°C. Some cultures may require initial low temperature (18°C: *Begonia × Cheimantha* hybrid tissues) for morphogenic response. A third factor of the culture environment that affects the performance of cultures during micropropagation is the constitution of the gas phase within the culture vessels. Ethylene, oxygen, carbon dioxide, ethanol and acetaldehyde are metabolically active gases with possible effects on morphogenesis and may promote unorganised growth of cells. The interaction of all these gases is complex and requires detailed studies to ascertain the real effects of the gas phase on micropropagation systems.

The inventions of temporary immersion system (TIS) for micropropagation of banana in liquid media was introduced by Alvard et al. (1993) and subsequently tested with success for clonal propagation of pineapples, *Hevea brasiliensis*, *Psidium guajava* and cacoa. *In vitro* plants growing in TIS bioreactors apparently do not show hyper hydration systems due to proper aeration (Altman and Hasegawa 2012).

16.5.4   Genotype

Screening and selection of genotypes among segregating populations could be a fruitful approach in the improvement of micropropagation capabilities of plant species which are recalcitrant in tissue culture. Bingham et al. (1975) first followed this approach for clonal propagation of alfalfa plants. By breeding different lines of alfalfa they screened the progeny for regeneration ability in cultures. Tran Thanh Van (1980) found that the response of different genotypes of *Nicotiana* species to their regeneration capabilities differed in cultures. A similar response was observed in closely related genotypes of three hexaploid oat species (Rines and McCoy 1981). Different propagation coefficients were achieved from various cultivars, lines and hybrids of grape cultured under similar culture conditions. Cultivars with vigorous germination and branching capacity (example: Jin Zhao Jin) propagated rapidly, while those with low rates of germination and weak branching capability (example: Seedless Block, Seedless Red) showed poor response in cultures (Cao 1990). Genotypic effects of this nature are frequent and underscore the fact that a micropropagation system developed for one particular cultivar will not automatically be applicable to another even within the same species.

16.6   Factors Affecting *in vitro* Rooting

Media having a low concentration of salts have proven satisfactory for rooting of shoots micropropagated at Stage II. For species in which induction of shoot multiplication requires a full-strength MS medium, reduction of the salt concentration to half or one-quarter has been found satisfactory for rooting. Roots are mostly induced in the presence of a suitable auxin (NAA or IBA 0.1 mg l^{-1}) in the medium although the shoots of some plants (gladiolus, *Narcissus* and strawberry) may readily root on a hormone-free medium. Riboflavin is reported to improve the quality of the root system in *Eucalyptus ficifolia*. Similarly, phloroglucinol has been observed to promote rooting in a number of rosaceous fruit trees (Jones 1983). However, the effectiveness of phloroglucinol in rooting has been a subject of controversy.

The micropropagation of hard-to-root species can be achieved if their juvenile explants are induced to develop shoots by suckering, coppicing, or hormone treatment. In some plants (*Eucalyptus citriodora*) shoots have been regenerated from adult material but *in vitro* multiplied shoots form roots only by passing them through a series of subcultures. The time required for *in vitro* rooting may vary from 10-15 days. Plants with roots around 5 mm in length have been found convenient for handling during transplantation. Longer roots may break in this process and lead to high mortality of transplanted plants. Rooting of cultured shoot in some cases is achieved *ex-vitro* (*see* Section 16.7).

16.7   Acclimatisation of Plants Transferred to Soil

Micropropagation on a large scale can be successful only when plants after transfer from culture to the soil show high survival rates and the cost involved in the process is low. Tissue culture plants generally show some structural and physiological abnormalities which include: (a) abnormal leaf morphology and anatomy,

(b) poor photosynthetic efficiency, (c) marked decrease in epicuticular wax, (d) malfunctioning of stomata, and (e) hypolignified stem. These characteristics as well as a heterotrophic mode of nutrition and a poor mechanism for water-loss control further render micropropagated plants vulnerable to transplantation shocks. Therefore, the transfer of individual plantlets to a potting mix and their acclimatisation under greenhouse conditions require the application of various methods to harden the plants/shoots for transplantation.

Plants are transferred to the soil usually after the *in vitro* rooting stage. However, the induction of *in vivo* rooting of cultured shoots (microcuttings) may be more economical besides producing good quality roots. Treatment of microcuttings (2.5-5 cm long) with root inducing growth substance (auxin) or commercial rooting powder may be necessary for rooting *in vivo* (Driver and Suttle 1987) Microshoots of tea were directly rooted by dipping the cutends in IBA (50 mg l^{-1}) for 10 min and subsequently transplanting in a soil: peat moss (1:1) mixture. Rooting *ex vitro* can also be done on stonewool substrate moistened with auxin solution (Das et al. 1990, Maene and Debergh 1987). Driver and Suttle (1987) also successfully induced rooting and acclimatization simultaneously in walnut and peach microshoots directly in the field. For acclimatization *ex vitro*, it is essential that the lower parts of tissue culture plants/shoots be washed thoroughly before their transfer to the potting mix (pumica, peat, vermiculite soil, sand, or their mixtures in different proportions). Transplanted plantlets or shoots are immediately irrigated with an inorganic nutrient solution and maintained under high humidity for the initial 10-15 days. This is required because plantlets during culture are adapted to almost 90-100% humidity. Manipulating sugar content, growth retardants and antirespirants in the medium could be effective in accumatization of tissue cultured plants (Hazarika 2003).

High humidity can be built up around transplanted plants by covering them with clean transparent plastic bags having a small hole for air circulation. The size of the hole can be enlarged after two weeks in order to reduce the humidity. This enables shoots to adapt well to greenhouse conditions and to establish functional roots. Partial defoliation of plantlets and application of transplants (1% Acropol, v/v) in the initial stages of transpiration reportedly improve the survival frequency due to reduction in water loss by plantlets. Most commercial laboratories these days have computerised hardening rooms with controlled conditions of light, temperature and humidity.

Direct transplantation of cultured plants (rice, tobacco) to the field has also been tried. Transplants survived with high frequencies provided a thin film comprising 50% aqueous glycerol and grease or paraffin (melting point 52-54°C) in an equal amount of diethylether, is applied on the surface of leaves with a brush before transplantation.

A major shock to plants, following transplantation, is the change from a substrate rich in organic nutrients to a substrate providing mostly inorganic nutrients. Attempts have been made to harden the shoot system by inducing autotrophism and development of surface wax on *in vitro* formed leaves. Increase in the epicuticular wax deposition could be induced either by exposing cultures (cabbage) to $CaCl_2$ or covering the medium with a thin layer of lanolin (chrysanthemum, cauliflower). These treatments also helped in reducing humidity. However, $CaCl_2$ or lanolin may prove detrimental to overall growth and development of plants. The relative humidity (RH) can also be reduced by opening the culture tubes inside a desiccator with $CaSO_4$ as the desiccant. This approach resulted in hardening of the tissue culture plants of carnation *in vitro* by development of wax after seven days. Addition of paclobutrazol (0.5-4 mg l^{-1}), an anti-gibberellin, to the rooting medium enhanced desiccation tolerance of micropropagated chrysanthemum, rose and grapevine (Roberts et al. 1992). Deleting sucrose from the medium, increasing CO_2 level (to 350 ppm) around the plant, and increased light intensity (200 µmol m^{-1}s^{-1}) also reduce RH and promote plant growth and photoautotrophic behaviour. These plants then acclimatise easily and have a higher survival rate (96%).

Tissue culture plants of tree-legume species (*Leucaena leucocephala*) show high survival rates provided a pretransplant is introduced. In this stage, plants are transferred to screw-cap bottles containing sterilised quartz sand irrigated with an inorganic nutrient solution carrying an efficient strain of *Rhizobium* (NCR 8). Bottles are initially kept closed for two weeks and subsequently the caps removed to maintain plants under controlled conditions of light and temperature (25 ± 2°C and 18 Wm^{-2}) for another two weeks before their final transfer to the field.

Storage organs have been induced in cultured shoots of several species. These structures do not require hardening and can be directly transplanted to the soil with survival rates comparable to a similar type of organs formed *in vitro*. The advantage of *in vitro* tuberisation is that the additional step of rooting the shoots is altogether eliminated. A well-known example is the production of aerial tubers by cultured shoots of potato under the influence of chloramequat. Akita and Takayama (1993) devised two-step culture scheme for mass propagation of potato tubers in jar fermentor. Other examples in which *in vitro* tuberisation can be induced are *Dioscorea bulbifera*, *D. alata* and *D. rotundata*. Formation of bulblets in cultures of *Muscari armeniacum* and *Narcissus tazetta* was observed to occur in the presence of activated charcoal, while a high sucrose concentration appeared critical for *in vitro* cormlet formation in *Gladiolus* shoots. Transplantation of micropropagated plants of *Gladiolus* has been a serious problem. Corms developed *in vitro* can circumvent this problem as they show high germination (75-80%) under field conditions (for details and references, *see* Ziv 1995, Bhojwani and Razdan 1996). Acclimatisation units devised with computer controlled light, temperature, humidity and CO$_2$ concentrations are cost prohibitive. Therefore

one step hardening technique (see Fasella and Hussain 2014) of tissue culture raised orchid seedlings is both cost effective and with high degree of efficiency in hardening.

16.8 Automation

Limiting manual labour in various steps involved in micropropagation process could be achieved by automation. Vitron^{TM} 501 developed by ForBio comprises a box-loader that unloads a stack of vessels, the lids of which are individually opened by pneumatic cap opener. A set of grippers and scissors cut the tips and lower nodes of the plant. These are then placed in media containing vessels and with the help of CCD camera the growth of cuttings is monitored. It is estimated that the processing rate of micropropagation is nearly size millions plants per year which seems a moderate volume considering the capital investment for the apparatus. Robotic systems for *in vitro* micropropagation demonstrate inverse relationship between volume of production versus cost of propagule and this happens to the main constraint. Moreover, Vitron 501 is merely suitable for excission and culture of intermodes. Further other automated tissue culture systems need to be developed for plants without distinguishable nodes e.g., bulbous plants, banana meristem culture, etc. For the descriptive details of these automated systems refer to Sluis (2006), Altman and Hasegawa (2012).

16.9 Clonal Multiplication of Woody Species

Woody perennials comprise valuable crops such as fruit and nut trees, plantation species (palms of various types, rubber, coffee, tea etc.), timbers (hardwood and softwood) and important trees of social forestry. Due to the long-term life cycle of these tree crops, progress in the improvement of woody perennials through the application of conventional methods has been extremely slow. The difficulty in rooting the cuttings from elite tree species (*Quercus, Fagus, Eucalyptus* and most conifers) has further complicated the process of their clonal propagation. According to Jones (1983), vegetative propagation of many monocotyledonous palms and forest species is virtually impossible. Conventionally, forests are multiplied by the growth of natural seedlings and by collection of seeds from randomly pollinated elite trees. *Cryptomeria japonica* is an exception which has been clonally propagated for centuries in Japan. Recently, attempts were made to establish clones of Norway spruce (*Picea abies* cv.) on a large scale in the forests of Sweden and the former Federal Republic of Germany (*see* Paranjothy et al. 1990). Members of Ericaceae (*Kalmia, Rhododendron* and *Vaccinium*) are some of the earliest woody shrubs micropropagated on commercial scale (Smith 2013). Some perennial crops that can be propagated in tissue culture and media used at the various stages of their micropro-pagation are listed in Table 16.2.

16.9.1 Hardwoods and Fruit Trees

During the past many years considerable success has been achieved with respect to cloning of temperate fruit species and other trees through tissue culture. The response of forest trees, as *Populus, Eucalyptus* and *Tectona,* to micropro-pagation has been highly positive. Tissue culture of nut trees also holds some promise (Zimmer-man 1986). Noteworthy success in tree tissue culture is the controlled flowering of *in vitro* propagated bamboo (Nadgauda et al. 1990). Other examples on *in vitro* propagation as well as conservation of woody trees are described in two volumes (Razdan and Cocking 1997, 2000).

Explants that have shown positive response in cultures for regeneration, however, are largely restricted to juvenile material (Fig. 16.6). Clonal multiplication of woody perennials *in vitro* on a commercial scale requires enormous efforts directed towards establishing cultures from adult explants. In this context Chaturvedi et al. (2004b) reported a highly reproducible and recurrent method of clonal propagation of a 50-year-old neem tree through forced axillary shoot branching from explants *in vitro*. Further, effective treatments to induce

efficient rooting of *in vitro* multiplied shoots and quality improvement of somatic embryos to achieve high frequency conversion to plantlets must be found. Shoot tips from seedlings of *Citrus,* jackfruit and black plums have readily responded in culture. Immature embryos of avocado have proved equally effective clonal materials in cultures. Since nucellar seedlings are naturally rejuvenated clones, they have been utilised as stock plants for micropropagation of several polyembryonic *Citrus* cultivars and mangosteens (Barlass and Skene 1986, Goh et al. 1988). Monoembryonic species of *Citrus,* varieties of mango, loquat and *Syzygium* spp. regenerate by somatic embryogenesis using nucellar explants (Litz et al. 1986). Also, many tropical and subtropical fruit crops are reported to regenerate from somatic embryos induced on explants from mature trees.

The nature of the explant, its orientation on the culture medium and genotype influence the response of adult materials *in vitro*. Shoot tips or nodal segments from tropical adult trees are reliable sources of explants. These explants can be excised from newly developed vegetative buds of jackfruit (Amin 1987) at the base of the main stem during the onset of flowering or during vigorous vegetative growth of guava (Jaiswal and Amin 1987). In avocado, hard pruning of the main "stem stimulates the growth of lateral buds, thus providing enough meristematic material for clonal propagation (Pliego-Alfaro et al. 1987). The conditioning of stock plants improves the culture response of shoot tips and nodal segments and so do warm and humid months (guava; Jaiswal and Amin 1987). Sometimes horizontal placement of shoot-tip explants on the medium may prove superior to vertical orientation.

Tropical fruit species which could be commercially propagated are banana, papaya and pineapple. The planting of extensive orchards of vegetatively propagated clones of some tropical and subtropical fruits has remained limited due to loss through pathogen attacks. Most *Citrus*-producing countries have

Table 16.2 Media composition (µM) for various stages of micro propagation in some perennial species[a]

Species	Media and stages	Species	Media and stages
Acacia auriculaeformis	1: ER, BA 13.2, IAA 8.6	*Morus alba*	1: MS, IAA 5.7-11, BA 4.4-8.8
	2: Same as above		2: MS, IAA 2.9-5.7, BA 2.2-4.4
	3: 1/2 ER, ZEA 9.1, NAA 2.7		3: MS, BA 2.2, glutamic acid 13.6, lysine 13.7
Carica papaya	1: MS, BA 2, NAA 1.1	*Murraya exotica*	1: MS, KIN 4.6, NAA 2.7, ZEA 2-3, BA 2.2
	2: Same as above 3: MS, NAA 5.4		2: same as above
Citrus microcarpa	1: 1/2 MS, NAA 0.5, KIN 9.3		3: 1/2 MS, ZEA 0.9, BA 2.6
	2: MS, IAA 2.9, KIN 9.3	*Rhododendron hybridum**	1: MS with $(NH_4)_2SO_4$, ratio of NH_4 : NO_3 changed to 1 : 1, KIN 2.3, NAA 0.5-1.25 or add low concentrations of ZEA
	3: MS (hormone free)		
C. sinensis	1: Murashige-Tucker (MT), KIN 0.5, IAA 5.7		
Coffea arabica	1: MS, BA or ZEA 4.6-9.1, NAA 1.1	*Rosa chinensis**	1: MS, NAA 0.5, BA 2.2
	2: Same as above		2: MS, BA 4.4
	3: 1/2 MS, IBA 1.0	*Salix matsudana × S. alba*	1: MS, BA 0.4, NAA 1
Cryptomeria japonica	1: MS		2: same as above
	2: MS, NAA 5.4		3: MS, NAA 1
	3: 1/3, MS, NAA 5.4	*Santalum* sp.	1: MS, BA 8.8, NAA 5
Elaeis citriodora	1: MS (liquid), KIN 1, BA 1.3, culture on roller drum		2: MS, BA 8.8
	2: Same as above but add agar		3: MS, IBA 4.9
	3: WH or MS	*Tectona grandis*	1: MS (Liquid), BA 0.4, KIN 0.5
E. guineensis	1: WH, 2,4-D 4.5, KIN 2.3		2: MS, BA 4.4, KIN 2.3
	2: Heller, 2,4-D 4.5, KIN 2.3		3: WH, IAA 13.9, IBA 12.3, IPA
	3: WH or Heller, IAA 5.7	*Vanilla planifolia**	1: MS, thiamine 1.5, pyridoxine 2.4, nicotinic Kononowicz and acid 4, glycine 26.6, inositol 555, LH 1000 mg/l, sucrose 87.6 mM
*Leucaena leucocephala**	1: MS, NH_4NO_3 changed to $(NH_4)_2SO_4$ 5.7 mM, Ba 4.4, NAA 0.05, CH 300 mg/l, sucrose 146 mM		
	3: same culture medium as above, IAA 2.9-5.7, sucrose 58.4 mM	*Vitis* sp.*	1: MS, 2,4-D 4.5, BA 0.4, LH 500 mg/L
*Malus domestica**	1: MS, BA 22		2: MS, NAA 10.7, BA 0.4
	2: MS, BA 13.2-22		

[a] For references and abbreviations used in media composition, refer to Chen and Evans (1990).
* Media constituents for all stages not mentioned.

established programmes for release of clonal material that has been indexed for freedom from *Citrus tristeza* virus after micrografting of meristems onto seedling rootstocks *in vitro* (Blitters et al. 1970, Duran-Villa 1997). Other diseases of fruit trees, such as root rot of avocado caused by *Phytophthora cinnamoni*, could be controlled *in vitro* if clonally propagated disease-resistant root-stocks were readily available.

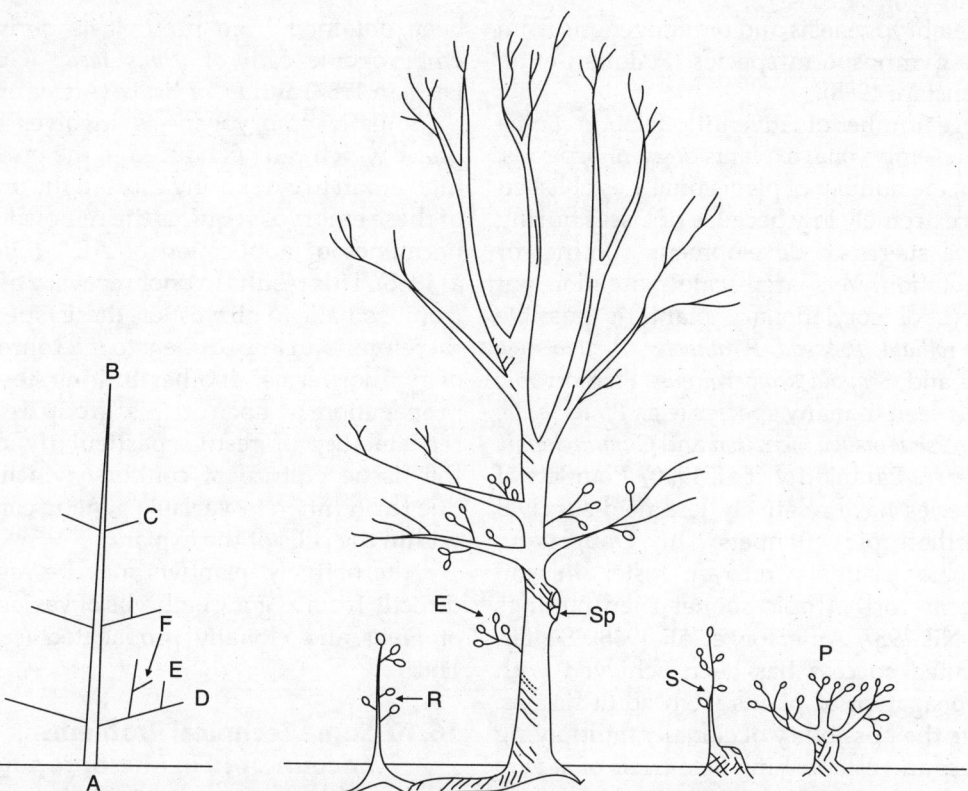

Fig. 16.6 Illustrations representing juvenile gradients in tree species. Left: juvenility in a regularly shaped conifer. The degree of juvenility is inversely proportional to the distance (along trunk and branches) between the root-shoot junction (A) and meristems. When considering the distances AB, AC, AE, AF, the meristem B is most mature and F the most juvenile. Right: juvenility in several of the hardwoods. Density of cross-hatching indicates degree of juvenility. Epicorm shoots (E), spheroblasts (Sp), root shoots (R), stump shoots (S) and severely pruned trees (P) are juvenile. Note: the characteristic features of the juvenile zone are the single trunk, retention of leaves close to trunk in winter and obtuse branch angles; the mature zone is characterised by the forked trunk and acute branch angles (after Bonga 1982; drawn from Pierik 1989).

16.9.2 Softwoods

Gymnosperms account for most of the world's produce of timber (softwood). Clonal propagation of gymnospermous species is desirable because of their long lifespan and 'open pollination'. Vegetative multiplication by cuttings has drawbacks as cuttings lack the vigour of seedlings and their growth declines with increasing age. The *in vitro* techniques are expected to help overcome these problems and regenerate plants at faster rates.

Tissue culture studies with gymnosperms commenced with Gautheret (1934) who cultured cambial tissues of *Pinus pinaster* and *Abies alba*. La Rue (1936) cultured embryos of several gymnosperms into normal-looking seedlings. Buds were also obtained from callus cultures of *Sequoia sempervirens* (Ball 1950, 1987). Despite these early successes little headway was made on tissue culture of gymnosperms until 1975 when reports appeared on complete regeneration of *Pinus palustris* plantlets (Sommer et al. 1975). Since then plantlets have been obtained via somatic embryogenesis and organogenesis from about 35 gymnosperm species (Hakman et al, 1985, Dunstan, 1988).

A large number of .adventitious buds can be induced on embryonal explants of gymnosperms. However, the number of plants finally established in soil is extremely low because of losses during successive stages of development, rooting, or transplantation. Yet, large-scale plantation out of embryo or cotyledonary plants is possible in *Pinus radiata, P. taeda, P. pinaster, Pseudotsuga menziesii* and *Sequoia sempervirens*. Plagiotropic growth is seen in many species (e.g., *Pinus taeda, P. radiata, Pseudotsuga menziesii* and *Cunninghamia lanceolate; see* Paranjothy et al. 1990). Plantlets of these species have relatively less field-survival than orthotropic plantlets. In Douglas-fir, plagiotropic plantlets recover faster in soil and become orthotropic sooner than cuttings (Abo-El-Nil 1987, Amerson et al. 1988). So far, only limited success has been achieved with micropropagation of plants from adult tissues. However, the possibility of clonally multiplying older trees and establishing field trials of *in vitro* derived plants has been demonstrated in species such as *Sequoia sempervirens, Pinus pinaster and P. radiata.*

Induction of somatic embryogenesis using immature embryo cultures *of Picea abies* (Hakman et al. 1985) and female gametophyte of *Larix decidua* (Nagmani and Bonga 1985) has been a major breakthrough in the tissue culture of gymnospermous species yielding softwoods. Other reports of similar nature are *of Pseudotsuga menziesii* (Durzan and Gupta 1987), *Larix* sp. (Klimaszewska 1989), *Pinus taeda* (Gupta and Durzan 1987) and *P. strobus* (Finer et al. 1989). In all these cases precotyledonary embryos (2-3 weeks after fertilisation) were used since older embryos yielded non-embryogenic calli. The possibility of obtaining somatic embryos from a mature zygotic embryo culture has been demonstrated in *Picea abies* (Von Arnold and Hakman 1986), *P. glauca* (Kartha et al. 1988) and *Pinus lambertiana* (Gupta and Durzan 1986). Well-developed somatic embryos have also been obtained from protoplasts derived from embryogenic calli of *Pinus taeda* (Gupta and Durzan 1987) and *Picea glauca* (Attree et al. 1987).

Somatic embryogenesis involves a callus-phase which can be induced in the presence of a high auxin to cytokinin ratio, but the maturation of these embryos requires the removal of phytohormones or application of ABA (Dunstan et al.1988). This results in poor recovery of plantlets from somatic embryos of these species and, therefore, is an impediment to micropropagation of gymnosperms. Another limiting step in clonal propagation of taxa of this group is the non-repeatability of results particularly regarding the tissue culture of conifers, which may be due to an inherent variable genetic constitution within the cells of the explant.

Alternatively, plantlets may be regenerated directly from zygotic embryos of various species of *Pinus* and clonally propagated (*see* Section 11.8.5).

16.10 Some Technical Problems Encountered in Micropropagation

16.10.1 Contamination of Cultures

During large-scale micropropagation of some plants certain types of slow-growing microbial (bacterial) contaminants persist even after initial surface sterilisation of explants (*see* Chapter 4). Such contaminants (*Pseudomonas* sp., *Erwinia* sp. and *Bacillus* sp.) may persist for many generations without being noticed and cause reduction in vigour or chlorosis in propagated plantlets (Knauss and Miller 1978). Addition of antibiotics or fungicides to the culture medium may control the infection by microbial contaminants (*see* Chapter 4) but for propagation of plants normally infected with latent or symptomless types of viruses or mycoplasmas, it is necessary to maintain stock of plants of them which are disease-free, by adopting the requisite procedures given in Chapter 15.

16.10.2 Browning of the Medium

A problem more frequently associated with

micropropagation of woody perennials is the accumulation of inhibiting substances in the growth medium during initiation of cultures. Explants from adult tissues of these species often produce excessive amounts of phenolic substances which rum the medium dark brown. Such a medium is toxic to the tissues and inhibits their growth. Quick transfer of explants to a fresh medium at short intervals could alleviate this problem in orchid bud cultures and nodal explant cultures of *Juglans regia*. Browning of the medium may also be prevented by dissecting explant tissues under the surface of liquids or by incorporating ascorbic acid or citric acid in culture media. For teak tissue cultures, polyvinyl pyrrolidone has proved an effective amendment to the culture medium. However, it may be emphasised that phenolic compounds such as phloroglucinol (PG) have been found essential for shoot multiplication and rooting in a number of rosaceous cultivars (Jones 1983). PG (found in the xylem sap of apple) is known to promote shoot multiplication and rooting in Pissardi plum and *Cinchona* (*see* Bhojwani and Razdan 1983, Mantell et al. 1985, George et al. 2008). Addition of AA (ascorbic acid) on the surface of culture medium reportedly prevents its browning by explants of banana cv. Formosana (Ko et al. 2009).

16.10.3 Hyperhydration

The term *hyperhydration* refers to morphological, physiological and metabolic dearrangements occurring during intensive shoot multiplication in culture. Various terms, viz. glassiness, glauciness, translucency and vitrescence have been used to describe these malformations although the more general term used was vitrification (Gaspar 1991, Ziv 1991). Vitrification is now referred to a process in cryopreservation (*see* Chapter 18). Hyperhydrated shoots generally show poor growth, become necrotic and, finally, die. Some preventive measures to reduce hyper-hydration include (a) increasing concentration of agar, (b) overlaying of medium with paraffin, (c) using desiccant such as $CuSO_4$ (d) bottom cooling of the culture vial to improve aeration, (e)

lowering cytokinin level or replacing one kind of cytokinin by another, and (f) manipulating NH_4^+ or salt concentrations in the culture medium.

16.11 Applications and Limitations of Micropropagation

The technique of micropropagation, an alternative to conventional methods of vegetative propagation, is applied with the objective of enhancing the rate of multiplication. Through tissue culture over a million plants can be grown from a small, even microscopic, piece of plant tissue within 12 months. Such a prolific rate of multiplication cannot be expected by any of the *in vivo* methods of clonal propagation. Further the advantage in propagation through tissue culture is that shoot multiplication usually has a short cycle (2-6 weeks) and each cycle results in logarithmic increase in the number of shoots. Additionally, tissue cultures give propagules such as minitubers or minicorms for plant multiplication throughout the year irrespective of the season. The small size of the propagules and their ability to proliferate in soil-free environment facilitates their storage on a large scale and also allows their large-scale dissemination by suitable means of transport across international boundaries. Using these methods stocks of germplasm can be maintained for many years.

Clonal propagation *in vitro* appears to have a permanent advantage in cases in which serious problems with disease occur. This is because of the fact that through *in vitro* methods more pathogen-free plants can be raised and maintained economically (*see* Chapter 15). Multiplication by cloning in dioecious species is extremely important when the seed progeny yields 50% males and 50% females and plants of one of the sexes are desired commericaly. For example, male plants of *Asparagus officinalis* are more valuable than the female plants of this species. Clonal propagation of male *Asparagus* by stem cuttings has not been successful but can be achieved through tissue culture. On the contrary, in seed-raised orchards, micropropagating female plants would save the losses suffered due to discarding a large number of naturally arising male plants.

A major advantage of micropropagation happens to be the minimum growing space required in commercial nurseries. Several thousand million plants can be maintained inside the culture vials on a shelf space built into a room of about 3 m × 3 m × 5 m. This makes possible the propagation of clones on a commercial scale for large number of horticultural species (African violet, banana, eucalyptus, ferns, orchids, gloxinia, gerbera and rhododendrons) in a single nursery. Some other advantages of *in vitro* propagation over conventional breeding systems are enumerated in Table 16.3. A comprehensive list of plants that can be donally propagated *in vitro* is given by Bhojwani and Razdan (1983, 1996), Mascarenhas et al. (1989), Bhaskaran and Prabhudesai (1989) and Sagawa and Kunisaki (1990).

Although commercial micropropagation has been possible for quite a number of plants there are many tree crops, including gymnosperms,

Table 16.3 Comparison of conventional and *in vitro* breeding system[a]

Factor	Conventional breeding	*In vitro* breeding
Growth cycle	Several decades	25–26 hr
Size	One to several dozen metres	50-100 μm in diameter
Space (10^6 offspring per year)	Several hectares	10 l of suspended cells
Quantities required for mutation rate investigation	10^7 trees	10^7 cells
Quantities required for character investigation	10^3 trees	10^3 cells
Time required for seed production	5–15 years	One month produces 10^6 somatic embryos for use as artificial (synthetic) seeds
Predictability of seed production	Seed-setting has on- and off-years and is affected by natural factors	Seed production is completely under artificial control
Genetic uniformity	Much variation occurs during sexual propagation and pollination, and selection should be controlled	Less variation in somatic embryos
Multiplication	Cutting or grafting at slow speed	Propagation coefficient of *in vitro* culture can increase by millions
Ploidy	Difficult to obtain haploid and homozygous lines; diploid materials	Easy to get haploid and pure lines
Breeding methods	Selection is most common; hybridisation and mutagenisation are available in some species; self-cross and back-cross are difficult or impossible	Combination of hybridisation and anther culture can accelerate breeding process and increase selection efficiency; mutation and mutant screening are conducted at cellular level with high efficiency; pure lines can be achieved through haploid culture and chromosome doubling; genetic engineering methods can be used

[a] After Durzan 1980.

that could not be multiplied under both *in vivo* and *in vitro* techniques. Considerable research is needed to overcome this limitation. In systems in which aseptic cultures have been established, *in vitro* propagation through many generations may lead to bulking up of off-types rather than clones. This problem could be offset by a conservative approach, namely adopting enhanced axillary branching. Another approach would be to not go beyond the fourth cycle and to raise only a few thousand plants from a selected explant. Generally, during repeated cycles *of in vitro* shoot multiplication a percentage of cultures show water-soaked or almost translucent leaves. Such shoots exhibit a decline in rate of growth and may ultimately die. This phenomenon, hyperhydration (earlier called *vitrification),* may be prevented in shoot cultures of globe artichoke *(Cyana scolymus)* by raising the agar concentration from 0.6% to 1% in the culture medium. However, this type of treatment is likely to drastically reduce the rate of propagation for most of the tissues. Other preventive measures of hyperhydration are described in Section 16.9.3.

In conclusion, the investment in commercial tissue culture business will depend to a large extent on cost of the laboratory set-up, type of plant to be propagated and the skill involved (Table 16.4). The production cost in developing countries, would be low in view of the fact that 63% components listed in this table involve manpower and labour charges which in these countries are generally low. These countries could provide an incentive in attracting alien investment (particularly from developed countries) in establishing laboratories for commercial plant micropropagation. For example, flowering of *Dendrobium* hybrid seedlings could be significantly shortened by 5 months *in vitro* as against the optimal time (over 30 months) of the field-grown plants. This would thus reduce cost of labour and optimize the space required for normal orchid breeding (Sim et al. 2007).

Table 16.4 Relative cost components associated with micropropagation of a typical crop[a,b]

Cost component	Per cent
Laboratory, direct labour	15
Utilities	9
Depreciation	7
Supervision (laboratory and greenhouse)	23
Planting, direct labour	9
Other production costs	37
Total	100

[a] Twyford International, Inc. Study (Source: Chu and Kurtz 1990).
[b] The overall production cost would vary in different countries; reasons are given in the text.

It is advisable that a commercial nursery man start with those crop species for which published methods are available (Appendices 16.1 to 16.8). Micropropagation protocols are further available for potato, banana, pear, walnut, gladiolus, syngonium, lilacs, rose, *Dieffenbachia* (*see* details in Bhojwani and Razdan 1996, Trigiano and Gray 2000, 2011) and other perennial species listed in Table 16.2. In addition to this the details of *in vitro* propagation of Ficus, African Violet, kalanchoe, cacti, insectivorous plants, ferns and other endangered species are given in Smith (2013). It is essential that growers have some training in tissue culture and plant husbandry for which short courses could be organised from time to time. Another approach followed nowadays is to automate micropropagation at its various stages. In this connection bioreactors are being used for large-scale multiplication of somatic embryos, shoots and bulbs (Aitken-Christie et al. 1995). Robots have also been used for subculture of shoots during stages III and IV of micropropagation (Sluis 2006, Altman and Hasegawa 2012). Automation or robotisation reduces the cost of the labour component in micropropagation. The synthesis of some useful compounds from cell and tissue cultures using bioreactors is comprehensively described in Chapter 17.

APPENDIX 16.1

List of Orchids which have been Clonally Propagated *in vitro*[a]

Plant	Explant	Plant	Explant
Anacamptis pyramidalis	Shoot tip	*Neottia nidus-avis*	Root
Aranda	Shoot tip, axillary bud	*Odontioda*	Shoot tip
(*Arachnis hookeriana* × *Vanda lamellata*)	Shoot tip	*Odontoglossum*	Shoot tip
		Odontonia	Shoot tip
Aranthera	Shoot tip	*Oncidium*	Shoot tip
Arundina bambusifolia	Shoot tip (from young seedling)	*Oncidium papilio*	Flower-stalk segment (with dormant apical buds)
Ascofinetia	Inflorescence segment (with flower primordia)	*Phajus*	Shoot tip
Brassocattleya	Axillary bud	*Phalaenopsis*	Flower-stalk segment
Calanthe	Shoot tip		Shoot tip
Cattleya	Shoot tip, axillary bud, lateral bud, leaf base, leaf tip		Leaf segment, stem segment, root segment
Cymbidium	Shoot tip	*Pleione*	Shoot tip
Dendrobium	Shoot tip	*Rhynchostylis gigantea*	Shoot tip, lateral buds
	Nodal segment	*Schomburgkia superbiens*	Lateral bud
	Flower stalk segment (with vegetative buds)	*Vanda* (Terete-leaf)	Shoot tip
			Stem section
Epidendrum	Leaf tip	*Vanda* (Strap-leaf)	Shoot tip
Laelia	Axillary bud	*Vanda* hybrid	Shoot tip
Laeliocattleya	Axillary bud	(*V. teres* × *V. hookeriana*)	Axillary bud, root segment
Lycaste	Shoot tip		
Miltonia	Shoot tip	*Vascostylis*	Inflorescence segment (with flower primordia)
Neostylis	Inflorescence segment (with flower primordia)	*Vuylstekeara*	Shoot tip

[a] For references refer to Arditti (1977).

APPENDIX 16.2
Composition of Media for Micropropagation of *Cattleya* Shoot Tip

Constituents	Media (amounts in mg l^{-1})[a]		
	Initiation	Maintenance[b]	Rooting
Inorganic nutrients			
$MgSO_4 \cdot 7H_2O$	120	120	250
KH_2PO_4	135	135	250
$Ca(NO_3)_2 \cdot 4H_2O$	500	500	1000
$(NH_4)_2 \cdot SO_4$	1000	1000	500
KCl	1050	1050	—
KI	0.099	0.099	—
H_3BO_3	1.014	1.014	0.056
$MnSO_4 \cdot 4H_2O$	0.068	0.068	7.5
$ZnSO_4 \cdot 7H_2O$	0.565	0.565	0.331
MoO_3	—	—	0.016
$CuSO_4 \cdot 5H_2O$	0.019	0.019	—
$CuSO_4$	—	—	0.040
$AlCl_3$	0.031	0.031	—
$NaCl_2$	0.017	0.017	—
$FeSO_4 \cdot 7H_2O$	—	—	25
$FeC_6H_5O_7 \cdot 3H_2O$	5.4	5.4	—
Organic nutrients			
Inositol	—	18.00	—
Nicotinic acid	—	1.22	—
Pyridoxine HCl	—	0.21	—
Thiamine HCl	—	0.34	—
Folic acid	—	4.4	—
Biotin	—	0.024	—
Calcium-d-pantothenate	—	0.48	—
Glutamic acid	—	15.0	—
Asparagine	—	13.0	—
Guanylic acid	—	182.0	—
Cytidylic acid	—	162.0	—
Growth regulators			
Kinetin	0.2	0.22	—
NAA	0.1	0.18	—
GA_3	—	0.35	—
Complex nutrients			
Coconut water	150 ml l^{-1}	50-150 ml l^{-1} } either one	—
Casein hydrolysate	—	100 ml l^{-1}	—
Sucrose	0.5%	2%	2%
Agar	—	—	1.2%-1.5%

[a] Taken from Arditti (1977).
[b] Maintenance medium is for protocorm multiplication and shoot development.

APPENDIX 16.3
Composition of Media for Micropropagation of *Cymbidium* Shoot Tip[a]

Constituents	Media (amounts in mg l^{-1})*		
	Wimber	Fonnesbech[b]	Modified Knudson C
Inorganic nutrients			
KNO_3	525	—	—
$MgSO_4 \cdot 7H_2O$	250	250	250
KH_2PO_4	250	250	250
K_2HPO_4	—	212	—
$(NH_4)_2SO_4$	500	300	500
$Ca(NO_3)_2 \cdot 4H_2O$	—	400	1000
$CaHPO_4$	200	—	—
H_3BO_3	—	10	0.056
$MnSO_4 \cdot 4H_2O$	—	25	7.5
$ZnSO_4 \cdot 7H_2O$	—	10	0.331
$Na_2MoO_4 \cdot 2H_2O$	—	0.25	—
$CuSO_4 \cdot 5H_2O$	—	0.025	0.040
MoO_3	—	—	0.016
$Fe_2(C_4H_4O_6)_3$	300	—	—
$FeSO_4 \cdot 7H_2O$	—	27.9	25
Na_2 EDTA	—	37.8	—
Organic nutrients			
Inositol	—	100	—
Nicotinic acid	—	1	—
Pyridoxine HCl	—	0.5	—
Thiamine HCl	—	0.5	—
Glycine	—	2	—
Growth regulators			
Kinetin	—	0.215	—
NAA	—	1.86	—
Complex nutrients			
Casamino acid	—	2000-3000 } either one	—
Tryptophan	2000	3000-4000	—
Coconut water	—	100-150 ml l^{-1}	—
Banana, ripe	—	—	15%
Sucrose	2%	3-4%	2%
Agar	—	0.8%	1.2-1.5%

[a] Adopted from Arditti (1977).
[b] After Fonnesbech (1972).
* Any one of the media can be used for all three stages of micropropagation.

APPENDIX 16.4
Composition of Media for Micropropagation of Strawberry[a,b]

Constituents	Media (amounts in mg l^{-1})*		
	Initiation	Shoot multiplication[c]	Rooting
Inorganic nutrients			
KNO_3	250	250	250
$MgSO_4 \cdot 7H_2O$	250	250	250
KH_2PO_4	250	250	250
$Ca(NO_3)_2 \cdot 4H_2O$	1000	1000	1000
KI	0.83	0.83	0.83
H_3BO_3	6.2	6.2	6.2
$MnSO_4 \cdot 4H_2O$	16.9	16.9	16.9
$ZnSO_4 \cdot 7H_2O$	8.6	8.6	8.6
$Na_2MoO_4 \cdot 2H_2O$	0.25	0.25	0.25
$CuSO_4 \cdot 5H_2O$	0.025	0.025	0.025
$CoCl_2 \cdot 6H_2O$	0.025	0.025	0.025
$FeSO_4 \cdot 7H_2O$	27.8	27.8	27.8
Na_2 EDTA	37.3	37.3	37.3
Organic nutrients			
Inositol	100	100	100
Nicotinic acid	0.5	0.5	0.5
Pyridoxine HCl	0.5	0.5	0.5
Thiamine HCl	0.1	0.1	0.1
Glycine	2	2	2
Growth regulators			
BAP	0.1	1	—
IBA	1	1	1
GA_3	0.1	0.1	—
Glucose	4%	4%	4%
Agar	0.8%	0.8%	0.8%

[a] Explant: shoot tip.
[b] After Boxus (1974).
[c] Mode of shoot multiplication; enhanced axillary branching.

APPENDIX 16.5
Composition of Media for Micropropagatlon of Red Raspberry[a,b,c]

Constituents	Media (amounts in mg l^{-1})	
	Initiation and shoot multiplication	Rooting
Inorganic nutrients		
NH_4NO_3	400	400
KNO_3	480	480
$CaCl_2 \cdot 2H_2O$	440	440
$MgSO_4 \cdot 7H_2O$	370	370
$NaH_2PO_4 \cdot H_2O$	380	380
KI	0.30	0.30
H_3BO_3	6.2	6.2
$MnSO_4 \cdot 4H_2O$	16.9	16.9
$ZnSO_4 \cdot 7H_2O$	8.6	8.6
$Na_2MoO_4 \cdot 2H_2O$	0.25	0.25
$CuSO_4 \cdot 5H_2O$	0.025	0.025
$COCl_6 \cdot 6H_2O$	0.025	0.025
$FeSO_4 \cdot 7H_2O$	55.7	55.7
Na_2 EDTA	74.5	74.5
Organic nutrients		
Inositol	100	100
Thiamine HCl	0.4	0.4
Adenine sulphate dihydrate	80	—
Growth regulators		
BAP	2	—
IBA	0.5	1
Activated charcoal	—	600
Sucrose	3%	3%
Agar	0.6%	0.6%

[a] Explant: shoot tip.
[b] After Anderson (1980).
[c] Mode of shoot multiplication: enhanced axillary branching.

APPENDIX 16.6
Composition of Media for Micropropagation of Blueberry[a,b]

Constituents	Media (amounts in mg l^{-1})	
	Initiation and shoot multiplication[c,d]	Rooting
Inorganic nutrients		
NH_4NO_3	160	
KNO_3	202	
$MgSO_4 \cdot 7H_2O$	370	
KH_2PO_4	408	
$(NH_4)_2SO_4$	198	
$Ca(NO_3)_2 \cdot 4H_2O$	708	
KI	0.83	Rooted *in vivo*
H_3BO_3	6.2	by dipping the
$MnSO_4 \cdot 4H_2O$	16.9	basal end in
$ZnSO_4 \cdot 7H_2O$	8.6	0.1% IBA on
$Na_2MoO_4 \cdot 2H_2O$	0.25	talc and
$CuSO_4 \cdot 5H_2O$	0.025	Planting in
$CoCl_2 \cdot 6H_2O$	0.025	potting mix
$FeSO_4 \cdot 7H_2O$	55.7	
Na_2 EDTA	74.4	
Organic nutrients		
Inositol	100	
Thiamine HCl	0.4	
Adenine sulphate dihydrate	80	
Growth regulators		
IAA	4	
2-ip	15	
Sucrose	3%	
Agar	0.55%	

[a] Explant: shoot tip.
[b] After Zimmerman and Broome (1980).
[c] Mode of shoot multiplication: enhanced axillary branching and adventitious shoots.
[d] Before rooting, the shoot clusters are maintained on a shoot elongation medium which differs from the shoot multiplication medium only in that the hormone content 2-ip is either omitted or its concentration reduced to 1-5 mg l^{-1}.

APPENDIX 16.7
Composition of Media for Micropropagation of Rhododendron[a,b,c]

Constituents	Media (amounts in mg l^{-1})	
	Initiation and shoot multiplication	Rooting[d]
Inorganic nutrients		
NH_4NO_3	400	133.3
KNO_3	480	160
$CaCl_2 \cdot 2H_2O$	440	146.6
$MgSO_4 \cdot 7H_2O$	370	123.3
NaH_2PO_4	380	126.6
KI	0.83	—
H_3BO_3	6.2	2.06
$MnSO_4$	16.9	5.63
$ZnSO_4 \cdot 7H_2O$	8.6	2.86
$Na_2MoO_4 \cdot 2H_2O$	0.25	0.083
$CuSO_4 \cdot 5H_2O$	0.03	0.01
$CoCl_2 \cdot 6H_2O$	0.03	0.01
$FeSO_4 \cdot 7H_2O$	55.70	18.56
EDTA	74.50	24.83
Organic nutrients		
Inositol	1 00	33.3
Thiamine HCl	0.40	0.133
Adenine sulphate	80	26.86
Growth regulators		
2-ip	5.00	—
IAA	1.00	—
Activated charcoal	—	600
Sucrose	3%	3%
Agar	0.6%	0.6%

[a] Explant: nodal cutting and terminal bud.
[b] After Kyte and Briggs (1979).
[c] Mode of shoot multiplication: enhanced axillary branching.
[d] In vitro formed shoots also rooted in a medium containing only sucrose (3%), 2-ip (5 mg l^{-1}) and activated charcoal (800 mg l^{-1}).
Note: Woody Plant Medium (WPM) has also been used to propagate *Rhodedendron* sp. by axillary bud proliferation (see Lloyd and McCown 1980).

APPENDIX 16.8
Composition of Media for Micropropagation of Date Palm[a,b]

Constituents	Media (amounts in mg l^{-1})	
	Callus induction	Somatic embryogenesis
Inorganic nutrients		
NH_4NO_3	1650	1650
KNO_3	1900	1900
$CaCl_2$	322	322
$MgSO_4 \cdot 7H_2O$	181	181
KH_2PO_4	170	170
$NaH_2PO_4 \cdot 2H_2O$	170	170
KI	0.83	0.83
H_3BO_3	6.2	6.2
$MnSO_4 \cdot H_2O$	16.9	16.9
$ZnSO_4 \cdot H_2O$	8.6	8.6
$Na_2MoO_4 \cdot 2H_2O$	0.25	0.25
$CuSO_4$	0.016	0.016
$CoCl_2 \cdot 6H_2O$	0.025	0.025
$FeNa \cdot EDTA$	36.7	36.7
Organic nutrients		
Inositol	100	100
Thiamine HCl	0.4	0.4
Growth regulators		
2,4-D	100	—
2-ip	3	—
Activated, neutralised charcoal	3000	3000
Agar	0.8%	0.8%
Sucrose	3%	3%

[a] Explant: lateral buds and shoot tips.
[b] After Tisserat (1981).

PART V
General Applications

Industrial Applications: Secondry Metabolite Production

The potential use of plant tissue culture for the production of compounds valuable to the industry has long been a subject of interest for biotechnologists. Culture practices involved in the growth and maintenance of micro-organisms for industrial level production of drugs, breweries and milk products are well established. Higher plants too are valuable sources of industrially important natural products, which include flavours, fragrances, essential oils, pigments, sweeteners, feedstocks, antimicrobials and pharmaceuticals. In most instances these chemical compounds belong to a metabolic group collectively referred to as *secondary metabolites.* These substances do not participate in vital metabolic functions of the host-plant tissues in the same manner as amino acids, nucleic acids, or other primary metabolites, but appear to serve as a chemical interface between the producing plant and its surrounding environment. It is presumed that secondary metabolites are produced by plants to ward off predators and to attract pollinators. To some extent, secondary substances may also help in combating infectious diseases (*see* Whitaker and Hashimoto 1986) and act as anticancer agents (Wink et al. 2005).

Plant cell cultures have advantages in metabolite production over intact plants due to the fact that (a) the rate of cell growth and phytochemical biosynthesis in cultures initiated from a very small amount of plant material is quite high, and the final product may be produced in a considerably short period of time. This is in contrast to large amounts of mature plant tissues processed to obtain a small quantity of a drug, (b) Plant cell cultures are maintained under controlled environmental and nutritional conditions which ensure the continuous yields of metabolites. On the contrary, continual decline in the natural habitats due to ecologically disturbed conditions may make the availability of source plants uncertain unless they are clonally propagated, (c) Suspension culture offers a more effective mechanism of incorporating precursors into cells than is found in whole plants, (d) New routes of synthesis can be recovered from deviant and mutant cell lines which can lead to production of novel compounds not previously found in whole plants, (e) Some cell cultures have the capacity for biotransformation of specific substrates to more valuable products by means of single- or multiple-step enzyme activity, and (f) Culture of cells may be more economical for those plants which take long periods to achieve maturity (e.g., *Papaver bracteatum*, the source of thebanine, takes two to three seasons to reach maturity).

Considering the high economical and pharmacological importance of secondary metabolites, industries are deeply interested in utilising plant tissue culture technology for large-scale production of these substances (Misawa 1994, Sato et al. 2001, Lee et al. 2010). About 25% prescription medicines and various raw materials used in the industries are obtained from plants. Further, the number of patents on pigments, cosmetics, perfumes and food additives have increased in recent years. Natural sources for the manufacture of these products are not enough to meet the consumers' demand and efforts have to be made to develop technology for their production at an industrial level. Tissue culture of natural materials may not be as expensive compared to the synthetic production of various commercial substances, or pharmaceuticals (Fujita 1990, Scragg 1994, Cai et al. 2012).

17.1 Advances in Tissue Culture for Synthesis of Useful Compounds

Application of plant tissue cultures for the production of useful secondary compounds has been recognised since the early 1950s. The pioneering attempts achieved little success since plant tissue culture technology was not well developed then. Plant cells generally produce only small amounts of useful compounds. The successful selection of cells producing high amounts of secondary compounds has been made possible because of the heterogeneity associated with cultured plant tissues or cells. Techniques are now available to induce and select stable genetic variants arising from the cultured tissues (*see* Chapter 14) and these can be applied for producing high amounts of vitamins, pigments, alkaloids, food flavours and useful metabolites in plant species (Collin and Watts 1983, Constabel and Vasil 1987).

The impact of utilising a specific tissue culture system for producing increased levels of a certain secondary metabolite over the intact plant has been observed in cell lines producing high nicotine in *Nicotiana tabacum*, anthocyanin in *Euphorbia millii*, anthraquinones in *Morinda citrifolia*, *Crocus sativa*, and in about

18 species from genera *Asperula, Galium* and *Sherardia*, shikonin derivatives in *Lithospermum erythrorhizon*, phenolic compounds *in Acer pseudoplatanus*, α-hydroxyquiphenylalanine (L-dopa) in *Stizolobium hassjoo*, ferruginol in *Salvia miltorrhiza*, sanguinarine in *Papaver somniferum*, ajmalcine and serpentine in *Catharanthus roseus*, rosmarinic acid in *Coleus blumei*, diosgenin in *Dioscorea deltoides* (*see* Whitaker and Hashimoto 1986, Sakuta and Komamine 1987, Giles and Songstad 1990, Isa and Ogasawara 1991).

In spite of the fact that several useful matabolites can be accumulated in cell cultures (Table 17.1), the exciting promise of plant tissue culture for the production of valuable secondary metabolites on a commercial scale is still restricted to a few products. Mitsui Petrochemical Industries (Tokyo) has achieved success in commercially producing shikonin, used both as a dye and a pharmaceutical, from *Lithospermum erythrorhizon* cell suspension cultures. Similarly, it has been possible to establish cell cultures for industrial production of berberine and ginseng extracted from roots of *Coptis japonica* and *Panax ginseng* (Sato and Yamada 1984, Ushiyama et al. 1986), respectively. Some more phytochemicals produced on large scale are mentioned under section 17.8.

To propel tissue culture production of secondary metabolites into the realm of economic feasibility one has to encounter two technical challenges. First, the establishment of a tissue culture system for synthesising the desired secondary product and subsequently selecting variant cultures that synthesise very high levels of this compound. Second, the challenge of designing bioreactor systems that can manipulate the genetic and biochemical capabilities of plant cell cultures to raise production levels even higher. Now that much basic and useful information on cell biochemistry is available, which is fundamental to all work on natural product synthesis, it appears that compounds (phenolics and flavonoids) which are widespread in plants and happen to be biologically less complex, could be exploited

Table 17.1 Accumulation of secondary metabolites reported in cell and tissue cultures[a]

Secondary metabolite	Type of culture[b]	Source plant
Anthocyanins		
Cyanidin	C	Dimorphothica auriculata
	C	Haplopappus gracilis
Delphinidin	C	D. auriculata
Anthraquinones		
Alizartin	C	Morinda citrifolia
Digitolutein	C	Digitalis lanata
4-hydroxydigitolutein	C	D. lanata
3-methyl purpurin	C	D. lanata
Rhein	C	Cassia angustifolia
Carotenoids		
Antheraxanthin		
Beta-carotene		
Lutein		
Lutein-5,6 epoxide	C	Ruta graveolens
Neoxanthin		
Voilaxanthin		
Zeaxanthin		
Chalcones and Deoxyflavones		
Daidzein	C, SC	Phaseolus aureus
	C, CS	Glycine max
2',4,4'-trihydroxy chalcone	C	P. aureus
Coumestanes and Coumarinochromans		
Coumestrol	C, SC	P. aureus
Pisatin	C	Pisum sativum
Soyagol	C, SC	P. aureus
Flavones and Flavanoids		
Apigenin	SC	G. max
	C	Solanum dulcamara
	C	Trigonella corniculata
Artimissinine	C	Artemisia scoparia
Chrysoerion	SC	Petroselinum hortense
Isorhamnetin	SC	Argemone mexicana
		P. hortens
Kaempferol	C	Agave wightii
	C	Allium sativum
	C	A. scoparia
	C	Dolichos lablab
	C	G. max

Contd...

Table 17.1 Contd.

Secondary metabolite	Type of culture[b]	Source plant
	C	*P. sativum*
	C	*Lycopersicon esculentum*
Luteolin	C	*Datura pinnata*
	SC	*Papaver hortense*
Negretin	SC	*Solanum tuberosum*
Nobeletin	C	*Citrus aurantium*
Quercetin	C	*A. wightii*
	SC	*Crotalaria juncea*
	C	*Datura metel*
	C	*D. tatula*
	C	*L. esculentum*
	C	*P. hortense*
	C	*S. aviculare*
Sinensetin	C	*C. aurantium*
Naphthoquinones		
Plumbagin	C	*Plumbago zeylanica*
Sapogenins		
Chlorogenin		
Hecogenin		
Sarsasapogenin	C	*Yucca aloifolia*
Smilagenin		
Tiggenin		
Diosgenin	C	*S. aviculare*
	C	*S. dulcamara*
	C	*S. nigrum*
	C	*S. xanthocarpum*
Gitogenin		
Hecogenin	C	*A. wightii*
Tiggenin	SC	*S. nigrum*
Sesquiterpenes		
Cryophyllenebisabolene	SC	*Andrographis paniculata*
	C	*Lindra strychinifolia*
Lindenenol	C	*L. strychinifolia*
Linderane	C	*L. strychinifolia*
Steroidal alkaloids		
Solasonine	C	*S. xanthocarpum*
Tomatin	C	*L. esculentum*

Contd...

Table 17.1 Contd.

Secondary metabolite	Type of culture[b]	Source plant
Sterols and Triterpenes		
Beta-amyrin	C	Paul's Scarlet Rose
	C	*Tylophora indica*
Beta-sitosterol	C	*Nicotiana tabacum*
Campesterol	C	*Withania somnifera*
	C	*D. metel*
	C	*Helianthus annuus*
	C	Paul's Scarlet Rose
	C	*Trigonella foenum-graceum*
	C	*T. indica*
Cholesterol	C, SC	*H. annuus*
	C, SC	*S. indicum*
Cycloartinol	C, SC	*N. tabacum*
Isofucosterol	C, SC	*H. annuus*
Lanosterol	C, SC	*S. nigrum*
Obtusifoliol	C, SC	*N. tabacum*
Stigma sterol	C	*Dioscorea tokora*
Tannins	C	*Camellia sinensis*
Catechin	SC	Paul's Scarlet Rose
Epicatechin	C	*C. sinensis*

[a] Adopted from Purohit and Mathur (1990).
[b] C = Callus; SC = Suspension culture.

for large-scale synthesis using tissue culture technology (Neumann et al. 1985, Constabel and Vasil 1987).

Various evidences suggest that the regulation of secondary metabolism is linked with induction of morphological differentiation in plants of some species. For example, the cardiac glycosides of *Digitalis* are principally found in leaf cells, quinine and quinidine in the bark of *Cinchona* trees, and tropane alkaloids in the roots of many solanaceous species. This shows that cells not only undergo a morphological specialisation during plant growth and maturation, but also differentiate in their capacity to produce specific chemicals. Investigations on cell differentiation versus secondary product synthesis using celery tissue culture revealed that cultures initiated from less differentiated tissues (globular and

heart-shaped embroys) demonstrated no flavour compounds, while those initiated from better differentiated tissues (torpedo-shaped embryos, petiole tissues) possesed the characteristic celery phthalide flavour compounds (Al-Abta and Collin 1978, Al-Abta et al. 1979). It was further assumed that presence of chlorophyll in differentiated tissues may be someway associated with the synthesis of flavour compounds.

A detailed and comprehensive review on the role of morphological and cellular differentiation in the synthesis of flavour compounds has been presented by Collin and Watts (1983). The most elegant work done subsequently was a study of the regulation of phenylpropanoid (flavonoid and coumarin) metabolism in parsley cell cultures (Hahlbrock and Scheel 1989). Two groups of enzymes are associated with induction and synthesis of these two types of compounds in cultured cells.

17.2 Techniques of Selecting Cell Lines for High Yields of Secondary Compounds

Most callus and liquid cell suspension cultures metabolising secondary compounds generally produce low yields of these compounds. This is attributed to the lack of organ differentiation in cell or callus cultures. In undifferentiated tissues and in cells, there is no sufficient compartment-alisation of enzymes required for synthesis and accumulation of secondary metabolites. On the contrary, organ-differentiating callus or tissue cultures have the necessary enzymes for the synthesis of secondary metabolites (*see* review by Butcher 1977). To increase the yield of secondary compounds by undifferentiated tissue or cell cultures, therefore, the application of some special techniques may be necessary for exploring the heterogeneity within cell populations in order to select lines capable of accumulating higher levels of the desired metabolite.

17.2.1 Cell Cloning

Cloning of cells can be used as a tool to explore natural heterogeneity within cell cultures and in the selection of high-yielding lines. The simplest procedure is to take single cells from suspension cultures and grow them individually into populations on a suitable medium (*see* Chapter 5). Each cell population is then screened for altered phenotype. Only those showing increased yields are finally selected and maintained by subcloning.

Alternatively, suspension cultures may be screened for phytochemical differences by passing through sieves of definite mesh size followed by density gradient centrifugation. Ozeki and Komamine (1985) obtained two populations of carrot suspension cultures, the heavier of which gave embryos and the lighter accumulated anthocyanin provided auxin was absent in the medium.

Examples in which it might be possible to enhance the yield using the cell cloning technique are *Daucus carota* (anthocyanin), *Lithospermum erythrorhizon* (shikonin), *Nicotiana tabacum* and *N. rustica* (nicotine), *Catharanthus roseus* (serpentine and ajmalicine), *Anchusa officinalis* (rosmarinic acid), *Coptis japonica* (berberine), *Lavandula vera* (biotin) and *N. tabacum* (ubiquinone-10). Cells used for cloning in most of these cases were plated as units varying from a single cell to a group of small aggregates. The use of aggregates did not seem to invalidate the generation of clones since the cells in these groups had only marginal phenotypic variations (Dougall 1987).

17.2.2 Visual or Chemical Analysis

A straightforward approach for identifying those cell lines which are high producers of desired secondary metabolites is to either visualise or chemically measure the approximate levels of these compounds in individual cell or cell colonies. Visual analysis involves microscopic estimations provided the compounds are coloured. This procedure has the advantage of being non-destructive while most chemical analytical procedures result in cell death. Visual selection also proves useful in cell cloning as it is possible to identify cells or callus with high concentrations of pigments (β-carotene in carrot and shikonin in *Lithospermum* cultures) and isolate them for cloning.

The procedures for chemical analysis utilise analytical techniques that are rapid, inexpensive, specific and very sensitive. An early example of such analysis was described by Zenk et al, (1977) who tested a portion of colonies derived from single-cell cultures of *Catharanthus roseus* for serpentine and ajmalicine, using a sensitive radioimmunoassay technique. More than 200 samples could be screened each day for selection of high-yielding strains. Hall and Yeoman (1986) used microspectrophotometry to demonstrate that cells accumulating anthocyanin differ from the remaining cells in *Catharanthus* cultures with respect to absorbance spectra. Similarly, differences in UV absorbance spectra by single cells of *Coleus blumei* and *Anchusa officinalis* suspension cultures were correlated to variations in their rosmarinic acid content.

Radioimmunoassay techniques employed in the selection of high-yielding lines of *Catharanthus* and *Anchusa* cells could be made more efficient if they were automated using flow-cytometric sorting systems. One might also devise staining procedures wherein an added compound would form a complex with the desired compound and enable its detection. Another possibility is to use a fluorescent antibody to detect the desired compound on the cell surface. However, these methods, if applied must be non-destructive allowing the cells to grow in cultures. Destructive methods using stains or antibodies are used only when replica plating procedures can be successfully applied and replicas of colonies obtained from single cells. In such a situation, one replica can be analysed destructively to ascertain the high-producing quality and the other reserved to allow the rescue of the high-producing clones after identification.

The 'cell squash method' has also been tried for chemical analysis of tobacco cells. In this method, portions of cell clones are squashed on filter paper and then sprayed with stain specific to a compound in order to visualise the high-producing colonies. Knoop and Beiderbeck (1985) plated suspension cultures on a cellophane sheet attached to activated carbon-coated filter paper which, in turn, was placed on the nutrient medium. Compounds released by the cells were absorbed in localised areas of the charcoal-coated filter paper. After removing the cellophane, the filter paper was oberved under UV light to assess the quality of metabolite released from each colony.

17.2.3 Selection for Resistance

Selection for resistance to toxic compounds can produce mutant cell lines which are able to accumulate secondary metabolites. Such mutants can be selected from large cell populations by their ability to grow in the presence of an inhibitor such as the toxic analogue of the required primary metabolite. Usually cells overproducing the primary metabolite dilute the toxic effect of its analogue, thus incorporating the trait resistance in them. Primary metabolites are usually precursors of secondary compounds. Overproduction of a primary metabolite could therefore lead to synthesis and accumulation of secondary compounds in cultured cell (Widholm 1987).

Cell lines selected for resistance to the tryptophan analogue 5-methyltryptophan (5 MT) produce strains which reportedly overproduce tryptophan in a number of tissue cultures (*see* also Chapter 14). This is due to mutation in the feedback tryptophan biosynthetic control enzyme anthranilate synthase (AS). Some of these overproducing lines were shown to synthesise 10—40 times the levels of natural auxin IAA, which is a tryptophan derivative (Sung 1979).

Another amino acid analogue, *p*-fluorophenylalanine (PFP), has been used to select a number of resistant cell lines (tobacco, carrot, sycamore, *Nicotiana glauca*). The resistant mechanisms include (a) overproduction of phenylalanine and (b) increased activity of phenylalanine ammonia lyase (PAL) which can detoxify PFP by converting it to a secondary metabolite p-fluorocinnamic acid (Berlin et al. 1982). Increased PAL activity was associated with 2- to 12-fold synthesis of phenolics derived from phenylalanine. Selection of *Catharanthus roseus* cells for resistance to 4 MT resulted in enhanced tryptophan decarboxylase (TDC) activity responsible for accumulation of the indole alkaloid ajmalicine (Sasse et al. 1983). Reports have also appeared about selection of tobacco and *Aspergillus officinalis* suspension cultures resistant to methionine analogue ethionine (ETH). In these cells, free methionine is overproduced which, in turn, leads to the synthesis of 5-methylmethionine and 5-methylcysteine at higher levels. From these various accounts it is apparent that overproduction of a primary compound by selection for resistance to its analogue leads to accumulation of secondary compounds, and that such selection procedures could be applied to cell cultures of other plant species for synthesis of desired compounds.

17.2.4 Auxotrophy

The procedure applied for selection of auxotrophs might be useful for increasing secondary compound production since a block in the necessary pathway would result in the build up of intermediates. In case such intermediaries happen to be precursors or have some positive effect on the production of secondary compounds, a greater synthesis of resultant secondary substances is expected to occur. The disadvantage in applying this procedure is that auxotrophic plants or cells have been selected only in a few cases. Secondly, selection must be done with haploid cells since auxotrophs are mostly mutants for recessive traits. The induction of haploidy in cell cultures is generally not easy.

17.3 Mass Cultivation of Plant Cells

Soon after the selection of cell lines for high yields, follow-up efforts (Fig. 17.1) are required for mass cultivation of these cells in order to achieve industrial level production of the desired metabolites. There are some biological and technological barriers in the mass culture of cells. The biological barriers include the low growth rate and genetic instability of cultured cells. The technological barriers are shear sensitivity due to the rheological nature of cells, oxygen transfer, cell aggregation and wall growth by the adhesion of cells. Therefore, several fundamental properties of plant cells are taken into account while considering the suitability of cell cultures for their possible mass growth and metabolite production. Plant cells (generally 20 μm-150 μm in dia) are 10-100 times larger than bacterial or fungal cells. Moreover, cells in culture exhibit volume changes and a wide variety of shapes as well as sizes. Cell walls have high tensile strength but low shear resistance. All these aspects contribute to the conventional use of microbial fermentation vessels unfavourable for mass culture of plant cells. Vessels used in microbial fermentation are mainly designed to develop high levels of shear and incorporate impellers or stirring paddles to aid gas transfers.

17.3.1 Free-cell Suspension Culture

Cell suspension culture is the most convenient mode of mass cultivation of free-cells and large-scale production of secondary compounds. Suspension cultures of plant cells should have adequate growth rates for efficient product formation which can be possible only by providing necessary culture requirements.

(A) Application of Bioreactors

In a normal culture vial, a cell suspension grows to relatively low densities because of the minimal quantity of nutrients available in the medium within the limited space of the vessel. Therefore, a wide variety of culture vessels ranging in size from 2 1 to 20,000 1 have been designed to grow high densities of plant cells (*see* review by Fowler 1982, Eibl and Eibl 2008). Israeli Biotech company Protali designed plastic bioreactors for production of glucocerebrosidase by plant cells. Two main problems encountered during the cultivation of cells at high density are the supply of oxygen and hydrodynamic stresses generated by the aeration-agitation system within the culture vessel. Taking into account these parameters of mass scale cell culture some basic types of airlift bioreactors have been designed (Fig. 17.2). These airlift systems can support biomass levels of 30 g l^{-1} dry weight without the development of stagnant zones since one of the important considerations with these systems is that the mixing capacity is linked with a proportionate outlet of gases. Otherwise the accumulation of gases at high levels could be inhibitory to growth and metabolic formation. Tanaka (1987) reported that establishment of high-density cultures for large-scale production of shikonin derivatives was possible by using

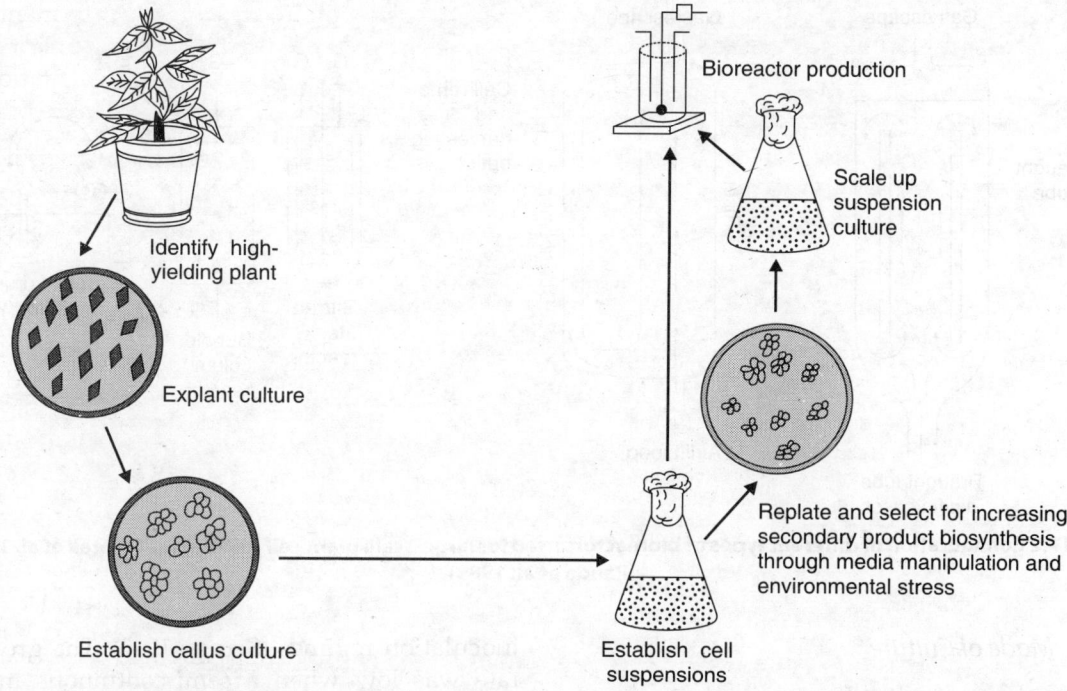

Fig. 17.1 Mass cultivation of cells producing high yields of desired metabolites (modified from Whitaker and Hashimoto 1986).

the bioreactor with a modified paddle type impeller or a rotating drum. Both these types of bioreactors are used to successfully cultivate shikonin cells upto 75 g l^{-1} dry weight. A high yield of berberine (3 gl^{-1}) has been achieved in a reactor with a hollow paddle-type stirrer (*see* Matsubara et al. 1989). Stirred-tank reactors with modified impellers that import improved mixing under low shear have been advocated for large-scale cultivation of fragile plant cell suspension cultures. Hooker et al. (1990) studied the effect of power consumption, flow pattern, the type of impeller, its size and position relative to .air-liquid interface and agitation speed on cell growth in batch cultures. A large flat-bladed turbine impeller (7.6 cm dia. 14.0 cm width) has been found to be optimal for phenolic production from tobacco cell cultures in a 5-litre reactor. Such impellers with high width-to-diameter ratio impart more gradual velocity gradients than the regular flat-bladed disc turbine thus providing well-distributed mixing patterns without dead

zones at low shear. However, the use of different agitation speeds during different stages of growth would be a more rational approach to successful cultivation a plant cells while using bioreactors.

Oxygen plays a major role in the energetics of plant cells in cultures. Some cultures require CO_2 in addition to O_2 to improve cell growth and metabolite synthesis. It is important that the level of these gases be controlled for their optimal utilisation in a bioreactor. Smith et al. (1990) described a bioreactor control system for controlling the dissolved concentrations of both oxygen and carbon dioxide simultaneously. Using a mathematical model, a strategy was developed that allowed the concentrations to be controlled without changing either the total gas-flow rate or the agitation speed, thereby imparting constant shear conditions in the reactor. This strategy has been shown to be applicable for both short and long-term cultivation of *Catharanthus roseus* cells.

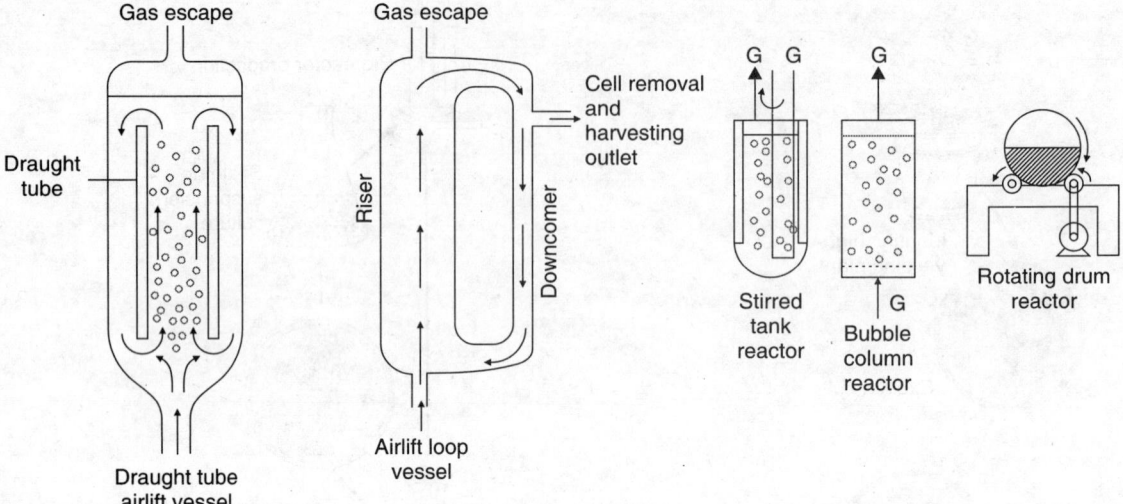

Fig. 17.2 Configuration of different types of bioreactors used for large-scale plant cell culture (*see* Mantell et al. 1985, Panda et al. 1989).

(B)   *Mode of Culture*

To maximise product formation various modes of cell culture have been followed. The fed-batch mode is used in cases in which the addition of a high concentration of substrate or sucrose (*see* Chapter 5) affects the growth of the suspension culture. Fermenter culture design for *Morinda citrifolia* cell lines directing anthraquinone accumulation is shown in Fig. 17.3. A semi-continuous or draw-fill mode of repeated batch cultures provides a suitable approach towards continuous cultivation of plant cells when the rate of product synthesis parallels the rate of growth. Two-stage culture is the obvious choice when cells are first propagated in a growth medium and then transferred to a production medium for synthesis of secondary compounds (e.g. *Catharanthus roseus* suspension cultures for producing serpentine. *Digitalis lanata* for diacetyl lantoside C. production).

The batch mode of cultivation proved successful in scaling up unstable but rapidly growing *Helianthus annuus cell* cultures to a volume of 80 1 in an airlift bioreactor. Increase in cell growth was also achieved in shake flasks (100 ml) from three 10-day cultures using 1:10 inoculation regimes (Scragg 1990). The growth rate was low when a semi-continuous mode was employed in a 7-litre airlift bioreactor and lasted only three cycles before the cells lost their viability. This shows that the mode of cell culture is dependent on the physiological state of the cells to be used for secondary metabolite production.

17.3.2   *Immobilised Plant Cell Culture*

The development of a new technique, known as surface immobilised plant cell (SIPC), is the most significant contribution in secondary metabolite production. SIPC efficiently retains the inoculum and growing biomass even at higher mixing rates. Cells are protected from high-liquid shear stresses and there is better cell-to-cell contact. The product formation is optimised by decoupling this phase from the growth phase (Rosevear 1988).

Methods of immobilisation are widely based on gel or membrane entrapment (Table 17.2). Alginate gel-trapped immobilised systems effectively enhance the production of secondary metabolites in a number of species. Employment of *elicitors* in these systems to stimulate

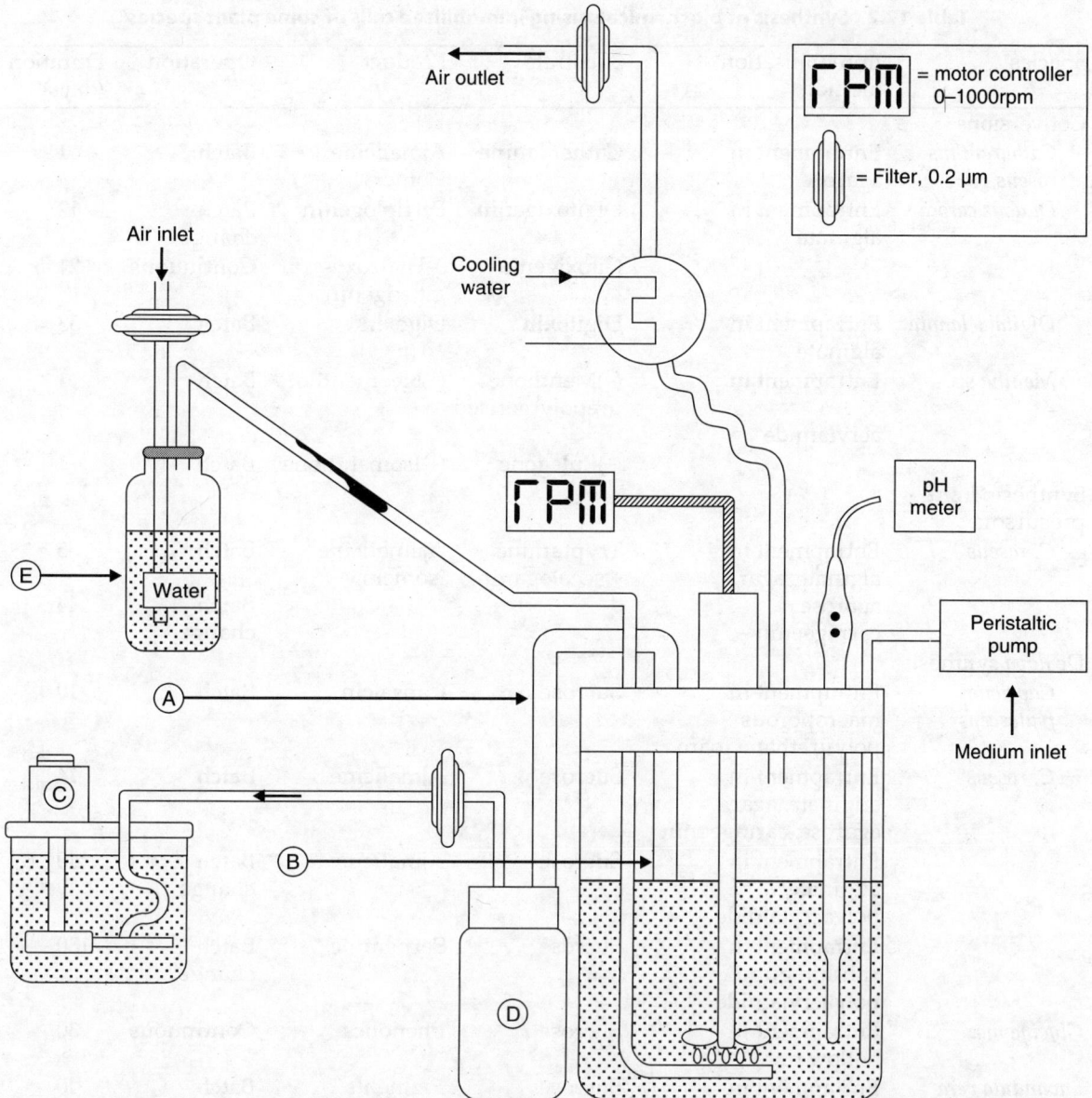

Fig. 17.3 Schematic drawing of the fermenter cultures. Fresh medium is pumped into the culture vessel at a constant rate. The overflow/sample is removed by a vacuum pump (C) through an enlarged tube (A) that is placed at a certain height to keep the culture volume constant (B). The overflow is collected in a sample bottle (D). The incoming air is moistened by leading it through a bottle filled with sterile water (E). (Adopted from Hagendoorn et al. 1999).

Table 17.2 Synthesis of biochemicals using immobilised cells of some plant species[a]

Species	Immobilisation method	Substrate	Product	Operation	Duration (days)
Conversions					
Catharanthus roseus	Entrapment in agarose	Cathenamine	Ajmalicine isomers	Batch	1
Daucus carota	Entrapment in alginate	Digitoxigenin	Petriplogenin	Batch change	12
		Gitoxigenin	5-Hydroxy-gitoxigenin	Continuous	21
Digitalis lanota	Entrapment in alginate	Digitoxin	Digoxin	Batch	33
Mentha sp.	Entrapment in acrylamide	(–)Menthone prepolymerised	(+)Neomenthol	Batch	1
		(+)Pulegone	(+)Isomenthone	Batch	1
Synthesis from precursors					
C. roseus	Entrapment in alginate, agar, agarose, carrageenin	Tryptamine + secologanin	Ajamalicine isomers	Batch	5
				Batch change	14
De nova synthesis					
Capsicum frutescens	Entrapment in macroporous polyurethane foam	Sucrose	Capsaicin	Batch	10
C. roseus	Entrapment in aliginate, agar, agarose, carrageenin	Sucrose	Ajmalicine	Batch	14
	Entrapment in aliginate/polyacrylamide	Sucrose	Ajmalicine	Batch change	40
	Entrapment in xanthin gum/polyacrylamide	Sucrose	Serpentine	Batch change	180
Glycine max	Entrapment in hollow fibre	Sucrose	Phenolics	Continuous	30
Lavandula vera	Entrapment in urethane	Sucrose	Pigments	Batch	10
Morinda citrifolia	Entrapment in alginate	Sucrose	Anthraquinone	Batch	21
Petunia hybrida	Entrapment in hollow fibre	Sucrose	Phenolics	Continuous	21
Solanum aviculare	Co-valant attachment to polyphenylene oxide beads	Sucrose	Steroid glycosides	Continuous	11

[a] Detailed references are given in Brodelius (1986).

secondary metabolism and application of a resin for adsorption of the product have further helped in releasing a higher quantity of secondary products. A further significance of these chemical 'helpers' is that the potential for enhancing the secondary compound production (normally stored intracellularly in vacuoles) can be achieved without using permeability agents, or inducing cyclic changes in pH, that might have deleterious effects on cells (Bisaria and Panda 1991).

(A) Entrapment in Gels

Procedures currently employed for the immobilisation of cultured plant cells are similar to those used for micro-organisms. Entrapment of cells or protoplasts is carried out in some kind of gel (alginate, carrageenin, agar, agarose and so forth) or combination of gels which are allowed to ploymerise around them. The first report using this approach was described by Brodelius et al. (1979) who immobilised *Catharanthus roseus, Digitalis lanata* and *Morinda citrifolia* cells by their entrapment in a matrix of calcium alginate (Fig. 17.4). Since then a number of reports have appeared on the application of gels for plant cell immobilisation. The protocol for entrapment of plant cells in various kinds of gels is given in Appendix 17.1.

(B) Entrapment in Nets or Foam

Lindsey et al. (1983) demonstrated that cells of various species growing actively in a liquid nutrient medium can be readily immobilised in blocks of polyurethane foam in a one-step process. The foam (reticulate polyurethane of various pore sizes, or polyester Declon; Corby, Northants, England) may be used in blocks of 1 cm^3. However, it is likely that in fixed-bed bioreactors sheets will be most effective (Yeoman 1987). Blocks of the gel are immersed in the cell suspension and flasks agitated on a rotary shaker. Initially the free cells and cell aggregates are washed in and out of the blocks but soon the aggregates become trapped deep

within the inert matrix. The component cells divide within the compartments of the foam and form large aggregates. After two weeks, or when the entire block has filled with cells, the loaded cubes are transferred to a low-growth medium which supports metabolic activity but not active cell division. Generally, the absence or presence of a lower concentration of nitrate, or phosphate, in the medium used for a particular cell culture, imparts a characteristic of low growth. Therefore, immobilised cells placed in a low-growth medium do not proliferate further and cells remain confined to the foam without being released into the medium.

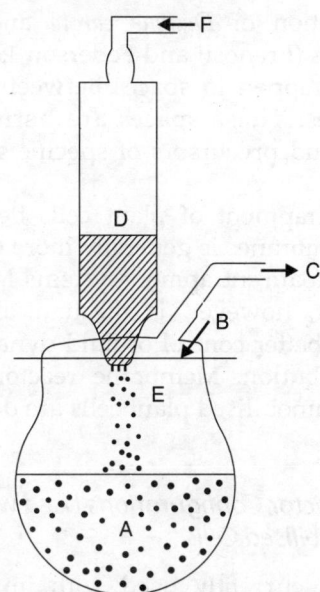

Fig. 17.4 Device for preparation of immobilised cells by their entrapment in a matrix of calcium alginate. A: medium in a container; B: lid; C: air outlet; D: reservoir of aliginate-cell suspension; E: six nipples (diameter 1 mm) for bead formation; F: sterile air inlet (after Brodelius 1984).

Immobilisation in a polyurethane matrix does not appear to affect cell viability. Cell suspensions of almost any degree of aggregation can be accommodated by meshes of polyurethane foam. Immobilised cells of *Capsicum frutescens* produced more capsaicin than freely suspended cells cultured under similar conditions (Lindsey and Yeoman 1984). Further, the reticulate polyurethane offers no barrier to the inward and outward diffusion of metabolites.

(C) Entrapment in Hollow-Fibre Membranes

Hollow-fibre membranes have also been used for the immobilisation of intact plant cells. Tubular fibres made of cellulose acetate silicone polycarbonate, and organised in parallel bundles within a reactor vessel, are used for the culture of animal cells as well as microorganisms (Yeoman 1987). Shuler (1981) described the entrapment of *Glycine max* cells in a similar structure. Hollow-fibre cartridges were also used for immobilisation of *Daucus carota* and *Petunia hybrida* cells (Prenosil and Pederson 1983). Cells can be entrapped in spaces between the fibre membranes. These spaces are permeable to nutrients and precursors of specific secondary products.

The entrapment of plant cells between or within membranes is generally more expensive than gel or foam entrapment systems. Membrane entrapment, however, is mechanically stable and offers better control of fluid dynamics and flow distribution. Membrane reactor systems used for immobilised plant cells are depicted in Fig. 17.5.

(D) Bioreactor Configurations Used with Immobilised Cells

Bioreactors currently used with immobilised plant cells entrapped in gels or foam may be placed in *fluidised-bed* and *fixed-bed* categories. In the fluidised-bed reactor the immobolised cells are agitated either by a flow of air (Fig. 17.6a) or by pumping the medium through the bioreactor (Fig. 17.6b). Contrarily, immobilised cells in the fixed-bed bioreactor are not agitated but held stationary and perfused with an aerated liquid culture medium at a relatively slow rate. The medium can be recirculated unchanged or may be modified either by removal or addition of constituents. With both types of bioreactors, the cells are protected from shear though fluid

mixing appears more effective with fluidised-bed reactors

17.3.3 Two-phase System Culture

Cultivation of plant cells in an aqueous two-phase system is a novel way of obtaining secondary products. In this method, cells are kept apart from the product by a separation phase (liquid) placed in a reactor. The advantage in this process is that phase-partitioning of the cells from the product facilitates continuous removal of the product from the reactor. Various polymers have been used for phase separation. Hooker and Lee (1990) cultured tobacco cells in an aqueous two-phase comprising dextran and PEG for the production of phenolic compounds. Silicon fluid has been used for phase separation of *Eschscholtzia californica* cell cultures. This fluid enhanced the production of benzophenanthridine alkaloids three-fold compared to single-phase culture (Byun et al. 1990). However, more detailed investigations are required in terms of phase separation and solubility of products before this technique can be applied on a wider scale.

17.3.4 Hairy Root Culture

Establishment of hairy root cultures by genetic transformation of plant tissues by the pathogenic bacterium *Agrobacterium rhizogenes* (*see* Chapter 13) is one of the most recent organ culture systems employed for large-scale production of secondary metabolites. Hairy roots produced by plant tissues have lateral branching and each branch has metabolite features similar to the normal root. Hairy root cultures can be useful especially for the production of root-associated metabolites because of their high-growth rate, genetic stability and high biosynthetic capabilities as compared to normal roots.

Conventional stirred-tank reactors are not suitable for mass cultivation of hairy roots because the contact of the impeller will not only damage the growing root tips, but also induce callus formation leading to decrease in

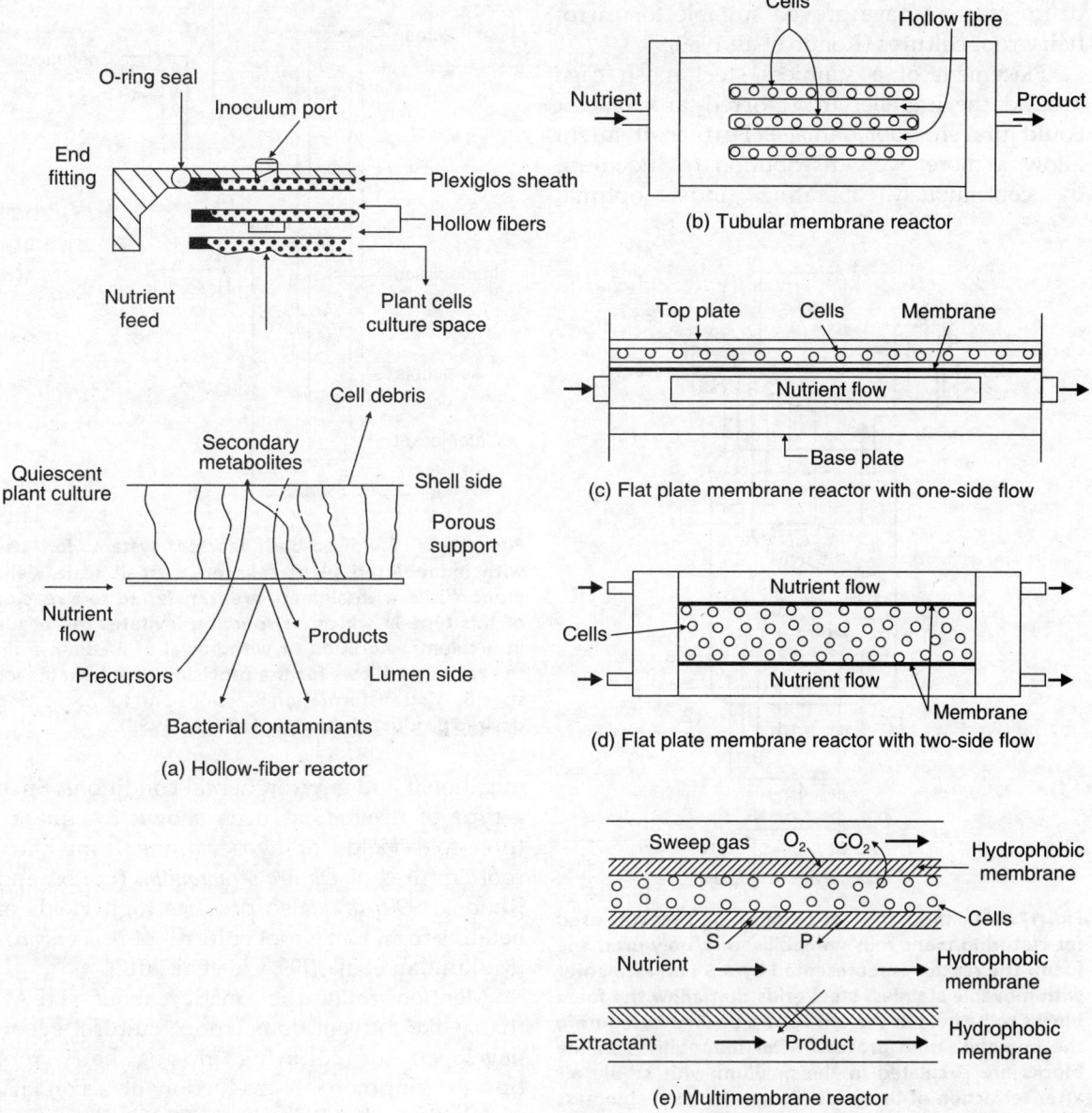

Fig. 17.5 Configuration of various membrane reactor systems used for immobilised plant cells (after Mantell et al. 1985, Panda et al. 1989).

productivity. Airlift turbine blade and rotating drum reactors have proved suitable for carrot hairy root cultures (Kondo et al. 1989).

Placement of a stainless steel mesh cage around the impeller in a stirred-tank reactor could prevent root damage. Further, it might allow a more even distribution of inoculum by continuously operating under optimal

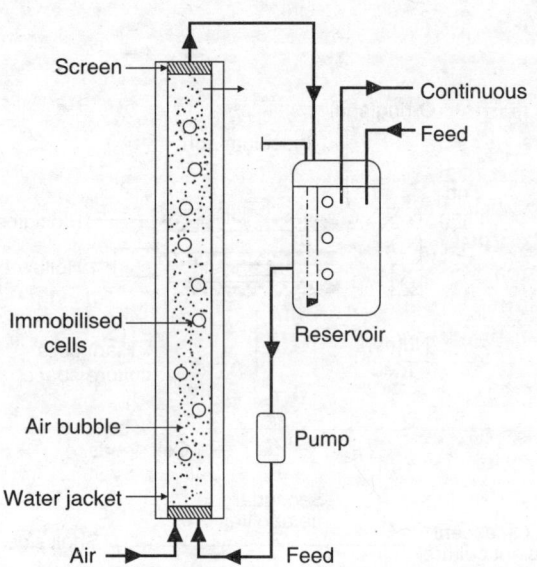

Fig. 17.6b Fluidised-bed reactor system for use with immobilised plant cells on a small scale. Cells immobilised with alginate are transferred to a reactor of this type in which air-sparging levitates the beads in a column. Addition or withdrawal of medium from the reservoir allows for the provision of substrates for specific biotransformation reactions and for harvesting desired products (after Mantell et al. 1985).

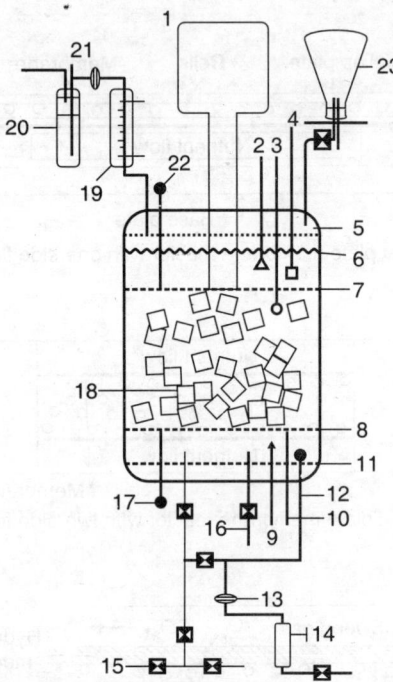

Fig. 17.6a Configuration of fluidised-bed reactor used for culturing plant cells immobilised in poly-urethane foam. The reactor is represented by a 5 l LH fermenter with movable stainless steel grids that allow the foam blocks to be held in the centre as a packed bed during the immobilisation process. After immobilisation the blocks are circulated in the medium with sterile air after retraction of the grids. Components: 1—biomass sampler vessel; 2—antifoam probe; 3—pH probe; 4—oxygen probe; 5—grid (raised position); 6—liquid level; 7—grid (lowered position); 8—grid (raised position); 9—heater; 10 temperature sensor; 11—grid (lowered position); 12—air input through sparge; 13—filter; 14—rotameter; 15—steam point; 16—liquid sample point; 17—handle to move grid; 18—foam particles; 19—grid (raised position); 20—condenser; 21—filter; 22—handle to move grid; 23—inoculation vessel (after Yeoman 1987).

nutritional and environmental conditions. Such a type of reactor has been shown to enhance three-fold yields of hyoscyamine from hairy root cultures of *Datura stramonium* (Hilton and Rhodes 1990) and also produce high yields of betalins from hairy root cultures of *Beta vulgaris* (Mukundan et al. 1999, Cai et al. 2012).

Mention may be made about HLAR (hairy-like adventitious root) culture sytem developed for culturing ginsing hairy root tips for commercial production of secondary metabolites. HLAR grown in 20 L balloon type bubble reactor has design suitable for nutrient availability, oxygen exchange, mixing and shear sensitivity to enable increase in biomass 30 times after 42 days of culture, thereby resulting in high yield of genoside production under 20°C/13°C and 16 h/8 h day night cycle (c.f. Bhojwani and Dantu 2013).

17.4 Elicitor-induced Accumulation of Products

Antimicrobial active compounds accumulated by a plant in response to an attack by microorganisms are referred to as *phytoalexins*. A wide variety of plant species are known to accumulate phytoalexins since these compounds impart chemical resistance, an important factor in plant defence. While studying the accumulation of phytoalexins, compounds of pathogen origin were found to induce secondary metabolism in plants. Such pathogen-origin compounds that induce the accumulation of secondary products in plants are known as *elicitors*. Product accumulation that may occur under stress caused by physical or chemical agents (UV, cold or heat exposure, ethylene, fungicides, antibiotics, salts of heavy metals, high salt concentrations, etc.) are called abiotic elicitors. At the Sixth International Congress for Plant and Cell culture, in 1986, the term 'elicitor' was precised to mean compounds of biological origin only and other treatments designated as *abiotic stress* (*see* Eilert 1987).

17.4.1 Mode of Interactions in Elicitor Formation

Four different types of plant-microbe interaction lead to the formation of elicitors. These include:

i) Direct release of the elicitor by the microorganism and its recognition by a receptor site located on the plasma membrane of the plant cell (example: soybean).

ii) Microbial enzymes responsible for development of cell-wall components act as elicitors in respective plant cells (examples: endo-polygalacturonic acid lyase from *Erwinia carotovara* and *Rhizopus stolonifer*).

iii) Plant enzymes release cell-wall components from the microorganisms which, in turn, induce phytoalexin formation in plant cells. These phytoalexins then form elicitor-active components from the plant cell walls (examples: chitasan from *Fusarium* cell walls, α-1,3-endoglucanase from *Phytophthora* cell walls).

iv) Elicitor compounds, endogenous and constitutive in nature, are formed in response to various stimuli (examples: oligosaccharides and acid hydrolysates released by digestion of cultured cells induce furanocoumarin formation in parsley cultures).

17.4.2 Methodology of Elicitation

(A) Selection of Microorganisms

There is no way of predicting an organism which could become an elicitor. The microorganisms selected need not be pathogens, they may also be saprophytes. A variety of microbes should be investigated in individual plant cell culture systems for elicitor formation. Preparations of virus, bacteria, blue-green algae and fungi have been successfully used in triggering elicitation. The quantity of microbial inoculum can be an important factor for elicitor formation in cultured plant cells. Concentrations of the fungal inoculum can be manipulated by using conidiospores, other asexual spores, or sporangia. However, their elicitation properties may differ from those of mycelium.

(B) Co-Culture

Since tissue cultures are maintained under aspetic conditions the deliberate infection with a living microorganism can sometimes be of interest with particular reference to: (a) culture of obligate parasites, (b) study of various processes of infestation, (c) general tests of susceptibility or resistance and (d) stimulation of secondary metabolism in cultured cells. The techniques concerning a-c are extensively described in the handbooks edited by Helgeson (1983) and Dixon (1985).

To stimulate secondary metabolism, it is necessary that cultures be transferred to a fresh medium prior to inoculation with a microorganism. Generally, callus cultures are transferred 1–4 weeks and suspension cultures 3–14 days before the date fixed for inoculation. Suspension culture cells can be elicited simply by the addition of a microbial inoculum. The advantage in using a suspension culture is that the cell material is more or less homogeneous and the concentration of inoculum can be more easily controlled.

Co-culturing cells with microbes can result in both synergistic and inhibitory effects. Filamentous fungi easily grow out of proportion, resulting in severe damage to plant cells. Use of complex sterile-culture homogenates avoids this problem while maintaining a wide range of possible interactions. Complex elicitor preparations are readily obtained by culturing a selected microorganism preferably on a tissue culture medium, followed by homogenisation and autoclaving of the entire culture. The process of autoclaving could lead to the release

or formation of elicitor compounds. For heat labile elicitors, the culture homogenates should be filter-sterilised.

Direct contact of plant cells and microbes in co-culture can be prevented by entrapment of either partner in gels as followed for immobilisation of cells (Section 17.3.2). In such procedures only non-mobile bacteria, yeasts, or slow growing fungi should be selected. Plant cell-microbial interaction occurs by diffusion of compounds in the medium and *via* the gaseous phase. Some products accumulated in cultured cells in response to elicitation are listed in Table 17.3.

17.5 Biotransformation using Plant Cell Cultures

Plant cells can also be used to accomplish certain changes in the structure and composition of industrially important chemical compounds.

Table 17.3   Elicitor-induced product accumulation as presented at the Sixth IAPTC Congress, Minneapolis 1986[a]

Elicitor	Preparation	Cell culture	Products
Aspergillus niger	Homogenate	*Cinchona ledgeriana, Rubia tinctoria* or *Morinda citrifolia*	Anthraquinones
Botrytis sp.	Homogenate	*Papaver somniferum*	Sanguinarine
Dendryphion sp.	Extract	*Papaver somniferum*	Sanguinarine
		Papaver bracteatum	Sanguinarine
Yeast	Carbohydrate preparation	*Glycine max*	Glyceollin
		Thalictrum rugosum	Berberine
Nigeran	—	*Solanum melongena*	Polyacetylenes
		Vigna angularis	Isoflavones
Pythium aphanidermatum, Eurotium rubum, Micromucor isabellina, or *Chrysosporium palmorum*	Filtrates and extracts	*Catharanthus roseus*	*Tryptamine ajmalicine* and catharanthine
Pythium aphanidermatum	Homogenate	*Catharanthus roseus*	*Strictosidine ajmalicine,* tabersonine, lochnericine and catharanthine

[a] For more secondary metabolites accumulated by elicitation, *see* Eilert (1987).

The conversion of a small part of a chemical molecule by means of biological systems is termed *biotransformation*. One of the promising fields in the biotechnological application of plant cell culture has been its use in biotransformation of some less important substrates to medicinal products. In the steroid industry, cell culture methods are particularly preferred since the synthetic processes are not only difficult but even uneconomical. Certain species of bacteria and fungi raised in fermenters have long been used to accomplish the specific alterations in the phenanthrene ring structure and side chain composition of several plant-derived steroids (e.g., diosgenin), in order to produce steroid sex hormones and anabolics (Aharonowitz and Cohen 1981).

Stohs and Staba (1964) were the first to report the biotransformation of steroid digitoxigenin to 3-dehydrodigitoxigenin during a 7-day incubation in *Digitalis lanata* and *D. purpurea* cell cultures. In a similar manner *D. purpurea* suspension cultures could be used for biotransformation of digitoxin to gitoxin purpurea glycoside A and B (*see* Purohit and Mathur 1990). The production of digoxin (cardiovascular drug) from digitoxin by *D. lanata* cultures is the classical example of utilising plant cell cultures for achieving a specific biotransformation on a large scale. Although undifferentiated cell cultures of *D. lanata* do not produce cardiac glycosides, they nevertheless perform special biotransformations on cardenolide glycosides when added to the medium (Alfermann et al. 1983). The production of digoxin has been achieved in airlift bioreactors of the down-draft type using immobilised cells of *Digitalis*.

Biotransformation requires selection of those cell types which have the enzymatic capabilities for catalysing the specific reaction of interest. Since the heterogeneity within a large population of cell is considerable it might be possible to devise procedures for selection of cell cultures responding positively to the transformation of a precursor into the respective product. Plant cells have been shown to perform more than one biotransformation with a given substrate (Barz and Ellis 1980): hence a thorough knowledge of the enzymology of the desired reaction needs to be incorporated into the cell selection process. Some examples of biotransformation using cell cultures are listed in Table 17.4. The selectivity in the biotransformation by various plant cell suspension cultures are summarised in Overview by Hamada and Furuya (1999).

17.6 Secondary Metabolite Production using Genetically-Engineered Plant Cell Cultures

Genetic manipulation of cell cultures has great potential for altering the metabolic profile of plants, thereby making them profitable to the industry. Transgenic cell cultures hold promise for production of flavour components, food additives and colours. Identification of key regulatory enzymes of secondary metabolic pathways is necessary in order to manipulate

Table 17.4   Some examples of biotransformation by plant cell cultures[a]

Substrate biotransformed	Plant cell culture	Product
Carvoxine	*Nicotiana tabacum*	Cavaxone
Codeinone	*Papaver somniferum*	Codeine
Ellipticine	*Choisya ternata*	5-formylellipticine
Solavetivone	*Solanurn tuberosum*	Hydroxylated derivatives
2-Succinylbenzoate	*Galium mollugo*	Anthraquinone
Valencene	*Citrus* spp.	Noothatone

[a] For detailed references *see* Whitaker and Hashimoto (1986).

them in cultured cells. Dixon et al. (1999) generated cell suspension cultures from transgenic plants over-expressing bean PAL and alfalfa C_4 transgenes in tobacco which correspondingly increased the levels of phenolic compounds with time in culture. Genetic engineering of metabolic pathways in plants are particularly applicable to such cases where elicitation fails to provide the necessary specificity. Attempts at altering phenylpropanoid pathway flux by genetic engineering using plant cell cultures (Dixon and Paiva 1995) and strategies to improve secondary product synthesis by metabolic engineering of crop plants are reviewed in an edited volume by Fu et al. (1999); also see Sato et al. (2001).

17.7 Factors Limiting Large-scale Production of Useful Compounds

The scale-up of plant cell cultures for the production of useful compounds presents a number of technical challenges. The large size and striking number of physiological differences among various cell types preclude the use of a microbial type fermentation for plant cell culture. Oxygen transfer is also a problem in large-scale cultures since plant cells have a tendency to aggregate. Moreover, secondary products generally are not excreted but retained inside the cell vacuole. This necessitates a destructive harvest to release the intracellular compounds. Recycling of cell cultures is hardly possible and overall cost production would increase. Thus, necessary parameters must be adopted to ensure high yields without destroying cells.

Studies on plant cell excretion have suggested that accumulation and excretion of secondary products may be conveniently modulated by strict control of external pH. Secondary products can also be excreted through non-destructive permeabilisation of immobilised plant cells (Brodelius and Nilsson 1983). However, specific excretion and accumulation parameters have to be worked out for individual cell lines (*see* Cai et al. 2012); such parameters would prove valuable in designing cost-effective bioreactors for scaling up the production of cell cultures.

17.7.1 Genotype of Mother Plant

A range of variability in the amount and type of secondary metabolites has been observed in cell cultures raised from different mother plants and sometimes even from the same mother plant. This is attributed to the differences between the genotype of the mother plant and relative productivity of cell cultures. Cell lines of tobacco from high-yielding mother plants tended to give high-yielding nicotine cultures, while low-yielding mother plants produced low-yielding cultures with respect to nicotine (Kinnersley and Dougall 1980). Further, a significant variation was observed in mean levels of metabolites even within a high-yielding strain under similar culture conditions. Cell cultures are known to show a gradual loss of productivity and very little is known about the factors that inhibit the secondary metabolite synthesis in tissue cultures or cause its re-emergence upon redifferentiation. Methods for enhancing productivity in cell lines were described earlier in this chapter.

High-yielding lines are rather easily distinguished from low-yielding lines where cells synthesize metabolites that are visibly discernible, such as anthocyanins, quinones, betacyanins and carotenoids. For other compounds, a quantitative analysis of cellular accumulation patterns of specific metabolites (rosemarinic acid and cinnamyl putrescines) may be attempted using microspectrophotometry.

17.7.2 Age of Cell Cultures

Asynchronously dividing batch cultures consist of a heterogeneous mixture of cells at different stages of biochemical and morphological differentiation. Cells divide rapidly during the initial and middle phases of culture, but soon the substrates in the medium become more limiting. This results in the decline of the proportion of cells undergoing divisions. Eventually, with the advent of the stationary phase, the majority of the cells become quiescent and the ageing process commences in more and more of them.

The accumulation of metabolites at any particular instant in the cell culture is the result of a dynamic balance between rate of its biosynthesis and biodegradation. A wide variety of metabolism is involved in the biosynthesis and biodegradation of the many secondary compounds. In view of this, there is no strict particular time in the cell cycle for harvesting the cells in order to obtain the maximum yield of secondary compounds. Most secondary metabolites accumulate during the late stages of growth in tobacco batch cultures although under some conditions certain metabolites in *Catharanthus* cell cultures accumulate even during the lag phase (Fig. 17.7).

17.7.3 Production Medium

An important limiting factor in the application of tissue culture for synthesis of metabolites at the industrial level is the composition of the medium (production medium) that promotes the production of a desired product. The production

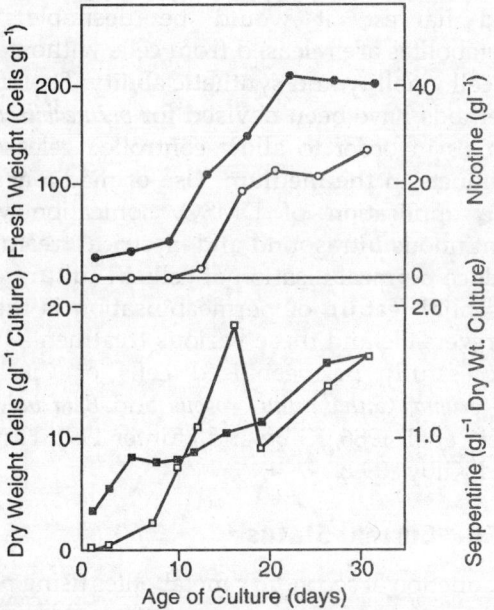

Fig. 17.7 Accumulation of secondary metabolites in batch cultures of *Nicotiana tabacum* (top) and *Catharanthus roseus* (bottom; *see* Mantell et al. 1985).

medium does not generally support the level of growth required to obtain an appropriate bio-mass. Therefore, a two-step approach is desirable whereby one medium is utilised solely for bio-mass growth and a second for secondary product accumulation. The presence or absence of phosphate in a medium greatly affects the expression and accumulation of some secondary products. About 50% increase in anthraquinone accumulation was reported in cell cultures of *Morinda citrifolia* by enhancing the phosphate concentration to 5 gl^{-1} (Zenk et al. 1975, Hagendoorn et al. 1999); paradoxically the overall accumulation of tryptamine and indole alkaloids could occur only by shifting *Catharanthus roseus* cells to a medium devoid of phosphate (Knobloch et al. 1981). Similar sensitivity to phosphate concentrations was also observed in *Nicotiana tabacum* and *Catharanthus roseus* cell cultures with respect to PAL activity and subsequent accumulation of cinnamoyl putrescine (Knobloch 1982, Knobloch and Berlin 1983).

Sources of carbon, nitrogen, vitamins, and metal ions have all played significant role in altering the expression of secondary pathways. The addition of sucrose, NO_3^-, Cu^{++}, or SO_4^{-2} in culture media above the optimum level has a profound effect on shikonin biosynthesis. Low concentration of sugar increased the production of ubiquinone-10 in tobacco cell cultures and anthraquinones in *Morinda* suspension cultures, while the carbon-nitrogen (C: N) ratio regulated the yields of catechol tannins in sycamore cell suspension cultures (Whitaker and Hashimoto 1986, Purohit and Mathur 1990, Hagendoorn et al. 1999).

Studies on the effect of growth regulators in nicotine production by tobacco cell cultures revealed that it could be best achieved by following a two-phase approach. The growth phase I was achieved on an MS medium containing 10^{-5} mmol l^{-1} NAA and 10^{-6} mmol l^{-1} kinetin. For growth phase II, the same medium was used but containing reduced levels of the two types of growth substances and no CH (Mantell et al, 1985). 2,4-D is reported to have both inhibitory (for shikonin synthesis in *Lithospermum* cultures, nicotine in tobacco cultures) and stimulatory (in *L*-dopa synthesis) effects. The proportion of the cytokinin and gibberellin in the medium is critical for secondary compound production, as in other areas of plant tissue culture discussed in previous chapters. The concentration of the two types needs to be decided depending on the individual cell culture system.

Secondary metabolism can also be triggered by treatment of cell cultures with specific elicitors (*see* Section 17.4) and other chemical factors to enhance production. Sometimes feeding specific precursors to the culture medium becomes necessary should these be considered the limiting factors for synthesis of particular secondary products. Feeding of specific precursors to media may even increase the yield of useful secondary compounds manifold. Cinnamic acid or a-phenylalanine are precursors fed to increase the synthesis of flavonoids and tropic acid for tropane alkaloids. Shikonin production increases three-fold when *Lithospermum* cells are fed with α-phenylalanine, while the addition of hydroquinone to *Datura* sp. suspension cultures reportedly enhances arbutin synthesis (*see* Purohit and Mathur 1990). Paclitaxel (Taxol) yields in cell cultures of *Taxus cuspidata* improved considerably by feeding these cultures with phenylalanine and other potential precursors like benzoic acid, serine etc. Similarly, phenylalanine fed into suspension cell cultures of *Salvia officinalis* stimulates rosmaric acid production (Maheshwari et al. 2008).

17.7.4   *Environmental Stress*

While exploring mechanisms for increasing secondary product biosynthesis through media manipulations, an unmistakable correlation between specific environmental conditions, or stresses, and the production of secondary compounds is becoming increasingly apparent (*see* also Section 17.7.5). The mRNA encoding enzymes phenylalanine ammonia lyase, and 4-coumarate : CoA ligase have been induced in parsley *(Petroselinum hortense)* cultures either by irradiation with UV light or treatment with cell-wall fraction of the fungal pathogen *Phytophthora megasperma* (Kuhn et al. 1984). These enzymes are involved in biosynthesis of phenylpropanoids which provide the precursors for the accumulation of flavonoids and furanocoumarins. In uninduced cells, neither of the secondary products are synthesised.

Generally, cell cultures are maintained at constant illumination for accumulation and obtaining increased yields of polyphenols, plastoquinones, flavonoids, carotenoids, volatile oil content, medicinal and other secondary compounds. However, the synthesis of a few compounds, such as shikonin, is inhibited by light.

17.7.5   *Permeabilisation of Cells*

Secondary metabolite synthesis occurs intra-cellularly and these products can be extracted from the cells by destructive (fractionation) methods. To ensure that each culture be used for repeated secondary metabolite production, and harvest, it would be desirable that metabolites are released from cells without loss of cell viability and synthetic ability. Therefore, methods have been devised for *permeabilisation of cells* in order to allow controlled release of products in the medium. Use of media at low pH, application of DMSO, sonication with continuous ultrasound and electrical treatment induce permeabilisation of cells in culture. The essential feature of permeabilisation is that it is reversible and these various treatments have successfully permeabilised cells of *Cinchona ledgeriana*, *Catharanthus roseus* and *Beta vulgaris* (Parr et al. 1986, Kilby and Hunter 1991, Hunter and Kilby 1999).

17.8   Current Status

Production of secondary metabolites using plant cell and tissue cultures apparently could have an advantage over *in vivo* synthesis of these products in whole plants. Climatic, geographic, or political factors which act as barriers for cultivation of plantation crops are altogether overcome since *in vitro* facilities can be created anywhere and plant propagation generated continuously all through the year with no constraints. Intensive cell culture is also a practical alternative for conservation of rare, nearly extinct, or variant (somaclonal/transgenic) plant germplasm (*see* Chapter 18). Large-scale culture of plant cells is now technically feasible using bioreactors, yet, in most cases, yields of secondary metabolites is too low for commercialisation. Industrial production of secondary metabolites from plant cell and tissue cultures initially remained restricted only to three products: shikonin, ginseng saponins, and berberine (Trigiano and Gray 2000). Shikonin is obtained from reddish purple roots of *Lithospermum erythrorhizon*, but due to uncontrolled mass collection of this plant from nature, it is on threshold of extinction in countries (Japan, China, Korea) where it grows wild. Mitsui Petrochemical Co., Japan, however, developed a commercial tissue culture process for production of shikonin (an antiseptic) using cell cultures. Ginseng saponins are extracted from roots of *Panax ginseng* (ginseng). Plant growth of this species is restricted to 30-40° North latitude. Over the years the demand for this plant has increased considerably and this spurred Meiji Seika Kaisha, Japan, to commercially develop mass scale root cell cultures of *Panax ginseng* for extraction of the product which is used as additive to tonic drinks, health foods, hebal liquors, etc. Similarly, berberine is produced from colored cell cultures of *Coptix japonica* for industrial purposes.

Phyton Biotech in Germany produced Taxol, an anticoncerous drug, by establishing suspension cultures from cambial meristematic cells (instead of dedifferentiated cells) of *Taxus* *cuspidate* and subsequently increased Taxol production on commercial scale using airlift bioreactors. Polysaccharides produced in cell cultures of *Polyanthes tuberosus* and Saiko saponin extracts from cultured roots of *Bupleurum falcatum* are other secondary metabolites reported to have been produced commercially (Gupta and Ibaraki 2006, Lee et al. 2010).

For numerous other secondary products (*see* Table 17.1, Bhojwani and Razdan 1996, Fu et al. 1999, Fasella and Hussain 2014), the *in vitro* processes have yet to achieve industrial level status. Growth of differentiated cells, elicitation and biotransformation are various approaches which have been attempted to increase secondary product activity in cells (Verpoorte et al. 1998, 1999), but differentiated cells are difficult to cultivate whereas elicitation and biotransformation have limitations because of their specificity to a class of chemical compounds specific to a plant species which may not be of interest. Metabolic engineering could be another approach for procuring higher yields but this technology is still in its infancy (Sato et al 2001). Also, ethical and safety concerns dominate statutory licensing of commercial transgenic products. Many countries have developed safety assessment measures to ensure human health friendly production of food flavours, and its ingredients, through *in vitro* processes (Hallagan et al. 1999). Thus, plant cell and tissue cultures have significant scope for synthesizing valuable secondary products of market value. Dow AgroSciences, USA was granted first regulatory approval by USDA for marketing plant cell produced vaccine and efforts are underway for commercial productions of recombinant human proteins for pharmaceutical purposes (biotherapautics) by plant cells (Tekoah et al. 2015).

APPENDIX 17.1

Protocol for Entrapment of Plant Cell in Various Gels[a]

Alginate

1. Dissolve 2 g alginate (protanol HF) in 90 ml of an appropriate medium by heating and constant stirring.
2. Sterilise the alginate solution at 121°C for 20 min.
3. Mix cells with alginate solution at room temperature in the ratio 1:4 (w/w). The fresh weight of cells and the acutal weight of the alginate solution should be taken into account while mixing them.
4. Fill a plastic syringe with this mixture of cells and alginate solution. Let it subsequently drip into a flask containing 50 ml medium fortified with 50 mmol $CaCl_2$ under continuous stirring.
5. Leave the beads formed for 30 min and then collect them by filtration.
6. Wash the beads with a medium containing 5 mmol $CaCl_2$.
7. The immobilised cell preparation is now ready for use.

Agarose/Carrageenin

1. Dissolve 5 g agarose (Sigma type VII) or carrageenin (NJAL 798 from FMC) in 95 ml of an appropriate medium* by heating and constant stirring.
2. Sterilise the agarose solution by autoclaving at 121°C for 20 min.
3. Equilibrate the agarose solution at 35°C.
4. Mix 2 g fresh weight cells (collected by filtration) with 8 g agarose solution at 35°C.
5. Pour the agarose/cell suspension into a 100 ml beaker containing 40 ml vegetable oil under stirring at 35°C
6. Stir until droplets of appropriate size have formed.
7. Place the vessel containing the mixture in an ice bath and cool to 10-15°C until the agarose has solidified.
8. Transfer the oil/bead suspension to Falcon centrifuge tubes containing 10 ml medium each and centrifuge at 100× g for 2 min.
9. Remove the oil and most of the aqueous phases by using an aspirator (with a trap to collect the oil).
10. Resuspend the beads in the medium and centrifuge at 100× g for 2 min.
11. Repeat steps 9 and 10 until the preparation is free from oil.
12. The immobilised cell suspension is now ready for use.

---

[a] Adopted from Brodelius (1986).
* Add 0.09% NaCl if carrageenin is used for entrapment.

18

Germplasm Conservation

Ever since primitive man learned the art of farming and realised the economic utility of plants, he started saving selected seeds or vegetative propagules from one season to the next. In effect, he was practising a type of germplasm conservation and management. Conservation of forest resources was taught and decreed in parts of India and China as far back as 700 B.C. (Swaminathan 1983). The primitive cultivars and their wild relatives constitute a pool of genetic diversity (germplasm or gene bank) which is invaluable for breeding programmes. However, the population and needs of man have steadily increased over time, leading to persistent exploitation of natural resources. The pressures created by increasing human population on the demand for more food, achievement of greater yields of major crop plants, developing pathogen resistance and utilisation of more marginal lands have led to a search for replacement of conventionally grown plants by varieties with superior traits. The introduction of superior varieties is made possible either by exploring the vast resources of wild species, or through the evolution of new cultivars as a result of the application of various scientific processes. Thus, continual introduction of new cultivars has paved the way for gradual exclusion of traditionally used other genotypes. Indiscriminate clearing of forests and conversion of agricultural lands

resulting from industrialisation, urbanisation and measures taken for economic development affect the ecosystem adversely. In these processes, depletion of naturally occurring plant genetic resources is reported to the extent that nearly 2000-60,000 higher plant species became endangered, rare and on threshold of extinction (Bajaj 1987, 1995, Staritsky 1997). Many of these species happen to be the agricultural crops (FAO 2009). Indubitably, some of the valuable gene pools might be lost unless co-ordinated efforts are made towards the conservation of genetic stocks all over the world. Realising the danger of erosion of genetic resources the UN Conference on Human Environment held in Stockholm, in 1972, recommended conservation of the habitats that are rich in genetic diversity. Two years later, in 1974, the Consultative Group on International Agricultural Research (CGIAR) established the International Board for Plant Genetic Resources (IBPGR), subsequently International Plant Genetic Resources Institute (IPGRI), with the objective of providing necessary support to the collection, conservation and utilisation of the plant genetic resources anywhere in the world.

18.1 Modes of Conservation

The principle of any technology designed for germplasm conservation should be to preserve the maximum possible genetic diversity of a particular plant or genetic stock for future use. Diversity in plants is discernible at the level of species, varieties and individuals. Special genetic stocks may include the material (mutant or breeder lines with identified gene or gene combinations) developed and used in ongoing breeding programmes. It has been estimated that the survival of approximately 9,000 wild species of crop plants from tropical regions are some way threatened (*see* Grout 1990). This further highlights the need for a positive approach to conservation of endangered plants. When new cultivars replace the primitive or conventionally used agricultural crops, it becomes especially important that the displaced crop be properly documented and conserved. According to UNEP (United Nations Environment Programme, 1995) over 15000 threatened species are conserved internationally in botanic gardens. Global climatic changes and natural hazards also affect the natural plant habitats, thereby contributing to relatively rapid changes in agricultural strategies. Access to maximum genetic diversity in this situation would obviously make a major difference to the speed and appropriateness of responses.

18.1.1   *In situ Conservation*

This method of conservation mainly aims at preservation of land races of plants with wild relatives in which genetic diversity exists and/or in which the weedy/wild forms present hybridise with related cultivars. These are evolutionary systems that are difficult for plant breeders to simulate and should not be knowingly destroyed.

The *in situ* conservation of habitats has received high priority in the world conservation strategy programmes launched since 1980. Institutional arrangements, especially in countries of the developing world, have been emphasised. This mode of conservation has some limitations, however. There is a risk of the material being lost due to environmental hazards. Further, the cost of maintaining a large proportion of available genotypes in nurseries or fields may be extremely high.

18.1.2   *Ex situ Conservation*

Ex situ conservation is the chief mode for preservation of genetic resources, which may include both cultivated and wild material. Generally seeds or *in vitro* maintained plant cells, tissues and organs are preserved under appropriate conditions for long-term storage as gene banks. This requires considerable knowledge of the genetic structure of populations, sampling techniques, methods of regeneration and maintenance of varietal gene pools, particularly in cross-pollinated plants.

Germplasm conservation in the form of seeds is most convenient since seeds occupy a relatively small space. Their transport to various introduction centres and gene banks is also economical. Initially, the IPGRI's activities on biodiversity conservation focused on the storage of germplasm as seedbanks which include storage of seeds under different conditions, namely as (a) *Base Collections* for securing conservation at low temperature, (b) *Active Collections* for evaluation, distribution and other use, (c) *Working Collections* for materials used by breeders, and (d) *Field Genebanks* for materials difficult to store as seeds. To meet these objectives, guidelines for "Collecting Plant Genetic Diversity" were published by IPGRI (*see* Guardino et al. 1995) and around 10,000 accessions maintained worldwide for food and agriculture (FAO 2009)

Seeds are characterised as *orthodox, recalcitrant* or *intermediate* (Ellis 1984) depending on their storage behaviour. Mature orthodox seeds are desiccation tolerant with low moisture content (20% or less on a wet basis) and can further be dried (5% or less moisture content) without loss of viability. These seeds remain viable for many years and their longevity can be further increased by storing at low temperature or cryopreservation. Recalcitrant seeds are sensitive to desiccation and cannot be dried below critical moisture level. About 118 species from 46 families of flowering plants are known to produce recalcitrant species (Pence 1995).

Seeds of some species (*Coffea arabica, Carica papaya, Elaies guineensis* and *Oreodoxy regia*) are desiccation tolerant but sensitive to low temperature. Such seeds, termed *intermediate* are, therefore, unsuitable for long-term storage by conventional drying (Ellis et al. 1991).

Drawbacks in conservation by seeds are: (a) loss of viability over the passage of time (recalcitrant and intermediate seeds) and susceptibility to insect or pathogen attack, (b) inability to maintain distinct clones except for inbred and apomict species and (c) non-applicability to vegetatively propagated crops (*Dioscorea, Ipomoea*, potato etc.)

Contrarily, *in vitro* methods are the most useful for conservation of vegetatively propagated plants, species with recalcitrant seeds and genetically engineered materials (Bajaj 1995, Krishnapillay 2000). The potential advantages of *in vitro* conservation include: (a) requirement for little space for preservation of a large number of clonally multiplied plants, (b) maintenance of the material in an environment free of pests or pathogens, (c) protection against dangers of natural environmental hazards, (d) availability of nucleus stock to propagate a large number of plants rapidly whenever necessary and (e) minimising the obstacles generally imposed by quarantine systems on the movement of live plants across national boundaries since they are raised and maintained in an aseptic environment. The present chapter deals only with the application of plant cell and tissue cultures in germplasm conservation.

18.2 Materials Used for Conservation

The materials stored *in vitro* may be isolated protoplasts, cells from suspension or callus cultures, meristem tips, propagules at various stages of development or organised plantlets. Even seeds are the materials stored *in vitro*. The cultured cells or shoots can be maintained by serial subcultures at 4-8 weekly intervals for virtually unlimited periods.

The storage of germplasm by repeated cultures has some disadvantages. These include the risk of material loss due to human error or the failure in maintenance of *in vitro* security resulting in invasion by a pathogen. Genetic instability is also affected during serial subculturing of the plant material. This is particularly observed in materials in an undifferentiated phase (cell or callus cultures). The genetic changes may range from additions or deletions of gene sequences (*see* Chapter 14).

Maintenance of the material as plantlets and subsequent propagation from their nodal cuttings reduces the risk of genetic instability. Prolonged storage of somatic embryos also ensures a high degree of genetic stability. A basic requirement for practical feasibility of a plant tissue culture method in germplasm conservation is to reduce the frequency of subcultures to the bare minimum. This can be mainly achieved by cold storage, freeze-preservation (currently termed *cryopreservation*), low-pressure and low-oxygen storage. Applications and limitations of various plant source materials in *in vitro* conservation are discussed in Razdan and Cocking (1997, 2000).

18.3 Methods of *in vitro* Conservation

The methods used for conservation of plant germplasm *in vitro* basically fall into two categories: (a) slow-growth systems, (b) cryopreservation (Monette 1995). Application of these methods invariably depends on the plant material to be used for storage.

18.3.1 Slow-growth Systems

Cultures grown under modified conditions enabling them to be stored for longer periods before transfer to fresh medium constitute slow-growth systems. Modifications in culture environment can be brought about by temperature reduction or manipulation of chemical constituents (using ABA or high sucrose content) of growth media. Slow-growth strategies are generally applicable for "medium-term storage" (Reed 1995, Reed and Chang 1997). Cold storage (reduced temperature under refrigerated conditions) is the most predominant method of medium-term storage and provides a back-up collection for use in plant distribution. This method is described in detail under Section 18.4. The disadvantage in slow growing cultures is the accumulation of variant cells over the passage of time which could be prevented using synthetic seed technology (*see* Section 18.6).

18.3.2 Cryopreservation

The principle underlying cryopreservation basically involves bringing the plant cell and tissue cultures to a non-dividing or zero metabolism state by subjecting them to superlow temperature in the presence or absence of cryoprotectants. In this technique the plant material is frozen and maintained at the temperature of liquid nitrogen (LN), which is around –196°C or –150°C in the vapour phase. Cryopreservation is an established method of storing sperm cells in order to use them for artificial insemination during animal breeding programmes. The progress made in the area of plant cryobiology since 1975, has shown that entire plants can now be regenerated from cells, meristems and embryos frozen and stored in for almost indefinite lengths of time. This procedure has been successfully applied for germplasm conservation of a wide range of plant species, such as cassava, pea, chick-pea, rice, wheat peanut, coconut, oilpalm, strawberry and sugarcane and other agrihorticulture crops (*see* Bajaj 1987, 1995; Razdan and Cocking 1997, 2000). Zygotic embryo or embryonic axes excised from orthodox, intermediate, or recalcitrant seeds of almost a hundred species growing in temperate and tropical climates from a range of crop, fruit and tree species can be successfully cryopreseved (Engelmann 2011). These studies have generated much enthusiasm to employ the cryogenic method as a meaningful tool for long-term conservation of desirable and diverse genetic stocks. Cryopreservation is most useful for long-term storage of plant germplasm since cells at ultralow temperature do not divide and remain genetically stable.

Cryopreservation of living plant cells demands considerable technical skill. It is essential that cells during freeze-preservation be protected against cryogenic injuries. Formation of ice crystals inside the cell can cause the rupture of organelles and the cell itself. Increase in concentrations of intracellular solutes to toxic levels is another potential source of cell damage. Cells may also suffer the loss of vital solutes through leakage during the freezing process. The physiological conditions of the material, prefreezing treatments and application of cryoprotectants considerably affect the viability of cells frozen to cryogenic temperatures. Cryopreservation broadly involves four steps: (a) *freezing*, (b) *storage*, (c) *thawing and* (d) *reculture.*

18.3.3 Freezing

Numerous protocols for Cryopreservation of plant cells have been described. The sensitivity of cells to low temperature varies with the species (*see* Finkel and Ulrich 1983). However, the general practice is to suspend the material in the culture medium and after treating with a suitable cryoprotectant transfer it to sterile polypropylene ampoules with screw-caps and freeze by one of the following methods:

(A) Slow-Freezing Method

In this method the material is frozen at slow cooling rates of 0.5-4°C min^{-1} starting from 0°C until the temperature reaches –100°C, and finally transferred to LN. The slow-freezing procedure has proved especially successful with cells from suspension cultures (Kartha 1987).

(B) Rapid-Freezing Method

This method is simple as the vials containing the plant material are directly lowered into a tank filled with LN. The temperature decreases rapidly, at the rate of –300°C to 11000°C min^{-1}. Dry ice (CO_2), used instead of LN, exerts a similar effect (Nitzsche 1983).

Seibert 1976 and Reed 1992, used the rapid-freezing procedure in which isolated shoot apices were frozen by first pouring LN in the freezing vial and then dipping it in a flask of LN. Isolated shoot apices of *Dianthus caryophyllus* and strawberry, maintained at −196°C for 2 months, remained viable and regenerated whole plants when cultured. The survival of shoot tips following this ultrarapid cooling procedure has been explained on the basis that the critical temperature zone for ice crystal growth is passed so rapidly that ice crystals of a lethal size are not formed inside the cells. The rapid-freezing method has been successfully applied for germplasm conservation of a large number of species (*see* Bhojwani and Razdan 1983, Withers 1985, Bajaj 1990a, b).

The water content of the samples is critical in deciding whether to use the slow- or rapid-freezing method. Small specimens with a low water content are appropriate materials for rapid-freezing method. Shoot apices of strawberry and *Solanum goniocalyx* are materials with small cells potentially able to regenerate plants after rapid freezing. The main drawback in following this procedure is that some plant materials do not survive rapid freezing by direct immersion in LN and, therefore, step-wise freezing becomes necessary.

(C) Step-wise Freezing Method

This method combines the advantages of both slow- and rapid-freezing procedures. The plant material is cooled step-wise (ca 1°C-5°C min⁻¹) to an intermediate temperature, maintained at that temperature for 30 min, and then rapidly cooled by plunging it into LN. The step-wise freezing method can be used for freeze-preservation of a wide range of plant materials, which include shoot apices, buds and suspension cultures. In the initial slow freezing, ice is formed outside the cells and the unfrozen protoplasm of the cells loses water due to the vapour deficit pressure between the supercoiled protoplasm and the external ice. Compared to rapid freezing method, strawberry shoot tips showed enhanced (60-80%) survival rate using this method (Sakai 1997).

(D) Dry-Freezing Method

Materials dehydrated by drying in an oven or under vacuum demonstrate remarkable resistance to cryogenic damage. The basic idea for the dry-freezing method originated from the fact that non-germinated dry seeds are able to survive freezing at superlow temperatures in contrast to water-imbibing seeds which show susceptibility to cryogenic injuries. Similarly, dehydration of cells under vacuum also leads to a better survival of plant organs after freezing at −196°C. However, there is a dehydration optimum which varies according to species and tissues.

Different types of freezing units have been used to freeze plant materials (Fig. 18.1). In particular, programmed freezing units (Palmer R 201 and R 202; Palmer Products, Windmill Road, Sunbury-on-Thames, England) have been recommended to achieve greater control on the rates of freezing for handling a large number of specimens at one time. A simple set-up of LN Dewar flask may also work satisfactorily besides being economical. In this, the cooling rates can be determined at various distances above the level of LN inside the flask. Ampoules containing plant material can then be exposed to a predetermined distance in the flask to achieve the desired cooling rate. Once the prefreezing temperature has been achieved, the ampoules can be lowered into the LN and stored. A routine cryopreservation procedure developed by Withers and King (1980) is described in Fig. 18.2.

(E) Vitrification

Water content in cells of the material to be preserved is critical in deciding whether to use slow- or rapid-freezing methods. To maintain viability of cells with more water content after cooling to the temperature of LN, it is essential to avoid intercellular freezing. This can be accomplished either by *drying or vitrification*. Materials dehydrated by drying in an oven or under vacuum demonstrate resistance to cryogenic damage. The drying process is not

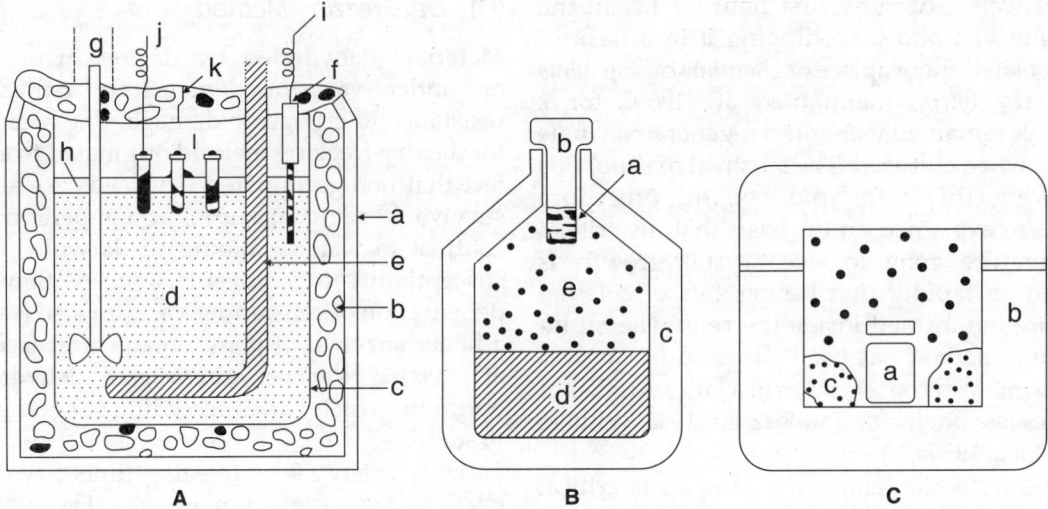

Fig, 18.1 Improvised freezing units designed for preservation of cell and tissue cultures. A: the unit consisting of an electrically cooled alcohol bath on the surface of which the specimens are floated (a—plastic bin; b—polystyrene packing material; c—glass beaker or metal cannister; d—alcohol such as methylated spirit; e—dip cooler; f—thermostat to cooler; g—stirrer; h—polystyrene raft; i—specimen ampoule; j—thermometer for recording specimen temperature; k—bag of polystyrene packing material). B: Unit used to achieve slow cooling by exposing the specimens to cold nitrogen vapour above the level of liquid nitrogen. The specimen ampoules (a) are attached to the plug (b) of a Dewar vessel (c). The vessel contains liquid nitrogen (d), which generates a cold temperature (e). C: A simple unit in which specimens are kept in a vacuum flask (a), later placed inside an insulated cabinet (b), containing solid CO_2 (c). An electrically cooled refrigerator running at ca –70°C would also be suitable (adopted from Withers 1984).

always successful as there is a dehydration optimum which varies according to species. Another potential procedure that prevents cells from cryogenic injuries and successfully applied to most of the plant materials for *in vitro* conservation is vitrification.

Vitrification is an effective mechanism against freezing wherein a highly concentrated cryoprotective solution supercools to very low temperatures and solidifies into metastable glass without undergoing crystallisation (Fahy et al. 1984). The metastable glass is exceedingly viscous and stops molecular diffusion inside the cells. In the vitrification method, cells or meristems are sufficiently dehydrated with a highly concentrated vitrification solution at 25° or 0°C without causing injury prior to immersion in LN. Vitrification solutions PVS2 (Sakai et al. 1990) and PVS3 (Nishizawa et al. 1993) have been developed which are glycerol-based and

less toxic. The complete vitrification procedure involves: (a) loading cells or meristems with an intermediate concentration of suitable cryoprotectant, (b) dehydrating cells and meristems by exposure to highly concentrated vitrification solution (e.g., PVS2 or PVS3). (c) transfer of cells and meristems to minicryotubes (1.8 ml) followed by proper sealing and immersion in LN, (d) rapid thawing in water bath at 40°C, and (e) removal of vitrification solution and reculture for shoot or plantlet regeneration.

The vitrification procedure has been particularly successful with the cryopreservation of species and cultivars in a range of herbaceous and woody plants (*see* Fig. 18.3).

(F) Encapsulation/Dehydration

Cryopreservation by vitrification requires careful control of the procedures for dehydration and cryoprotectant permeation in order to prevent

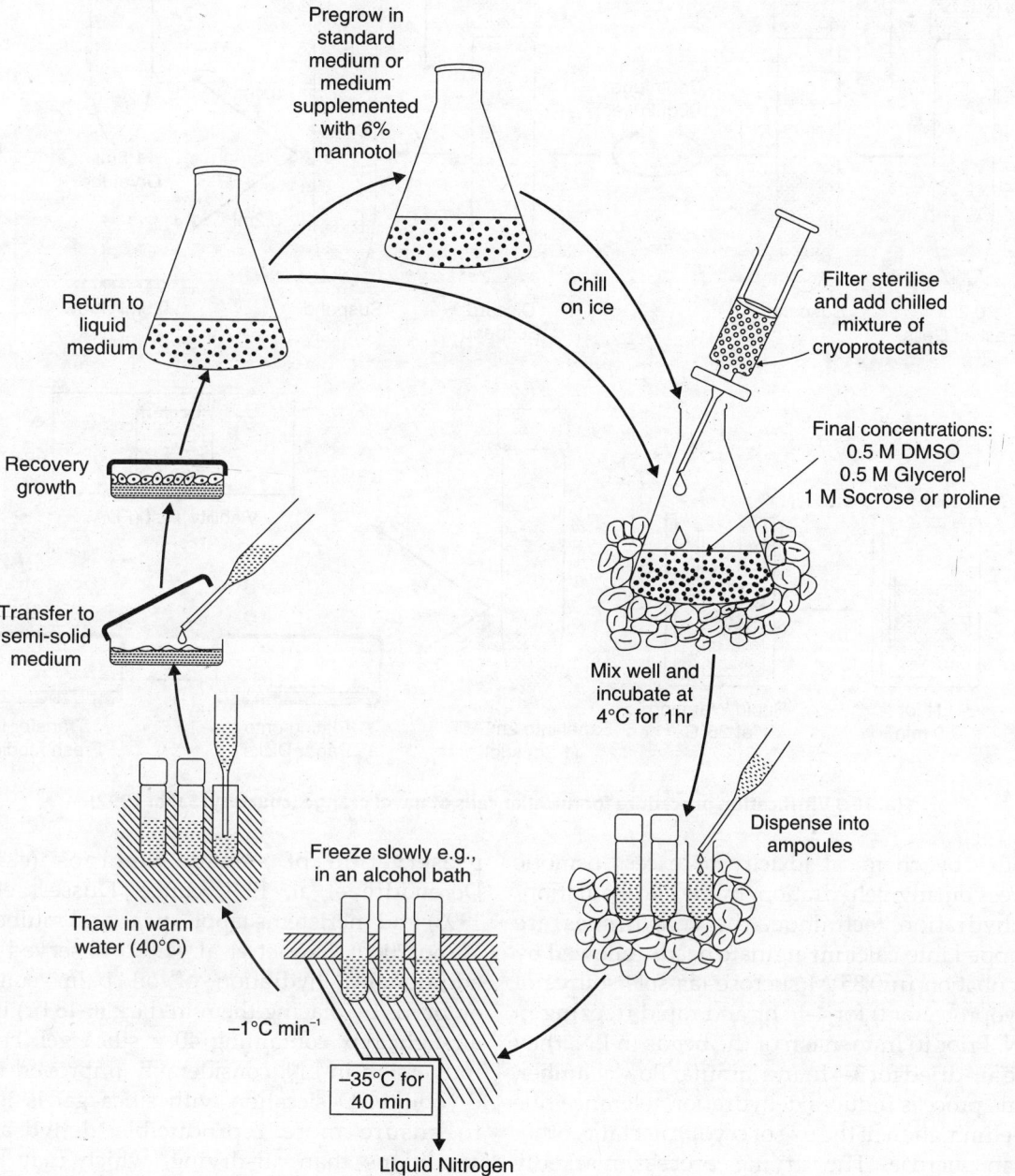

Fig. 18.2 Routine procedure for cryopreservation of cell cultures developed by Withers and King (1980). The appropriate freezing unit to maintain the frozen material is shown in Fig. 18.1 A.

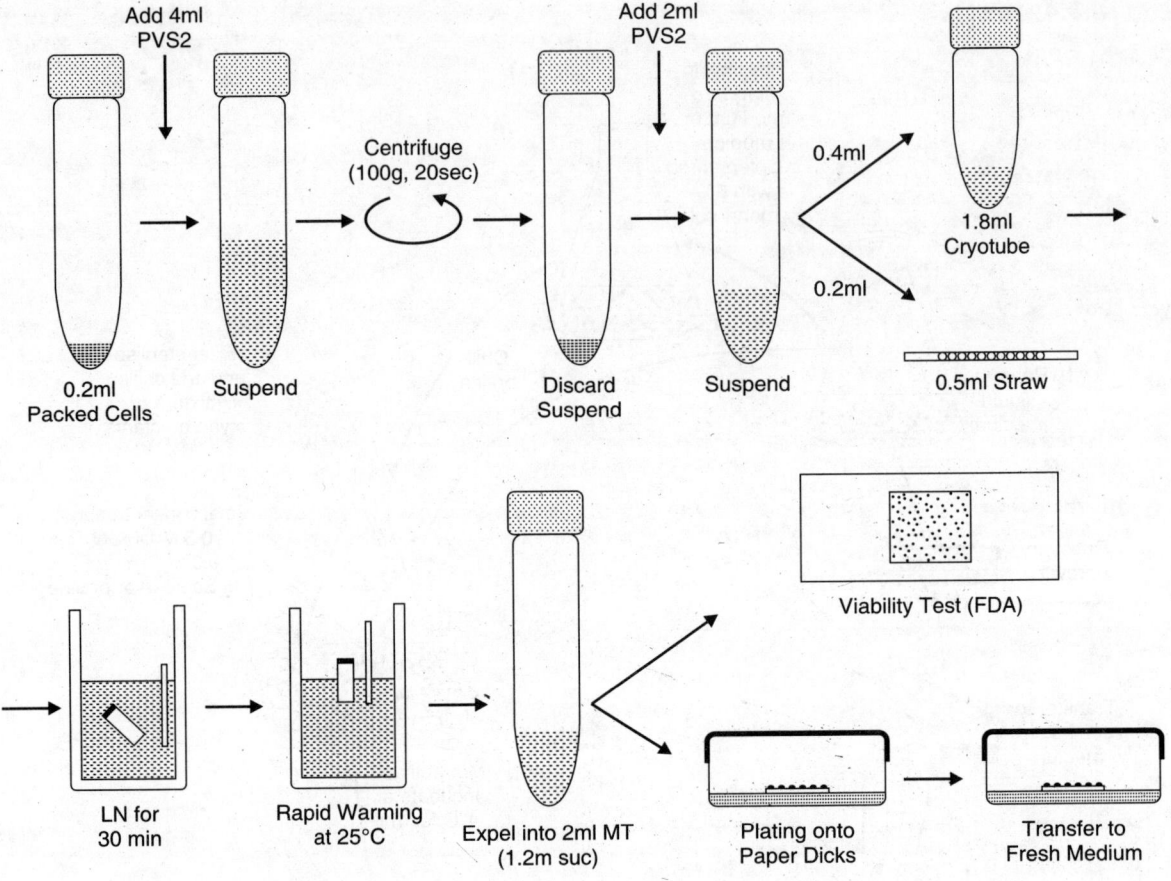

Fig. 18.3 Vitrification procedure for nucellar cells of navel orange (courtesy: Sakai 1997).

injury by chemical toxicity or excess osmotic stress during dehydration. In the encapsulation/dehydration technique, the cells/tissues are trapped into calcium alginate beads followed by incubation in 0.85 M sucrose (as sole source of cryoprotectant) for 4-16 hr and rapid freezing in LN. Prior to immersion of the beads in LN, they are air-dried for 3-4 hr in a laminar flow chamber. This process induces dehydration tolerance due to elimination of the use of cryoprotectants other than sucrose. The drying process markedly increases, the sugar concentration in the beads, which results in the glass transition during cooling to -196°C (Dereuddre et al. 1991).

The encapsulation/dehydration technique has found increasing application in cryo-preservation of somatic embryos (carrot, Dereuddre et al. 1991; coffee, Dussert et al. 1993) and meristems (apple, pear and mulberry; Niino 2000). Dumet et al. (1997) observed that additional dehydration of oil palm somatic embryos by placing them in dark (6-18 hr) in an air-tight box containing 40 g silica gel, before immersion in LN, considerably improved their cryobility. Desiccation with silica gel is likely to ensure more reproducible dehydration conditions than air-drying, which may vary depending on humidity and speed of airflow inside the chamber.

The encapsulation/dehydration method for Cryopreservation of coffee apices is given in Fig. 18.4.

18.3.4 Factors Affecting Freezing

(A) Physiological Condition of Material

A close correlation is found between the cell cycle of the plant material and its freezing potential. The relatively small and cytoplasmically rich (meristematic) cells of the late lag phase or early exponential phase have the highest freeze tolerance. It is generally recommended that materials should be subcultured and cryopreserved only when most of the cells are in the late lag or exponential phase. This can be readily determined in the case of cell suspensions. In comparatively larger tissues (shoot apices, young embryos and plantlets), highly vacuolated cells are severely damaged and regrowth occurs from the actively dividing meristematic cell component. Cells of the mature zygotic embryos do not seem to recover and their survival may require special treatment (Withers 1979, Dumet et al. 1997).

(B) Prefreezing Treatments and Cryoprotection

Various studies have shown that freshly harvested cells or tissues may not survive supercooling and require to be conditioned by their brief culture (pre-growth) before freezing. Prefreezing treatment of this type proved beneficial for potato and banana shoot apices only when they were cultured for 48 hr in the presence of cryoprotectant 5% dimethylsulphoxide (DMSO, Grout and Henshaw 1978, Thorpe and Yeung 2011). The process of hardening is also important as a prefreezing treatment in tissue culture. Plants grown at low temperature (4°C) for 3 days to a week before taking shoot apices for cryopreservation are reported to considerably enhance the survival frequency of the excised tissues. During the hardening process some molecular alterations occur in the plasma membrane which include changes in linkage groups within proteins, sugars and similar substances (Nitzsche 1983).

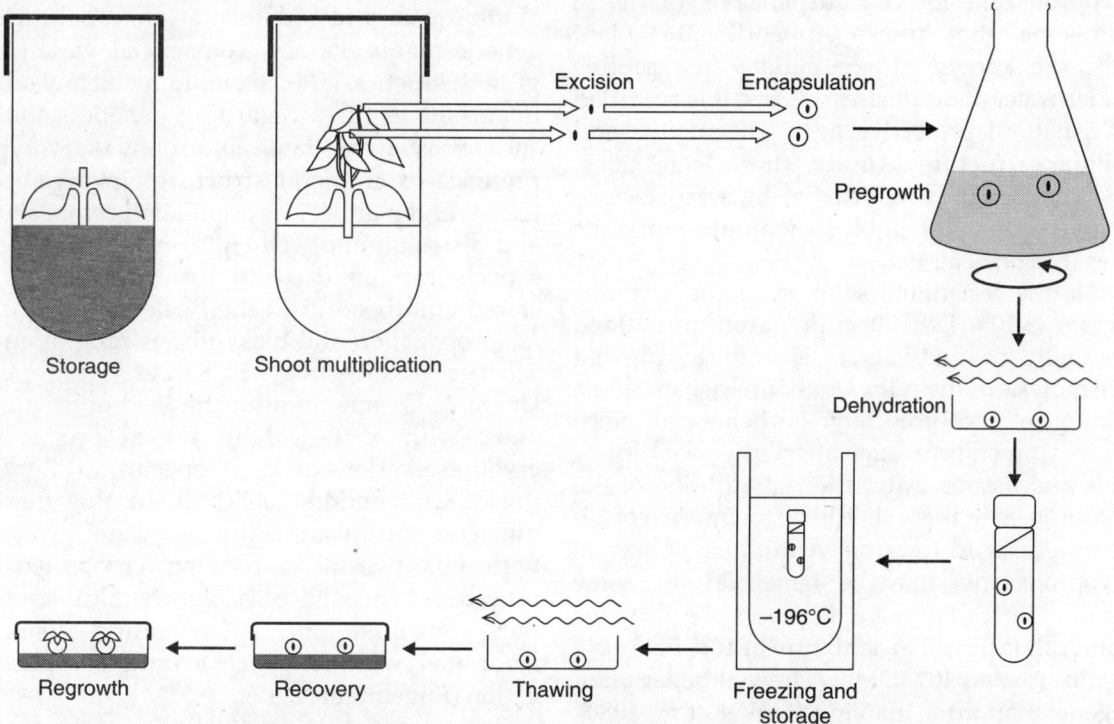

Fig. 18.4 Cryopreservation of coffee apices, as described by Mari et al. (1995).

Dehydration also increases the chances of a material to survive freezing. Fully hydrated materials, on the other hand, require rapid thawing to ensure cellular survival. The process of dehydration, however, increases the intracellular concentration of solutes and thus makes them toxic to cells. To protect cells from this toxic solution effect, *cryoprotectants* are added to the freezing mixture. Substances which in the solid state are amorphous, instead of crystalline, and dissolve in water with or without sugar generally prevent the formation of large ice crystals inside the cell due to freezing of the protoplasm. Such substances are mostly used as cryoprotectants (Uemura and Sakai 1980). Several of the cryoprotectants (Appendix 18.1) tried have different efficiencies. For plants, the most frequently used cryoprotectant is DMSO. The other compounds sometimes used successfully, either alone or in combination with DMSO, are sugar and glycerol. Withers and King (1979) found proline to be a superior cryoprotectant for cell suspensions of maize and some other species (Appendix 18.2). Plant cells are known to accumulate free proline under water and salt stresses, and it is regarded as a natural protective agent of enzyme and cellular structures under these conditions. 1,2-Propanediol is reported to be a satisfactory cryoprotectant for imbibed walnut embryonic axes (Brison et al. 1992).

Normally a dilute solution of the cryoprotectant (5-10% DMSO) is prepared and added gradually at intervals of 5 min to prevent plasmolysis of the cells. Use of an ice bath while adding a cryoprotectant is beneficial since room temperature may affect the viability of cells and tissues. After the last addition of the cryoprotectant there should be an interval of 20-30 min prior to freezing. A mixture of several cryoprotectants may be beneficial for some tissue cultures. For example, pieces of embryonic palm callus frozen in a mixture of 10% PEG, 0.44 M glucose and 10% DMSO showed better plant regeneration after thawing (Finkel et al. 1980). Prefreezing and cryoprotection treatments applied to some cell cultures are listed in Table 18.1. Droplet freezing and droplet vitrification have also proved successful for a number of species (Engelmann 2012).

18.3.5 Maintenance of Cryogenic Cultures

Maintenance of the frozen material at the desired temperature is very important. Temperatures above −130°C may allow ice-crystal growth inside the cells and, consequently, reduce their viability. Long-term storage of material frozen at −196°C will require an LN refrigerator that can accommodate about 4000 ampoules of 2 ml size. It has been estimated that 20-251 LN are consumed per week in the refrigerator and hence a regular supply of LN is necessary to store the frozen material.

While storing large numbers of germplasm a proper documentation should be made. The documented information may include exact taxonomic classification of the material, culture history, morphogenic potential, synthetic or biotransformation capabilities, genetic manipulations, somaclonal variations, growth kinetics, culture medium and any other important features regarding genetic stability such as isozyme data, autotrophy, karyotype, presence of chimeral structures in organised tissues and plantlets, or population composition and distribution of cell cultures. Data on these aspects can be used in future exchange of stored cultures and in their sale as feedstocks for propagation and biosynthesis programmes. Mention may be made of EUCOST (European Union of Co-operation in the field of Scientific and Technical Research) funded project on cryopreservation of crop species in Europe during the period 2007-2010. In this period nineteen European countries were involved with objective of improving cryoprotection procedures through physiobiochemical studies that contribute in understanding changes associated with tolerance toward cryopreservation (Engelmann 2012).

It is essential that the viability of the germplasm be tested periodically on some samples

Table 18.1 Examples of pregrowth and cryoprotection treatments applied to cell suspension cultures[a,b]

Species	Pregrowth	Cryoprotection
Daucus carota	Cells harvested 2 weeks after transfer to maintenance medium containing 10 mM myoinositol and 0.1% casein hydrolysate	10% DMSO applied at room temperature; no period of equilibration
Daucus carota, Nicotiana tabacum	Cells harvested after 4 days growth in standard medium and in exponential growth	10% DMSO plus 10% glycerol applied at 0°C; no period of equilibration
N. tabacum	Cells harvested after 3 days growth in presence of 6% mannitol and in exponential growth	0.5 M DMSO plus 0.5 M glycerol plus 1 M sucrose applied at ice temperature; 1 hr equilibration
N. sylvestris	Cells harvested after 3-5 days growth in presence of 6% sorbitol	0.5 M DMSO plus 0.5 M glycerol applied at ice temperature; 1 hr equilibration
Zea mays and many other species	Cells harvested after 4-7 days growth in standard medium or in presence of 6% mannitol or 10% proline	0.5 M DMSO plus 0.5 M glycerol plus 1 M sucrose or proline applied at room temperature or on ice; 1 hr equilibration
Catharanthus roseus	Cells harvested after 4 days growth in standard medium and cultured for 20 hr in presence of 1 M sorbitol	1 M sorbitol plus 0.5 M DMSO applied at ice temperature; 1 hr equilibration

[a] The treatments are listed in order of increasing modification in conditions of pregrowth and cryoprotection.
[b] After Withers (1985).

during long-term storage and properly recorded. These details would reduce the risk of exposing other samples to ambient temperature during sample removal.

18.3.6 Thawing

Cryopreserved materials are plunged into warm water at 37-40°C which gives a rapid thawing rate of 500-750°C min^{-1}. After about 90s the material is transferred to either to an ice bath or water bath at room temperature and maintained there until ready for reculture. Rapid thawing protects cells from the damaging effects of ice-crystal formation. However, if the water content of the cells had been reduced to an optimum level before freezing, the thawing rate is less critical. Measuring thermal events during cooling and re-warming using DSC (Differential Scanning Calorimetry) proved beneficial in establishing efficient cryopreservation protocol for zygotic embryos of *Parkia speciosa* and various citrus

species growing in Australia (Nadarajam et al. 2008, Hamiton et al. 2009).

18.3.7 Reculture

Thawed materials are washed several times to remove the cryoprotectants so as to avoid any deplasmolytic injury to the cells. The washed material is then recultured in a fresh culture medium following the standard procedures. Some workers have strongly recommended that thawed material not be washed. This may be because certain vital substances leached from the cells during the freezing process are lost by thorough washing. In fact, the frozen and thawed cells of *Zea mays*, and somatic embryos as well as plantlets of carrot, showed faster recovery and higher survival rate when they were cultured in the presence of the surrounding medium (Withers and King 1979). An alternative means of achieving post-thaw recovery while avoiding deplasmolysis and cryoprotectant injury was explored by Chen et al. (1984). They layered thawed cells of various *Catharanthus roseus* genotypes onto a filter paper disc overlying a semi-solid medium. After 4-5 hr the filter paper bearing the cells was plated on a fresh medium where the latter resumed growth. Culturing thawed cells without washing has been found more suitable for *Digitalis lanata* as well (Diettrich et al. 1982).

Recultured frozen and thawed plant materials exhibit special requirements for better survival. Grout et al. (1978) observed that shoot tips from frozen seedlings of tomato developed directly into plantlets on the addition of GA_3 to the culture medium. In the absence of GA_3 apices formed calli which later differentiated adventitious shoots. Normally the control (non-frozen) shoot apices do not require GA_3 to directly develop into plants. Activated charcoal is also reported to enhance the overall survival of freeze-preserved and thawed plants of carrot (Withers 1979). Anthony et al. (1996, 2000) observed that supplementation of culture media with surfactant Pluronic® F-68 (0.01-0.2%, w/v) increased the post-thaw growth of cryopreserved suspension cultured cells of *Oryza sativa* cvs. Taipei 309 (Japonica rice), tarom and ryegrass *(Lolium trifolium).* Similar beneficial effects were noticed with regard to the post-thaw growth of cryopreserved cells of Indica rice variety Pusa Basmati 1 on incorporation of commercial haemoglobin solution (Erythrogen™) into the culture medium. Other factors governing post-thaw recovery of cryopreserved cells are media composition (especially NH_4^+ concentration), use of activated charcoal, and method of removing cryoprotectant during thawing (Watanabe 2000).

18.3.8 *Viability of Cryopreserved Cells and Tissues*

The techniques to determine the viability of cryopreserved cells and tissues are the same as used for cell cultures (*see* Chapter 5). Generally, the cryopreserved materials are stained with FDA, Evan's blue, or 2,3,5-triphenyltetrazolium chloride (TTC). Initially, most of the cells in a population may appear to be viable. However, often, a small proportion recovers completely and divides to regenerate complete plants. The best indicator for survival of frozen and thawed cells, therefore, is their entry into the division phase and regrowth of shoot apices/embryos in culture. The viability is then synonymous with the survival and represented as:

$$\frac{\text{No. of cells/organs growing}}{\text{No. of cells/organs thawed}} \times 100$$

18.4 Cold Storage

Germplasm conservation by storing materials in cultures at low and non-freezing temperatures (1-9°C) has also been practised for some plants (Table 18.2). At low temperatures the ageing of the plant material is slowed down but not completely stopped as in cryopreservation. The advantage in this method is that cells and tissues are not subjected to cryogenic injuries. Further subculture of the plant material is necessary but only very infrequently. Medium-term cold storage (Section 18.3.1) is not only simple, it also gives a high rate of plant material survival.

Cold storage of *in vitro* derived shoots/plants has been particularly successful in fruit-tree species. Virus-free strawberry plants could be maintained for 6 years at 4°C provided a few drops of the liquid medium were added to the cultures after every three months (Mullin and Schlegel 1976). About 800 cultivars of grape plants have been stored for over 15 years at 9°C by yearly transfer to fresh medium (*see* Morel 1975). The use of ABA and high levels of sucrose may help to prolong the interval between two subcultures. Germplasm of a wide range of species can now be stored for 6-24 months by maintaining their shoot cultures in slow growth at low or cold temperatures (Table 18.3). Plantlets at slow growth are generally stored in glass tubes or plastic boxes (Gunning and Largerstedt 1985), although use of "gas permeable-heat-sealable" polythene bags is also recommended (Reed 1991).

Table 18.2 Some examples of shoots/plants stored *in vitro* at 1-9°C[a]

Species	Period in storage (months)	Viability (%)
Fragaria × ananassa	72	100
F. virginiana	72	100
F. vesca	72	100
Lolium multiflorum	10-11	88-100
Lotus corniculatus	1	90
Malus domestica	12	100
Medicago sativa	15-18	94-95
Rubus sp.	14	95
Trifolium pratense	15-18	70-86
T. repens	15-18	89-92
	11	90-100
Vitis vinifera	12	?

[a] See details in Bhojwani and Razdan (1983).
? = Not known.

18.5 Low-pressure and Low-oxygen Storage

Attempts have been made to develop low-pressure storage (LPS) and low-oxygen storage (LOS) as feasible techniques for future conservation of cultured plant materials. It is also envisaged to use these procedures (Fig. 18.5) as alternatives to cryopreservation and cold storage. In LPS, the atmospheric, pressure surrounding the tissue cultures is reduced, resulting in partial decrease of the pressure created by gases in contact with the plant material. On the other hand, in LOS the atmospheric pressure (760 mm Hg) is not reduced but the inert gases (particularly nitrogen) are combined with oxygen to create low-oxygen pressure. The results obtained from plant tissue culture experiments with LPS and LOS have been found comparable regardless of the plant or species used. These are summarised in Appendix 18.3.

Low-pressure or hypobaric systems have been found useful for both short-term and long-term storage. Short-term storage is generally used to extend the shelf life of plant materials such as fruits, cut flowers, vegetables, potted plants and cuttings. Long-term storage is used for materials grown in cultures. The principles on which low-pressure systems are based include: (a) maintaining materials in an atmosphere of controlled temperatures, (b) reducing atmospheric pressure in such a manner that the partial pressures of each gas within the storage and around the cultured material are

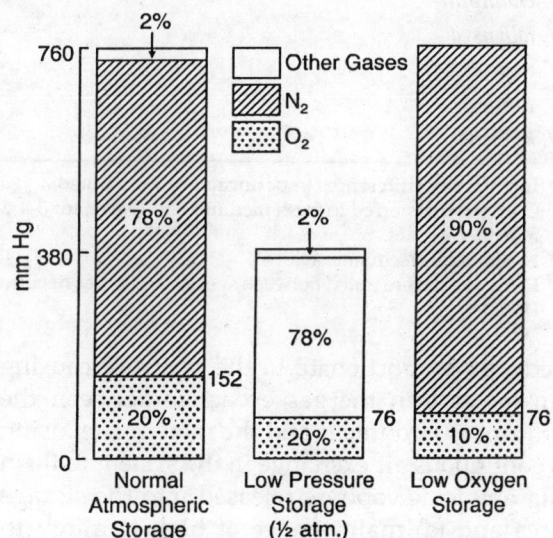

Fig. 18.5 Graphic representation of tissue culture storage under normal atmospheric pressure, low pressure and low oxygen (adopted from Bridgen and Staby 1983).

Table 18.3 Conditions for storage of shoot cultures of various species at reduced temperatures[a]

Species	Normal culture conditions	Storage conditions	Storage period
Dioscorea alata	26°C; Standard	16-20°C; 3%	8-24 months
D. rotundata	medium	mannitol	
Ipomoea batatas			
Solanum stenotomum	22°C; 16-hr	12°C /16-hr day	12 months[a]
S. goniocalyx	photoperiod;	6°C/8-hr night;	
S. chaucha	3% sucroce	8% sucroce	
S. juzepejukii			
S. tuberosum			
S. curtilobum			
Solanum species	22°C; Standard	22°C; 20 mg/litre ABA;	12 months[a]
(as above)	medium; 20 ml medium	60 ml medium	
Solanum species	22°C; Standard	22°C; 6%	12 months[a]
(as above)	medium	mannitol;	
S. tuberosum	20-22°C; 16-hr	10°C; 16-hr	24 months[b]
	photoperiod 4000lx;	photoperiod 2000 lx;	
	Standard medium	50 mg/litre B-nine[c]	
Trifolium repens	25°C; 16-hr photoperiod	5°C; dark	10 months[d]
Vitis amurensis	26°C/15-hr day,	9.5°C; 15-hr	6 months[d]
V. berlandieri	20°C/9-hr night	photoperiod	
V. caribea			
V. champini			
V. labrusca			
V. rupestris			
V. vinifera			
V. hybrids			

[a] Interspecific differences in performance were noted.
[b] Cultures transferred to fresh medium and placed for 3-4 weeks at 20-22°C could be returned to slow-growth conditions for a further 2 years.
[c] N-dimethylsuccinamic acid.
[d] Differences were noted between species and types of culture (proliferating shoots or single-rooted shoots). Source: Withers (1985).

reduced proportionate to the pressure causing an increase in the gas exchange between the culture environment and the material, (c) using a continuous air exchange in the system to flush out any toxic vapours released into the storage area and (d) maintenance of high humidty to preclude shrinkage, weight loss and desiccation of the material.

LPS has an added advantage of reducing the microbial activity in cultures in which the cells or tissues carry them systemically. Subatmospheric pressures also have a fungistatic effect by inhibiting spore germination, mycelial growth and sporulation of various fungi.

The partial pressures of oxygen below 50 mm Hg reduce organised and unorganised plant tissue growth. This is because low oxygen intake by cultured materials results in a decrease in CO_2 production during LOS. Consequently, the photosynthetic process is reduced in these tissues, thereby inhibiting their further growth. The idea that low partial pressures of oxygen may be advantageous for the storage of plant tissue cultures was propounded by Caplin (1959). His experiments with carrot tissue cultures demonstrated that the amount of growth under mineral oil (liquid petroleum widely used for conservation of cultures of various micro-organisms) is controlled by the supply of oxygen to the tissue. Since mineral oil prevented the diffusion of gases, therefore the level of oxygen below the oil reduced gradually due to intake by cultured tissues. This ultimately resulted in suspension of tissue growth.

The major limitation in using LOS for germplasm conservation is the long-term effects of low partial pressures of oxygen on plants. Studies on tobacco and chrysanthemums revealed that tissue or plant growth was not inhibited by either LPS or LOS treatments for the initial 6-week period, but later there was considerable difference in the growth of stored materials. The medium desiccation observed in these treatments would further limit the storage time of cultures unless there is a control mechanism for monitoring the relative humidity within the storage chamber. A comprehensive review of LPS and LOS treatments has been given by Bridgen and Staby (1983).

18.6 Synthetic Seed Technology for Short-Term Storage

Synthetic seeds are proficient tools to facilitate short-term germplasm conservation with particular objective for exchange in trade such as in floriculture, forestry, etc. Synseeds provide readily available germplasm source of mass propagation as they are capable of getting convertible into plants like conventional seeds. A high percentage (ca 80%) of germination was observed in excapsulated synthetic seed on short-term storage upto 150 days in petridishes or screw capped polythene tube (without super cooling) in case of orchid *Dendrobium* 'White Fairy' (Siew et al. 2014). More importantly, plants regenerated from synseed of the desert date palm

(*Balamites aegypticum* Del) with this procedure were found genetically stable using ISSR (Inter Simple Sequence Repeat) system (Varshney and Anis 2014). For synthesis of synseeds see Section 7.5.2.

18.7 *In vitro* Genebanks

In addition to highlighting the potential applications of various germplasm resources in this chapter, *in vitro* storage ensures international exchange of germplasm which is disease-free. Aspects of establishing *in vitro* genebanks and exchange of conserved germplasms at the international level are discussed in detail by Withers (1988). It is now possible to exchange genetic resources maintained *in vitro* for cassava, potato, sweet pea, yam, and several other species. Like seedbanks (Section 18.1.2), the types of *in vitro* collections are designated as: (a) *In vitro* Base Collections (long-term storage of materials by cryopreservation and not for distribution), (b) *In vitro* Active Collections (storage of materials in slow growth for a relatively short period), which can be used for multiplication, evaluation, indexing and distribution, and (c) *In vitro* Genebanks (comprising *in vitro* base and active collections) for exchange of germplasm at the national/international level.

The principle of *in vitro* gene banks should be to preserve the maximum possible genetic diversity of a particular genetic stock for future use. Genetic stocks may include materials (mutant or breeder lines with identified gene or gene combinations) specially developed for use in the ongoing breeding programmes. When new cultivars replace the primitive or conventionally used agricultural groups, it becomes all the more important that germplasm of displaced crops be properly conserved either *in situ* or *in vitro*. Global climatic changes also affect the natural plant habitats or bring about rapid changes in agricultural strategies. Thus, the materials which would otherwise be discarded due to pressures on resources can be stored *in vitro* without reservation (*see* Staritsky 1997).

18.8 Applications and Limitations

Progress in the development of plant cell and tissue culture techniques for long-term germplasm conservation has been quite significant, especially in view of the level of activity and small number of workers involved. The techniques of cell culture in particular have been so refined that only a few species are likely to be recalcitrant to known procedures. During the past decade studies on the cryopreservation of cell cultures have trended from basic aspects (cryoprotection, freezing, thawing, injury, recovery etc.) to applied aspects.

The evidence available so far indicates that cryopreservation is the most reliable approach to the long-term preservation of cell cultures which possess the biosynthetic capacity for synthesis and accumulation of secondary metabolites (Kartha 1987, Staritsky 1997). A brief outline of the protocol used for the cryopreservation of secondary compound-producing cell cultures is given in Table 18.4. Accumulation of secondary metabolites requires a certain degree of structural differentiation, often accompanied by increase in cellular and vacuolar volume as well as lengthening of the generation period. Therefore, the identification of appropriate growth stages most conducive to freezing becomes critical in devising an appropriate cryopreservation strategy of cells producing secondary metabolites. Recent developments regarding reculture techniques using various additives (*see* Section 18.3.7) and judicious use of cryoprotectants should enable cryopreservation of many secondary metabolite-producing cell cultures. Demonstration of post-freezing stability in terms of ploidy, growth rate, mitotic index and, most importantly, the biosynthetic activity in preserved cells (*see* Kartha 1987) further supports the prospect of using cryogenic techniques for the long-term preservation of such specialised cells.

Germplasm conservation using cell, tissue and organ cultures earlier reported in 60 species including various crops (*see* Withers 1985; Bajaj 1987, 1990a,b) has widened. From the information available, conservation is possible in both monocotyledonous and dicotyledonous families. Temperate species figure more prominently than tropical species (*see* Razdan and Cocking 1997, 2000).

In biotechnology laboratories, Cryopreservation and medium-term storage of plant materials can be of enormous value in maintenance of stock cultures and samples that await screening for future use. Almost an infinite number of replicates from germplasm *of* a large number of plant species can be stored *in vitro* using little space in contrast to the thousands of hectares of land that would be required for the same number of plants if maintained *in situ*. Thus, the material which would otherwise be discarded due to pressures on resources can certainly be stored without reservations. The principal objectives of germplasm conservation using cell and tissue cultures have been: (a) conservation of somaclonal and gametoclonal variation in cultures, (b) maintenance of recalcitrant seeds, (c) conservation of cell lines producing medicines, (d) storage of pollen for enhancing longevity, (e) conservation of rare germplasm arising through somatic hybridisation or other methods of genetic manipulations, (f) delaying the process of ageing, (g) storage of meristem culture for micropropagation, micrografting and production of disease-free plants, (h) conservation of plant material from endangered species, (i) establishment of germplasm banks and (j) exchange of information as well as germplasm at the international level. Additionally, cryopreservation has potential use in cryotherapy (*see* Chapter 15).

The expensive equipment needed to provide controlled and variable rates of cooling/warming temperatures can, however, be a limitation in the application of *in vitro* technology for large-scale germplasm conservation. It may be necessary to develop a low-cost technology to obtain the freezing equipment necessary once a protocol has been developed for preservation of a particular cell line, tissue, or plantlet. Further,

Table 18.4   Protocol used for cryopreservation of secondary metabolite-producing cell cultures[a]

Species	Methodology
Catharanthus roseus	Four-day-old cultures precultured in liquid medium with 1 M sorbitol for 6-2 hr. Frozen using 1 M sorbitol + 5% DMSO. Optimal cooling rate 0.5°C/min to –35°C; stored in liquid nitrogen (LN). Regrowth on filter paper placed over nutrient medium. No post-thaw wash.
Daucus carota	Two-week-old cultures treated with equal volume of medium containing 10% DMSO and frozen at 1°C/min to –70°C followed by storage in LN. Cells recultured after washing.
Digitalis lanato	Cells precultured for 1 week in medium with 3% mannitol; cryoprotectants: sucrose, glycerol and DMSO. Cooling rate 0.5-2.0°C/min to –60°C followed by storage in LN. Regrowth on solid medium. No post-thaw wash.
D. lanata	Six-day-old cultures pregrown in medium with 6% mannitol for 3 days. Cryoprotectants: DMSO, glycerol and sucrose. Cooling rate 1°C/min to –35°C and stored in LN. Regrowth on solid medium. No post-thaw wash.
Dioscorea deltoides	Cells precultured in medium containing asparagine (20 mM), or alanine (50 mM), or proline (20 mM), or serine (10 mM). Cryoprotectant: 7% DMSO. Cooling rate 0.5°C/min to –30°C, 9°C/min from –30 to 70°C, followed by storage in LN. Post-thaw wash.
Lavandula vera (callus)	Green callus pieces suspended in liquid medium. Cryoprotectants: 5% DMSO and 10% glucose. Cooling rate 1°C/min to –40°C followed by storage in LN. Reculture on solid medium after washing.
Panax ginseng	Cold hardening of cells by gradually increasing sucrose concentration from 3 to 25% and simultaneous decrease of culture temperature to 2°C. Cryoprotectants: sucrose-glycerine-DMSO. Cooling rate 0.5°C/min to –30°C, 9°C/min from –30 to –70°C, followed by storage in LN. Reculture after post-thaw wash.

[a] For details, refer to Kartha (1987).

distribution techniques and information networks need to be developed for materials conserved in cultures. It is hoped that in the near future the value of unique cell cultures will become increasingly recognised for conservation of higher genotypes, as is currently done with microbial cultures.

Progress with reference to the development of *in vitro* genebanks has been slow but steady although some prominent centres of *in vitro* Active Genebanks (IVAG) have been established which meet the standards of IBPGR. Notable among these are CIAT, CIP, IITA, IVIA, NCGR and IPGRI. A general tendency toward germplasm storage in seedbanks has contributed little to understanding the application of *in vitro* methods

for conservation. Moreover, the efforts made for the establishment of *In vitro* Base Genebanks (IVBG) of cryopreserved materials are far from satisfactory. The expensive equipment needed to provide controlled and variable rates of cooling/warming temperatures can also be a limitation in the application of *in vitro* technology for large-scale germplasm conservation. Establishment of *in vitro* genebanks would be profitable only by creating the facilities and infrastructure necessary for worldwide movement of germplasm. The *in vitro* gene banks and crops mostly they conserve are: CIP (International Potato Centre, Peru) – potato, and tubers; CIAT (Centro de Agriculture Tropical, Colombia) – Manihot; Gatersleben Gene Bank, Germany

– potato, onion spp.; CENARGEN (National Centre for Genetic Resources and Biotechnology, Brazil) – cassava, yam, asparagus, stevia, strawberry, banana, potato and sweet pea; International Institute of Tropical Agriculture, Nigeria – *Dioscorea* spp; NBPGR (National Bureau of Plant Genetic Resources, India) – *Musa, Fragaria, Colocasia, Dioscorea, Ipomoea, Curcuma,* ginger, medicinal and Aromatic Plants; NCGR (National Clonal Germplasm Storage, USA) – *Humulus, Mentha, Pyrus, Ribes, Fragaria, Rubus, Vaccinium*; INIBAP (International Network for Improvement of Banana and Plantain, France) – Musa; Institute of Plant Breeding, University of Philippines – cassava, sweet potato, potato, yam, banana, garlic; BGRC (Braunschweig Genetic Resources Centre, Germany) – Potato (c.f. Razdan and Cocking 1997, 2000; Bhojwani and Dantu 2013). Gene Banks especially for fruits has been established w.e.f Jan 1st, 2003 at Fruit Breeding Institute, Dresden – Pillnitz, Germany, which closely integrates with national decentralized network of the Fruit Gene Bank as a collection and holding partner (Manek and Hanke 2014).

APPENDIX 18.1

Cryoprotectants Used in Germplasm Conservation Using Cell and Tissue Cultures[a]

Acetamide
Diethylene glycol
Dimethylacetamide
Dimethylsulphone
Dimethylsulphoxide
Ethylene glycol
Formamide
Glucose
Hexamethylene tetramine
Mannose
Monoacetine
Polyvinyl pyrrolidone
Propylene glycol
Pyridoxine-n-oxide
Ribose

APPENDIX 18.2

Protocol for Cryopreservation of Cells from Suspension Cultures[b]

1. Subculture the actively growing suspension culture into fresh medium supplemented with 10% proline.
2. Wash cells after 3–4 days in proline-free medium and again suspend them in fresh medium.
3. Chill the cell suspension on an ice bath and add an equal volume of chilled-culture medium containing 20% proline gradually in four aliquots over a period of 1 hr.
4. Transfer chilled cells sterile to polypropylene ampoules with screw caps (1 ml into each 2 ml ampoule).
5. Place the ampoules in a controlled freezing apparatus and cool to –30°C at a rate of about 1°C min^{-1}.
6. After 30–40 min transfer the ampoules to the LN container or refrigerator.
7. For thawing, take out ampoules from the storage container and agitate them in water at 40°C for 1–2 min.
8. Transfer the thawed cells from the ampoules onto an agar medium along with the surrounding liquid medium for regrowth.

---

[a] Source: Nitzsche (1983); survival efficiency of the plant material reported to be in the range of 20-80%, depending on the cryoprotectant.
[b] After Withers and King (1979).

APPENDIX 18.3

Conclusions Based on Experiments with Plant Tissue Cultures Subjected to Low-pressure and Low-oxygen Treatments[a]

1. Partial pressures of oxygen below 50 mm Hg reduce the rate of growth of plants stored *in vitro*.
2. Low partial pressures of gases in LPS or LOS exert similar effects on plant tissue cultures.
3. Low partial pressures of oxygen affect both organised and unorganised tissues.
4. No phenotypic growth differences are observed after reculture of tissues removed from either type of storage.

---

[a] Compiled from Bridgen and Staby (1983).

19 Intellectual Property Rights of New Plants, Products or Processes, Obtained by Biotechnology or Conventional Methods

The past two decades have witnessed a rapid advancement in the production of new improved plants using biotechnology through application of plant tissue culture and gene transformation methods. Agriculturally important new plant varieties have also been obtained conventionally through selection, crossings, mutation and polyploidisation. Many of the economically useful new plant varieties, products and processes, obtained or developed by either of these procedures have been found of value at commercial scale. With these advances the agricultural biotechnology has developed as fast expanding industry with scope to offer remarkable economic, environmental and social opportunities. Notwithstanding the need for increase in the production of food, fibre, ornamental and medicinal plants there is considerable focus on development of genetically improved bioenergy crops which can contribute to eco-friendly and costeffective production of biofuels (Harfouche et al. 2012). Genome sequencing of several plant species provide building blocks for plant improvement and are of use to biotech industries; they thus help in expansion of the existing scientific knowledge (Singh et al. 2009).

19.1 Intellectual Property Protection and Rights

Concurrent to the increasing application of plant biotechnology and conventional methods of producing new plant varieties, products,or processes on commercial scale issues like legal characterisation of Intellectual Property (IP) and Intellectual Property Protection (IPP) have arisen, which can be addressed by giving them 'Rights' such as Intellectual Property Rights (IPR). Prominent examples of IPR in areas of plant biotechnology and agriculture science include recombinant DNA created through genetic engineering techniques using restriction enzymes, transgenics, *in vitro* regenerated new improved plant varieties, trade related biochemical processes as well as their products, and the plant breeder's rights of protection to newly developed plant hybrids or cultivars. The term 'property' more often refers to physical object (land, housing and household goods). It is tangible and protected with ownership rights guaranteed as per law of the country. The 'intellectual property' on the other hand is intangible that includes patents, trade secrets, copy rights, or trademarks. Patents provide its holder legal right to use described invention commercially. Trade secrets comprise private information about specific technical processes or formulations that a company protects from others, whereas Copyrights protect the authorship of published works from unauthorised use. Trademarks are words or symbols that identify a product or process of the particular company. The intellectual property rights (IPR) in respect of new improved plants, products or processes, have become essential in order to prohibit competitors from making, copying, using or selling their propriety subject matter. It may be interesting to note that number of biotechnological companies in USA (Ca.2100) and Europe (Ca.2200) till 2008 made huge investments in stock market. While some of these companies made profits others met with failures and even faced closure due to lack of proper implementation of IPR guidelines as they were not familiar about the products and processes that required protection in biotechnology or plant science research (Fasella and Hussain 2014). Since no biotechnology company would initiate high-risk long-term projects without knowing whether products of its research efforts can legally be protected, it thus becomes more pertinent that inventors be given exclusive rights to novel products or processes that they develop. It is a well-known fact that inventions like novel products, processes, discovery or development of new plant varieties in agriculture have greatly benefitted the mankind; the society therefore has a stake in promoting such innovations at industry level. Keeping this in view it is being increasingly recognized that the strategy to cover both society and industry concerns can best be achieved by formulating guidelines of intellectual property rights (IPR) for biotechnological inventions including newly obtained plant varieties on the basis of patent system.

19.2 Patenting of Biotechnological Inventions/New Plant Varieties

A patent is a legal document that gives patent holder exclusive rights to implement the described invention commercially. Point to note is that patent holder can develop other products from his original invention but competitors would need to obtain license for the right to use this invention for developing a product based on it. Biological products or processes are patentable subject to satisfying undermentioned basic requirements (c.f. Glick et al. 2010):

- Patent must in effect be 'novel', i.e. not used elsewhere other than the country where patent application was submitted.
- Patent must have inventive step that was not obvious earlier to other workers in the field.
- Patent whether it is a process, instrument, biochemical molecule, microorganism, or multicellular organism, must be useful for R&D and in some way economically as well as commercially viable.
- Patent application must provide descriptive account of the invention sufficient enough for implementation by other person knowledgeable in the same field.

Duration of the patent under IPR usually is 20 years from the date of filing any type of invention. A provisional specification may be filed to establish priority of the invention if it is in conceptual stage and submission of full description may need more time. It should be noted that the provisional specification is a scientific-cum-legal document of the patent in which no amendment is allowed.

19.2.1 Patent Categories

Patenting of inventions arising out of basic research on plants, or life forms, even though economically significant was initially considered unethical based on notion that a patent cannot be granted for products of nature. For a society it would be inappropriate to give monopoly to an individual or a company for something that occurs naturally and was merely discovered. As patent laws vary in different countries the concept of not granting patent based on life forms has undergone a change particularly since the discovery of oil-eating bacteria (*Pseudomonas*) in 1980 by A. Chakrabarty who at that time worked for General Electric Corporation in USA. This was first case of a life form granted patent in USA to a Multinational Corporation (Fasella and Hussain 2014). Another example of intellectual property of life form is the development of new crop varieties in the field which are protected through plant breeder's rights (PBR). Under PBR the plant breeder who developed a new variety enjoys exclusive rights for marketing that variety with exemption to farmer (farmer's exemption) who is generally a permitted replantation of the variety by saving seeds. To deprive farmers this advantage some countries, however, have allowed grant of 'utility patents' for such plant genetic materials so that patented material can neither be used for further breeding nor permit farmers to save seeds for further cultivation without paying fee to the patent holder. Utility patents can be granted for plant inbred new varieties, hybrids, plant parts (e.g., seeds, pollen, fruits, flowers) besides genes, biotechnological processes,and many other plant-derived products (Adams et al. 2009).

Mention may be made about 'product patents' and 'process patents' (Table 19.1). Product patents comprise homogeneous substances, complex mixtures and various devices; whereas process patents include preparative procedures, methodologies, and actual uses.

19.2.2 Granting of a Patent

System for granting of a patent may vary according to a country. However, general procedure is to prepare an application with the help of an expert, normally a patent lawyer, and organize it in a defined pattern. The application must have title, an abstract describing the nature of invention, a section on the background stating current 'state-of-the-art' in the field of invention with comprehensive summary, figures or schematic representations that depict functioning of the invention, list of claims about the invention, and finally its use. The application is submitted to Patent and Trademark Office (PTO), where it is reviewed by an examiner to adjudicate novelty, utility, non-ambiguity, feasibility and acceptability as a patentable invention. Upon examiner agreeing that invention satisfies all laid criteria of patentability

Table 19.1 Examples of biotechnology inventions of product patents and process patents.*

Category	Examples
Product patents	
Substance	Cloned genes, recombinant proteins, monoclonal antibodies, plasmids, promoters, vectors, cDNA sequencing, antigens, peptides, RNA constructs, antisense oligonucleotides, peptide nucleic acids, ribozymes, and fused proteins.
Complex mixtures	Multivalent vaccines, biofertilisers, bioinsecticides, microorganisms, transformed cell lines, transgenic organisms.
Devices	Pulse-field gel electrophoresis apparatus, DNA-sequencing units, and microprojectile gene gun.
Process patents	
Process of preparation	DNA isolation, synthesizing double-stranded DNA, vector-insert construction, PCR applications, and purifications of recombinant proteins.
Method of working	Nucleic acid hybridisation essays, diagnostic procedures and mutation detecting systems using PCR.
Use	Applying biofertilisers and bioinsecticides, fermentation by genetically modified organisms, nontherapeutic treatment systems.

* adopted from Glick et al. (2010).

a patent will be awarded. If patent application is rejected there is a provision for an applicant to file an appeal to a Patent Appeals Board. In case this appeal is turned down then decision can be challenged legally. It is necessary that filing of an application for a patent should be completed at the earliest possible date to register the priority as delay in filing patent application may entail risks because other inventors might apply for grant of the said patent and forestall rights of the first inventor. To safeguard such eventuality a provisional specification (*see* Section 19.2) can be filed. However, complete specification of the patent must be submitted within 12 months of filing the provisional specification. The period may further be extended by 3 months. Patent application with complete specification will be published 18 months after its date of filing. A "benchside"guide to patents and patenting is provided by Johnson (1996).

Most importantly the grant of a patent does not give right to sell invention or use it for production purposes until all statutory regulations have been followed. To quote an example, for a patent granted to genetically engineered microorganism the manufacturer must satisfy recombinant DNA regulations for its multiplication, distribution, and release. Also protection of patent rights being responsibility of patent holder, therefore, if any other person is presumed to be infringing on a patent he/she will invite lawsuit against him/her which can only be decided by the courts and not the patent office. In similar way if a person or company feels that awarded patent is inappropriate then the legitimacy of that patent can be challenged in a court of law (Glick et al. 2010).

19.2.3   *Terminator Technology Patent*

A patent under 'control of gene expression' dubbed as terminator technology in the media was issued in USA to Delta and Pine Land Company (D and PL) and US Department of Agriculture (USDA), in 1998, as it was claimed that genetic system in the patent worked well with tobacco and cotton. The combination of genes described in patent provides a very broad protection of plant components like plant cells, tissues, seeds and whole plants of any species (both transgenic and conventional varieties). This encouraged inventors of this technology to apply for patents in at least 78 countries. USDA, D and PL also decided to make terminator technology ideally available to seed companies under licensing agreements. The possible advantages of using this technology for gene protection advocated by Monsanto include: (i) minimum outcrossing with related species, (ii) enhanced choice of seed varieties for farmers, (iii) provision for producing consistently high quality seeds, (iv) yearly improvement in these seeds to produce best quality, and (vi) maintenance of desirable characteristics in varieties grown for more than one year. The basic objective of Inventors was to develop this technology as 'technology protection system' for corporate breeders against free use of transgene technology. For example, the technology could be used with purpose of desired trait expressing in one situation, but not in the other, particularly with regard to traits like male sterility, drought or insect resistance, flower development,and time of seed germination. It was also proposed to protect the rights of seed producers/breeders by disallowing farmers to use their harvested seeds for replantation, thereby ensuring adequate returns to owner holding patent of the crop variety. Agriculturists as well as others by and large expressed displeasure for such stipulations and considered terminator technology as 'traitor technology'. In the document prepared in 1998-99 on behalf of the conference of parties (CoP) of the convention of biological diversity (CBD) the terminator technology was named as v-GURT and t-GURT (v = variety, t = trait; GURT = genetic use restriction technology) and considered it as anti-poor farmer and also working against the concept of sustainable use of biodiversity (*see* for details Fasella and Hussain 2014).

19.3   Patent Rights in Different Countries

Patent rights are generally extended throughout

a country where application has been filed. To protect use of an invention in other countries it is advisable to file patent application separately in each country since patent offices in different countries often reach quite different conclusions about the same patent (Holman 2007). For example, Genentech applied for a patent in the United Kingdom for the production of human tissue plasminogen activator (tPA) in 1989, using recombinant DNA processes, which the company developed as therapeutic agent for the prevention and treatment of coronary thrombosis. A total of 20 claims were presented in original patent application of which some were broad and others narrow in scope. The patent was rejected by UK's patent office and subsequently the company appealed to the United Kingdom's Patent Appeals Board which after considerable deliberations invalidated all claims for variety of reasons. Paradoxically the same company Genentech was awarded patent for the same protein human tPA in USA giving them exclusive rights for marketing. Further Japanese version of Genentech's tPA patent has limited amino acid sequence which not only the company was given permission to clone but even other companies could sell variant forms of human tPA in Japan. Thus, the application for same patent was rejected as well as approved involving three different patent offices. Not everyone seems to favour patenting particularly when opponents argue that awarding patents encourage monopoly, restrict competition, encourage prices soar, and promote large corporates at the expense of individual inventors and small companies. With all such concerns still patent system has become an accepted norm in different countries as it does not appear to prevent fundamental research or act as an impediment to innovations. To substantiate this fact U.S. patent granted to Stanley Cohen and Herbert Boyer in 1980 for recombinant DNA technology for use of both plasmid and viral vectors to make constructs or for cloning of foreign genes has not proved

hindrance to further research on recombinant DNA technology (Marshall 1997, Evans 2001).

In Europe, the Act: European Parliament and Council Directive 98/44/EC of 6 July 1998 attempts to clarify the distinction between what is patentable and not patentable. In the context of plants, inventions which have value for industrial application are patentable. Plant or biological material which is isolated from its natural environment or produced by means of technical process may be subject of invention, but exploitation of such invention if found contrary to public policy or morality is not patentable. European Commission's Group on Ethics, Science and New Technologies has primary responsibility to evaluate all ethical aspects of biotechnology. Further plant propagating or breeding material sold to farmer by patent holder can be used by farmer only for his own agricultural purposes. However, where a breeder is not able to acquire plant variety right, and to avoid infringing a prior patent, he has aprovision to apply for compulsory licence for nonexclusive use of the invention protected by that patent. Fleck and Baldock (2003) have given a descriptive account of Intellectual property protection for plant-related inventions in Europe.

In several developing countries plant variety protection (PVP) is in the process of legal status. In India, the patent law under Indian Patent Act 1970 did not allow product patents on foods, chemicals, drugs, and pharmaceuticals. Only process patents were granted in these areas provided the same product was produced by alternate process. Should there be a dispute about product not being produced by alternate process the burden of proving this lay with original patent holder who lodged the complaint. With the adoption of TRIPs agreement (*see* Section 19.4.2) under the aegis of GATT (General Agreement of Trade and Tariffs), and later pressure from WTO (World Trade Organisation) and private industry, Indian Patent Act 1970 was amended by the Indian Parliament in 2001.

It was approved that new plant varieties can be protected in India under 'New Plant Variety and Farmer's Rights Protection Act 2001'. The protection is provided to new plant varieties not through patents but by granting farmers rights to use seeds from his variety for planting the next crop. There are also provisions for benefit sharing among other farmers, penalty for marketing spurious propagation material, and protecting extant varieties. The total period of protection is 10 years from the date of registration. *Distinctiveness, uniformity, stability, novelty* and *denomination* are five main criteria to arrive at a decision whether a plant variety is new or not. *Distinctiveness* of a new variety is apparent by its clearly distinct features in comparison to any other related variety whose existence is a matter of common knowledge at the time of filing the IPP application. *Uniformity* is determined by relevant characteristic features of the variety being transmitted uniformly during propagation; these characteristics if remain unchanged during repeated propagations point to the fact that the variety is *stable*. Similarly, the variety will be deemed *novel* or new if it's propagating or harvesting material at the time of filing the application for protection under PBR has not been sold or otherwise disposed to others. The *denomination per se* refers to the generic designation of new variety which is a must to enable a variety to be identified.

For plant improvement using biotechnology, or conventional procedures, the essential pre-requisite is that natural plant diversity is preserved vis-à-vis current concerns about degradation in environment due to climate change and other factors. There is thus need for sustainable maintenance of biodiversity and to ensure this India has enacted Biodiversity Act 2002. The Act has specific provisions about ownership of intellectual property rights associated with exploitation of biodiversity. For example, industries have to obtain prior consent of the National Biodiversity Authority for exploring biodiversity in India and in the event of benefits accrued through R&D based on use of the biodiversity there is provision for sharing of these benefits with local community. The Union Govt. will ensure that benefits reach through State Governments to the local community.

19.4 Harmonisation of Patent Laws

Considering the fact that number of cross-border disputes and multiple infringement suits related to patents is increasing firstly because the scope and coverage of patent protection differs from one country to the other and secondly the intellectual property right has become a big component of global trade, there seems an urge for harmonisation of the patent laws so that such an effort would improve the world's capacity to innovate as a whole. Harmonisation of patent laws would prove beneficial to liberalised technology transfer as well as foreign direct investment to developing and underdeveloped countries. A step in this direction has been Substantive Patent Law Treaty (SPLT) proposed by the World's Intellectual Property Organisation (WIPO) based in Geneva which also included plant intellectual property rights (Koo et al. 2004). Prior to this, Paris Convention of 1988 sought international cooperation for enforcement of uniform patent laws underlying the basic principles of equal treatment for domestic and foreign inventors. In Paris convention 100 member states participated and decided to allow inventors to claim international priority upon filing of a patent application even in one member state. European Patent Convention (EPC) 1973 is another international group which incidentally has the distinction of introducing the first patent statute with specific provision for biotechnological inventions (Bera 2009). There are certain other prominent international agreements that include provisions for IPP and IPR of new plant varieties and biotechnology inventions.

19.4.1 Intellectual Property Protection Agreement UPOV

While making tentative efforts for harmonisation of patent laws and considering the fact that intellectual property protection of plants varies widely between countries, the UPOV (Union International pour la Protection des Obtentions Végétales) came into existence as an international agreement outlining minimum standards for the protection of plant varieties. Member countries that have signed the UPOV convention are obliged to guarantee that standards set are enforced within their own legislation. UPOV guidelines were enacted into law in 1961 and subsequently the Act revised in 1972, 1978 and 1991 with the objective of adding important changes that enable more protection for a plant variety which is distinct from known existing varieties. Further the definition of the propagating material has been specified and provisions of permitting farmer's rights to harvest and sow seeds for personal use well defined so that they are acceptable to many countries (Sechley and Schroeder 2002).

UPOV Act of 1961, 1972 preserved the breeder's right to use seed of a new plant variety to generate further new varieties, but the 1978 version protected, in addition to breeder, the farmer's privilege of keeping his harvest apart for subsequent sowing. In the 1991 Act, significant amendments were added and number of plant genera and species that were given protection selectively in the UPOV Act 1978 ultimately increased to all plants. Also the period of protection for most plants was extended from 15 to 20 years; for trees and vines 25 years. Importantly, farmer's privilege to sell or exchange seeds of a protected variety with other farmers has been withdrawn, but option is given to a member state of UPOV (if it so desires) to restore this farmer's privilege by enacting necessary legislation in the respective country under UPOV Article 15, 1991. Another important amendment under Article 14, 1991 includes ability to cover essentially derived varieties. A 'derived variety' comprises the characteristics of protected variety with only a minor change. Derived varieties may be spontaneous or induced mutants, a somaclonal variant, or a variant obtained by backcrossing or genetic engineering. Essential criteria for derived variety being that minor change that occurs in the initial variety must be genetic and expressed significantly through plant phenotype. About 68 countries had become members of UPOV which included USA and most countries of the European Union (Sterckx 2010).

19.4.2 Trade Related Intellectual Property Rights Agreement (TRIPS)

One of the notable steps taken towards harmonisation of patent laws is the agreement on Trade Related Aspects of Intellectual Property Rights (TRIPS) that came into effect on January 1, 1995. It is most comprehensive multilateral agreement intended to contribute to promotion of technological innovations including transfer as well as dissemination of technology for mutual benefit of its producers and users in a manner conducive to social and economic welfare (Article 7, Article 8, 1996). To balance rights and obligations the Act 1.1 leaves member states "free to determine the appropriate method of implementing the provisions of this agreement within their own legal system and practice". The November 2005 decision of the Council of TRIPS even allowed least-developed country members to postpone many TRIPS obligations until 2013. Countries have provision to refuse patent for those inventions that are required to protect *ordre public*, morality and human health under Act. 27 (2). Nations may tailor their patent system according to the domestic needs, development, and potential for growth (Act. 27.1). With regard to IP rights of plants, TRIPS states that member countries must protect plant varieties by patenting as per *sui generis* (one of a kind, unique) systemor by a combination thereof as mentioned in Article 27(3) b, 1996. Options given in issuing a patent for new plants obtained through biological processes (such as crossing or selection) by a member country appear flexible under Article 27(2), 1996. Comprehensive details are given in document 'Agreement on Trade Related Aspects of Intellectual Property Rights (TRIPS)' published by World Intellectual Property Office (WIPO), Publication No. 223(E), 1996. ISBN 92-805-0640-4.

19.4.3 Treaty on Plant Genetic Resources

The treaty was adopted by consensus during 31st session of the FAO conference on November 3, 2001 and actually came into effect on June 29, 2004 after necessary ratification by 120 member countries. The treaty ensures public availability of the plant genetic resources that serve as raw materials for plant breeders to produce new crop varieties and infact replaces previous FAO agreement "The International Undertaking on Plant Genetic Resources (IU) 1983" that was voluntary and not legally binding. The basic objective of the treaty on plant genetic resources happens to protect farmer's privilege and develop Multilateral System of Access and Benefit Sharing (MLS). The MLS gives easy access to genetic resources like major food and feed crops and shares benefits arising from exploitation of these materials in accordance with multilaterally agreed terms and conditions. Genetic resources are accessed free of charge by member countries of the treaty, but for others fee is charged which does not exceed the minimal cost (*see* details in Smolders 2005).

19.4.4 Other Agreements

(a) Material Transfer Agreements: Generally in research laboratories of university, Govt., or privately funded scientific institutes, the academic research is conducted in a relatively free environment with unrestricted access to basic knowledge and use of products previously developed for a concerned scientific project (Bentwich 2010). Under patent laws no exemption exists for academic use of patented materials and to come over this impediment administrators

at publicly funded research institutions were prompted to execute multiple material transfer agreements (MTAs) permitting them to use patented materials in laboratory research and breeding programs. MTAs are preferred for transfer of intangible property like plasmid vector for making constructs, by inserting gene of interest, or plant germplasm in countries where IPR is not appropriate (Kowalski et al. 2002). In centres of the Consultative Group on international Agricultural research (CGIAR), MTAs are a sort of legal arrangement toutilise IP for biotechnological research. Germplasm exchange of all members in TRIPS and CGIAR system are now covered w.e.f. January 2007 by standardised MTA known as Uniform Biological MTA (UBMTA). It may be noted that this agreement stipulates that materials exchanged be used only for research purposes and a separate agreement would be required on patent terms if the product of research is used for commercial purposes. Further material(s) may not be transferred without prior- written approval of original provider (Mowery and Ziedonis 2007).

(b) Technology transfer agreement: Universities and research institutions in many countries face tremendous pressure from government agencies for undertaking collaborative research in partnership with industry in order to be eligible for public funding. This would thus lighten the burden on the public exchequer and propel new inventions to be utilised commercially. Countries in EU and the U.S. provide good examples of successful technology transfer (TT) between the universities and industry of inventions generated with commercial value. In the U.S., the establishment of Technology transfer offices (TTOs) and the biotechnology industry are mostly linked to Bayh-Dole Act of 1980. Since IP ownership rights were with federal government Bayh-Dole Act signalled the U.S. government support for commercialization of research products and further created a belief that universities are a source of innovation that can be used to strengthen the country's economy.

This paved way allowing universities for the first time under 'University and Small Business Patent Procedures Act 35 U.S.C.§2003(1980)' to take ownership rights of patent based on discoveries or inventions made as a result of federal funding (Berman 2008).

(c) Patent Cooperation Treaty (PCT): Envisaged for accelerating IP protection in developing countries China is fifth among other developing countries that signed this treaty. PCT is an agreement by which patents can be simultaneously lodged in different countries although guideline policies still need improvement vis-á-vis other agreements (Harfouche et al. 2012).

19.5 IPR of Trade Secrets in Agricultural Biotechnology

Trade secrets in the area of agricultural biotechnology include private propriety information on aspects such as: (i) Hybridisation conditions; (ii) Cell lines; (iii) Corporate merchandising plans; and (iv) Customer lists. Also the duration of trade secrets is unlimited and may not require satisfying the difficult conditions as laid down for filing patent applications. However, disclosure of trade secret or its use without authorisation entails punishment in a court of law and owner may be allowed appropriate compensation.

Maintenance of trade secrets at times becomes untenable particularly when the results are published or discussed in conferences and exchange of information becomes inevitable. Yet the reliance on trade secrets appears more prudent in comparison to patents which have specified period of protection.

19.6 IPR of Copyright and Trademarks

Authored books, edited books, audio and video cassettes are best Examples of copyright which cannot be produced without permission of the person who holds copyright, i.e. author, editor or publisher. Copyright rules are modified time to time considering that computer software and other materials are now included under copyright. Another example that may be mentioned of copyright in biotechnology is the published DNA sequence data coding for certain protein. In research, however, alternate sequence coding for same protein may be prepared using wobble in genetic code to ensure that the copyright is not infringed. Computer data bases and photomicrographs of DNA constructs may also be copyrighted. It may be interesting to note that an instruction manual can be copyrighted and even protected as trade secret although it cannot be patented. Copyright rules appear to have limitations. As per copyright rules photocopy of a book on biotechnology, plant sciences, or on any other subject is not allowed but ideas given in the book can be used for research or dissemination of the knowledge.

Trademark in biotechnology or plant science research is a word or symbol that a manufacturer or merchant uses to identify his goods to distinguish them from those manufactured or sold by other. Laboratory equipment or instruments used for this purpose are known by their trademarks. A plant biotechnology product given a brand name is protected as IP by registering it as trademark with appropriate patent office and so are certain vectors available for recombinant DNA research. Bollgard® is a registered trade mark of Monsanto Company for Bollgard® cotton. Other trademarks such as Roundup Ready® or LibertyLink® refers to specific traits imparted by transgene and may be of value in marketing to farmers. A significant aspect of the agricultural biotechnology variety having agronomic traits of commercial value is its trademark identification and reputation. Trademarks are awarded by national governments and protection is valid only in countries issuing them. For more details on Section 19.5 and 19.6 refer to Friedman (2009), Harfouche et al. (2012), Fasella and Hussain (2014).

19.7 Freedom-to-Operate in Agricultural Biotechnology

A product developed from biotechnology on commercial scale is both financial as well as time intensive. Further access to ownership of IP in such products being complex there are situations where no single public or private enterprise is able to provide full IP rights to guarantee Freedom-to-Operate (FTO) products of a certain technology within a particular jurisdiction. For example, products improved using genetic engineering technology may entail know-how of many proprietary processes each of which may be protected under IP/tangible property (TP) law. It becomes imperative, therefore, to conduct FTO analysis to identify rights of product's IP and TP (Krattiger 2007). FTO means that for a given product, at a given point of time, no IP rights of third party is fringed in respect to a particular market or geography. This concept, however, does not look as a practicable productive strategy for companies and public sector institutions. Analysing the FTO to find out that a product under development does not infringe third party IP rights, therefore, is a vital step which may only be planned and executed prior to launching a new product because of the prevalence of national jurisdiction over patent and IP laws that could vary between countries. A country in which a product will be developed and commercialized invariably would involve engaging an attorney for FTO analysis to cover market risks (Chi-Ham et al. 2010). For instance a transgenic plant-product FTO analysis scheme involves three main steps: (i) *product deconstruction*, (ii) *product clearance,* and (iii) *FTO decisions*. The *product deconstruction* involves identification of all components (e.g. germplasm: cv, spontaneous variety; expression vectors; promoters; terminators; GOIs; selection markers; excision markers), processes or methods (e.g. *Agrobacterium-* mediated genetic transformation; particle bombardment; EXACT™ technology, etc.) involved in the development of a product. For *Product clearance* IP and TP landscape of the product are analysed using extensive patent and literature searches on IP laws of different countries. MTAs and license agreements related to proposed product are also examined. Lastly, the data obtained from *product deconstruction*

and *product clearance* studies are subjected to make strategic *FTO decisions* that help reduce the risk of IP litigation (*see* for details Harfouche et al. 2012).

International and national agricultural research centres have far greater FTO in research related to food crops, particularly in the developing world, which involves public-private partnership. African Agricultural Technology Foundation (AATF) has a program known as Water-Efficient Maize for Africa (WEMA). Under this program AATF in partnership with International Wheat and Maize Improvement Center (CIMMYT), and Monsanto, have plans to develop African varieties with long-term goal of making drought-tolerant and royalty-free maize available to small scale African farmers (Pardey et al. 2003).

19.8 Concluding Remarks on IPR in Plant Science or Agriculture

There seems a general impression that Intellectual property rights may act as barriers to access innovations in agriculture or plant science research. On the contrary IPR in these areas seems to help accelerate economic and social development as well as reduce market risks for a company or an industry in making and selling of innovative products, processes, or genetically improved plant varieties. Problems of delayed or blocked access to needed research materials are often attributed to absence of adequate MTAs considering that there is variability of patent laws among different countries (Lei et al. 2009). While the primary focus of the private sector is on classical crash crops, because of their large market potential, the public sector such as universities, research institutions, and other non-profit organisations paradoxically use biotechnology or conventional procedures mainly for improvement of specialty crops which may have small market but are quite important from the perspective of developing countries. A belief that is attracting attention emphatically is based on public-private sector collaboration as this could directly address to IPR issues by including data sharing and patent pooling-an equivalent of an IP clearing house (Graff and Zilberman 2001). This approach is likely to promote sharing access with enabling technologies (e.g. gene delivery methods, marker-gene excision techniques, selectable markers, promoters, etc.) which have key role in the production of transgenic varieties, and provide FTO provision in the development of these new crops. Plant biologists have been advocating for open-source biology under a forum 'Biological Innovations for Open Society (BIOS)' with perception that an open access patent database could be used to collect IP data from several national patent offices in order to enable filter IP that could avoid overlapping of competing rights (Jefferson 2007). This would need worldwide harmonization of IP rights, identification of areas with common interests between the public and private sectors, and a substantial increase in R&D investments.

References

Abel, P.P., Powell, P., Nelson, R.S., De, B., Hoffmann, N., Rogers, S.G., Fraley, R,T. and Beachy, R.N. 1986. Delay of disease development in transgenic plants that express.the tobacco mosaic virus coat protein gene. Science 232: 738-743.

Abo El-Nil, M. 1987. Tissue culture of Douglas-fir and Western North American conifers. *In*: J.M. Bonga and D.J. Durzan (Eds.), Cell and Tissue Culture in Forestry, Vol. III. Martinus Nijhoff Publ., Dordrecht, pp. 80-100.

Adams, A.N. Barbara, D.J., Morton, A. and Darby, P. 1996. The experimental transmission of hop latent viroid and its elimination by low temperature treatment and meristem culture. Ann. Appl. Biol. 128: 37-44.

Adams, D.J., Beniston, L.J. and Childs, P.R.N. 2009. Promoting creativity and innovations in biotechnology. Trends Biotechnol. 27: 445-457.

Adeyemi, O., Aremu, O., Bairu, M.W., Delezal, K., Finnie, J.F. and Van Staden, J. 2012. Topolins: a panacea to plant tissue culture challenges? Plant Cell Tiss. Organ Cult. 108:1-16.

Afonso, C.L., Harkins, K.R., Thomas-Compton, M., Krejci, A.E. and Galbraith, D.W. 1985. Selection of somatic hybrid plants in *Nicotiana* through fluorescence-activated sorting of protoplasts. Biotechnology 3: 811-816.

Agrawal, A., Sanayaima, R., Tandon, R. and Tyagi, R.K. 2010. Cost-effective in vitro conservation of banana using alternatives of gelling agent (isabgol) and carbon source (market sugar). Acta Physiol. Plant. 32: 703-711.

Aharonowitz, V. and Cohen, G. 1981. The microbiological production of pharmaceuticals. Sci. Amer. 245: 106-119.

Ahkong, Q.F., Howell, J.I., Lucy, J.A., Safwat, F., Davey, M.R. and Cocking, E.C 1975. Fusion of hen erythrocytes with yeast protoplasts induced by polyethylene glycol. Nature (London) 255: 66-67.

Ahloowalia, B.S. and Sherington, J. 1985. Transmission of somaclonal variation in wheat. Euphytica 34: 525-537.

Aitken-Christie, J., Kozai, T. and Takayama, S. 1995. Automation in plant tissue culture—general introduction and Overview. *In*: Aitken-Christie et al. (Eds), Automation and Environmental Control in Plant Tissue Culture. Kluwer, Dordrecht, pp. 1-18.

Aitken-Christie, J., Singh, A.P., Morgan, K.J. and Thorpe, T.A. 1985. Explant developmental state and shoot formation in *Pinus radiata* cotyledons. Bot. Gazette 146: 196-203.

Akita, M. and Takayama, S. 1993. Effects of medium surface level control on the mass propagation of potato tubers using a jar fermentor culture technique. Plant Tissue Cult. Lett. 10: 242-248.

Al-Abta, S. and Collin, H.A. 1978. Cell differentiation in embryoids and plantlets of celery tissue cultures. New Phytol. 80: 517-521.

Al-Abta, S., Galpin, I.J. and Collin, H.A. 1979. Flavour compounds in tissue cultures of celery. Plant Sci. Lett. 16: 129-134.

Alfermann, A.W., Bergmann, W., Figur, C., Helmbold, U., Schwantag, D., Schuller, I. and Reinhard, E. 1983. Biotransformation of methyl digitoxin to methyl digoxin by cell cultures of *Digitalis lanata*. *In*: S.H. Mantell and H. Smith (Eds.), Plant Biotechnology. Cambridge Univ. Press, Cambridge, pp. 67-74.

Al-Mallah, M.K., Davey, M.R. and Cocking, E.C. 1989. Formation of nodular structures on rice seedlings by *Rhizobia*. J. Exp. Bot. 40: 473-478.

Altman, A. and Hasegawa, P.M. 2012. Plant Biotechnology and Agriculture: Prospects for the 21st Century. Elsevier, London, U.K., 586 pp.

Altman, A., Nadel, B.L., Falash, Z. and Levin, W. 1990. Somatic embryogenesis in celery: induction, control and changes in polyamines and proteins. *In*: N.J.J. Nijkamp et al. (Ed's.), Plant Cellular and Molecular Biology. Kluwer, Dordrecht, pp. 454-459.

Alvard, D., Cote,F. and Teisson,C. 1993. Comparison of methods of liquid medium culture for banana micropropagation. Plant Cell Tiss. Organ Cult. 32:55-60.

Amerson, H.V., Frampton, L.J., Mott, R.L. and Spaine, P.C. 1988. Tissue culture of conifers using loblolly pine as a model. *In*: J.W. Hanover and D.E. Keathley (Eds.), Genetic Manipulation of Woody Plants. Plenum Press, N.Y., pp. 117-137.

Amin, M.N. 1987. *In Vitro* clonal propagation of guava (*Psidium guajava* L.) and jackfruit (*Arthocarpus heterophyllus* Lamk.). Ph.D. Dissert ation, BanarasHindu Univ., Varanasi, India.

Amin, M.N. and Jaiswal, V.S. 1987. Rapid clonal propagation of guava through *In Vitro* shoot proliferation on nodal explants of mature trees. Plant Cell Tissue Organ. Cult. 9: 235-243.

Ammirato, P.V. 1983. Embryogenesis. *In*: D.A. Evans et al. (Eds.), Handbook of Plant Cell Culture, Vol. 1. Macmillan, N.Y., pp. 82-123.

Ammirato, P.V. 1984. Induction, maintenance, and manipulation of development in embryogenic cell suspension cultures. *In*: I.K. Vasil (Ed.), Cell Culture and Somatic Cell Genetics of Plants, Vol. 1: Laboratory Procedures and their Applications. Academic Press, Orlando, pp. 139-151.

Amoo, S.O., Finnie, J.F. and Van Staden, J. 2011. The role of metapolins in alleviating micropropagation problems. Plant Growth Regul. 63:197-206.

An, G., Watson, B.C., Stachel, S., Gordon, M.P. and Nester, E.W. 1985. New cloning vehicles for transformation of higher plants. EMBO J. 4: 277-284.

Anderson, W.C. 1980. Tissue culture propagation of red raspberries. *In*: Proceedings of Conference on Nursery Production of Fruit Plants Through Tissue Culture—Applications and Feasibility. Agric. Res. Sci. Educ. Admin., U.S.D.A., Beltsville, pp. 27-34.

Anthony, P. Jelodar, N.B., Lowe, K.C., Power, J.B. and Davey, M.R. 1996. Pluronic F-68 increases the post-thaw growth of cryopreserved plant cells. Cryobiology 33: 508-514.

Anthony, P., Davey, M.R., Azakanandam, Power, J.B. and Lowe, K.C. 2000. Cryopreservation of plant germplasms: new approaches to enhanced post-thaw recovery. *In*: M.K. Razdan and E.C. Cocking (Eds.) Conservation of Plant Genetic Resources *In Vitro*. Vol. 2: Applications and Limitations. Science Publishers Inc., Enfield, USA, pp. 21-37.

Arditti, J. 1977. Clonal propagation of orchids by means of tissue culture — a manual. *In*: J. Arditti (Ed.), Orchid Biology—Reviews and Perspectives. Cornell Univ. Press, U.S.A., pp. 203-293.

Asad, S., Mukhtar, Z., Nazir, F., Hashmi, J.A., Mansoor, S. and Zafar, Y. 2008. Silicon-carbide whisker-mediated embryogenic callus transformation of cotton (*Gossypium hirsutum* L.) and regeneration of salt tolerance plants. Mol. Biotech. 40:161-169.

Attree, S.M., Bekkaoui, F., Dunstan, D.I. and Fowke, L.C. 1987. Regeneration of somatic embryos from protoplasts isolated from an embryogenic suspension culture of white spruce (*Picea glauca*). Plant Cell Rep. 6: 480-483.

Aviv, D., Fluhr, R., Eddman, M. and Galun, E. 1980, Progeny analysis of the interspecific somatic hybrids. *Nicotiana tabacum* (CMS) + *N. sylvestris* with respect to nuclear and chloroplast markers. Theor. Appl. Genet. 56:145-150.

Badr, Y.A., Kereim, M.A., Yahia, M.A., Fouad, D.O. and Baheilden. A. 2005. Production of fertile transgenic wheat plants by laser micropuncture. Photochem. Photobiol. Sci. 4:803-807.

Baer, G.Y., Yemets, A.I. Stadnichuk, N.A., Rakhmetov, D.B. and Blume, Y.B. 2007. Somaclonal variability as a source for creation of new cultivars of finger millet (Eleusine *coracana*(l.) Gaertn.) Cytol. Genet. 41:204-208.

Bagga, S., Rajasekhar, V.K., Guha-Mukherjee, S. and Sopory, S.K. 1985. Enhancement in the formation of shoot initials by phytochrome in stem callus cultures of *Brassica oleraceae* var. *botrytis*. Plant Sci. 38: 61-64.

Bajaj, Y.P.S. 1987. Cryopreservation of plant cell cultures and its prospects in agricultural and forest biotechnology.*In*:S.Nateshetal.(Eds.),Biotechnology in Agriculture. Oxford & IBH Publishing Co. Pvt. Ltd., New Delhi, India, pp. 109-131.

Bajaj, Y.P.S. (Ed.). 1989a. Biotechnology in Agriculture and Forestry, Vol. 8: Plant Protoplasts and Genetic Engineering I. Springer-Verlag, Berlin, pp. 1-444.

Bajaj, Y.P.S. (Ed.). 1989b. Biotechnology in Agriculture and Forestry, Vol. 9: Plant Protoplasts and Genetic Engineering II. Springer-Verlag, Berlin, pp. 1-499.

Bajaj, Y.P.S. (Ed.). 1990a. Biotechnology in Agriculture and Forestry, Vol. 10: Legumes and Oilseed Crops I. Springer-Verlag, Berlin, pp. 1-682.

Bajaj, Y.P.S. (Ed.). 1990b. Biotechnology in Agriculture and Forestry, Vol. 11: Somaclonal Variation in Crop Improvement I. Springer-Verlag, Berlin, pp. 1-685.

Bajaj, Y.P.S. (Ed.). 1995. Biotechnology in Agriculture and Forestry 32: Cryopreservation of Plant Germplasm 1. Springer, Berlin, 512 pp.

Baker, R. and Kinnaman, H. 1973. Elimination of pathogens from shoot-tip cultures. *In*: P.F. Kruse, Jr. and M.K. Patterson (Eds.), Tissue Culture: Methods and Applications. Academic Press, N.Y., pp. 735-739.

Baker, R. and Phillips, D.J. 1962. Obtaining pathogen-free stock by shoot-tip culture. Phytopathology 52:1242-1244.

Ball, E. 1946. Development in sterile culture of stem tips and adjacent regions of *Tropaeolum majus* L. and of *Lupinus albus* L. Am. J. Bot. 33: 301-318.

Ball, E.A. 1950. Differentiation in a callus culture of *Sequoia sempervirens*. Growth 14: 295-325.

Ball, E.A. 1987. Tissue Culture multiplication of *Sequoia*. *In*: J.M. Bonga and D.J. Durzan (Eds.), Cell and Tissue Culture in Forestry, Vol. III. Martinus Nijhoff Publ., Dordrecht, pp. 146-157.

Ball, S.G. 1990. Molecular basis of somaclonal variation. *In*: Y.P.S. Bajaj (Ed.), 1990b. Biotechnology in Agriculture and Forestry, Vol. 11: Somaclonal Variation in Crop Improvement I. Springer-Verlag, Berlin, pp. 134-152.

Barbara, D.J. and Clark, M.F. 1986. Immunoassays in plant pathology. *In*: T.L. Wang (Ed.), Immunology in Plant Science. Cambrige University Press, Cambridge, pp. 197-219.

Barbier, M. and Dulieu, H.L. 1980. Genetic changes observed on tobacco (*Nicotiana tabacum*) plants regenerated from cotyledons by *In Vitro* culture. Ann. Amelior. Plant. 30: 321-344.

Barlass, M. and Skene, K.G.M. 1986. Citrus (*Citrus* species). *In*: Y.P.S. Bajaj (Ed.), Biotechnology in Agriculture and Forestry; Vol. I: Trees I. Springer-Verlag, Berlin, pp. 207-219.

Barton, K.A., Binns, A.N., Matzke, A.J.M. and Chilton, D.D. 1983. Regeneration of intact tobacco plants containing full length copies of genetically engineered T-DNA, and transmission of T-DNA to Ri progeny. Cell 32: 1033-1043.

Barwale, U.B., Kerns, H.R. and Widholm, J.M. 1986, Plant regeneration from callus cultures of several soybean genotypes via embryogenesis and organogenesis. Planta 167: 473-481.

Barz, W. and Ellis, B.E. 1980. Plant cell culture and their biotechnological potential. Ber. Deutsch. Bot. Ges. 94: 1-26.

Basile, D.V., Wood, H.N. and Braun, A.C. 1973, Programming of cells for death under defined experimental conditions: Relevance to the tumor problem. Proc. Natl. Acad. Sci., U.S.A. 70: 3055-3059.

Bates, G.W. 1989. Electrofusion: The technique and its application to somatic hybridization. *In*: Y.P.S, Bajaj (Ed.), Biotechnology and Forestry, Vol. 9: Protoplasts and Genetic Engineering II. Springer-Verlag, Berlin, pp. 241-256.

Batra, V., Prakash, S. and Shivanna, K.R. 1990. Intergeneric hybridization between *Diplotaxis scifolia*, a wild species and crop brassicas. Theor, Appl. Genet. 80: 537-541.

Bayliss, M.W. 1980. Chromosome variation in plant tissues in culture. *In*: I.K. Vasil (Ed.), Perspectives in Plant Cell and Tissue Culture, Int. Rev. Cytol, Suppl., 11B. Academic Press, N.Y., pp. 113-144.

Bechtold, N., Ellis, J. and Pelletier, G. 1993. In planta Agrobacterium-medicated gene transfer by infiltration of adult *Arabidopsis thaliana* plants. C.R. Acad. Sci., Paris, Life Science 316: 1194-1199.

Belliard, G., Pelletier, G., Vedel, F. and Quetier, F. 1978. Morphological characteristics and chloro-plast DNA distribution in different cytoplasmic parasexual hybrids of *Nicotiana tabacum*. Mol. Gen. Genet. 165: 231-237.

Belliard, G., Vedel, F. and Pelletier, G. 1979. Mito-chondrial recombination in cytoplasmic hybrids of *Nicotiana tabacum* by protoplast fusion. Nature (London) 281: 401-403.

Bentwich, M. 2010. Changing rules of the game: Addressing the conflict between free access to scientific discovery and intellectual property rights. Nature Biotechnology 28: 137-140.

Bera, R.K. 2009. Harmonisation of patent laws. Curr. Sci. 96: 457-458.

Bergmann, L. 1960. Growth and division of single cells of higher plants in *In Vitro*. J. Gen. Physiol. 43: 841-851.

*Bergner, A.D. (1921). Cited by Han, H. 1987.

Berlin, J., Witte, L., Hammer, J., Kuboschke, K.G., Zimmer, A. and Pape, D. 1982. Metabolism of p-fluorophenyla-lanine in p-fluorophenylalanine sensitive and resistant cell cultures. Planta 155: 244-250.

Berman, E.P. 2008. Why did universities start patenting? Institution-Building and the road to the Bayh-Dole Act. Social Studies Sci. 38: 835-871.

Bevan, M. 1984. Binary *Agrobacterium* vectors for plant transformation. Nucl. Acid Res. 12: 8711-8721.

Bhaskaran, S. 1987. Somaclonal variation, *In*: S. Natesh et al. (Eds.), Biotechnology in Agriculture. Oxford & IBH Publishing Co. Pvt. Ltd., New Delhi, pp. 101-108.

Bhaskaran, S. and Prabhudesai, V.R. 1989. Tissue culture in plantation crops. *In*: V. Dhawan (Ed.), Applications of Biotechnology in Forestry and Horticulture. Plenum Press, N.Y., pp. 87-118.

Bhat. J.G. and Murthy, N.N. 2007. Factors affecting in vitro gynogenic haploid production in niger (Guizotia abyssimica (L.A.) Cass. Plant Growth Regul. 52:241-248.

Bhojwani, S.S. 1990. Plant Tissue Culture: Applications and Limitations, Elsevier, Amsterdam, pp. 1-461.

Bhojwani, S.S. and Bhatnagar, S.P. 1990. Embryology of Angiosperms. Vikas Publishing Co., New Delhi, India, pp. 1-370.

Bhojwani, S.S. and Dantu, P.K. 2013. Plant Tissue Culture: An Introductory Text. Springer, India, 309 pp.

Bhojwani, S.S. and Johri, B.M. 1971. Morphogenetic studies on cultured mature endosperm of *Croton bonplandianwn*. New Phytol. 70: 761-766.

Bhojwani, S.S. and Razdan, M.K. 1983. Plant Tissue Culture: Theory and Practice. Elsevier Science Publishers, Amsterdam, pp. 1-502.

Bhojwani, S.S. and Razdan, M.K. 1996. Plant Tissue culture Theory and Practice, a Revised Edition. Elsevier, Amsterdam, 467 pp.

Bhojwani, S.S. and Soh, W.Y. (Eds.). 2001 Current Trends in Embryology of Angiosperms. Elsevier, Amsterdam, 521 pp.

Bhojwani, S.S. and Thomas, D.T. 2001. *In Vitro* gynogenesis. *In*: S.S. Bhojwani and W.Y. Soh (Eds.), Current Trends in Embryology of Angiosperms. Elsevier, Amsterdam, pp. 489-508.

Bürstrom, H. 1953. Physiology of root growth. Ann. Rev. Plant Physiol. 4: 237-252.

Bingham, E.T., Hurley, L.V., Kaatz, D.M and Saunders, J.N. 1975. Breeding alfalfa which regenerates from callus tissue in culture. Crop Science 15: 719-721.

Biotechnology in Agriculture and Forestry 32: Cryopreservation of Plant Germplasm I. Springer, Berlin, pp. 321-331.

Bisaria, V. and Panda, A. 1991. Large-scale plant cell culture: methods, applications and products. Current Opinion in Biotechnology 2: 370-374.

Blitters, W.P., Murashige, T., Rangan, T.S. and Nauer, E. 1970. Investigations on established virus-free plants through tissue culture. Calif. Citrus Nursery Soc. 9: 27-30.

Boll, W.G. and Street, H.E. 1951. Studies on the growth of excised roots. I. The stimulatory effect of molybdenum on the growth of excised tomato roots. New Phytol. 50: 50-75.

Bolwell, G.P. and Robertson, D. 1997. Cited from Fukuda 1997.

Bonner, J. 1937. Vitamin B_1, a growth factor for higher plants. Science 85: 183-184.

Bonnett, H.T. and Torrey, J.G. 1966. Comparative anatomy of endogenous bud and lateral root formation in *Convolvulus arvensis* roots cultured *In Vitro*. Am J. Bot. 53: 496-507.

Bornman, C.H. 1993, Maturation of somatic embryos. *In*: K. Redenbaugh (Ed.), Synseeds: Applications of Synthetic Seeds to Crop Improvement. CRC Press, Boca Raton, Fl., pp. 105-113.

Borotto-Fernandez, E.G., Sommerbauer, T., Popowich, E., Scharl, A. and Laimer, F. 2009. Somatic embryogenesis from anthers of the autochthonous *Vitis vinifera* cv. Domina leads to Arabis mosaic virus-free plants. Eur.J.Plant Pathol. 124:171-174.

Boudet 1995 (cited from Fukuda 1997).

Bourgin, J.P. and Nitsch, J.P. 1967. Obtention de Nicotiana haploides à partir d'étamines cultivées *In Vitro*. Ann. Physiol. Veg. 9: 377-382.

Boxus, P. 1974. The production of strawberry plants by *In Vitro* micropropagation. J. Hortic. Sci. 49: 209-210.

Boxus, P. 1999. Micropropagation of strawberry via axillary shoot proliferation. *In*: R.D. Hall (Ed.). Plant Cell Culture Protocols, Humana Press, Inc., Totowa, N.J., pp. 103-125.

Boxus, P., Quorin, M. and Laine, J.M. 1977. Large scale propagation of strawberry plants from tissue culture. *In*: J. Reinert and Y.P.S. Bajaj (Eds.), Applied and Fundamental Aspects of Plant Cell, Tissue and Organ Culture. Springer-Verlag, Berlin, pp. 130-143.

Brand, M.H. and Lineberger, R.D. 1992. Micropropagation of American sweetgum (*Liquidambar styraciflua* L.). *In*: Y.P.S. Bajaj (Ed.), Biotechnology in Agriculture and Forestry, Vol. 18. High -Tech and Micropropagation II. Springer, Berlin, pp. 3-24.

Brar, D.S. and Jain, S.M. 1998. Somaclonal variation: Mechanism and application in crop improvement. *In*: S.M. Jain et al. (Eds.), Somaclonal Variation and Induced Mutations in Crop Improvement. Kluwer Academic Publishers, Dordrecht, pp. 15-37.

Braun, A.C. 1959. A demonstration of the recovery of the crown-gall rumor cell with the use of complex tumors of single-cell origins. Proc. Natl. Acad. Sci., U.S.A. 45: 932-938.

Braun, A.C. and Wood, H.N. 1976. Suppression of the neoplastic state with the acquisition of specialized functions in cells, tissues and organs of crown-gall teratomas of tobacco. Proc. Natl. Acad. Sci., U.S.A. 73: 496-500.

*Bravo, J.E. and Evans, D.A. 1985. Somaclonal variation in *Nicotiana alata*. Cited by Evans, D.A. and Sharp, W.R. 1986a, p. 115.

Brettell, R.I.S., Dennis, E.S., Scowcroft, W.R. and Percock, J.W. 1986. Molecular analysis of a somaclonal mutant of maize alcohol dehy-drogenase. Mol. Gen. Genet. 202: 235-239.

Bridgen, M.P. and Staby, G.L. 1983. Protocols of low-pressure storage. *In*: D.A. Evans et al. (Eds.), Handbook of Plant Cell Culture: Vol. 1: Techniques for Propagation and Breeding. Macmillan Publishing Co., N.Y., pp. 816-827.

Bright, S.W.J., Miflin, B.J. and Rognes, S.E. 1982. Threonine accumulation in the seeds of a barley mutant with an altered aspartate kinase. Biochem. Genet. 20: 229-243.

Bright, S.WJ, Kneh, J.S.H. and Rognes, S.E. 1983. Lysine transport in two barley mutants with altered uptake to basic amino acids in the roots. Plant Physiol. 72: 821-824.

Brison, M., Paulus, V., De Boucaud, T. and Dospa, F. 1992. Cryopreservation of walnut and plum shoot tips. Cryobiology 29: 738.

Brisson, N., Paszkowski, J., Penswick, J.R., Gronenborn, B., Potrykus, I. and Hohn, T. 1984. Expression of a bacterial gene in plants using a viral vector. Nature (London) 310: 511-514.

Brodelius, P. 1984. Immobilization of cultured plant cells and protoplasts. *In*: I.K. Vasil (Ed.), Cell Culture and Somatic Cell Genetics of Plants, Vol. 1: Laboratory Procedures and Their Applications. Academic Press, Orlando, pp. 535-546.

Brodelius, P. 1986. Immobilized plant cells and protoplasts. *In*: D.A. Evans et al. (Eds.), Handbook of Plant Cell Culture, Vol. 4: Techniques and Applications. Macmillan Publishing Co., N.Y., pp. 287-315.

Brodelius, P. and Nilsson, K. 1983. Permeabilization of immobilized plant cells, resulting in release of intracellulary stored products with preserved cell viability. Eur. J. Appl. Microbiol. Biotechnol. 17: 275-280.

Brodelius, P., Deus, B., Mosbach, K. and Zenk, M.H. 1979. Immobilized plant cells for the production and transformation of natural products. FEBS Lett. 103: 93-97.

Brodersen, P. and Vionett, O. 2006. The diversity of RNA silencing pathways in plants. Trends Genet. 22: 268-280.

Brown, P.T.H. and Lörz, H. 1986. Molecular changes and possible origins of somaclonal variation. *In*: F. Semal (Ed.), Somaclonal Variations and Crop Improvement. Martinus Nijhoff, Publ., Dodrecht/Lancaster, pp. 148-156.

Bürstrom, H. 1953. Physiology of root growth. Ann. Rev. Plant Physiol. 4: 237-252.

Butcher, D.N. 1977. Secondary products in plant tissue culture. *In*: J. Reinert and Y.P.S. Bajaj (Eds.), Plant Cell, Tissue and Organ Cultures. Springer-Verlag, Berlin, pp. 668-693.

Buvat, R. 1944. Recherches sur la dédifferenciation des cellulales végétales. Ann. Sci. Nat. Bot. Biol. Veg. 5: 1-130.

Buvat, R. 1945. Recherches sur la dédifferenciation des cellules végétales. Ann. Sci. Nat. Bot. Bioi. Veg. 6: 1-119.

Byun, S.Y., Pederson, H. and Chen, C.K. 1990. Two-phase culture for the enhanced production of benzopheran thridine alkaloids in cell suspension of *Eschscholtzia california*. Phytochemistry 29: 3129-3135.

Caboche, M. and Lurquin, P.P. 1987. Liposomes as carriers for the transfer and expression of nucleic acids into higher plant protoplasts. *In*: L. Packer and R. Douce (Eds.), Methods in Enzymology. Academic Press, New York, 148: 39-45.

Cai, Z., Kastell, A. and Knorr, D. 2012. Exudation: an expanding technique for continuous production and release of secondary metabolites from plant cell suspension and hairy root cultures. Plant Cell Rep. 31: 461-477.

Cao, Z. 1990. Grape: micropropagation. *In*: Z. Chen et al. (Eds.), Handbook of Plant Cell Culture, Vol. 6: Perennial Crops. McGraw-Hill Publishing Co., N.Y., pp. 312-328.

Caplin, S.M. 1959. Mineral-oil overlay for conservation of plant tissue cultures. Am. J. Bot. 46:324-329.

Caplin, S.M. and Steward, F.C. 1948. Effect of coconut milk on the growth of explants from carrot root. Science, 108:655-657.

Carlson, P.S. 1970. Induction and isolation of auxotrophic mutants in somatic cell cultures of *Nicotiana tabacum*. Science, N.Y., 168: 487-489.

Carlson, P.S. 1973. Methionine sulfoximine-resistant mutants of tobacco. Science 180: 1366-1368.

Carlson, P.S., Smith, H.H. and Dearing, R.D. 1972. Parasexual interspecific plant hybridization. Proc. Natl. Acad. Sci., U.S.A. 69: 2292-2294.

Cassells, A.C. 1989. Uptake of viruses by plant protoplasts and their use as transforming agents. *In*: Y.P.S. Bajaj (Ed.), 1989b, Biotechnology in Agriculture and Forestry, Vol. 9: Plant Protoplasts and Genetic Engineering II. Springer-Verlag, Berlin, pp. 388-405.

Cassells, A.C., Minas, G. and Long, R. 1980. Culture of pelargonium hybrids from meristems and explants: Chimeral and beneficially infected varieties. *In*: D.S. Ingram and J.P.S. Helgeson (Eds.), Tissue Culture Methods for Plant Pathologists. Blackwell Scientific Publishers, Oxford, U.K., pp. 125-130.

Castelli, S., Manzocchi, L. A. and Torti, G. 1988. Callus induction in endosperm from maize mutants with different IAA content. *In*: M.S.S. Pais et al. (Eds.), Plant Cell Biotechnology. Springer-Verfag, Berlin, pp. 63-67.

Cazzulino, D.L., Pedersen, H., Chin, C.K. and Styer, D. 1990. Kinetics of carrot somatic embryo development in suspension culture. Biotechnology and Bioengineering 35: 781-786.

Chabaud, M., Passiatore, J.E., Cannon, F. and Buchanon-Wollaston, V. 1988. Parameters affecting the frequency of kanamycin resistant alfalfa obtained by *Agrobacterium tumefaciens* mediated transformation. Plant. Cell Rep. 7: 512-516.

Chaleff, R.S. 1986. Selection for herbicide-resistant mutants. *In*: D. Evans et al. (Eds.), Handbook of Plant Cell Culture, Vol. 4: Techniques and Applications. Macmillan Publishing Company, N.Y., pp. 133-149.

Chaleff, R.S. and Carlson, P.S. 1975. *In Vitro* selection for mutants of higher plants. *In*: L. Ledoux (Ed.), Genetic Manipulation and Plant Material. Plenum Press, N.Y., pp. 351-363.

Chatterjee, G., Sikdar, S.R., Das, S. and Sen, S.K. 1988. Intergeneric somatic hybrid production through protoplast fusion between *Brassica juncea* and *Diplotaxis muralis*. Theor. Appl. Genet. 76:915-922.

Chaturvedi, R. 2001. Morphogenesis in Haploid, Diploid and Triploid Tissue Cultures of *Azadirachta indica* A. Juss. Ph.D Thesis, University of Delhi, India, 133 pp.

Chaturvedi, R., Razdan, M.K. and Bhojwani, S.S. 2003a. Production of haploids of neem (*Azadirachta indica* A.Juss.) by anther culture. Plant Cell Rep. 21:531-537.

Chaturvedi, R., Razdan, M.K. and Bhojwani, S.S. 2003b. An efficient protocol for the production of triploid plants from endosperm callus of neem, *Azadirachta indica* A. Juss. J. Plant Physiol. 160: 557-564.

Chaturvedi, R., Razdan, M.K. and Bhojwani, S.S. 2004a. *In vitro* morphogenesis in zygotic embryo cultures of neem (*Azadirachta indica* A. Juss.). Plant Cell Rep. 22:801-809.

Chaturvedi, R., Razdan, M.K. and Bhojwani, S.S. 2004b. *In vitro* clonal propagation of an adult tree of neem (Azadirachta indica A. Juss.) by forced axillary branching. Plant Sci. 166:501-506.

Chauhan, H. and Khurana, P. 2011. Use of doubled haploid technology for development of stable drought tolerant bread wheat (*Triticum aestivum*) transgenics. Plant Biotechnol. J. 9: 408-411.

Chen, K. and Gao, C. 2013. TALENs: customizable molecular DNA scissors for genome engineering of plants. J. Genet. Genomics 40: 271-279.

Chen, T.H.H., Kartha, K.K., Leung, N.L., Kurz, W.G.W., Chatson, K.B. and Constabel, F. 1984. Cryopreservation of alkaloid producing cell cultures of periwinkle (*Catharanthus roseus*). Plant Physiol. 75: 726-731.

Chen, Z. 1990. Haploid induction in perennial crops. *In*: Z. Chen et al. (Eds.), Handbook of Plant Cell Culture, Vol. 6: Perennial Crops. McGraw-Hill Publishing Company, N.Y., pp. 62-75.

Chen, Z. and Evans, D.A. 1990. General techniques of tissue cultures in perennial crops. *In*: Z, Chen et al. (Eds.), Handbook of Plant Cell Culture, Vol. 6: Perennial Crops. McGraw-Hill Publishing Company, N.Y., pp. 22-61.

Chi-Ham, C.L., Clark, K.L. and Bennet, A. 2010. The intellectual property landscape for gene suppression technologies in plants. Nature Biotechnology. 28: 32-36.

Cho, M.S. and Zapata, F.J. 1990. Plant regeneration from microspores of indica rice. Plant Cell Physiol. 31: 881-885.

Chourey, P.S. and Kemble, R.J. 1982. Transposition event in tissue cultured cells of S-cms genotype of maize. Maize Genet. Coop. Newsl. 56: 70.

Christensen, A.H., and Quail, P.H. 1996. Ubiquitin promoter-based vectors for high level expression of expression of selectable and/ or screenable marker genes in monocotyledonous plants. Transgenic Res. %:213-218.

Christou, P., Swain, W.F., Yang, N.S. and Mc Cabe, D.E. 1989. Inheritance and expression of foreign genes in transgenic soybean plants. Proc. Natl. Acad. Sci., USA. 66: 7500-7504.

Chu, C.C. 1978. The N_6 medium and its applications to anther culture of cereal crops. *In*: Proceedings of Symposium on Plant Tissue Culture. Science Press, Peking, pp. 43-50.

Chu, I.Y.E. 1986. The application of tissue culture of plant improvement and propagation in the ornamental horticultural industry. *In*: R.H. Zimmerman et al. (Eds.), Tissue Culture as a Plant Production System for Horticultural Crops. Martinus Nijhoff, Dordrecht, The Netherlands, pp. 15-33.

Chu, I.Y.E. and Kurtz, S.L. 1990. Commercialization of plant micropropagation. *In*: P.V. Ammirato et al. (Eds.), Handbook of Plant Cell Culture, Vol. 5. Ornamental Species. McGraw-Hill Publishing Company, N.Y., pp. 126-164.

Chuang, C, Ouyang, T., China, H., Chou, S. and Ching, C. 1978. A set of potato media of wheat anther culture. *In*: Proceedings of Symposium on Plant Tissue Culture. Science Press, Peking, pp. 51-56.

Chugh, A., Amundsen, E. and Eudes, F. 2009. Translocation of cell-penetrating peptides and delivery of their cargoes in triticale microspores. Plant cell Rep. 28: 801-810.

Church, D.L. 1993. Tracheary element differentiation in *Zinnia* mesophyll cultures. Plant Growth Regl. 12: 179-188.

Clapham, D. 1973. Haploid *Hordeum* plants from anthers *in vitro*. Z. pflanzenzucht. 69: 142-145.

Clough, S.J. and Bent, A.F. 1998. Floral dip: a simplified method for Agrobacterium-mediated transformation of *Arabidopsis thaliana*. Plant J. 16: 735-43.

Cocking, E.C. 1979. Parasexual reproduction in flowering plants. New Zealand J. Bot. 17: 665-671.

Cocking, E.C. and Davey, M.R. 1987. Gene transfer in cereals. Science 236: 1259-1262.

Cocking, E.C. and Peberdy, J.F. (Eds.). 1974. The Use of Protoplasts from Fungi and Higher Plants as Genetic Systems. Dept. Botany, Univ. Nottingham, UK.

Cocking, E.C. and Pojnar, E. 1969. An electron microscopy study of the infection of isolated tomato fruit protoplasts by tobacco mosaic virus. J. Gen. Virol. 4: 305-312.

Cocking, E.C. 1960. A method for the isolation of plant protoplasts and vacuoles. Nature (London) 187: 927-929.

Collins, H.A. and Edwards, S. 1998. Plant Cell Culture. Bios Scientific, Oxford, UK, 158 pp.

Collin, H.A. and Watts, M. 1983. Flavor production in culture. *In*: D.A. Evans et al. (Eds.), Handbook of Plant Cell Culture, Vol. 1: Techniques for Propagation and Breeding. Macmillan Publishing Co. N.Y., pp. 729-747.

Collins, G.B. and Grosser, J.W. 1984. Culture of Embryos. *In*: I.K. Vasil (Ed.), Cell Culture and Somatic Cell Genetics of Plants, Vol. 1: Laboratory Proceedings and their Applications. Academic Press, Orlando, pp. 241-257.

Comai, S. and Henikoff, S. 2006. TILLING: practical single nucleotide discovery. Plant. J. 45:684-694.

Conia, J., Bergounioux, C., Perennes, C., Brown, S., Müller, Ph. and Gadal, P. 1987. Chromosomes from protoplasts of *Petunia hybrida*, flow-analysis and sorting. Bio Cell 58 (Suppl.): 5a.

Constabel, F. and Vasil, I.K. (Eds.). 1987. Cell Culture and Somatic Cell Genetics of Plants, Vol. 4: Cell Culture in Phytochemistry. Academic Press, Inc., San Diego, pp. 1-314.

Constabel, F., Kurz, W.G.W. and Kutney, P. 1982. Variation in cell cultures of periwinkle *Catharanthus roseus*. *In*: A. Fujiwara (Ed.), Plant Tissue Culture. Maruzen, Tokyo, pp. 301-304.

Cordewener, J.H.G., Busink, R., Traas, J.A., Custer, J.B.M., Dons, J.J.M and Van Lookeren Campagne, M.M. 1994. Induction of microspore embryo-genesis in *Brassica napus* is accompanied by specific changes in protein synthesis. Planta 195: 50-56.

Corredoira, E., Ballester, A., Ibarra, M. and Vieitiz, A.M. 2015a. Induction of somatic embryogenesis in explants of shoot cultures established from adult *Eucalyptus globulus* and *E.stigma* X *E. maidenii* trees. Tree Physiol. 35: 678-690.

Corredoira, E., Valladeres, S., Vieitiz, A.M. and Ballester, A. 2015b. Chestnut, European (*Castanea sativa*). Methods Mol. Biol. 1224: 163-176.

Cowen, N.M., Johnson, C.D., Armstrong, K., Miller, M., Woosley, A., Pescitelli, S., Skakut, M., Belmar, S. and Petolino, J.F. 1992. Mapping gene conditioning In Vitro androgenesis in maize using RFLP analysis. Theor, Appl. Genet. 84: 720-724.

Crossway, A., Hauptli, H., Houck, C.M., Irvine, J.M., OaV. es, J.V. and Perani, L.A. 1986. Micromani-pulation techniques in plant biotechnology. Biotechniques 4: 320-334.

Croughan, S.S., Quisenbery, S.S., Eichchorn, M.M., Gloyer, P.D. and Brown, T.F. 1994. Registration of Brazos-R3 bermudagrass germplasm. Crop Sci. 34: 542.

Cseplo, A., Medgyesy, P. and Marton, L. 1986. *In Vitro* induction, isolation and transfer of chloroplast mutations in *Nicotiana*. Proc. Nuclear Techniques and *In Vitro* Culture for Plant Improvement. IAEA, Vienna, pp. 137-146.

Cuming, A.C., Tuvet, M. and Butler, W. 1996. Gene expression and embryonic maturation in cereals. *In*: T.C. Wang and A. Cuming (Eds.), Embryo-genesis the Generation of a Plant-Environment Plant Biology Series. BIOS Scientific Publishers Limited, Oxford, pp. 113-132.

Curtis, W.R. 1993. Cultivation of roots in bioreactors. Curr. Opinion Biotechnol. 4: 205-210.

Cutter, E.G. 1965. Recent experimental studies of the shoot apex and shoot morphogenesis. Bot. Rev. 31: 7-113.

D'Amato, F. 1977. Cytogenetics of differentiation in tissue and cell cultures. *In*: J. Reinert and Y.P.S. Bajaj (Eds.), Applied and Fundamental Aspects of Plant Cell, Tissue and Organ Culture. Springer-Verlag, Berlin, pp. 343-357.

D'Hallulin, K., Bonne, E., Bossut, M. and Page, R.L. 1999. Transformation of maize via tissue electro-poration. *In*: D. Hall (Ed.), Methods in Molecular Biology, Vol. III. Plant Cell Culture Protocols. Humana Press Inc., Totowa, N.J., pp. 367-374.

Daghma, D.E.S., Hansel, G., Rutten, T., Meltzer, M. and Kumlehn, J. 2014. Cellular dynamics during early barley pollen embryogenesis revealed by time-lapse imaging front. Plant Sci. 5:675.

Darbani, B., Eimanifar, A., Stewart, C.N. Jr. and Camargo, W. 2007. Methods to produce marker-free transgenic plants. Biotechnol. J. 2:83-90.

Das, S.C., Barman, T.S. and Singh, R. 1990. Plant regeneration and establishment in nursery. Assam Rev. Tea News 79: 24-27.

Datta, S.K. and Wenzel, G. 1987. Isolated microscope derived plant formation via embryogenesis in *Trithanus aestivum* L. Plant Sci. 48: 49-54.

Davey, M.R., Cocking, E.C., Freeman, J., Pearce, N. and Tudor, I. 1980. Transformation of *Petunia* protoplasts by isolated *Agrobacterium* plasmids. Plant Sci. Lett. 18: 307-313.

Davey, M.R., Frearson, E.M. and Power, J.B. 1976. Polyethylene-induced transplantation of chloro-plasts into protoplasts: an ultrastructural assessment. Plant Sci. Lett. 7: 7-16.

De Buyser, J., Henry, Y., Lonnet, P., Hertzog, R. and Hespel, A. 1987. Florin: a doubled haploid wheat variety developed by the anther culture method. Plant Breed. 98: 53-56.

De Fossard, R.A., Myint, A. and Lee, E.C.M. 1974. A broad spectrum tissue culture experiment with tobacco. (*Nicotiana tabacum*) pith tissue callus. Physiol. Plant. 30: 125-130.

De Guzman, E.V., del Rosario, A.G. and Eusebio, E.C. 1971. The growth and development of coconut "makapuno" embryos *In Vitro*. III. Resumption of root growth in high sugar media. Philipp. Agr. 55: 566-579.

De Jong, A.J., Schmidt, E.D.L and De Vries, S.C. 1993. Early events in higher plant embryogenesis. Plant Mol. Biol. 22: 367-377.

De La Peña, A., Lörz, H. and Schell J. 1987. Transgenic rye plants obtained by injecting DNA into young floral tillers. Nature 325: 274-276.

De Laat, A.M.M. and Blaas, J. 1984. Flow cytomerric characterization and sorting of plant chromosomes. Theor. Appl. Genet. 67: 463-167.

De Laat, A.M.M. and Blaas, J. 1987. An improved method for protoplast microinjection suitable for transfer of entire plant chromosomes. Plant Sci. 50: 161-169.

De Paepe, R., Prat, D. and Knight J. 1983. Effects of consecutive androgeneses on morphology and fertility in *Nicotiana sylvestris*. Can. J. Bot. 61: 2038-2046.

De Ropp, R.S. 1955. The growth and behaviour *In Vitro* of isolated plant cells. Proc. Roy. Soc. Lond. B. 144: 86-93.

De Vries, S.C., Booj, H., Janssens, R., Vogels, R., Saris, L., Lo Schiavo, F., Terzi, M. and van Kammeu, A. 1988. Carrot somatic embryogenesis depends on the phytohormone-controlled presence of correctly glycosylated extracellular proteins. Genes Dev. 2: 462-476.

De Wet, J.R., Wood, K.V., Helinski, D.R. and Deluca, M. 1985. Cloning of firefly luciferase cDNA and the expression of active luciferase in *Escherichia coli*. Proc. Natl. Acad. Sci., USA 82: 7870-7873.

Deak, M., Kiss, G.B., Konez, C. and Dudits, G. 1986. Transformation of *Medicago* by *Agrobacterium* mediated gene transfer. Plant Cell Rep. 5: 97-100.

Deambrogio, E. and Dale, P.I. 1980. Effect on 2,4-D on the frequency of regenerated plants in barley (*Hordeum vulgare*) cultivar 'AKKa' and on genetic variability between them. Cereal Res. Commun. 8: 417-424.

Debergh, P.C. and Maene, L.J. 1981. A scheme for commercial propagation of ornamental plants by tissue culture. Sci. Hortic. 14: 335-345.

Deblaere, R., Reynaerts, A., Hofte, H., Hernalsteens, J.P., Leemans, J. and Van Montagu, M. 1987. Vectors for cloning in plant cells. Methods Enzymol. 153: 277-292.

Dejong, A.J., Cordewener, J., Schiavo, F.L., Terzi, M., Vandekerckhove, J., Van Kammen, A. and De Vries,

C.C. 1992. A carrot somatic embryo mutant is rescued by chitinase. Plant Cell 4: 425-433.

Demura, T. and Fukuda, H. 1994. Novel vascular cell-specific genes whose expression is regulated temporally and spatially during vascular system development. Plant Cell. 6: 967-981.

Dereuddre, J., Hassen, N., Blandin, S. and Kaminski, M. 1991. Resistance of alginate-coated somatic embryos of carrot (*Daucus carota* L.) to desiccation and freezing in liquid nitrogen 2. Thermal analysis. Cryo-Lett. 12s 135-148.

Deshayes, A., Herrera-Estrella, L. and Caboche, M. 1985. Liposome mediated transformation of tabacco mesophyll protoplasts by and *Escherichia coli*. EMBO J. 4: 2731-2735.

Dhaliwal, S. and King, P.J. 1978. Direct pollination of *Zea mays* ovules *In Vitro* with Z. *mays*, *Z. mexicana* and *Sorghum bicolor* pollen. Theor. Appl. Genet. 53: 43-46.

Dhillon, S.S., Wernsman, E.A. and Miksche, J.P. 1983. Evaluation of nuclear DNA content and heterochromatin changes in anther-derived dihaploids of tobacco (*Nicotiana tabacum* W. Coker 139). Can. J. Genet. Cytol. 28: 169-173.

Dietrich, K. 1924. Über kultur von Embryonen ausserhalb des Sameus. Flora (Jena) 117: 379-417.

Diettrich, B., Popov, A.S., Pfeiffer, B., Neumann, D., Butenko, R. and Luckner, M. 1982. Cryopreservation of *Digitalis lanata* cell cultures. Planta Med. 46: 82-87.

Dijak, M., Smith, D.L., Wilson, T.J. and Brown, C.W. 1986. Stimulation of direct embryogenesis from mesophyll protoplasts of *Medicago sativa* L. Plant Cell Rep. 5: 468-470.

Dix, P.J. 1977. Chilling resistance is not transmitted sexually in plants regenerated from *Nicotiana sylvestris* cell lines. Z. Pflanzenphysiol. 84:223-226.

Dixon, R.A. 1985. Plant Cell Culture — A Practical Approach. IRL Press Ltd., Oxford, pp. 1-236.

Dixon, R.A. and Paiva, N,L. 1995. Stress-induced phenylpropanoid metabolism. Plant Cell 7: 1085-1097.

Dixon, R.A., Howles, P.J., Lamb, C, Korth, K., He, X.H., Sewalt, V.H. and Rasmussen, S. 1999. Plant secondary metabolism: control points and prospects for genetic manipulation of phenylpropanoid biosynthesis. *In*: T.J. Fu, et al. (Eds.), Plant Cell and Tissue Culture for the Production of Food Ingredients. Kluwer, Dordrecht, pp. 7-22.

Doi, H., Hoshi, N., Yamada, E., Yokoi, S., Nishihara, M., Hikage, T. and Takhata, Y. 2013. Efficient haploid and doubled haploid production from unpollinated ovule culture of gentianas (*Gentiana* spp.). Breed Sci. 63(4): 400-406.

Dong, J., Bowra, S. and Vincze, E. 2010. The development and evaluation of single cell suspension from wheat and barley as model system; a first step towards functional genomics. BMC Plant Biol. 10:239-250.

Dougall, D.K. 1987. Cell cloning and the selection of high yielding strains. *In*: F. Constabel and I.K. Vasil (Eds.), Cell Culture and Somatic Cell Genetics of Plants, Vol. 4: Cell Culture in Phytochemistry. Academic Press, Inc., San Diego, pp. 117-124.

Draper, J. and Scott, R. 1991. Gene transfer to plants. *In*: D. Grierson (Ed.), Plant Genetic Engineering, Vol. 1. Blackie, Glasgow and London, pp. 38-81.

Driver, J. A. and Suttle, G.R.L. 1987. Nursery handling of propagules. *In*: J.M. Bonga and D.J. Durzan (Eds.). Cell and Tissue Culture in Forestry, Vol. 2. Specific Principles and Methods: Growth and Development. Martinus Nijhoff, Dordrecht, pp. 320-335.

Ducos, J.P., Labbe, G., Lambot, C. and Pétiard, V.L. 2007. Pilot scale process for the production of pre-germinated somatic embryos of selected robusta (*Coffea canephora*) clones. In Vitro Cell Dev. Biol. Plant. 43: 652-659.

Dudits, D., Bogre, L. and Gyorgyey, J. 1991. Molecular and cellular approaches to the analyses of plant embryo development from somatic cells *in vitro*. J. Cell Sci. 99: 475-484.

*Duhamel du Monceau, H.L. 1756. La Physique des Arbres, ou II Est Trait ê de I'Anatomie des plantes et de l'economie Végétale pour Servir d'Introduction au Trait ê complet des Bois et des Forêsts. P.H.L. Guêrin Pub. (cited by Gautheret 1985, p. 52).

Dumet, D., Berjak, P. and Engelmann, F. 1997. Cryopreservation of zygotic and somatic embryos producing recalcitrant and intermediate seeds. *In*: M.K. Razdan and E.G. Cocking (Eds.), Conservation of Plant Genetic Resources *In Vitro*. Vol. 1: General Aspects, Science Publishers Inc., Enfield, USA, pp. 153-172.

Dunsmuir, P. and Suslow, T. 1989. Structure and regulation of organ and tissue-specific genes in plants. *In*: J. Schell and I.K. Vasil (Eds.), Cell Culture and Somatic Cell Genetics of Plants, Vol. 6: Molecular Biology of Plant Nuclear Genes. Academic Press, Inc., San Diego, pp. 215-227.

Dunstan, D., Bekkaoui F., Pilon, M., Fowke, L.C. and Abrahams, S.R. 1988. Effects of abscisic acid and analogues on the maturation of white spruce (*Picea glauca*) somatic embryos. Plant Sci. 58: 77-84.

Dunstan, D.I. 1988, Prospects and progress in conifer biotechnology. Can. J. For. Res. 18: 1497-1506.

Dunwell, J.M. 1985. Haploid cell cultures. *In*: R.A. Dixon (Ed.), Plant Cell Culture: A Practical Approach. IRL Press, Oxford, UK, pp. 21-36.

Dunwell, J.M. 1999. Transformation of maize using silicon whiskers. *In*: D. Hall (Ed.). Methods in Molecular Biology Vol. III. Plant Cell Culture Protocols. Humana Press Inc., Totowa, N.J., pp. 375-382.

Dunwell, J.M. and Wetten, A. (Eds.). 2012. Transgenic Plants: Methods and Protocols. Humana Press, N.Y., 497 pp.

Dupuis, J.M., Millot, C., Teufel, E., Arnault, J.L. and Preyssinet, G. 1999. Germination of synthetic seeds. *In*: W.Y. Soh and S.J. Bhojwani (Eds.), Morphogenesis in Plant Tissue Cultures. Kluwer, Dordrecht, pp. 419-441.

Duran-Vila, N. 1997. Conservation of citrus genetic resources. *In*: M.K. Razdan and E.G. Cocking (Eds.)- Conservation of Plant Genetic Resources *In Vitro*, Vol. 1: General Aspects. Science Publishers, Inc., Enfield (USA), pp. 175-199.

Durzan, D.J. 1980. Progress and promise in forest genetics. Proc. 50th Anniv. Conf. Paper Sci. Technol: The Cutting Edge. May 8-10. Inst. Paper Chemistry, Appleton, Wisconsin, pp. 31-60.

Durzan, DJ. and Gupta, P.K. 1987. Somatic embryogenesis and polyembryogenesis in Douglas fir cell suspension cultures. Plant Sci. 52: 229-235.

Dussert, S., Engelmann, F., Chabrillange, N., Anthony, F., Woirot, M. and Hamon, S. 1997. *In Vitro* conservation of coffee (*Coffea* spp.) germplasm. *In*: M.K. Razdan and E.C. Cocking (Eds.), Conservation of Plant Genetic Resources *In Vitro*. Vol. 1: General Aspects. Science Publishers Inc., Enfield, USA, pp. 287-305.

Dwivedi, S.L., Britt, A.B., Tripathi, L., Sharma, S., Upadhaya, H.D. and Oritz, R. 2015. Haploids: Constraints and opportunities in plant breeding. Biotechnol. Adv. 33:812-829.

Edallo, S., Zuccinalli, C., Perenzin, M. and Salamani, F. 1981. Chromosomal variation and frequency of spontaneous mutation associated with *In Vitro* culture and plant regeneration in maize. Maydica 26: 39-56.

Eibl, R. and Eibl, D. 2008. Design of bioreactors suitable for plant cell and tissue cultures. Phytochem. Rev. 7: 593-598.

Eilert, U. 1987. Elicitation: methodology and aspects of application. *In*: F. Constabel and I.K. Vasil (Eds.), Cell Culture and Somatic Cell Genetics of Plants, Vol. 4: Cell Culture in Phytochemistry. Academic Press, Inc., San Diego, pp. 153-196.

Ellis, R.H. 1984. Revised table of seed storage characteristics. Plant Genet. Resour. Newsl., 58: 16-33.

Ellis, R.H., Hong, I.D., Roberts, E.H. and Soetisna, U. 1991. An intermediate category of seed storage behaviour II. Effects of provenance, immaturity, and imbibition on desiccation tolerance in coffee. J. Exp. Bot. 42: 653-657.

Elliston, K. and Messing, J. 1989. The molecular architecture of plant genes and their regulation. *In*: S. Kung and C.J. Arntzen (Eds.), Plant Biotechnology. Butterworths, Boston, pp. 115-140.

Engelmann, F. 2011. Cryopreservation of embryos: an overview. *In*: T.A.Thorpe and E.C. Yeung (Eds.), Plant Embryo Culture: Methods and Protocols. Methods in Molecular Biology Series Vol. 710, Humana Press, N.Y., pp. 155-186.

Engelmann, F. 2012. Germplasm collection, storage and conservation. *In*: A. Altman and M. Hasegawa (Eds.), Plant Agriculture: Prospects for the 21st Century. Elsevier, London, 255-267.

Engelmann, F. 1994. Cryopreservation for the long-term conservation of tropical crops of commerical importance *In*: Proc. Int. Symp. Application of Plant *In Vitro* Technology. Serdang, Malaysia, pp. 33-67.

Eriksson, T. 1965. Studies on the growth requirements and growth measurements of cell cultures of *Haplopappus gracilis*. Physiol. Plant. 18: 976-993.

Eudes, F., Shim, Y.S. and Jiang, F. 2014. Engineering the haploid genome microspores. Biocatal. Agric. Biotechnol. 3: 20-23.

Evans, D.A. 1988. Application of somaclonal variation. *In*: A. Mizrahi (Ed.), Advances in Biotechnological Processes, Vol. 9: Alan-R. Liss, Inc., New York, pp. 204-220.

Evans, D.A. 1989. Somaclonal variation—Genetic basis and breeding applications. Trends in Genetics 5: 46-50.

Evans, D.A. and Sharp, W.R. 1986a. Somaclonal and gametoclonal variation. *In*: D.A. Evans et al. (Eds.), Handbook of Plant Cell Culture, Vol. 4: Techniques and Applications. MacMillan Publishing Company, N.Y., pp. 97-132.

Evans, D.A., and Sharp, W.R. 1986b. Applications of somaclonal variation. Biotechnology 4: 528-532.

Evans, D.A., Sharp, W.R. and Bravo, J.E. 1987. Plant somaclonal variation and mutagenesis. Nestle Research News. 1986/87: 63-71.

Evans, D.A., Sharp, W.R. and Medina-Filho, H.P. 1984. Somaclonal and gametoclonal variation. Am. J. Bot 71: 759-774.

Evans, D.A., Sharp, W.R., Ammirato, P.V. and Yamada, Y. (Eds.). 1983. Handbook of Plant Cell Culture, Vol. 1: Techniques for Propagation and Breeding. Macmillan Publishing Co., New York, 970 pp.

Evans, P. 2001. Impact of US PTO guidelines on gene patents. Genet. Eng. News 21: 6-44.

Fahleson, J., Eriksson, I., Landgren, M., Stymne, S. and Glimelius, K. 1994. Intertribal somatic hybrid between *Brassica napus* and *Thlaspi perfoliaturn* with high content of *T. perfoliatum*-specific nervonic acid. Theor. Appl. Genet. 87: 795-804.

Fahy, G.M., Macfarlane, D.R., Angell, C.A. and Meryman, H.T. 1984. Vitrification as an approach to cryopreservation. Cryobiol. 21: 407-426.

Falah, M., Mozafari, J., Sokhandan, B.N. and Hashemi, M. 2009. Elimination of a DNA virus associated with yellow leaf curl disease in tomato using an electrotherapy technique. Acta Hortic. 808: 157-162.

FAO. 2009. Draft second report on World's plant genetic resources for food and agriculture. Food and Agriculture Organization of the United Nations, Rome, pp.1-12.

Fasella, P. and Hussain, A. 2014. Plant Biotechnology. Scientific Inti. Pvt. Ltd., New Delhi, 499 pp.

Faustmann, O., Kern, R., Sanger, H.L. and Muhlbach, H.P. 1986. Potato spindle tuber viroid (PSTV) RNA oligomers of (+) and (–) polarity are synthesized in potato protoplasts after liposome-mediated infection with PSTV. Virus Res. 4: 213-227.

Fayos, O., Valies, M.P., Gareês-Claver, A., Maller, C., and Castillo, A.M. 2015. Doubled haploid production from Spanish onion (*Allium cepa* L.) germplasm: embryogenesis induction, plant regeneration and chromosome doubling. Front. Plant Sci. 6: 684.

Feuillet et al. 1995. (cited from Fukuda 1997).

Finer, J.J., Kriebel, H.B. and Beewar, M.R. 1989. Initiation of embryogenic callus and suspension cultures of eastern white pine (*Pinus strobus* L.). Plant Cell Rep. 8: 203-206.

Finkel, B. and Ulrich,). 1983. Protocols of cryopreservation. *In*: D.A. Evans et al. (Eds.), Handbook of Plant Cell Culture, Vol. 1: Techniques for Propagation and Breeding. Macmillan Publishing Co., N.Y., pp. 806-815.

Finkel, B.J., Ulrich, J.M. and Tisserat, B.B. 1980. Regeneration of date palm trees from callus stored at –196°C. Plant Physiol. 65 (Suppl.): 36.

Fischer-Iglesias, C. and Neuhans, G. 2001. Zygotic embryogenesis. *In*: S.S. Bhojwani and W.Y. Soh (Eds.). Current Trends in the Embryology of Angiosperms. Kluwer, Dordrecht, pp. 223-247.

Fish, N., Steele, S.H. and Jones, M.G.K. 1988. Field assessment of dihaploid *Solatium tuberosum* and *S. brevidens* somatic hybrids. Theor. Appl. Genet. 76: 880-886.

Fleck, B. and Baldock, C. 2003. Intellectual property protection from plant-related inventions in Europe. Nature Biotechnology 4: 834-838.

Flick, C.E. 1983. Isolation of mutants from cell culture. *In*: D. Evans et al. (Eds.), Handbook of Plant Cell Culture, Vol. 1: Techniques for Propagation and

Breeding. Macmillan Publishing Co., N.Y., pp. 393-441.

Fluhr, R. 1989. Structure and regulation of light-induced genes: Genes involved in photosynthesis. *In*: J. Schell and I.K. Vasil (Eds.), Cell Structure and Somatic Cell Genetics of Plants, Vol. 6: Molecular Biology of Plant Nuclear Genes. Academic Press Inc., San Diego, pp. 134-153.

Fonnesbech, M. 1972. Organic nutrients in the media for propagation of Cymbidium *In Vitro*, Physiol. Plant. 27: 360-364.

Foroughi-Wehr, B. and Mix, G. 1979. *In Vitro* response of *Hordeum vulgare* L. anthers cultured from plants grown under different environments. Environ. Exp. Bot. 19: 303-309.

Forsberg, J., Landgren, M. and Glimelius, K. 1994. Fertile somatic hybrids between *Brassica napus* and *Arabidopsis thaliana*. Plant Sci. 95: 213-223.

Fosket, D.E. 1968. Cell division and the differentiation of wound-vessel members in cultured stem segments of *Coleus*, Proc. Natl. Acad. Sci., U.S.A. 59: 1089-1096.

Fowler, M.W. 1982. The large scale cultivation of plant cells. Prog: Ind. Microbiol. 17: 207-229.

Fraley, R.T., Rogers, S.G., Horsch, R.B., Eichholtz, D.A., Flick, J.S., Fink, C.L., Hoffmann, N.L. and Sanders, P.R. 1985. The SEV system, a new disarmed Ti plasmid vector system for plant transformation. Biotechnology 3: 629-635.

French, R., Janda, M. and Ahlquist, P, 1986. Bacterial gene inserted in an engineered RNA virus: Efficient expression of monocotyledonous plant cells. Science 231: 1294-1297.

Friedman, Y. 2009. Intellectual property and biotechnology innovation: To protect or not protect? J.Comm. Biotechnol. 15: 285-286.

Fromm, M.E., Taylor, L.P. and Walbot, Y. 1985. Expression of genes electroporated into monocot and dicot plant cells. Proc. Natl. Acad. Sci., U.S.A. 82: 5804-5828.

Fu, T.J., Singh, G. and Curtis, W.R. (Eds.). 1999. Plant Cell and Tissue Culture for the Production of Food Ingredients. Kluwer, Dordrecht, 285 pp.

Fujimura, T. and Komamine, A. 1979a. Involvement of endogenous auxin in somatic embryogenesis in a carrot cell suspension culture. Z. pflanzenphysiol. 95: 13-19.

Fujimura, T. and Komamine, A. 1979b. Synchronization of somatic embryogenesis in a carrot suspension culture. Plant Physiol. 64:162-164.

Fujimura, T. and Komamine, A. 1980. Mode of action of 2,4-D and zeatin on somatic embryogenesis in a carrot suspension culture. Z. Pflanzenphysiol. 99: 1-8.

Fujita, Y. 1990. The production of industrial compounds. *In*: S.S. Bhojwani (Ed.), Plant Tissue Culture: Applications and Limitations. Elsevier, Amsterdam, pp. 259-275.

Fukuda, H. 1992. Tracheary element formation as a model system of cell differentiation. Int. Rev. Cytol. 136: 289-332.

Fukuda, H. 1997. Xylogenesis: Initiation, progression, and cell death. Annual Review of Plant Physiology and Plant Molecular Biology, 47:299-325.

Fukuda, H. 2004. Signals that control plant vascular differentiation. Nature Rev. Mol. Cell Biol. 5:379-391.

Fukuda, H. and Komamine, A. 1985. Cytodifferentiation. *In*: I.K. Vasil (Ed.), Cell culture and Somatic Cell Genetics of Plants, Vol. 2: Cell Growth, Nutrition, Cytodifferentiation, and Cryopreservatlon. Acad. Press, Inc., Orlando, pp. 213-253.

Fukui, K. 1983. Sequential occurrence in a growing rice callus. Theor. Appl. Genet. 65: 225-230.

Galbraith, D.W. 1989. Flow cytometric analysis and sorting of somatic hybrid and transformed protoplasts. *In*: Y.P.S. Bajaj (Ed.), Biotechnology in Agriculture and Forestry, Vol. 9: Plant Protoplasts and Genetic Engineering II. Springer-Verlag, Berlin, pp. 304-327.

Galiba, G. and Sutka, J. 1988. Frost resistance of somaclones derived from *Triticum aestivum L.* winter wheat calli. Plant Breeding 102: 101-104.

Gamborg, O.L. and Phillipis, G.C. 1995. Plant Cell, Tissue and Organ Culture: Fundamental Methods. Springer-Verlag, Heidelberg, 348 pp.

Gambrog, O.L., Davis, B.D. and Stahlhut, R.W. 1983. Somatic embryogenesis in cell cultures of *Glycine* species. Plant Cell Rep. 2:209-212.

Gambrog, O.L., Miller, R.A. and Ojima, K. 1968. Nutrient requirements of suspension cultures of soybean root cells. Exp. Cell Res. 50:151-158.

Gaspar, T. 1991. Vitrification in micropropagation. *In*: Y.P.S. Bajaj (Eds.), Biotechnology in Agriculture and Forestry, Vol. 17. High-Tech and Micropropagation. I. Springer, Berlin, pp. 116-123.

Gautheret, R.J. 1934, Culture du tissu cambial. C.R. Acad. Sci. (Paris) 198: 2195-2196.

Gautheret, R.J. 1939. Sur la possibilite de réaliser la culture indefinie des tissus de tubercules de carotte. C.R. Hebd. Seances Acad. Sc. 208: 118-120.

Gautheret, R.J. 1942. Manual Technique de Culture de Tissus Végétaux. Masson Publ., Paris.

Gautheret, R.J. 1955. Sur la variabilité des proprietés physiologiques des cultures de tissus végétaux. Rev. Gen. Bot. 62: 5-112.

Gautheret, R.J. 1985. History of Plant Tissue and Cell Culture: A Personal Account. *In*: I.K. Vasil (Ed.), Cell Culture and Somatic Cell Genetics of Plants, Vol. 2: Cell Growth, Nutrition, Cytodifferentiation, and

Cryopreservation. Academic Press, Inc., Orlando, pp. 1-59.

Gavin, A.L., Conger, B.V. and Trigiano, R.N. 1989. Sexual transmission of somatic embryogenesis in *Dactylis glomerata*. Plant Breed. 103: 251-254.

Gengenbach, B.G. 1985. *In Vitro* pollination, fertilization, and development of maize kernels. *In*: I.K. Vasil (Ed.), Cell Culture and Somatic Cell Genetics of Plants, Vol. 1: Laboratory Procedures and Their Applications. Academic Press, Inc., Orlando, pp. 276-282.

Gengenbach, B.G., Green, C.E. and Donovan, C.M. 1977. Inheritance of selected pathotoxin resistance in maize plants regenerated from cell cultures. Proc. Natl. Acad. Sci., U.S.A. 74: 5113-5117.

Genovesi, A.D. and Collins, G.B. 1982. In vitro production of haploid plants of corn via anther culture. Crop Sci. 22:1137-1144.

George, E.F., Hall, M.A. and Deklerk, G.J. (Eds.). 2008. Plant Propagation by Tissue Culture 3rd Edition Vol.1. The Background. Springer, the Netherlands, 501 pp.

Giles, K.L. 1989. Chloroplast uptake and genetic complementation. *In*: Y.P.S. Bajaj (Ed.), Biotechnology in Agriculture and Forestry, Vol. 9: Plant Protoplasts and Genetic Engineering II. Springer-Verlag, Berlin, pp. 428-446.

Giles, K.L. and Songstad, D.D. 1990. Plant tissue culture in twenty-first century. *In*: S.S. Bhojwani (Ed.), Plant Tissue Culture: Applications and Limitations. Elsevier, Amsterdam, pp. 424-434.

Gleba, Y.Y. and Hoffmann, F. 1979. "Arabidobrassica", Plant-genome engineering by protoplast fusion. Natur Wissenchaften 66: 547-554.

Gleba, Y.Y. and Shlumukov, L.R. 1990. Somatic hybridisation and cybridization. *In*: S.S. Bhojwani (Ed.), Plant Tissue Culture: Applications and Limitations. Elsevier, Amsterdam, pp. 316-345.

Glick, B.R., Pasternak, J.J. and Patten, C.L. 2010. Molecular Biotechnology: Principles and Applications of Recombinant DNA (4th Edition). ASM Press, Washington D. C., 927 pp.

Glimelius, K. 1987. Enrichment of somatic hybrids with flow cytometry. Int. Assoc. Plant Tissue Cult. Newslett. 52: 2-10.

Glimelius, K., Eriksson, T., Grafe, R. and Müller, A.J. 1978. Somatic hybridization of nitrate-reductase-deficient mutants of *Nicotiana tabacum* by protoplast fusion. Physiol. Plant. 44: 273-277.

Gmitter, F.G. Jr., Ling, X.B. and Deng, X.X. 1990. Induction of triploid *Cirus* plants from endosperm calli *In Vitro*. Theor. Appl. Genet. 80: 785-790.

Gnanam, A. and Kulandaivelu, G. 1969. Photosynthetic studies with leaf cell suspensions from higher plants. Plant Physiol. 44: 1451-1456.

Goh, H.K.L., Rao, A.N. and Loh, C.S. 1988. *In Vitro* plantlet formation in mangosteen (*Garcinia mangostana* L.). Ann. Bot. 62: 87-93.

Goldberg, R.B. 1988. Plants: novel developmental processes. Science 240: 1460-1467.

Gonsalves, D. 1998. Control of papaya ringspot virus in papaya: A case study. Ann. Rev. Phytopathol. 36: 415-437.

Gordon-Kamm, W.J., Spencer, T.M., Mangano, M.L, Adams, T.R., Danies, RJ., O'Brien, J.V., Start, W.G., Adams, W.R., Chambers, S.A., Willetts, N.G., Krueger, R.W. Kausch, A.P., Mackey, CJ. and Lemeux, P.G. 1990. Transformation of corn using microprojectile bombardment. In Vitro Cell Dev. Biol. 26 (Part II): 69.

Gostimsky, S.A., Kokaeva, Z.G. and Konovalov, F.A. 2005. Studying plant genomic variation using molecular markers. Genetika 41: 480-492.

Graff, G. and Zilberman, D. 2001. Towards an intellectual property clearinghouse for agricultural biotechnology. IP Strategy Today 3:1-13.

Grappin, P., Bouinot, D., Sotta, B., Miginiac, E. and Jullien, M. 2000. Control of seed dormancy in *Nicotiana plumbaginifolia*: Post-imbibition abscisic acid synthesis imposes dormancy maintenance. Planta 210: 279-285.

Gray, D.J. 1987. Quiescence in monocotyledonous and dicotyledonous somatic embryos induced by dehydration. HortScience 22: 810-814.

Gray, D.J., Trigiano, R.N. and Conger, B.V. 1993. Liquid suspension culture production of orchardgrass somatic embryos and their potential for the breeding of improved varieties. *In*: K. Redenbaugh (Ed.), Synseeds: Application of Synthetic Seeds to Crop Improvement. CRC Press, Boca Raton, FL., pp. 351-356.

Gray, J.C. 1989. Genetic manipulation of the chloro-plast genome. *In*: S.W. Kung and CJ. Arntzen (Eds.), Plant Biotechnology. Butter-worths, Boston, pp. 317-335.

Green, C.E. and Phillips, R.L. 1974. Potential selection systems for mutants with increased lysine, threonine, and methionine in cereal crops. Crop Sci. 14: 827-830.

Green, C.E. and Phillips, R.L. 1975. Plant regeneration from tissue cultures of maize. Crop Sci. 15: 417-421.

Gresshoff, P.M., Skotnicki, M.L., Eadie, J.F. and Rolfe, B.C. 1977. Viability of *Rhizobium trifolii* bacterioids from clover root nodules. Plant Sci. Lett. 10: 299-304.

Grierson, D. (Ed.). 1991. Plant Genetic Engineering. Blackie, Glasgow, U.K., pp. 1-267.

Groose, R.W. and Bingham, E.T. 1984. Variation in plants regenerated from tissue culture of tetraploid alfalfa heterozygous for several traits. Crop Sci. 24: 655-658.

Grout, B.W.W. 1990. *In Vitro* conservation of germplasm. *In*: S.S. Bhojwani (Ed.), Plant Tissue Culture:

Applications and Limitations, Elsevier, Amsterdam, pp. 394-411.

Grout, B.W.W. 1999. Meristem-tip culture for propagation and virus elimination. *In*: R.D. Hall (Ed.). Methods in Molecular Biology, Vol. III. Plant Cell Culture Protocols. Humana Press Inc., Totowa, N.J., pp. 115-125.

Grout, B.W.W. and Henshaw, G.G. 1978. Freeze-preservation of potato shoot-tip cultures. Ann. Bot. (London) 42: 1227-1229.

Grout, B.W.W., Westcott, R.J. and Henshaw, G.G. 1978. Survival of shoot meristems of tomato seedlings frozen in liquid nitrogen. Cryobiology 15: 478-483.

Grover, A., Aggarwal, P.K., Kapoor, A., Katiyar-Agrewal, S., Agarwal, M. and Chandramouli, A. 2003. Addressing abiotic stresses in agriculture through transgenic technology. Curr. Sci. 89: 355-367.

Gu, S.-r., Gui, Y-I and Xu, T-Y. 1985. Induction of endosperm plantlets in *Lycium*. Acta. Bot. Sin. 27: 106-109.

Guardino, L., Ramantha Rao, V. and Reid, R. (Eds.). 1995. Collecting Plant Genetic Diversity—Technical Guidelines. IPGRI, CAB International, Walkingford, UK, 748 pp.

Guha, S. and Maheshwari, S.C. 1964. *In Vitro* production of embryos from anthers of *Datura*. Nature (London), 204: 497.

Guha, S. and Maheshwari, S.C. 1966. Cell division and differentiation of embryos in the pollen grains of *Datura In Vitro*. Nature 212: 97-98.

Gui, Y., Gu, S. and Xu, T. 1988. Differentiation of organs in endosperm culture of Chinese gooseberry. *In*: Genetic Manipulation in Crops. Cassell Tycooly, United Kingdom, pp. 364-366.

Gui, Y-I., Mu, X-J. and Xu, T-Y. 1982. Studies on morphological differentiation of endosperm plantlets of Chinese gooseberry *In Vitro*. Acta. Bot Sin. 24: 216-221.

Gunning, J. and Lagerstedt, H.B. 1985. Long-term storage techniques for *In Vitro* plant germplasm. Proc. Intl. Plant Prop. Soc. 35: 199-205.

Gupta, H.S., Rech, E.L., Cocking, E.C. and Davey, M.R. 1988. Electroporation and heat shock stimulate division of potoplasts of *Pennisetum squamulatum*. J. Plant Physiol. 133: 457-459.

Gupta, P.K. and Durzan, D.H. 1986. Somatic polyembryogenesis from callus of mature sugar pine embryos. Biotechnology 4: 643-645.

Gupta, P.K. and Durzan, D.H. 1987. Biotechnology of somatic polyembryogenesis and plantlet regeneration in loblolly pine. Biotechnology 5: 147-151.

Gupta, S. and Ibaraki, Y. (Eds.). 2006. Plant Tissue Culture Engineering. Springer, the Netherlands, 478 pp.

Guri, A and Sink, K.C. 1988. Interspecific somatic hybrid plants between eggplant *Solanum melongena* and *Solanum torvum*, Theor. Appl. Genet. 76: 490-496.

Haberlandt, G. 1902. Kulturversuche mit isolierten pflanzenzellen. Sitzungsber. Akad. Wiss. Wien., Math.-Naturwiss. KL, Abt. 1. III: 69-92.

Hagan, G., Gulfoyle, T.J. and Gray, W.M. 2004. Auxin signal transduction. *In*: P.J. Davies (Ed.), Plant Hormones. Kluwer, Dordrecht, pp. 282-303.

Hagendoorn, M.J.M., Jamar, D.C.L. and Van der Plas, L.H.W. 1999. Directing anthraquinone accumulation via manipulation of Morinda suspension cultures. *In*: R.D. Hall (Ed.), Methods in Molecular Biology, Vol. III. Plant Cell Culture Protocols. Humana Press Inc., Totowa, N.J., pp. 383-391.

Hahlbrock, K. and Scheel, D. 1989. Physiology and molecular biology of phenylpropanoid metabolism. Ann. Rev. Plant Physiol. 40. 347-369.

Hain, R., Stabel, P., Czernilofski, A.P., Steinbiss, H.H., Herrara-Estrella, L. and Schell, J. 1985. Uptake, integration and genetic transmission of a selectable chimeric gene by plant protoplasts. Mol. Gen. Genet. 199: 161-168.

Hakman, I., Fowke, L.C., Von Arnold, A. and Eriksson, T. 1985. The development of somatic embryos in tissue cultures initiated from immature embryos of *Picea abies* (Norway spruce). Plant Sci. 38: 53-59.

Halford, N.G. (Ed.). 2006. Plant Biotechnology: Current and Future Applications of Genetically Modified Crops. John Wiley & Sons Ltd., Chichester, West Sussex, England, 303 pp.

Hall, R.D. (Ed) 1999. Plant Culture Protocols. Humana Press, Totowa, 421 pp.

Hall, R.D. and Yeoman, M.M. 1986. Temporal and spatial heterogeneity in the accumulation of anthocyanin in cell cultures of *Catharanthus roseus* (L.) G. Don. Exp. Bot. 37: 48-60.

Hallagan, J.B., Putnam, J.M., Higley, N.A. and Hall, R.L. 1999. The safety assessment of flavor ingredients derived from plant cell and tissue culture. *In*: T.J. Fu et al. (Eds.) Plant Cell and Tissue Culture for the Production of Food Ingredients. Kluwer, Dordrecht, pp. 251-258.

Hamada, H. and Furuya, T. 1999. The selectivity by plant biotransformation. *In*: T.J. Fu, et al. (Eds.), Plant Cell and Tissue Culture for the Production of Food Ingredients. Kluwer, Dordrecht, pp. 113-120.

Hamilton, K.N., Ashmora, S.E. and Pritchard, M.W. 2009. Thermal analysis and cryopreservation of seeds of Australian wild *Citrus* species (Rutaceae): *Citrus australiaca, C.indora* and *C. garrawagi*. CryoLetters 30:268-279.

Han, Hu. 1987. Application of pollen-derived plants in crop improvement. *In*: S. Natesh et al. (Eds.), Biotechnology in Agriculture. Oxford & IBH Publishing Co. Pvt. Ltd., New Delhi, pp. 155-168.

Hannay, J.W. and Street, H.E. 1954. Studies on the growth of excised root. III. Molybdenum and manganese requirements of excised tomato roots. New Phytol. 2: 297-302.

Hannig, E. 1904. Zur Physiologic pflanzlicher Embryonen. I. Über die cultur von Crucifever-Embryonen ausserhalb des Embryosacks. Bot. Ztg. 62: 45-80.

Hansen, J. and Hildebrandt, A.C. 1966. The distribution of tobacco mosaic virus in plant callus culture. Virology 28: 15-21.

Harfouche, A., Meilan, R., Grant, K. and Shier, V.K. 2012. Intellectual property rights of biotechnologically improved plants. *In*: A. Altman and P.M. Hasegawa (Eds.), Plant Biotechnology and Agriculture: Prospects for 21st Century. Elsevier Inc., Oxford, U.K. pp. 525-539.

Harms, C.T. 1984. The multiple-drop-array (MDA) screening technique. *In*: I.K. Vasil (Ed.), Cell Culture and Somatic Cell Genetics of Plants, Vol. 1: Laboratory Procedures and Their Applications. Academic Press, Orlando, pp. 213-220.

Harms, C.T., Lörz, H., and Potrykus, I. 1979. Multiple-drop-array (MDA) technique for the large-scale testing of culture media variations in hanging microdrops cultures of single cell systems II. Determination of phytohormone combinations for optimal division response in *Nicotiana tabacum* protoplast cultures. Plant Sci. Lett. 14: 237-244.

Harrell, R.C. and Cantliffe, D.J. 1991. Automated evaluation of somatic embryogenesis in sweet potato by machine vision. *In*: I.K. Vasil (Ed.), Cell Culture and Somatic Cell Genetics of Plants, Vol. 8. Scale-up and Automation in Plant Propagation. Academic Press, San Diego, CA, pp. 179-195.

Hasezawa, S., Matsui, C, Nagata, T. and Syono, K. 1981. Transformation of *Vinca* protoplasts mediated by *Agrobacterium* spheroplasts. Cail. J. Bot. 61: 1052-1057.

Haslam, T.M. and Yeung, E.C. 2011. Zygotic embryo culture: an overview. *In*: T.A. Thorpe and E.C. Yeung (Eds.) Plant Embryo Culture: Methods and Protocols. Methods in Molecular Biology Series Vol. 710, Humana Press, N.Y., pp. 3-15.

Hayford, M., Medford, J., Hoffmann, N., Rogers, S. and Klee, H. 1988. Development of a plant transformation selection system based on expression of genes encoding gentamycin acetyl transferases. Plant Physiol. 86: 1216-1222.

Hazarika, B.N. 2003. Acclimatization of tissue-cultured plants. Curr. Sci. 85: 1704-1712.

Heath-Pagliuso, S. and Rappaport, L. 1990. Somaclonal variant UC-T3: the expression of Fusarium wilt resistance in progeny arrays of celery, *Apium graveolens L.* Theor. Appl. Genet. 80: 390-394.

Heberle-Bors, E. 1985. *In Vitro* haploid formation from pollen: A critical review. Theor. Appl. Genet. 71: 361-374.

Heinz, D.J. and Mee, G.W.P. 1971. Morphologic, cytogenetic enzymatic variation in *Saccharum* species hybrid clones derived from callus tissue. Am. J. Bot. 58: 257-262.

Heinz, D.J., Krishnamurthi, M., Nickell, L.G. and Maretzki, A. 1977. Cell, tissue and organ culture in sugarcane improvement. *In*: J. Reinert and Y.P.S. Bajaj (Eds.), Applied and Fundamental Aspect of Plant Cell, Tissue and Organ Culture. Springer-Verlag, Berlin, pp. 3-17.

Helgeson, J.P. 1983. Studies of host-pathogen interactions *In Vitro*. *In*: J.P. Helgeson and B.J. Deverall (Eds.), Use of Tissue Culture and Protoplasts in Plant Pathology. Academic Press, N.Y., pp. 9-38.

Hemenway, C.A., Turner, N.E,, Powell, P.A. and Beachy, R.N. 1989. Genetic engineering of plants for viral disesase resistance. *In*: J. Schell and I.K. Vasil (Eds.), Cell Culture and Somatic Cell Genetics of Plants, Vol. 6: Molecular Biology of Plant Nuclear Genes. Academic Press, Inc., San Diego, pp. 406-424.

Henry, R.J. 2013. Molecular markers in plants. Wiley-Blackwell, West Sussex, 197 pp.

Hess, D., Schnelder, G., Lörz, H. and Blaich, G. 1976. Investigations on the rumour induction in *Nicotiana giauca* by pollen transfer of DNA isolated from *Nicotiana langsdorffii*. Z. Pflanzenphysiol. 77: 247-254.

Heszky, L.E. and Simon-Kiss, I. 1992 'DAMA', The first plant variety of biotechnology origin in Hungary, registered in 1992. Hung. Agric. Res. I: 30-32.

Hibberd, K.A. and Green, C.E. 1982. Inheritance and expression of lysine plus threonine resistance selected in maize tissue culture. Proc. Natl. Acad. Sci., U.S.A. 79: 559-563.

Hilder, V.A., Gatehouse, A.M.R., Sheerman, S.E., Barker, R.F. and Boulter, D. 1987. A novel mechanism of insect resistance engineered in tobacco. Nature 300: 160-163.

Hilton, M.G. and Rhodes, M.J.C. 1990. Growth and hyoscyamine production in hairy root cultures of *Datura stramonium* in a modified stirred tank reactor. Appl. Microbiol. Biotechnol. 33: 132-138.

Himmel, M.E., Ding, S.H., Johnson, D.K., Adney, W.S., Nimolos, M.R., Brady, J.W. and Foust, T.D. 2007. Biomass recalcitrance: Engineering plants and enzymes for biofuel production. Science 315: 804-807.

Hollings, M. 1965. Disease control through virus stock. Ann. Rev. Phytopathol. 3: 367-396.

Hollings, M. and Stone, O.M. 1968. Techniques and problems in the production of virus tested planting material. Sci. Hortic. 20: 57-72.

Holman, C.M. 2007. Patent border wars: Defining the boundary between scientific discoveries and patentable inventions. Trends Biotechnol. 25: 539-543.

Holmes, F.O. 1948. Elimination of spotted wild from *Dahlia*. Phytopathology 38: 314.

Hooker, B.S. and Lee, J.M. 1990. Cultivation of plant cells in aqueous two-phase polymer system. Plant Cell Rep. 8: 546-549.

Hooker, B.S., Lee, J.M. and An, G. 1990. Cultivation of plant cells in stirred vessel. Effect of impeller designs. Biotechnol. Bioeng. 34: 502-508.

Horsch, R.B., Fry, J.B., Hoffmann, N.L., Wallroth, M., Eichholtz, D., Rogers, S.G. and Fragley, R.T. 1985. A simple and general method for transferring genes into plants. Science 227: 1229-1231.

Hoshino, Y., Miyashita, T. and Thomas, T.D. 2011. *In vitro* culture of endosperm and its applications in plant breeding: approaches to polyploidy breeding. Sci. Hortic. 130: 1-8.

Hu, C. and Wang, P. 1986. Embryo Culture: Technique and Application. *In*: D.A. Evans et al. (Eds.), Handbook of Plant Cell Culture, Vol. 4. Techniques and Applications. Macmillan Publishing Company, N.Y., pp. 43-96.

Hu, H. 1983. Genetic stability and variability of pollen-derived plants. *In*: *S.K.* Sen and K.L. Giles (Eds.), Plant Cell Culture in Crop Improvement. Plenum Press, N.Y., pp. 145-157.

Hu, H. and Guo, X. 1999. *In Vitro* induced haploids in plant genetics and breeding. *In*: W.Y. Soh and S.S. Bhojwani (Eds), Morphogenesis in Plant Tissue Cultures. Kluwer, Dordrecht, pp. 329-361.

Huang, B., Bird, S., Kemble, R., Simmonds, D., Keller, W. and Miki, B. 1990. Effects of culture density, conditioned medium and feeder cultures on microspore embryogenesis in *Brassica napus* L. cv. Topas. Plant Cell Rep. 8: 594-597.

Hunter, C.S. and Kilby, N.J. 1999. Their accumulation and release *In Vitro*. *In*: R.D. Hall (Ed.). Methods in Molecular Biology, Vol. Ill, Plant Cell Culture Protocols. Humana Press Inc., Totowa, N.J., pp. 403-410.

Hwang, S.C. 1991. Somaclonal resistance in Cavendish banana to Fusarium wilt. Plant Prot. Bull. (Taiwan) 33: 124-132.

Isa, T. and Ogasawara, T. 1991. High yield production of anthraquinone by cell suspension cultures of *Crocus saliva L*, Plant Tissue Culture Letters 8:171-174.

Islam, S.M. and Tuteja, N. 2012. Enhancement of androgenesis by abiotic stress and other pretreatments in major crop species. Plant Sci. 182: 134-144.

Israeli, Y., Reuveni, O. and Lahav, E. 1991. Qualitative aspects of somaclonal variations in banana propagated by *In Vitro* techniques. Sci. Hortic. 48: 71-88.

Ito, M. and Maeda, M. 1974. Meiotic division and fusion of nuclei in multinucleate cells from the induced fusion of meiotic protoplasts of liliaceous plants. Bot. Mag. Tokyo 87: 219-228.

Izhar, S. and Power, J.B. 1979. Somatic hybridization in *Petunia*: a male sterile cytoplasmic hybrid. Plant Sci. Lett. 14: 49-55.

Jacobsen, E. and Sopory, S.K. 1978. The Influence of possible recombination of genotypes on the production of microspore embryoids in anther cultures of *Solanum tuberosum* and dihaploid hybrids. Theor. Appl. Genet. 52:119-123.

Jacoli, G.G. 1978. Sequential degeneration of myco-plasma-like bodies in plant tissue cultures infected with aster yellows. Can. J. Bot. 56: 133-140.

Jain, R. and Babbar, S.B. 2005. Guar gum and isabgol as cost-effective alternative gelling agents for *in vitro* multiplication of an orchid, *Dendrobium chrysotoxum*. Curr. Sci. 88: 292-295.

Jain, S.M., Brar, D.S. and Ahloowalia, B.S. (Ed.) 1998. Somaclonal Variation and Induced Mutations in Crop Improvement. Kluwer, Dordrecht, 615 pp.

Jaiswal, V.S. and Amin. M.N. 1987. *In Vitro* propagation of guava from shoot cultures of mature trees. J. Plant Physiol. 130: 7-12.

Jakše, M., Hirscheger, P., Bohanec, B. and Harvey, M.J. 2010. Evaluation of gynogenic responsiveness and pollen viability of selfed doubled haploid onion lines and chromosome doubling via somatic regeneration. J. Am. Soc. Hortic. Sci. 135: 67-73.

Jefferson, R. 2007. Science as social enterprise. The CAMBIA BiOS initiative. Innovations: Technology, Governance, Globalization 1: 13-44.

Jefferson, R.A. 1987. Assaying chimeric genes in plants. The GUS gene fusion system. Plant Molec. Biol. Rep. 5: 387-405.

Jeffs, R.A. and Northcote, D.H. 1967. The influence of indole-3-acetic acid and sugar on the pattern of induced differentiation in plant tissue culture. J. Cell Sci. 2: 77-88.

Johnson, E. 1996. A benchside guide to patents and patenting. Nature Biotechnol. 14: 288-291.

Johnson, L.B., Stuteville, D.L., Schlarbaum, S.E. and Skinner, D.Z. 1984. Variation in phenotype and chromosome number in alfalfa protoclones regenerated from nonmutagenized calli. Crop Sci. 24: 948-951.

Johnson, R., Narvaez J., An, G. and Ryan, C. 1989. Expression of protemase inhibitors I and II in transgenic tobacco plants: effects of natural defense against *Manduca sexta* larvae. Proc. Nad. Acad. Sci., USA, 86: 9871-9875.

Johri, B.M. and Bhojwani, S.S. 1965. Growth responses of mature endosperm in cultures. Nature (London) 208: 1345-1347.

Jonard, R. 1986. Micrografting and its applications to tree improvement. *In*: Y.P.S. Bajaj (Ed.), Biotechnology in Agriculture and Forestry, Vol. 1: Trees I. Springer-Verlag, Berlin, pp. 31-18.

Jones, L.E., Hildebrandt, A.C., Riker, A.J. and Wu, J.H. 1960. Growth of somatic tobacco cells in micro-culture. Am. J. Bot. 47: 468-475.

Jones, O.P. 1983. *In Vit ro* propagation of tree crops. *In*: S.H. Mantell and H. Smith (Eds.), Plant Biotechnology. Cambridge Univ. Press, Cambridge, U.K., pp. 139-159.

Joseph, R., Yeoh, H.M. and Loh, S.S. 2004. Induced mutations in cassava using somatic embryos and identification of mutant plants with altered starch and composition. Plant Cell Rep. 23: 91-98.

Joshi, P.C. and Ball, E. 1968. Growth of isolated palisade cells *of Arachis hypogaea*. *In Vitro*. Dev. Biol. 17: 308-325.

Jouanneau, J.P. 1971. Controle par les cytokinines de la synchronisation des mitoses dans les cellules de tabac. Exp. Cell Res. 67: 329-337.

Joy, R.W., McIntyre, D.D., Vogel, H.J. and Thorpe, T.A. 1996. Stage-specific nitrogen metabolism in developing carrot somatic embryos. Physiol. Plant. 97: 149-159.

Kadkade, P.G. and Seibert, M. 1977. Phytochrome regulated organogenesis in lettuce tissue culture. Nature 170: 49-50.

Kagan-Zur, V., Mills, D. and Mizrachi, Y. 1990. Callus fromation from tomato endosperm. Acta Hortic. 280: 139-142.

Kameya, T. and Hinata, K. 1970. Induction of haploid plants from pollen grains of *Brassica*. Jpn. J. Breed. 20: 82-87.

Kameya, T., Hinata, K. and Mizushima, U. 1966. Fertilization *In Vitro* of excised ovules treated with calcium chloride in *Brassica oleracea* L. Proc. Jpn. Acad. 42: 165-167.

Kando, Y., Fujita, T., Sujiyama, N. and Fukuda, H. 2015. A novel system for xylem differentiation in *Arabidopsis thaliana*.Molec. Plant 8: 6212-621.

Kane, M. 2000. Culture indexing for bacterial and fungal contaminants. *In*: R.N. Trigiano and D.G. Gray (Eds.), Plant Tissue Culture Concepts and Laboratory Exercises. CRC Press, Boca Raton, pp. 427-431.

Kanta, K. and Maheshwari, P. 1963a. Intra-ovarian fertilization in some angiosperms. Phytomorphology 13: 215-229.

Kanta, K. and Maheshwari, P. 1963b. Test-tube fertilization in some angiosperms. Phytomorphology 13: 230-237.

Kanta, K., Rangaswamy, N.S. and Maheshwari, P. 1962. Test-tube fertilization in flowering plants. Nature (London) 194: 1214-1217.

Kao, K.N. 1977. Chromosomal behaviour in somatic hybrids of soybean—*Nicotians glauca*. Mol. Gen. Genet. 150: 225-230.

Kao, K.N. and Michayluk, M.R. 1975. Nutritional requirements for growth of *Vicia hajastana* cells and protoplasts at a very low population density in liquid media. Planta 126: 105-110.

Kao, K.N. and Michayluk, M.R. 1989. Fusion of plant protoplasts—Techniques. *In*: Y.P.S. Bajaj (Ed.), Biotechnology in Agriculture and Forestry, Vol. 8: Plant Protoplasts and Genetic Engineering I. Springer-Verlag, Berlin, pp. 277-288.

Karmani, O., Aghavaisi, B. and Pour, A.M. 2009. Molecular aspects of somatic-embryogenic transition in plants. J. Chem. Biol. 2:177-190.

Karp, A. 1990. Somaclonal variation in potato. *In*: Y.P.S. Bajaj (Ed.), Biotechnology in Agriculture and Forestry, Vol. 11: Somaclonal Variation in Crop Improvement I. Springer-Verlag, Berlin, pp. 379-399.

Kartha, K.K. 1984. Elimination of viruses. *In*: I.K. Vasil (Ed.), Cell Culture and Somatic Cell Genetics of Plants, Vol. 1: Laboratory Procedures and Their Applications. Academic Press, Inc., Orlando., pp. 577-585.

Kartha, K.K. 1987. Cryopreservation of secondary metabolite-producing plant cell cultures. *In*: F. Constabel and I.K. Vasil (Eds.), Cell Culture and Somatic Cell Genetics of Plants, Vol. 4: Cell Culture in Phytochemistry. Academic Press, Inc., San Diego, pp. 217-227.

Kartha, K.K., Fowke, L.C., Leung, N.L., Casewell, K.L. and Hakman, I. 1988. Induction of somatic embryos and plantlets from cryopreserved cell cultures of white spruce (*Picea glauca*), J. Plant Physiol. 132: 529-539.

Kartha, K.K and Gamborg, O.L. 1975. Elimination of cassava mosaic disease by meristem culture. Phytopathology 65: 826-828.

Kartha, K.K., Gamborg, O.L., Constabel, F. and Shyfuk, J.P. 1974. Regeneration of cassava plants from apical meristems. Plant Sci. Lett. 2:107-113.

Kassanis, B. 1957. The use of tissue cultures to produce virus-free clones from infected potato varieties. Ann. Appl. Biol. 45: 422-427.

Katavic, V., Haughn, G.W., Reed, D., Martin, M. and Kunst, H. 1994. In planta transformation of *Arabidopsis thaliana*. Mol. Gen. Genet. 245:363-370.

Kato, A. 1978. The involvement of photosynthesis in inducing bud formation on excised leaf segments of *Helionopsis orientalis* (Liliaceae). Plant Cell Physiol. 19: 791-799.

Kato, A. and Geiger, H.H. 2002. Chromosome doubling of haploid maize seedlings using nitrous oxide at the flower primordial stage. Plant Breed. 121: 370-377.

Keller, W.A. and Melchers, G. 1973. The effect of high pH and calcium on tobacco leaf protoplast fusion. Z. Naturforschung 28: 737-741.

Kemeya T. and Hinata, K. 1970. Induction of haploid plants from pollen grains of *Brassica*. Japan J. Breed 20: 82-87.

Kent, N.F. and Brink, R.A. 1947. Growth *In Vitro* of immature *Hordeum* embryos. Science 106: 547-548.

Kessell, R.H.J., Goodwin, C. and Philp, J. 1977. The relationship between dissolved oxygen concentration, ATP and embryogenesis in carrot (*Daucus carota*) tissue cultures. Plant Sci. Lett. 10: 265-274.

Khayat, E. 2012. An engineering view to micro-propagation and generation of true to type and pathogen free plants. *In*: A. Altman and P. Hasegawa (Eds.), Plant Biotechnology and Agriculture: Prospects of 21st Century. Elsevier, London, pp. 229-241.

Khor, E. and Loh, C.S. 2005. Artificial seeds. *In*: V. Nedovi and R.W. Willaert (Eds.), Application of Cell Immobilisations Biotechnology. Springer, Berlin, pp. 527-537.

Kielly, G.A. and Bowley, S.R. 1992. Genetic control of somatic embryogenesis in alfalfa. Genome 35: 474-477.

Kilby, NJ. and Hunter, C.S. 1991. Towards optimization of vacuole-located secondary product from *in vitro* grown plant cells using 1.02-MHz ultrasound. Appl. Microbiol. Biotechnol. 33: 488-451.

Kim, Y.H. and Janick, J. 1989. ABA and polyoxencap-sulation or high humidity increases survival of desiccated somatic embryo of celery. Hort. Science 24: 674-676.

Kin, M.S., Fraser, L.G. and Harvey, C.F. 1990. Initiation of callus and regeneration of plantlets from endosperm of *Actinidia* interspecific hybrids. Sci. Hort. 44: 107-117

King, P.J., Cox, B.J., Fowler, M.W. and Street, H.E. 1974. Metabolic events in synchronised cell cultures of *Acer pseudoplatanus* L. Planta 117: 109-122.

Kinnersley, A.M. and Dougall, O.K. 1980. Correlation between the nicotine content of tobacco plants and callus cultures. Planta 149: 205.

Kim, Y.H. and Janick, J. 1989. ABA and polyoxencapsulation or high humidity increases survival of desiccated somatic embryo of celery. Hort. Science 24: 674-676.

Klee, H.J. and Rogers, S.G. 1989. Plant gene vectors and genetic transformation: plant transformation systems based on the use of *Agrobacterium tumefaciens*. *In*: J. Schell and I.K. Vasil (Eds.), Cell Culture and Somatic Cell Genetics of Plants, Vol. 6: Molecular Biology of Plant Nuclear Genes. Academic Press, Inc., San Diego, pp. 1-24.

Klein, T.M., Fromm, M., Weissinger, A., Tomes, D., Schalf, S., Sletten, M. and Sanford, J.C. 1988. Transfer of foreign genes into intact maize cells with high-velocity microprojectiles. Proc. Natl. Acad. Sci., U.S.A. 85: 4305-4309.

Klein, T.M., Wolff, E.D., Wu, R. and Sanford, J.C. 1987. High-velocity microprojectiles for delivering nucleic acids into living cells. Nature 327: 70-73.

Kleinhofs, A. and Behnki, R. 1977. Prospects for plant genome modification by nonconventional methods. Ann. Rev. Genet. 11: 79-102.

Klercker, J.A. 1892. Fine Methode zur Isoleiring lebender Protoplasten. Oefvers Vetenskaps Adad., Stockholm, 9: 463-471.

Klimaszewska, K. 1989. Plantlet development from immature zygotic embryos of hybrid larch through somatic embryogenesis. Plant Sci. 63: 95-103.

Klopmeyer, M.J. 2000. Indexing for plant pathogens. *In*: R.N. Trigiano and D.J. Gray (Eds.). Plant Tissue Culture: Concepts and Laboratory Exercises. CRC Press, Boca Raton, pp. 417-425.

Knauss, J.F. 1976. A tissue culture method for producing *Dieffenbachia* pieta. Proc. Fla. State Hort. Soc. 89: 293-295.

Knauss, J.F. and Miller, J.W. 1978. A contaminant, *Erwinia carotovara*, affecting commercial plant tissue cultures. *In Vitro* 14: 754-756.

Knobloch, K.H. 1982. Uptake of phosphate and its effect on phenylalanine ammonia lyase activity and cinnamoyl putrescine accumulation in cell suspension cultures of *Nicotiana tabacum*. Plant Cell Rep. 1: 128-130.

Knobloch, K.H. and Berlin, J. 1983. Influence of phosphate on the formation of indole alkaloids and phenolic compounds in cell suspension cultures of *Catharanthus roseus*. I. Comparison of enzyme activities and product accumulation. Plant Cell, Tissue, Organ Cult. 2: 333-340.

Knobloch, K.H., Bentnagel, G. and Berlin, J. 1981. Medium-induced formation of indole alkaloids and concomitant changes of interrelated enzyme activities in cell suspension cultures *of Catharanthus roseus* cell cultures. Z. Naturforsch. 36a: 40-43.

Knoop, B. and Beiderbeck, R. 1985. Adsorbent filter — a tool for the selection of plant suspension culture cells producing secondary substances. Z. Naturforsch. C. Biosci. 40C: 297-300.

Ko, S.-W., Wong, C.-K. and Woo, S.-C. 1983. A simplified method of embryo culture in rice of *Oryza sativa L.* Bot. Bull. Acad. Sin. 24: 97-101.

Ko, W.H., Su, C.C., Chen, C.L. and Chao, C.P. 2009. Control of lethal browning of tissue culture plantlets of Cavendish banana cv. Formosana with ascorbic acid. Plant Cell Tiss. Organ Cult. 96: 137-141.

Kohinoor Begum and Islam, M.O. 2015. Selection of salt tolerant somaclones for development of salt stress tolerant sugarcane varieties. J. Plant 3(4): 37-46.

Kohlenbach, H.W. 1966. Die Entwicklungspotenzen explantierter und isolierter Dauerzellen. I. Das Strechungs-und Tei lungs-wanchrstun isolierter Mesophyllzellon von *Macleaya cordata.* Z. Pflanzen-physiol. 55: 142-157.

Kohlenbach, H.W. 1978. Comparative somatic embryo-genesis. *In*: T.A. Thorpe (Ed.), Frontiers of Plant Tissue Culture. Univ. Calgary Press, Canada, pp. 59-66.

Kohlenbach, H.W. 1984. Culture of isolated mesophyll cells. *In*: I.K. Vasil (Ed.), Cell Culture and Somatic Cell and Genetics of Plants, Vol. 1: Laboratory Procedures and Applications. Academic Press, Inc., Orlando, pp. 204-212.

Kohlenbach, H.W. and Schopke, C. 1981. Cytodiffe-rentiation to tracheary elements from protoplasts of *Zinnia elegans,* Naturwissenschaften 68: 576-577.

Kohlenbach, H.W., Korber, M. and Li, L. 1982. Cytodifferentiation of protoplasts isolated from a stem embryo system of *Brassica napus* to tracheary elements. Z. Pflanzenphysiol. 107: 367-371.

Komamine, A. 2001. Mechanisms of somatic embryo-genesis in carrot suspension cultures. Phytomor-phology (Golden Jubilee Issue: Trends in Plant Science). 51: 277-288.

Komatsuda, T. and Ohyama, K. 1988. Genotypes of high competence for somatic embryogenesis and plant regeneration in soybean *Glycine max.* Theor. Appl. Gen. 75: 695-700.

Konar, R.N. and Nataraja, K. 1969. Morphogenesis of isolated floral buds of *Ranunculus sceleratus* L. *In Vitro.* Acta. Bot. Neerl. 18: 680-699.

Kondo, O., Honda, H., Taya, M. and Kobayashi, T. 1989. Comparison of growth properties of carrot hairy root in various bioreactors. Appl. Microbiol. Biotechnol. 32: 291-294.

Koo, B., Nottenberg, C. and Pardey, G. 2004. Plants and intellectual property: An international appraisal. Science 306: 1295-1297.

*Kott, and Kasha. (1985). Cited from Pierik, R.L.M. 1989. *In Vitro* Culture of Higher Plants. Martinus Nijhoff Publishers, Dordrecht, The Netherlands, p. 147.

Kott, L.S. and Beversdorf, W.D. 1990. Enhanced plant regeneration from microspore derived embryos of *Brassica napus* by chilling, partial desiccation and age selection. Plant Cell, Tissue, Organ Cult. 23: 187-192.

Kotte, W. 1922. Kultur versuche mit isolierten Wurzelspitzen. Beitr. Allg. Bot. 2: 413-434.

Kowalski, S.P., Ebora, R., Kryder, R.D. and Potter, R. 2002. Transgenic crops, biotechnology and ownership rights: What scientists need to know? The Plant J. 31: 407-421.

Kragh, K.M., Hendriks, T., de Jong, A. J., Schiavo, F.L., Bucherna, N., Nojrup, P., Mikkelsen, J.D. and de Vries, S.C. 1996. Characterization of chitinases able to rescue somatic embryos of the temperature-sensitive carrot variants tsll. Plant Mol. Biol. 31: 631-645.

Kranz, E. and Lörz, H. 1993. *In Vitro* fertilization with isolated, single gametes results in zygotic embryo-genesis and fertile maize plants. Plant Cell 5: 739-746.

Kranz, E. and Scholten, S. 2008. In vitro fertilization analysis of early post-fertilization development using cytological and molecular techniques. Sex. Plant Reprod. 21:67-77.

Kranz, K. 1999. *In Vitro* fertilization with isolated single gametes. *In*: R.D. Hall (Ed.), Plant Cell Culture Protocols. Humana Press, Totowa, pp. 259-267.

Krattiger, A. 2007. Freedom to operate, public sector research and product-development partnerships: Strategies and risk management options. *In*: A. Krattiger, R.T. Mahoney, L. Nelsen, J.A. Thompson, A.B. Bennet, K. Suryanarayana, G.D. Graff, C.D. Fernandez and S.P. Kowalski (Eds.), Intellectual Property Management and Agricultural Innovations: A Handbook of Best Practices. Oxford: MIHR and PIPRA, Davis, pp. 1312-1327. Available online at www.ipHandbook.org

Krens, F.A., Molendijk, L., Wullems, G.J. and Schilperoort, R.A. 1982. *In Vitro* transformation of plant protoplasts with Ti-plasmid DNA. Nature (London) 296: 72-74.

Krishnamurthi, M. and Tlaskal, J. 1974. Fiji disease resistant *Saccharum officinanum* var. Pinda subc-lones from tissue culture. Proc. Int.Soc. Sugarcane Technol. 15: 130-137.

Krishnapillay, B. 2000. Towards the use of cryopreser-vation as a technique for conservation of tropical recalcitrant seed species. *In*: M.K. Razdan and E.C. Cocking (Eds.), Conservation of Plant Genetic Resources *In Vitro*. Vol. 2: Applications and Limitations, Science Publishers, USA, pp. 139-166.

Krumbiegel, G. and Schielder, O. 1979. Selection of somatic hybrids after fusion of protoplasts from *Datura innoxia* and *Atropa belladonna*. Planta 145: 371-375.

Kruse, A. 1974. An *in vivo/vitro* embryo culture technique. Hereditas 77: 219-224.

Kuboi, T. and Yamada, Y. 1978a. Regulation of the enzyme activities related to lignin synthesis in cell aggregates of tobacco cell culture. Biochem. Biophys. Acta 542: 181-190.

Kuboi, T. and Yamada, Y. 1978b. Changing cell aggregation lignification in tobacco suspension cultures. Plant Cell Physiol. 19: 437-443.

Kuhlmann, V., Foroughi-Wehr, B., Graner, A. and Wenzel, G. 1991. Improved culture system for microspores of barley to become a target for DNA uptake. Plant Breed. 107: 115-168.

Kuhn, D.N., Chappell, J., Boudet, A. and Hahlbrock, K. 1984. Induction of phenylalanine ammonialyase and 4-Coumarate; CoA ligase mRNAs in cultured plant cells by UV light of fungal elicitor. Proc. Natl. Acad. Sci., U.S.A. 81: 1102-1106.

Kumar, P.P. and Loh, C.D. 2012. Plant tissue culture for biotechnology. *In*: Altman and P.M. Hasegawa (Eds.), Plant Biotechnology and Agriculture for 21st Century. Elsevier, London, pp.131-138.

Kumar, P.P., Raju, C.R., Chandrashekhran, M. and Iyer, R.D. 1985. Induction and maintenance of friable callus from the cellular endosperm of *Cocos nucifera* L. Plant Sci. 40: 203-207.

Kumar, U. and Rangaswamy, N.S. 1977. Regulation of seed germination and polarity in seedling development in *Orobanche aegyptica* by growth substances. Biol. Plant. 19: 353-359.

Küster, E. 1909. Über die Verschmelzung nachter Protoplasten. Ber. Dtsch. Bot. Ges. 27: 589-598.

Kyte, L. and Briggs, B. 1979. A simplified entry into tissue culture production. Proc. Int. Plant Prop. Soc. 29: 90-95.

La Rue, C.D. 1936. The growth of plant embryos in culture. Bul. Torrey Bot. Club. 63: 365-382.

La Rue, C.D. 1949. Cultures of the endosperm of maize. Am J. Bot. 36: 798 (Abstr.).

Laibach, F. 1925. Das Taubwerden von Bastardsmen und die Kunstliche Aufzucht fruh absterbender Bastardembryonen. Z. Bot. 17: 417-459.

Laibach, F. 1929. Ectogenesis in plants. Methods and genetic possibilities of propagating embryos otherwise dying in the seed. J. Hered. 20: 201-208.

Lakshmi Sita, G. 1987. Triploids. *In*: J.M. Bonga and D.J. Durzan (Eds.), Cell and Tissue Culture in Forestry,

Vol. 2: Specific Principles and Methods—Growth and Development. Martinus Nijhoff Publ., Dordrecht, The Netherlands, pp. 269-284.

Lakshmi Sita. 1997. Gynogenic haploids *In Vitro*. *In*: S.M.Jain et al. (Eds), *In Vitro* Haploid Production in Higher Plants, Vol. 5. Kluwer, Dordrecht, pp. 175-193.

Lalitha, N., Devi, L.M., Banerjee, R., Chattopadhyay, S., Saha, A.K. and Bindroo, B.B. 2014. Effect of plant derived gelling agents as agar substitute in micropropagation of mulberry (Morus indica cv. S-1635). Intl. J. Adv. Res. 2:683-690.

*Lamp, and Mills. (1933). Cited by La Rue C.D. 1936. The growth of plant embryos in culture. Bull. Torrey Bot. Club 63: 365-382.

Lampton, R.K. 1952. Developmental and Experimental Morphology of the Ovule and Seed of *Asimina triloba* Dunal. Ph.D. Thesis, Michigan University, Ann Arbor, MI, 104 pp.

Landley, R.B., Cenci, A., Gyot, R. Bertrand, B., Georget, F., Dechamp E., Herra, J-C. , Aribi, J., Lashermes, P. and Etienne, H. 2015. Assessment of genetic and epigenetic changes during cell culture ageing and relations with somaclonal variation in *Coffea arabica*. Plant Cell Tiss. Org. Cult. 122: 517-531.

Larkin, P.J. 1998. Introduction. *In*: S.M. Jain et al. (Ed.), Somaclonal Variation and Induced Mutation in Crop Improvement. Kluwer, Dordrecht, pp. 3-13.

Larkin, P.J. and Scowcroft, W.R. 1981. Somaclonal variation—a novel source of variability from cell cultures for plant improvement. Theor. Appl. Genet. 60: 197-214.

Larkin, P.J., Brettell, R.I.S. and Scowcroft, W.R. 1984. Heritable somaclonal variation in wheat. Theor. Appl. Genet. 67: 443-455.

Lawyer, A.L., Berlyn, M.B. and Zelitch. 1.1980. Isolation and characterization of glycine hydroxamate-resistant cell lines of *Nicotiana tabacum*. Plant Physiol. 66: 334-341.

Leduc, N., Douglas, G.C. and Monnier, M. 1992. Methods of non-stigmatic pollination in *Trifolium repens* (Papilionaceae): seedset with self- and cross-pollinations *in vitro*. Theor. Appl. Genet. 83: 912-918.

Lee, E.K., Jin,Y. W., Park, J.H., Yoo, Y.M., Hong, S.M., Amir,R., Yan, Z. , Kwon, E., Eflick, A., Tomlinson, S., Halbritter, F., Waibel, T., Yun, B.W. and Loake, G.J. 2010. Cultured cambial meristematic cells as source of plant natural products. Nature Biotechnol. 28:1213-1217.

Lei, Z., Juneja, R. and Wright, B.D. 2009. Patents versus patenting: Implications of intellectual property protection for biological research. Nature Biotechnol. 27: 36-40.

Lenee, P., Sangwan-Norreel, B.S. and Sangwan, R.S. 1987. Nuclear DNA contents and ploidy in somatic embryos derived from androgenic plants of *Datura innoxia* Mill. J. Plant Physiol. 130: 37-48.

Lesney, M.S., Callow, P.W. and Sink, K.C. 1986. A technique for bulk production of cytoplasts and miniprotoplasts from suspension culture-derived protoplasts. Plant Cell Rep. 5: 115-118.

Letham, D.S. 1963. Zeatin, a factor inducing cell division isolated from *Zea mays*. Life Sci. 2: 569-579.

Letham, D.S. 1974. Regulators of cell division in plant tissues. The cytokinins of coconut milk. Physiol. Plant. 32: 66-70.

Lin, X., Hwang, G.J. and Zimmerman, J.L. 1996. Isolation and characterization of a diverse set of genes from carrot somatic embryos. Plant Physiol. 112: 1365-1374.

Lindsey, K. and Yeoman, M.M. 1984. The viabilty and biosynthetic activity of cells of *Capsicum frutescens* Mill. cv. annuum immobilised in reticulate polyurethane foam. J. Exp. Bot. 35: 1684-1696.

Lindsey, K. and Yeoman, M.M. 1985. Dynamics of plant cell cultures, *In*: I.K. Vasil (Ed.), Cell Culture and Somatic Cell Genetics of Plants. Vol. 2: Cell Growth, Nutrition. Cytodifferentiation, and Cryopreservation. Academic Press, Inc., Orlando, pp. 61-102.

Lindsey, K., Yeoman, M.M., Black, G.M. and Mavituna, F. 1983. A novel method for the immobilization and culture of plant cells. FEBS Lett. 155: 143-149.

Linsmaier, E.M. and Skoog, F. 1965. Organic growth factor requirements of tobacco tissue cultures. Physiol. Plant. 18: 100-127.

Litz, R.E. and Canover, R.A. 1981. Effect of sex type, season and other factors on *in vitro* establishment and culture of *Carica papaya* L. explants. J. Amer. Soc. Hort. Sci. 106: 792-794.

Litz, R.E., Jarret, R.L. and Asokan, P. 1986. Tropical and subtropical fruits and vegetables. *In*: R.H. Zimmerman et al. (Eds.), Tissue Culture as a Plant Production System for Horticultural Crops. Martinus Nijhoff, Dordrecht, The Netherlands, pp. 237-251.

Litz, R.E., Mathews, H., Moon, P.A. Pliego-Alfaro, F., Yuragalevitch, C. and Dewald, S.G. 1993. Somatic embryos of mango (*Mangifera indica* L.). In K. Redenbaugh (Ed.), Syn seeds: Applications of Syn seeds to crop Improvement. CRC Press, Boca Raton, Fl., pp. 409-425.

Liu, C.M., Xu, Z.H. and Chua, N.H. 1993. Proembryo culture: *In Vitro* development of early globular-stage zygotic embryos from *Brassica juncea*. Plant J. 3: 291-300.

Liu, J., Su, Q., Au, L. and Yang, A. 2009. Transfer of a minimal linear marker-free and vector-free smGFP cassette into soybean via ovary dip transformation. Biotechnol. Lett. 31: 295-303.

Liu, M. and Chen, W. 1978. Tissue and cell culture as aids to sugarcane breeding II. Performance and yield of callus derived lines. Euphytica 27: 273-282.

Liu, M.C. 1981. *In Vitro* methods applied to sugarcane improvement. *In*: T.A. Thorpe (Ed.), Plant Tissue Culture: Methods and Applications in Agriculture. Academic Press, N.Y., pp. 299-323.

Liu, S., Gu, Y., Gu, S. and Xu, T. 1987. Induction of endosperm calluses and regeneration of endosperm plantlets of *Asparagus officinalis*. Acta Bot. Sin. 29: 373-376.

Liu, W., Xu, Z. and Chua, N. 1993. Auxin polar transport is essential for the establishment of bilateral symmetry during early plant embryogenesis. Plant Cell 5: 621-630.

Lloyd, G.B. and Mc Cown, B.H. 1980. Woody plant medium. Intl. Propg. Soc. 30:421-427.

Lo Schiavo, F., Giuliano, G. and Sung, Z.R. 1988. Characterization of temperature sensitive carrot cell mutant impaired in somatic embryogenesis. Plant Sci. 54: 157-164.

Logemann, J. and Schell, J. 1993. The impact of biotechnology on plant breeding or how to combine increases in agricultural productivity with an improved protection of environment. *In*: I. Chet (Ed.), Biotechnology in Plant Diseases. Wiley-Liss, New York, pp. 1-14.

Lonsdale, D.M. 1987. Cytoplasmic male sterility: A molecular perspective. Plant Physiol. Biochem. 25: 265-271.

Loo, S.W. 1945. Cultivation of excised stem tips of *Asparagus in vitro*, Am. J. Bot. 32: 13-17.

Lörz, H. 1984. Enucleation of protoplasts: Preparation of cytoplasts and mini protoplasts. *In*: I.K. Vasil (Ed.). Cell Culture and Somatic Cell Genetics of Plants, Vol. 1: Laboratory Procedures and Their Applications. Academic Press, Orlando, pp. 448-453.

Lörz, H. and Scowcroft, W.R. 1983. Variability among plants and their progeny regenerated from protoplasts Su/su heterozygotes of *Nicotiana tabacum*. Theor. Appl. Genet. 66: 67-75.

Lörz, H., Baker, B. and Schell, J. 1985. Gene transfer to cereal cells mediated by protoplast transformation. Mol. Gen. Genet. 199: 178-182.

Luckett, D.J. and Smithard, R.A. 1991. Doubled haploid production by anther culture for Australian barley breeding. Aust. J. Agric. Res. 43: 67-78.

Lurquin, P.P. 1989. Uptake and integration of exogenous DNA in plants. *In*: Y.P.S. Bajaj (Ed.), Biotechnology in Agriculture and Forestry, Vol. 9: Plant Protoplasts and Genetic Engineering II. Springer-Verlag, Berlin, pp. 54-74.

Lurquin, P.P. and Kado, C.I. 1977. *Escherichia coli* plasmid pBR313 insertion into plant protoplasts and into their nuclei. Mol. Gen. Genet. 154: 113-121.

Lynch, P.T., Isaac, S. and Collin, H.A. 1989. Uptake of fungal protoplasts by plant protoplasts. *In*: Y.P.S. Bajaj (Ed.), Biotechnology in Agriculture and Forestry, Vol. 9: Plant Protoplasts and Genetic Engineering II. Springer-Verlag, Berlin, pp. 406-427.

Lyznik, L.A. and Tsai, C.Y. 1989. Protein synthesis in endosperm cell cultures of maize (*Zea mays* L.). Plant Sci. 64: 105-114.

MacDonald, M.V., Hadwiger, M.A., Aslam, F.N. and Ingram, D.S. 1988. The enhancement of anther culture efficiency in *Brassica napus* ssp. *oleifera* Metzg. (Sinsk.) using low dose of gamma irradiation. New Phytol. 110: 101-107.

Maene, L.J. and Debergh, P.C. 1983. Rooting of tissue cultured plants under in vivo conditions. Acta Hortic. 131: 201-208.

Maheshwari, N. 1958. *In Vitro* culture of excised ovules of *Papaver somniferum*. Science, N.Y., 127: 342.

Maheshwari, P. and Kanta, K. 1964. Control of fertilization. *In*: H.F. Linskens (Ed.), Pollen Physiology and Fertilization. Elsevier, Amsterdam, pp. 187-193.

Maheshwari, P., Garg, S. and Kumar, A. 2008. Taxoids: biosynthesis and *in vitro* production. Biotechnol. Mol. Biol. Rev. 3: 71-87.

Majeskwa-Sawka, A. and Nothnagel, E.A. 2000. The multiple roles of arabinogalactan proteins in plant development. Plant Physiol. 122: 3-9.

Maliga, P. 1984. Isolation and characterization of mutants. *In*: D. Evans et al. (Eds.). Handbook of Plant Cell Culture, Vol. 4: Techniques and Applications. Macmillan Publishing Company, N.Y., pp. 133-149.

Maliga, P. 1985. Cell culture, isolation and characterization of agronomically useful mutants of higher plants. *In*: Biotechnology in International Agricultural Research; Inter-Centre Seminar on International Agricultural Research Centres and Biotechnology, Manila. Philippines, 1984, pp. 111-120.

Maliga, P., Lázár, G., Joo, F., Nagy, A.H. and Menczel, L. 1977. Restoration of morphogenic potential in *Nicotiana* by somatic hybridization. Mol. Gen. Genet. 157: 291-296.

Maliga, P., Lörz, H., Lázár, G. and Nagy, F. 1982. Cytoplast-protoplast fusion for interspecific chloroplast transfer in *Nicotiana*. Mol. Gen. Genet. 185: 211-215.

Maluszynski, M., Szarejko, I. and Sigurbjörnsson, B. 1996. Haploidy and mutation techniques. *In*: S.M. Jain, S.K. Sopory and R.E. Veilleux (Eds.), *In Vitro* Haploid Production in Higher Plants. Vol. 1. Fundamental Aspects and Methods. Kluwer. Dordrecht, pp. 67-93.

Manawadu, I.P. and Dahanayake, N. 2015. Effect of carbon sources on shoot regeneration of raddish (*Raphnus sativus* L. var. Beeralu Rabu. J. Agrisearch 2:277-280.

Manek, H. and Hanke, M. 2014. 10 years of fruit gene bank at Dresten- Pillmitz under Federal responsibility. J. Kulturpflanzen 66: 117-129.

Mantell, S.H., Matthews, J.A. and McKee, R.A. 1985. Principles of Biotechnology, Blackwell Scientific Publications, Oxford, U.K., pp. 1-269.

Manzocchi, L.A., Bianchi, M.W. and Viotti, A. 1989. Expression of zein in long term cultures of wild type and opaque-2 maize endosperms. Plant Cell Rep. 7: 639-643.

Maqsood, M., Mujib, A. and Siddiqui, Z.H. 2012. Synthetic seed development and conversion to plantlet in *Catharanthus roseus*. (l) G. Don. Biotechnol. 11: 37-43.

Maraschin, S.F., Priester, W., Spaink, H.P. and Wang, M. 2005. Androgenic switch: an example of plant embryogenesis from male gametophyte perspective. J. Exp. Bot. 56: 1711-1726.

Margara, J., Rancillac, M. and Bouniols, A. 1967. La neo formation *in vitro* de bourgeons infiorescentiels chez *Cichorium intybus* L. Etude methodologique. Colloq. Int. C.N.R.S. 167: 71-82.

Mari, S., Engelmann, F., Chabrillange, N., Huet, C. and Michaux-Ferrière, N. 1995. Histocytological study of coffee (*Coffea racemosa* and S. *sessiflora*) *In Vitro* plantlets during their cryopreservation using the encapsulation dehydration technique. Cryo-Lett. 16: 289-298.

Mariani, C., Gossele, V., De Beuckeleer, M., De Block, M., Goldberg, R.B., De Greef, W. and Leemans, J. 1992. A chimeric ribonuclease-inhibitor gene restores fertility to male sterile plants. Nature 357: 384-387.

Marshall, E. 1997. Companies rush to patent DNA. Science 275: 780-781.

Martin, S.M. 1980. *In*: E.J. Staba (Ed.), Plant Tissue Culture as a Source of Biochemicals. CRC Press, Boca Baton, Florida, pp. 149-166.

Marty, D. 1988. La tomate dans tons ses états. Biofutur 72: 43-48.

Mascarenhas, A.F., Khuspe, S.S., Nadgauda, R.S., Gupta, P.K., Murlidharan, E.M. and Khan, B.M. 1989. Biotechnological application of plant tissue culture to forestry in India. *In*: V. Dhawan (Ed.), Applications of Biotechnology in Forestry and Horticulture. Plenum Press, N.Y., pp. 73-86.

Masuda, H., Ozeki, Y., Amino, S. and Komazunik, A. 1984. Changes in cell wall polysaccharides during elongation in a 2,4-D free medium in a carrot suspension culture. Physiol. Plant. 62: 65-72.

Matsubara, K., Katani, S., Yoshino, T., Murimoto, T. and Fujita, Y. 1989. High density culture of *Coptis japonica* cells increases berberine production. J. Chem. Tech. Biotechnol. 46: 61-69.

Matsubara, S. 1962. Studies on the growth-promoting substance "embryo factor" necessary for the culture of young embryos of *Datura tatula in vitro*. Bot. Mag. (Tokyo) 75: 1-18.

McCabe, D.E., Swain, W.F., Martinell, B.J. and Chri-stou, P. 1988. Stable transformation of soybean (*Glycine max*) by particle acceleration. Bio. Technol. 6: 923-926.

McGranahan, G.H., Leslie, C.A., Uratsu, S.L. and Dandekar, A.M. 1990. Improved efficiency of the walnut somatic embryo gene transfer system. Plant Cell Rep. 8: 512-516.

McMullen, M.D. and Finer, J.J. 1990. Stable transformation of cotton and soybean embryogenic cultures via microprojectile bombardment. J. Cell. Biochem. 14E: 285.

Medgyesy, P., Coiling, R. and Nagy, F. 1985. A light sensitive recipient for the effective transfer of chloroplast and mitochondrial traits by protoplast fusion in *Nicotiana*. Theor. Appl. Genet. 70: 590-594.

Mehta, R.A., Cassol, T., Li, N. Ali, N., Handa, A.K. and Mattoo, A.K. 2002. Nature Biotechnol. 20(6): 613-618.

Melchers, G. and Labib, G. 1974. Somatic hybridization of plants by fusion of protoplasts. I, Selection of light resistant hybrids of 'haploid' light sensitive varieties of tobacco. Mol. Gen, Genet. 135: 277-294.

Melchers, G., Sacristan, M.D. and Holder, A.A. 1978. Somatic hybrid plants of potato and tomato regenerated from fused protoplasts. Carlsberg Res. Commun. 43: 203-218.

Melechko, A.V., Merkulov, V.I., McKnight, T.E., Guiloðn, M.A., Klein, K.L., Lowndes, D.H. and Simpson, M.L. 2005. Vertically aligned carbon nanofibers an related structures: controlled synthesis and directed assembly. J. Appl. Physics 97:41301.

Mellor, F.C. and Stace-Smith, R. 1977. Virus-free potatoes by tissue culture. *In*: J. Reinert and Y.P.S. Bajaj (Eds.), Applied and Fundamental Aspects of Plant Cell, Tissue and Organ Culture. Springer-Verlag, Berlin, pp. 616-635.

Menczel, L., Lázár, G. and Maliga, P. 1978. Isolation of somatic hybrids by cloning *Nicotiana heterokaryons* in nurse cultures. Planta 143: 29-32.

Merkle, S.A., Parrot, W.A. and Flinn, B.S. 1995. Morphogenetic aspects of somatic embryogenesis. *In*: T.A. Thorpe (Ed.), *In Vitro* Embryogenesis in Plants. Kluwer, Dordrecht, pp. 155-205.

Meybodi, D.E., Mozafera, J., Babaeiyan, N. and Rahiman, H. 2011. Application of electrotherapy for elimination of potato potyvirus. J.Agric. Sci. Tech. 13: 921-927.

Michayluk, M.R. and Kao, K.N. 1975. A comparative study of sugars and sugar alcohols on cell regeneration and sustained cell division in plant protoplasts. Z. Pflanzenphysiol, 75: 181-185.

Miller, A.R. and Roberts, L.W. 1984. Ethylene biosynthesis and xylogenesis in *Lactuca* pith explants cultured *in vitro* in the presence of auxin and cytokinin. The effect of ethylene precursors and inhibitors. J. Exp. Bot. 35: 691-698.

Miller, C.O. 1961. Kinetin related compounds in plant growth. Ann. Rev. Plant Physiol. 12: 395-408.

Miller, C.O., Skoog, F., Okumura, F.S., Von Saltza., M.H. and Strong, P.M. 1955. Structure and synthesis of kinetin. J. Am. Chem. Soc. 77: 2662-2663.

Miller. L.R. and Murashige, T. 1976. Tissue culture propagation of tropical foliage plants. *In Vitro* 12: 797-813.

Minocha, S.C. 1980. Callus and adventitious shoot formation in excised embryos of white pine (Pi strobus). Can. J. Bot. 58: 366-370.

Mir, R.R. and Varshney, R.K. 2013. Future prospectus of molecular markers in plants. *In*: R.J. Henry (Ed.), Molecular markers in plants. Wiley-Blackwell, West Sussex, U.K., 197 pp.

Misawa, M. 1994. Plant Tissue Culture: An Alternative for Production of Useful Metabolites. FAO Agriculture Services Bull., Rome, 87 pp.

Mishra, V.K. and Goswami, R. 2014. Haploid production in higher plants. ICJCBS Rev. Paper 1(1): 1-21.

Mizuno, K. and Komamine, A. 1978. A possible role of cyclic AMP on tracheary element formation in cultured-root slices. Bot Mag. 91: 213-219.

Mohamed-Yasseen, Y., Barringer, S., Schloupt. R.M. and Splittstoesser, W.E. 1995. Activated charcoal in tissue culture: An overview. Plant Growth Regulator Society of America, 23 (4): 206-213.

Molle, F., Dupis, J.M., Ducos, J.P., Ansehn, A., Crolus-Savidan, I., Peitiard, V. and Freyssinet, G. 1993. Carrot somatic embryogenesis and its applications to synthetic seeds. *In*: K. Redenbaugh (Ed.), Synseeds: Applications of Synthetic Seeds to Crop Improvement. CRC Press, Boca Raton, Fl., pp. 257-287.

Monette, P.L. 1995. Conservation of germplasm of kiwifriut (Actinidia species). *In*: Y.P.S. Bajaj (Ed.). Biotechnology in Agriculture and Forestry 32: Cryopreservation of Plant Germplasm I. Springer, Berlin, pp. 321-331.

Monnier, M. 1976. Culture *in vitro* de embryon immature de *Capsella bursa-pastoris* Moench (I). Rev. Cytol. Biol. Veg. 39: 1-120.

Monnier, M. 1978. Culture of zygotic embryos. *In*: T. A. Thorpe (Ed.), Frontiers of Plant Tissue Culture. University of Calgary Press, Canada, pp. 277-286.

*Monnier, M. 1980. Bull. Soc. Bot. Biol. 151: 391-392 (cited by Pierik, 1989, p. 141).

Morel, G. 1948. Recherches sur la culture associee deparasites obligatiories et de tissus végétaux. Ann. Epiphyt. 14: 123-234.

Morel, G. 1950. Sur la culture des tissues de deux monocotylédones. C.R. Hebd. Seances Acad. Sci. 230: 1099-1101.

Morel, G. 1960. Producing virus-free *Cymbidium*. Am. Orchid Soc. Bull. 29: 495-497.

Morel, G. 1963. La culture *in vitro* du meristéma apical de certaines orchidées. C.R. Hebd. Seances Acad. Sc. 256: 4955-4957.

Morel, G. 1975. Meristem culture techniques for the long-term storage of cultivated plants. *In*: O.H. Frankel and J.G. Hawkes (Eds.), Crop Genetic Resources for Today and Tomorrow. Cambridge Univ. Press, U.K., pp. 327-332.

Morel, G. and Martin, C. 1952. Guérison de dahlias atteints d'une maladie à virus. C.R. Acad. Sci., Paris 235: 1324-1325.

Morel, G. and Martin, C. 1955. Guérison de pommes de terre atteintics de maladies à virus. C. R. Acad. Agric. Fr. 41: 471-474.

Morel, G. and Müller, J.F. 1964. La lecture *In Vitro* du meristeme apical de la pomme de terre. C.R. Acad. Sci. Paris 258:5250-5252.

Mori, K. 1971. Production of virus-free plants by means of meristem culture. Jpn. Agric. Res. 6:1-7.

*Mori, K. 1977. Acta Hortic. 78: 389-396.

Morrison, I.M. 1972. A semimicro method for the determination of lignin and its use in predicting the digestibility of forage crops. J. Sci. Food Agric. 23: 455-463.

Morrison, R.A. and Evans, D.A. 1988. Haploid plants from tissue culture: new plant varieties in a shortened time frame. Biotechnology 6: 684-689.

Motomura, T., Hidaka, T., Moriguchi, T., Akihama, T. and Omura, M. 1995. Intergeneric somatic hybrids between *Citrus* and *Atalantia* or *Severinia* by electrofusion and recombining mitochondrial genomes. Breed. Sci., 45: 309-314.

Mowery, D.C. and Zeidonis, A.A. 2007. Academic patents and material transfer agreements: Substitutes or complements? J. Technol. Transfer 32: 157-172.

Mu, S.K., Fraser, L.G. and Harvey, C.F. 1990. Initiation of callus and regeneration of plantlets from endosperm of *Actinidia* interspecific hybrids. Scientia Hortic. 44: 107-117.

Muhammad, K., Gul, Z., Jammal, Z. Ahmad, M., Rahman, A.R. and Khan, Z.U. 2015. Effect of coconut water from fruit-maturity stages as natural substitute for synthetic PGR in *in vitro* potato micropropagation. Intl. J. Biosci. 6: 84-95.

Muir, W.H. 1953. Cultural conditions favouring the isolation and growth of single cells from higher plants *In Vitro*. Ph.D. Thesis, Univ. of Wisconsin, U.S.A.

Muir, W.H., Hildebrandt, A.C. and Riker, A.J. 1954. Plant tissue cultures produced from single isolated plant cells. Science 119: 877-878.

Mukundan, U., Bhide, V. and Dawda, H. 1999. Production of betalains by hairy root cultures of *Beta vulgaris* L. *In*: T.J. Fu, et al. (Eds.), Plant Cell and Tissue Culture for the Production of Food Ingredients. Kluwer, Dordrecht, pp. 121-127.

Mullin, R.H. and Schlegel, D.E. 1976. Cold storage maintenance of strawberry meristem plantlets. HortScience 11: 100-101.

Mullin, R.H., Smith, S.H., Frazier, N.W., Schlegel, D.E. and McCall, S.R. 1974. Meristem culture frees strawberries of mild yellow edge, pallidosis and mottle diseases. Phytopathology 64: 1425-1429.

Mundree, S.G., Iyer, R., Baker, B., Conard, N., Davis, E.J., Govender, K., Maredza, A.T. and Thomson, J.A. 2006. Prospects of using genetic modification to engineering drought tolerance in crops. *In*: N. Halford (Ed.), Plant Biotechnology: Current and Future. Applications of Genetically modified Crops, pp. 193-205.

Murakawa, T., Kajiyama, S.i., and Fukui, K. 2008a. Improvement of bioactive-beads-mediated transformation by concomitant application of electroporation. Plant Biotechnology 25: 387-390.

Murakawa, T., Kajiyama, S.i., Ikeuchi, T., Kawakami, S. and Fukui, K. 2008b. Improvement of transformation efficiency by bioactive-beads-mediated gene transfer using DNA-lipofectin complex as entrapped genetic material. J. Biosci. Bioengineer. 105: 77-80.

Murakishi, H.H. and Carlson, P.S. 1976. Regeneration of virus-free plants from dark-green islands of tobacco mosaic virus-infected tobacco leaves. Phytopathology 66: 931-932.

Murashige, T. 1974. Plant propagation through tissue cultures. Ann. Rev. Plant Physiol. 25:135-166.

Murashige, T. 1978a. Principles of rapid propagation. *In*: K.W. Hughes et al. (Eds.), Propagation of Higher Plants through Tissue Culture. U.S. Deptt. of Energy, Washington, D.C., pp. 14-24.

Murashige, T. 1978b. The impact of plant tissue culture on agriculture. *In*: T.A. Thorpe (Ed.), Frontiers of Plant Tissue Culture. Univ. of Calgary Press, Calgary, Canada, pp. 15-26.

Murashige, T. and Skoog, F. 1962. A revised medium for rapid growth and bioassays with tobacco tissue cultures. Physiol. Plant. 15: 473-497.

Murashige, T., Shabde, M.N., Hasegawa, P.M., Takakotri, F.H. and Jones, J.B. 1972. Propagation of *Asparagus* through shoot apex culture. I. Nutrient medium for formation of plantlets. J. Am. Soc. Hortic. Sci. 97: 158-161.

Murovec, J. and Bohanec, B. 2012. Haploids and doubled haploids in breeding. *In*: I. Abdurakhmonov (Ed.), Plant breeding. In Tech Europe, Croatia, pp. 88-106.

Murthy, B.N.S., Murch, S.J. and Saxena, P. 1998. Thidiazuron: A potent regulator of *in vitro* plant morphogenesis. In Vitro Cellular and Developmental Biology, Plant 34: 267-275.

Nabors, M.W., Gibbs, S.E., Bernstein, C.S. and Meis, M.E. 1980. NaCl tolerant tobacco plants from cultured cells. Z. Pflanzenphysiol. 97: 13-17.

Nadarajan, J., Mansor, M-, Krishnapillay, B., Staines, H.J., Bensen, E. E. and Harding, K. 2008. Application of differential scanning calorimetry in developing cryopreservation strategies for *Perkia speciosa*, a tropical tree producing recalcitrant seeds. CryoLetters 29: 95-110.

Nadgauda, R.S., Parshamani, V.A. and Mascarenhas, A.F. 1990. Precocious flowering and seedling behaviour in tissue-cultured bamboos. Nature 344: 335-336.

Nag, K.K. and Johri, B.M. 1971. Morphogenic studies on endosperm of some parasitic angiosperms. Phytomorphology 21: 202-218.

Nagai, N., Kitauchi, F., Okamoto, K., Kanda, T., Shiwasaka, M. and Okazaki, M. 1994. A transient increase of phenylalanine ammonia-lyase transcript in kinetin-treated tobacco callus. Biosci. Biotech. Biochem, 58: 558-559.

Nagata, T. and Takebe, I. 1971. Plating of isolated tobacco mesophyll protoplasts on agar medium. Planta 99: 12-20.

Nagmani, R. and Bonga, J.M. 1985. Embryogenesis in subcultured callus of *Larix decidua*. Can. J. For. Res. 15:1088-1091.

Nair, L., Shirgurkar, S. and Mascarenhas, A.F. 1986. Studies on endosperm culture of *Annona squamosa* Linn. Plant Cell Rep. 5: 132-135.

Navarro, L., Roistacher, C.N. and Murashige, T. 1975. Improvement of shoot tip grafting *in vitro* for virus-free *Citrus*. J. Am. Soc. Hortic. Sci. 100: 471-479.

Negrutiu, L, Mouras, A., Gleba, Y.Y., Sidorov, V., Hinnisdale, S., Famelaer, Y. and Jacobs, M. 1989. Symmetric/asymmetric fusion combinations in higher plants. *In*: Y.P.S. Bajaj (Ed.), Biotechnology in Agriculture and Forestry, Vol. 8: Plant Protoplasts and Genetic Engineering I. Springer-Verlag, Berlin, pp. 304-319.

Neuhaus, G., Spangenberg, G., Mittelsten, S.O. and Schweiger, H.G. 1987. Transgenic rapeseed plants obtained by microinjection of DNA into microspore derived embryoids. Theor. Appl. Genet. 75: 30-36.

Neumann, K.H., Barg, W. and Reinhard, E. (Eds.). 1985. Primary and Secondary Metabolism in Plant Cell Cultures. Springer-Verlag, Berlin, pp. 1-377.

Nielsen, L.W. 1960. Elimination of the internal cork virus by culturing apical meristems of infected sweet potatoes. Phytopathology 50: 840-841.

Niino, T. 2000. Cryopreservation of deciduous fruits and mulberry trees. *In*: M.K. Razdan and E.C. Cocking (Eds.), *In*: Conservation of Plant Genetic Resources *In Vitro*. Vol. 2: Applications and Limitations. Science Publishers Inc., Enfield, USA, pp. 195-223.

Nishimura, S., Terashima, T., Higashi, K. and Kamada, H. 1993. Bioreactor culture of somatic embryos for mass propagation of plants. *In*: K. Redenbaugh (Ed.), Synseeds: Application of Synthetic Seeds to Crop Improvement. CRC Press, Boca Raton, FL., pp. 175-181.

Nishizawa, S., Sakai, A., Amano, Y. and Matuzawa, T. 1993. Cryopreservation of Asparagus (*Asparagus officinalis* L.) embryogenic suspension cells and subsequent plant regeneration by vitrification method. Plant Sci. 88: 67-73.

Nitsch, C. 1974. La culture de pollen isole sur milieu synthetique. C.R. Acad. Sciences Paris. 278: 1031-1034.

Nitsch, C. 1977. Culture of isolated microspores. *In*: J. Reinert and Y.P.S. Bajaj (Eds.), Applied and Fundamental Aspects of Plant Cell, Tissue and Organ Culture. Springer-Verlag, Berlin, pp. 268-278.

Nitsch, C., Anderson, S., Godard, M., Neuffer, M.G. and Sheridan, W.F. 1982. Production of haploid plants of *Zea mays* and *Pennisetum* through androgenesis. *In*: E.D. Earle and Y. Demarley (Eds.), Variability in Plants Regenerated from Tissue Culture. Praeger Publishers, N.Y., U.S.A., pp. 69-91.

Nitsch, J.P. 1951. Growth and development *in vitro* of excised ovules. Am. J. Bot. 38: 566-577.

Nitsch, J.P. and Nitsch, C. 1969. Haploid plants from pollen grains. Science 163: 85-87.

Nitzsche, W. 1983. Germplasm Preservation. *In*: D.A. Evans et al. (Eds.), Handbook of Plant Cell Culture, Vol. 1: Techniques for Propagation and Breeding. Macmillan Publishing Co., N.Y., pp. 782-805.

Nobécourt, P. 1937. Cultures en série de tissus végétaux sur milieu artificial. C.R. Hebd. Seances Acad.Sci. 200: 521-523.

Nobécourt, P. 1939. Sur la pérennite et l'augmentation de volume des cultures de tissus végétaux. C.R. Seances Soc. Biol. Ses. Fit. 130: 1270-1271.

Nobécourt, P. 1955. Variations de la morphologic et de la structure de cultures de tissus végétaux. Ber Schweiz. Bot. Ges. 65: 475-480.

Nomura, K. and Komamine, A. 1999. Physiological and morphological aspects of somatic embryogenesis. *In*: W.Y. Soh and S.S. Bhojwani (Eds.), Morphogenesis in Plant Tissue Cultures. Kluwer, Dordrecht, pp. 115-130.

Norstog, K. 1973. New synthetic medium for the culture of premature barley embryos. *In Vitro* 8: 307-308.

Norstog, K. 1979. Embryo culture as a tool in the study of comparative and developmental morphology. *In*: W.R. Sharp et al. (Eds.), Plant, Cell and Tissue Culture: Principles and Applications. Ohio State University Press, Columbus, U.S.A., pp. 179-202.

Nower, A.A. 2014. In vitro propagation and synthetic seed production: an efficient method for *Stevia rebaudiana* Bertoni. Sugar Tech. 16: 100-108.

Ochatt, S.J. and Power, J.B. 1992. Plant regeneration from cultured protoplasts of higher plants. *In*: M. Moo. Young et al. (Eds.), Comprehensive Biotechnology (2nd Suppl.). Pergamon Press, New York, pp. 99-127.

Oglevee-O' Donovan, W. 1993. Culture indexing for vascular wilts on viruses. *In*: J. White (Ed.), Geraniums IV. Ball Publishing, West Chicago, IL, pp. 277-286.

Ohkawa, Y. 1988. Breeding technique of pollen cultivation. Nogyo. Oyobi. Engei. Vol. 63(1): 141-145.

Ohta, Y. 1986. High-efficiency genetic transformation of maize by a mixture of pollen and exogenous DNA. Proc. Natl. Acad. Sci., U.S.A. 83: 715-719.

Okamoto, T. 2011. *In vitro* fertilization of rice gametes: Production of zygotes and embryos in culture. *In*: T.A. Thorpe and E.C. Yeung (Eds.), Plant Embryo Culture: Methods and Protocols. Methods in Molecular Biology, Springer, N.Y., pp. 17-27.

Okamura, S., Miyasaka, K. and Nishi, A. 1973. Synchronization of carrot cell culture by starvation and cold treatment. Exp. Cell Res. 78: 467-470.

Onishi, N., Sakamoto, Y. and Hirosawa, T. 1994. Synthetic seeds as an application of mass production of somatic embryos. Plant Cell, Tissue, Organ Cult., 39: 137-145.

*Ooms, G. et al. (1985). Cited in Tranh Thanh Van, K. and Trinh, T.H. 1990.

Ooms, G., Bossen, M.E., Burrell, M. and Karp, A. 1986. Genetic manipulation of potato with *Agrobacterium rhizogenes*. Potato Res. 29: 367-379.

Otten, L.A. and Schilperoort, R.A. 1978. A rapid microscale method for the detection of lysopine and nopaline dehydrogenase activities. Biochim, Biophys, Acta 527: 497-500.

Ow, D.W. and Howell, S.H. 1989. Incorporation of the firefly luciferase gene. *In*: Y.P.S. Bajaj (Ed.), Biotechnology in Agriculture and Forestry, Vol. 9: Plant Protoplasts and Genetic Engineering II. Springer-Verlag, Berlin, pp. 376-387.

Ow, D.W., Wood, K.Y., Deluca, M., De Wet, J.R., Helinski, D.R. and Howell, S.H. 1986. Transient and stable expression of the firefly luciferase gene in plant cells and transgenic plants. Science 234: 856-859.

Ozeki, V. and Komamine, A. 1985. Induction of anthocyanin synthesis in relation to embryogenesis in carrot suspension culture—A model system for the study of expression and repression of secondary metabolism. *In*: K.H. Neumann et al. (Eds.), Primary and Secondary Metabolism in Plant Cell Cultures. Springer-Verlag, Berlin, pp. 100-106.

Padgette, S.R., Ciopa, G., Shah, D.M., Fraley, R.M. and Kishore, G.M. 1989. Selective herbicide tolerance through protein engineering. *In*: J. Schell and I.K. Vasil) (Eds.), Cell Culture and Somatic Cell Genetics of Plants, Vol. 6: Molecular Biology of Plant Nuclear Genes. Academic Press, Inc., San Diego, pp. 441-476.

Paek, K.Y., Hahn, E.J. and Park, S.O. 2011. Micropropagation of Phalaenopsis orchids via protocorm or protocorm- like bodies. *In*: T.A. Thorpe and E.C.Yeung (Eds.), Plant Embryo Culture: Methods and Protocols. Methods in Molecular Biology, Springer, N.Y., pp. 293-306.

Paine, J.A. et al. 2005. Improving the nutritional value of golden rice through increased pro-vitamin A content. Nature Biotechnol. 4: 482-487.

Panda, A.K., Mishra, S., Bisaria, V.S. and Bhojwani, S.S. 1989. Plant cell reactors—a perspective. Enzyme Microb. Technol. 11: 386-397.

Paranjothy, K., Saxena, S., Banerjee, M., Jaiswal, V.S. and Bhojwani, S.S. 1990. Clonal multiplication of woody perennials. *In*: S.S. Bhojwani (Ed.), Plant Tissue Culture: Applications and Limitations. Elsevier, Amsterdam, pp. 190-219.

Pardey, P.G., Wright, B.D., Nottenberg, C., Binenbaum, E. and Zambrano, P. 2003. Intellectual property and developing countries: Freedom to operate in agricultural biotechnology. Biotechnology and Genetic Resource Policies, Brief 3. Intl. Food Policy Res. Inst., Washington, DC. pp. 1-6.

Pareek, L.K. (Ed.) 2001. Trends in Plant Tissue Culture and Biotechnology. Agrobios, Jodhpur, 334 pp.

Parr, A.J., Robins, R.J. and Rhodes, M.J.C. 1986. Product release from plant cells grown in culture. *In*: M. Phillips et al. (Eds.), Secondary Metabolism in Plant Cell Cultures. Cambridge University Press, Cambridge, pp. 173-177.

Parrot, W.A., Williams, E.G., Hildebrandt, D.F. and Collins, G.B. 1989. Effects of genotype on somatic embryogenesis from immature cotyledons of soybean. Plant Cell, Tissue, Organ Cult. 16: 15-21.

Pasternak, T., Miskolizi, P., Ayaydin, F., Mészáros, T., Dudits, D. and Fehér, A. 2000. Exogenous auxin and cytokinin dependant CDKs and cell division in leaf protoplast-derived cells of alfa alfa. Plant Growth Regu. 32: 129-141.

Paszkowski, J. and Saul, M.W. 1986. Direct gene transfer to plants, *In*: H. Weissbach and A. Weissbach (Eds.), Methods in Enzymology, Vol. 118. Academic Press, N.Y., pp. 668-675.

Patel, K.R. and Berlyn, G.P. 1983. Cytochemical investigations on multiple bud formation in tissue cultures *of Pinus coulteri*. Can. J. Bot. 61: 575-585.

Pehu, E. 1991. RFLP analysis of organeller genomes in somatic hybrids. *In*: K. Lindsey (Ed.), Plant Tissue Culture Manual. Kluwer, Dordrecht, pp. D6: 1-8.

Pelletier, G. and Ilami, M. 1972. Les facteurs de l'androgenese In *Vitro* chez *Nicotiana tabacum*. Z. Pflanzenphysiol., 68: 97-114.

Pelletier, G. and Ramagosa, I. 1993. Somatic hybridization in plant breeding: principles and prospects. *In*: M.D. Hayward and N.O. Basemark (Eds.), Plant Breeding Series I. Chapman and Hall, London, pp. 93-106.

Pence, V.V. 1995. Cryopreservation of recalcitrant seeds. *In*: Y.P.S. Bajaj (Ed.), Biotechnology in Agriculture and Forestry 32: Cryopreservation of Plant Germplasm 1. Springer, Berlin, pp. 29-50.

Pennell, R.I., Janniche, L., Scofield, G.N., Booij, H., de Vries, S.C. and Roberts, K. 1992. Identification of a transitional cell state in the developmental pathway to carrot somatic embryogenesis. J. Cell Biol. 119: 1371-1380.

Perry, S.E., Lethi, M.D. and Fernandez, D.E. 1999. The MADS-domain protein AGAMOUS-like 16 accumulates in embryogenic tissues with diverse origins. Plant Physiol. 120: 121-130.

Petolino, J.F., Cowen, N.M., Thompson, S.A. and Mitchell, J.C. 1990. Gamete selection for heat stress tolerance in maize. J. Plant Physiol. 136: 219-224.

Phillips, R.L., Kaepler, S.M. and Olhoft, P. 1994. Genetic instability of plant tissue cultures: breakdown of normal controls. Proc. Natl. Acad. Sci., USA, 91: 5222-5226.

PhytaSource. 1994. Technical Bulletin. Sigma Chemical Company, St. Louis, MO 63178, USA.

Pierik, R.L.M. 1989. *In Vitro* Culture of Higher Plants. Martinus Nijhoff Publishers, Dordrecht, The Netherlands, pp. 1-344.

Pijnacker, L.P., Ferwerda, M.A., Puite, K.J. and Roest, S. 1987, Elimination of *Solanum phureja* nucleolar chromosomes in S. *taberosum* + S. *phureja* somatic hybrids. Theor. Appl. Genet. 73: 878-882.

Pliego-Alfaro, F., Encina, C.L. and Barcelo-Munoz, M. 1987. Propagation of avocado root stocks by tissue culture. S. Afr. Growers Yrb. 10: 36-39.

Potrykus, I. and Hoffmann, F. 1973. Transplantation of nuclei into protoplasts of higher plants. Z. Pflanzenphysiol. 69: 287-289.

Potrykus, I., Harms, C.T. and Lörz, H. 1977. Callus formation from stem protoplasts of corn (*Zea mays* L.). Mol. Gen. Genet. 156: 347-350.

Potrykus, I., Saul, M.W., Petruska, J., Paszkowski, J. and Shillito, R.D. 1985. Direct gene transfer to cells of a graminaceous monocot. Mol. Gen. Genet. 199: 183-188.

Power, J.B. and Cocking, E.G. 1968. A simple method for the isolation of very large numbers of leaf protoplasts using mixtures of cellulase and pectinase. Biochem. J. 111: 33.

Power, J.B. Frearson, E.M., Hayward, C. and Cocking, E.C. 1975. Some consequences of the fusion and selective culture *of Petunia* and *Parthenoclissus* protoplasts. Plant Sci. Lett. 5: 197-207.

Power, J.B., Berry, S.F., Chapman, J.V and Cocking, E.C. 1980. Somatic hybridization between sexually incompatible petunias. *Petunia parodii, Petunia parviflora*. Theor. Appl. Genet. 57: 1-4.

Power, J.B., Berry, S.F., Chapman, J.V., Cocking, E.C. and Sink, K.C. 1979. Somatic hybrids between unilateral cross-incompatible *Petunia* species. Theor. Appl. Genet. 55: 97-99.

Power, J.B., Cummins, S.E. and Cocking, E.C. 1970. Fusion of isolated plant protoplasts. Nature (London) 225: 1016-1018.

Power, J.B., Davey, M.R., Freeman, J.P., Mulligan, B.J. and Cocking, E.C. 1986. Fusion and transformation of plant protoplasts. *In*: H. Weissbach and A. Weissbach (Eds.), Methods in Enzymology, Vol. 118. Academic Press, N.Y., pp. 578-588.

Power, J.B., Frearson. E.M., Hayward, C., George, D., Evans, P.K., Berry, S.F. and Cocking, E.C. 1976. Somatic hybridization of *Petunia hybrida* and *P. parodii*. Nature 263: 500-502.

Prat, D. 1983. Genetic variability induced in *Nicotiana sylvestris* by protoplast culture. Theor. Appl. Genet. 64: 223-230.

Prenosil, J.E. and Pederson. H. 1983. Immobilized plant cell reactors. Enzyme Microbiol. Technol. 5: 323-331.

Prioli, L.M. and Sondahl, M.R. 1989. Plant regeneration and recovery of fertile plants from protoplasts of maize (*Zea mays* L.). Biotechnol. 7: 589-594.

Pritchard, H.W., Beeby, L.A. and Davies, R.I. 2000. Role of embryo culture in the seed conservation of palms and other species. *In*: M.K. Razdan and E.G. Cocking (Eds.), Conservation of Plant Genetic Resources *In Vitro*, Vol. 2. Science Publishers Inc., Enfield, USA, pp. 89-138.

Pullman, G.S. and Bucalo, K. 2011. Pine somatic embryogenesis using zygotic embryos as explants.

In: T.A. Thorpe and E.C.Yeung (Eds.), Plant Embryo Culture: Methods and Protocols. Methods in Molecular Biology, Springer, N.Y., pp. 267-291.

Pullman, G.S., Zeng, X., Copeland-Kemp, B., Lucerzi, J., May, S.W. and Bucalo, K. 2015. Conifer somatic embryogenesis: Improvements by supplementation of medium with oxidation-reduction agents. Tree Physiol. 35: 209-224.

Purohit, S.S. and Mathur, S.K. 1990. Fundamentals of Biotechnology. Agro-Botanical Publishers (India), Bikaner, India, pp. 1-164.

Quak, F. 1977. Meristem culture and virus-free plants. *In*: J. Reinert and Y.P.S. Bajaj (Eds.), Applied and Fundamental Aspects of Plant Cell, Tissue, and Organ Culture. Springer-Verlag, Berlin, pp. 598-615.

Quinn, J., Simon. J.E. and Janick, J. 1989. Recovery of g-linolenic acid from somatic embryos of borage. J. Am. Soc. Hort. Sci. 114: 511-515.

Racchi, M.L. and Manzocchi, L.A. 1988. Anthocyanin and proteins as biochemical markers in maize endosperm culture. Plant Cell Rep. 7: 78-81.

Raghavan, V. 1966. Nutrition, growth and morphogenesis of plant embryos. Biol. Rev. 41: 1-58.

Raghavan, V. 1978. Origin and development of pollen embryoids and pollen cultures in cultured anther segments of *Hyoscyamus niger* (henbane). Am. J. Bot. 65: 984-1082.

Rajasekaran, K. 2013. Biolistic transformation of cotton zygotic embryo meristem. *In*: B.H. Zhang, (Ed.), Transgenic Cotton: Methods and Protocols. Methods Mol. Biol. 958: 47-57.

Ramawat, K.G. 2000. Plant Biotechnology. S. Chand Co. Ltd., New Delhi, 230 pp.

Rangaswamy, N.S. 1961. Experimental studies on female reproductive structures of *Citrus microcarpa* Bunge. Phytomorphology 11: 109-127.

Rangaswamy, N.S. and Shivarma, K.R. 1971. Overcoming self-incompatibility in *Petunia auxillaris* II. Placental pollination *in vitro*. J. Indian Bot. Soc. 50A: 286-296.

Rao, A.Q., Baksh, A., Kaini, S. Shahzad, K., Shahid, A.A. Husnain, T. 2009. The myth of transformation. Biotechnol. Adv. 27: 753-763.

Rashid, A. 1983. Angiosperm pollen—A system for cell differentiation. Curr. Sci. 52: 964-967.

Rashid, A. and Reinert, J. 1981. High-frequency embryogenesis in *ab initio* pollen cultures of *Nicotiana tabacum*. Die Naturwissenschaften 68: 378.

Rathore, K.S and Goldsworthy, A. 1985. Electric control of shoot regeneration in plant tissue cultures. Biotechnology 3: 1107-1109.

Raveh, D., Huberman, E. and Galun, E. 1973. *In Vitro* culture of tobacco protoplasts: Use of feeder techniques to support division of cells plated at low densities. *In Vitro* 9: 216-222.

Ravi, M. and Chan, S.W.L. 2010. Haploid plants produced by centromere-mediated genome elimination. Nature 464: 615-618.

Razdan, A., Razdan, M.K., Rajam, M.V. and Raina, S.N. 2008. Efficient protocol for *in vitro* production of androgenic haploids of *Phlox drummondii*. Plant Cell Tiss. Organ Cult. 95: 245-250.

Razdan, M.K. 1980. A rapid method of isolating chlorophyll-deficient protoplasts for selection in the somatic hybridization. *In*: P.S. Rao et al. (Eds.). Plant Tissue Culture, Genetic Manipulation and Somatic Hybridization of Plant Cells. Proc. Symp. BARC, India, pp. 248-254.

Razdan, M.K. 1984-85. Fusion of higher plant protoplasts and selection of somatic hybrids. *In*: R.N. Gohil (Ed.), Recent Trends in Botanical Research. Scientific Publishers, Jodhpur, India, pp. 65-78.

Razdan, M.K. 2003. Introduction to Plant Tissue Culture (2nd Edition). Science Publishers, Enfield, USA, 375 pp.

Razdan, M.K. and Cocking, E.C. 1981. Improvement of legumes by exploring extra-specific genetic variation. Euphytica 30: 818-833.

Razdan, M.K. and Cocking, E.C. 1997. Conservation of Plant Genetic Resources *In Vitro*, Vol. 1: General Aspects. Science Publishers Inc., Enfield, USA; 314 pp.

Razdan, M.K. and Cocking, E.C. 2000. Conservation of Plant Genetic Resources *In Vitro*, Vol. 2: Application and Limitations. Science Publishers Inc., Enfield, USA, 315 pp.

Razdan-Tiku, A., Razdan, M.K. and Raina, S.N. 2014. Production of triploid plants from endosperm cultures of Phlox drummondii. Biologia Plantarum 58:153-158.

Rechinger, C. 1893. Untersuchungen über die Grenzen der Teil barkeit im Pflanzenreich. Abh. Zool. Ges. Wien. 43:310-334.

Redenbaugh, K. 1993. Synseeds: Applications of Synthetic Seeds to Crop Improvement. CRC Press, BocaRaton,Fl., 481 pp.

Redenbaugh, K. and Walker, K. 1990. Role of artificial seeds in alfalfa breeding. *In*: S.S. Bhojwani (Ed.), Plant Tissue Culture: Applications and Limitations. Elsevier, Amsterdam, pp. 102-135.

Redenbergh, K. 1990. Application of artificial seed to tropical crops. Hort. Sci. 25: 225-251.

Redenbergh, K., Paasch, B., Nichoi, J., Kossler, M., Viss, P. and Walker, K. 1986. Somatic seeds: encapsulation of asexual plant embryos. Biotechnology 4: 797-801.

Ree, J.F. and Guerra, M.F. 2015. Palm (Araceae) somatic embryogenesis. *In vitro* Cellu. Dev. Biol. Plant 51: 582-602.

Reed, B.M. 1989. *In Vitro* conservation of germplasm. *In*: H.T. Stalker and C. Chapman (Eds). IBPGR Training Courses: Lecture Series 2. IBPGR, Rome, pp. 23-30.

Reed, B.M. 1991. Application of gas-permeable bags for *In Vitro* cold storage. Plant Cell Rep. 10: 431-434.

Reed, B.M. 1992. Cold storage of strawberries *In Vitro*: a comparison of three storage systems. Fruit Var. J. 46: 98-102.

Reed, B.M. 1995. Factors affecting the *In Vitro* storage of strawberry. HortScience 30: 871.

Reed, B.M. and Chang, Y. 1997, Medium-and long-term storage of *In Vitro* cultures of temperate and fruit crops. *In*: M.K. Razdan and E.C. Cocking (Eds.), Conservation of Plant Genetic Resources *In Vitro*. Vol. 1: General Aspects. Science Publishers Inc., Enfield, USA, pp. 67-105.

Reinert, J. 1958. Morphogenese und ihre Kontrolle an Gewebckuluren aux Carotten. Naturwissenschaften 45: 344-345.

Reinert, J. 1959. Über die Kontroile der Morphogenese und die Induktion von Advientiveem bryonen an Gcwebekuluren aus Karotten. Planta 58: 318-333.

Reish, B., Duke, S.H. and Bingham, E.T. 1981. Selection and characterization of ethionine-resistant alfalfa (*Medicago sativa* L.) cell lines. Theor. Appl. Genet. 59: 89-94.

Reynaerts, A., Van de Wiele, H., De Sutter, G. and Janssens, J. 1993. Engineered genes for fertility control and their application in hybrid seed production. Sci. Hortic. 55: 125-139.

Reynolds, J.F. and Murashige, T. 1979. Plant cell lines. *In*: W.B. Jakoby and I.H. Pastan (Eds.), Methods in Enzymology, Vol. 58. Academic Press, N.Y., pp. 478-486.

Rhodes, C.A., Lowe, K. and Ruby, K. 1988. Plant regeneration from embryogenic maize cell cultures. Biotechnology 6: 56.

Rhodes, C.A., Pierce, D.A., Mettler, I.J., Mascarenhas, D. and Detmer, J.J. 1989. Genetically transformed maize (*Zea mays* L.) plants from protoplasts. *In*: Y.P.S. Bajaj (Ed.), Biotechnology in Agriculture and Forestry; Vol. 9: Plant Protoplasts and Genetic Engineering II. Springer-Verlag. Berlin, pp. 99-106.

Rines, H.W. and McCoy, T.J. 1981. Tissue culture initiation and plant regeneration in hexaploid species of oats. Crop Science 21: 837-842.

Robbins, W.J. 1922. Effect of autolysed yeast and peptone on growth of excised corn root tips in the dark. Bot. Gaz. (Chicago) 74: 59-79.

Roberts, A.V., Walker, S., Horan, I., Smith, E.F. and Mottleg, J. 1992. The effect of growth retardants, humidity and lighting at Stage III on Stage IV of micropropagation in chrysanthemum and rose. Acta Hortic. 319: 153-158.

Robertson, D., Beech, I. and Bolwell, G.P. 1995. Regulation of the enzymes of UDP-sugar metabolism during differentiation of French bean (*Phaseolus vulgaris* L.). Phytochemistry 39: 21-28.

Robertson, D. and Haigler, E. 1990. Cited from Robertson, D. et al. 1995.

Rode, A., Hartmann, C, De Buyser, J. and Henry, Y, 1989. Evidence for a direct relationship between mitochondrial genome organization and regeneration ability in hexaploid wheat somatic tissue cultures. Curr. Genet. 14: 387-394.

Ronald, P.C. 1997. Making rice disease resistant. Sci. Am. 77:100-105.

Rosevear, A. 1988. Immobilized plant cells. *In*: R.D. King and P.S. Cheetham (Eds.), Food Biotechnology. Elsevier Applied Science, London, pp. 83-116.

Rossini, L. 1972. Division of free leaf cells *of Calystegia sepium In Vitro*. Phytomorphology 22: 21-29.

Rowhani, A., Uyemoto, J.K., Golino, A. and Martelli, G.P. 2005. Pathogen testing and certification of *Vitis* and *Prunus* species. Ann. Rev. Phytopathol. 43: 261-268.

Ruffoni, B., Pistelli, L., Bertoli, A. and Pistelli, S. 2010. Plant cell cultures: bioreactors for industrial production. Adv. Exptl. Med.Biol. 698: 203-221.

Rukavtsova, E.B., Alekseeva, V.V. and Burianov, Y.I. 2010. The use of RNA interference for the metabolic engineering of plants. Russ. J. Bioorg. Chem. 36:146-156.

Saad, A.M. and Elshahed, I. A.M. 2012. Plant tissue culture media. *In*: A.Leva and M.R. Rinaldi (Eds.), Recent Advances in Plant *In Vitro* Culture. In-Tech., pp. 29-40.

Sagawa, Y. and Kunisaki, J.T. 1990. Micropropagation of floricultural crops. *In*: P.V. Ammirato et al. (Eds.), Handbook of Plant Cell Culture, Vol. 5: Ornamental Species. McGraw-Hill Publishing Company, N.Y., pp. 25-56.

Sakai, A. 1997. Potentially valuable cryogenic procedures for cryopreservation of cultured plant meristems. *In*: M.K. Razdan and E.C. Cocking (Eds.), Conservation of Plant Genetic Resources *In Vitro*. Vol. 1: General Aspects. Science Publishers Inc., Enfield, USA, pp. 53-66.

Sakai, A., Kobayashi, S. and Oiyama, 1.1990. Cryopreservation of nuclear cells of navel orange (*Citrus sinensis* osb. var. *brasiliensis* Tanaka) by vitrification. Plant Cell Rep. 9: 30-33.

Sakamoto, Y., Onishi, N. and Hirosawa, T. 1995. Delivery systems for tissue culture by encapsulation. *In*: J. Aitken-Christie et al. (Eds.), Automation and

Environmental Control in Plant Tissue Culture. Kluwer, Dordrecht, pp. 215-243.

Sakuta, M. and Komamine, A. 1987. Cell growth and accumulation of secondary metabolites. *In*: F. Constabel and I.K. Vasil (Eds.), Cell Culture and Somatic Cell Genetics of Plants, Vol. 4: Cell Culture in Phytochemistry. Academic Press, Inc., San Diego, pp. 97-114.

Sakuth, T., Schobert, C., Pecsvaradi, A., Eichholz, A., Komor, E. and Orlich, G. 1993. Specific proteins in the sieve tube exudate of *Ricinus communis* L. seedlings: separation, characterization and *in vivo* labelling. Planta 191: 207-213.

San Noeum, L.H. 1976. Haploides *d'Hordcum vulgare* L. par culture *in vitro* non fecondes. Ann. Amelior. Plantes 26: 751-754.

Sánchez-Diaz, R.A., Castillo, A.M. and Vallies, M.P. 2013. Microspore embryogenesis in wheat: new marker genes for early, middle and late stages of embryo development. Sex. Plant Reprod. 26: 287-296.

Sanford, K.K., Earle, W.R. and Likely, G.D. 1948. The growth *in vitro* of single isolated tissue cells. JNCL, Natl. Cancer Inst. 9: 229-246.

Sangthong, R., Chin, D.P., Supaibul-watana, K. and Mii, M. 2009. Gametoclonal hybridization between egg cell protoplasts and mesophyll protoplasts of *Petunia hybrida*. Plant Biotechnol. J. 26: 377-383.

Saravitz, J., C.H. and Boyer, C.D. 1987. Starch characteristics in culture of normal and mutant maize endosperm. Theor. Appl. Genet. 12: 489-495.

Sasse, F., Buchholz, M. and Berlin, J. 1983. Selection of cell lines of *Catharanthus roseus* with increased tryptophan decarboxylase activity. Z. Naturforsch., C: Biosci. 38 C: 916-922.

Sato, F. and Yamada, V. 1984. High berberine-producing cultures of *Coptis japonica* cells. Phytochemistry 23: 281-285.

Sato, F., Hashimoto, A., Hachiya, A., Tamara, K. Choi, K.B., Murashige, T., Fujimoto, H. and Yamada, Y. 2001. Metabolic engineering of plant alkaloids biosynthesis. Proc. Natl. Acad. Sci., USA 98: 367-372.

Saunders, J.W. and Bingham, E.T. 1972. Production of alfalfa plants from callus tissue. Crop Sci. 12: 804-808.

Saxena, P.K., Lui, V., and King, J. 1987. Nuclear transplantation into protoplasts; Optimal conditions for induction and determination of nuclear uptake. J. Plant Physiol. 128: 451-460.

Saxena, P.K., Mil, M., Crosby, W.L., Fowke, L.C. and King, J. 1986. Transplantation of isolated nuclei into plant protoplasts. Planta 168: 29-35.

Schardl, C.L., Byrd, A.D., Binzoin, G., Mitchell, A., Hildebrandt, D.F. and Hunt, A.G. 1987. Design and construction of a versatile system for expression of foreign genes in plants. Gene 61: 1-11.

Schell, J. and Vasil, I.K. (Eds.). 1989. Cell Culture and Somatic Cell Genetics of Plants, Vol. 6: Molecular Biology of Plant Nuclear Genes. Academic Press, Inc., San Diego, pp.1-494.

Schell, J. St. 1987. Transgenic plants as tools to study the molecular organization of plant genes. Science 237: 1176-1183.

Schenk, H. 1982. *Brassica napus*: successful resynthesis by protoplast fusion between *B. oleracea* and *B. campestris* V. Int. Congr. Plant Tissue Cell Cult. Tokyo, p. 109 (abstr.).

Schieder, O. 1978. Somatic hybrids of *Datura innoxia* Mill. + *Datura discolor* Benth. and of *Datura innoxia* Mill. + *Datura stramonium* L. var. *totula* L. I. Mol. Gen. Genet. 162: 113-119.

Scott, K.J., Chin, J.C. and Wood, CJ. 1978. Isolation and culture of cereal protoplasts. *In*: Proceedings of Symposium on Plant Tissue Culture. Science Press, Peking, pp. 298-315.

Scott, R., Dagless, E., Hodge, R., Wyatt. P., Soufleri, I. and Draper, J. 1991. Patterns of gene expression in developing anthers of *Brassica napus*. Plant Mol. Biol. 17: 195-209.

Scragg, A.H. 1990. Large scale cultivation of *Helianthus annus* cell suspension. Enzyme Microb. Technol. 12: 82-85.

Scragg, A.H. 1994. Secondary products from cultured cells and organs II. Large scale culture. *In*: R.A. Dixon and R.A. Gonzales (Eds), Plant Cell Culture: A Practical Approach. IRL Press, Oxford University Press, pp. 199-224.

Scragg, A.H. 1999. Alkaloid accumulation in *Catharanthus roseus* suspension cultures. *In*: R.D. Hall (Ed.), Methods in Molecular Biology, Vol. Ill, Plant Cell Culture Protocols Humana Press Inc., Totowa, N.J., pp. 393-400.

Sechley, K.A. and Schroeder, H. 2002. Intellectual property protection of plant biotechnology invention. Trends Biotechnol. 20: 456-461.

Sehgal, C.B. 1969. Experimental studies on maize endosperm. Beitr. Biol. Pflanz. 46:233-238.

Sehgal, C.B. and Khurana, S. 1985. Morphogenesis and plant regeneration from cultured endosperm of *Emblica officinalis* Gaertn. Plant Cell Reports. 4: 263-266.

Seibert, M. 1973. The effects of wavelength and intensity on growth and shoot initiation in tobacco callus. *In Vitro* 80: 435.

Seibert, M. 1976. Shoot initiation from carnation shoot apices frozen to −196°C. Science (N.Y.) 191: 1178-1179.

Semal, J. and Lepoivre, P. 1990. Application of tissue culture variability to crop improvement. *In*: S.S.

Bhojwani (Ed.), Plant Tissue Culture: Applications and Limitations. Elsevier, Amsterdam, pp. 301-315.

Senaratna, T., McKersie, B.D. and Bowley, S.R. 1990. Artificial seeds of alfalfa (*Medicago sativa* L.). Induction of desiccation tolerance in somatic embryos. *In Vitro* Cell. Dev. Biol. 26: 85-90.

Senda, M., Takeda, J., Abe, S. and Nakamura, T. 1979. Induction of cell fusion plant protoplasts by electric stimulation. Plant Cell Physiol. 20: 1441-1443.

Shabde, M. and Murashige, T. 1977. Hormonal requirements of excised *Dianthus caryophyllus* L. shoot apical meristem *in vitro*. Am. J. Bot. 64: 443-448.

Shah, D.M., Horsch, R.B., Klee, H.J., Kishore, G.M., Winter, J.A., Turner, N.E., Hironata, C.M., Sanders, P.R., Gasser, C.S., Ag kent, S., Siegel, N.R., Rogers, S.G. and Fraley, R.T. 1986. Engineering herbicide tolerance in transgenic plants. Science 233: 478-481.

Shahin, E.A. 1984. Isolation and culture of protoplasts: tomato. *In*: I.K. Vasil (Ed.), Cell Culture and Somatic Cell Genetics of Plants, Vol. 1: Laboratory Procedures and Their Applications. Academic Press, Inc., Orlando, pp. 381-390.

Sharp, W.R., Evans, D.A. and Sondahl, M.R. 1982. Applications of somatic embryogenesis to crop improvement. *In*: A. Fujiwara (Ed.), Plant Tissue Culture. Proceedings 5th International Congress of Plant Tissue and Cell Culture, Japan. Japanese Association for Plant Tissue Culture, pp. 759-762.

Shaw, C.H., Leemans, J., Shaw, C.H., Van Montagu, M. and Schell, J. 1983. A general method for the transfer of cloned genes to plant cells. Gene 23: 315-330.

Sheat. D.E.G., Flechter, B.H. and Street, H.E. 1959. Studies on the growth of excised roots VIII. The growth of excised tomato roots supplied with various inorganic sources of nitrogen. New Phytol. 58: 128-141.

Shepard, J.F. 1977. Regeneration of plants from protoplasts of potato virus X-infected tobacco leaves. Virology 78: 261-266.

Shepard, J.F. 1981. Protoplasts as sources of disease resistance in plants. Ann. Rev. Phytopathol. 19: 145-155.

Shepard, J.F., Bidney, D. and Shahin, E. 1980. Potato protoplasts in crop improvement. Science (N.Y.) 208: 17-24.

Shihkin, M., Fraser, L.G. and Harvey, C.F. 1990. Initiation of callus and regeneration of plantlets from endosperm of *Actinidia* interspecific hybrids. Scientia Horticulture 44: 107-117.

Shillito, R.D., Carswell, G.K., Johnson, C.M., DiMaio, J.J. and Harms, C.T. 1989. Regeneration of fertile plants from protoplasts of elite inbred maize. Biotechnology 7: 581-587.

Shininger, T.L. 1975. Is DNA synthesis required for the induction of differentiation in quiescent root cortical parenchyma? Dev. Biol. 45: 137-150.

Shuler, ML. 1981. Production of secondary metabolites from plant tissue culture—problems and prospects. Ann. N.Y. Acad. Sci. 369: 65-79.

Siew, W.L., Kwok, M.Y., Ong, Y.M., Liew, H.P. and Yew, B.K. 2014. Effective use of synthetic technology in regeneration of *Dendrobium* White Fairy orchid. J. ornamental Plants 4:1-7.

Sikdar, S.R., Chatterjee, G., Das, S. and Sen, S.K. 1990. "Erussica", the intergeneric fertile somatic hybrid developed through protoplast fusion between *Eruca saliva* and *Brassica juncea* (L.) Czern, Theor. Appl. Genet. 79: 561-567.

Sim, G.E., Loh, C.S. and Goh, C.J. 2007. High frequency in vitro flowering *Dendrobium* Madame Thong-in (Orchidaceae). Plant Cell Rep. 26: 383-397.

*Simon, S. 1908. Experimentelle unterschungen uber die differenzierung vorgaangen callus-gewebe von holzge wachsen. Jahrb. Wiss. Bot. 45: 351-478.

Simpson, J. and Herrera-Estrella, L. 1989. DNA recombinants and transformation of agricultural crops. *In*: Y.P.S. Bajaj (Ed.), Biotechnology in Agriculture and Forestry, Vol. 9: Plant Protoplasts in Genetic Engineering II. Springer-Verlag, Berlin, pp. 75-98.

Singh, A., Hallihosur, S. and Ranga, A. 2009. Changing landscape in biotechnology patenting. World Patent Information 31: 219-225.

Sjodin, C. and Glimelius, K. 1989a. *Brassica naponigra*, a somatic hybrid resistant to *Phoma lingam*. Theor. Appl. Genet. 77: 651-656.

Sjodin, C. and Glimelius, K. 1989b. Transfer of resistance against *Phome lingam to Brassica napus* by asymmetric hybridization combined with toxin selection. Theor. Appl. Genet. 78: 513-520.

Skálová, D., Maváratilová, B., Doleźalová, I., Vašut, R.J. and Lebeda, A. 2012. Haploid and mixoploid cucumber (*Cucumis sativus* L.) protoplasts- isolation, regeneration and fusion. J. Appl. Bot. Food Qual. 85: 64-72.

Skirvin, R.M. 1978. Natural and induced variation in tissue culture, Euphytica 27: 241-266.

Skirvin, R.M. and Janick, J. 1976. Tissue culture-induced variation in scented *Pelargonium* sp. J. Amer. Soc. Hortic. Sci. 101: 281-290.

Skoog, F. 1944. Growth and organ formation in tobacco tissue cultures. Am. J. Bot. 31: 19-24.

Skoog, F. and Miller, C.O. 1957. Chemical regulation of growth and organ formation in plant tissue cultured *In Vitro*. Symp. Soc. Exp. Biol. 11: 118-131.

Skoog, F. and Tsui, C. 1951. Growth substances and the formation of buds in plant tissues. *In*: F. Skoog (Ed.),

Plant Growth Substances. Univ. Wisconsin Press, Madison, Wisconsin, pp. 263-285.

Sladky, Z. and Havel, L. 1976. The study of the conditions for the fertilization *In Vitro* in maize. Biol. Plant. 18: 469-472.

Slater, A., Scott, N.W. and Fowler, M.R. (2nd Edition), 2008. Pant Biotechnology: The Genetic Manipulation of Plants. Oxford Univ. Press, Oxford, 376 pp.

Sluis, C.J. 2006. Integrating automation technologies with commercial micropropagation. *In*: S. DuttaGupta and I. Ibaraki (Eds.), Plant Tissue Culture Engineering. Springer, the Netherlands, pp. 231-251.

Smith, J.M., Davison, S.W. and Payne, G.F. 1990. Development of a strategy to control the dissolved concentrations of oxygen and carbon-dioxide at constant shear in a plant cell bioreactor. Biotechnol. Bioeng. 35: 1088-1101.

Smith, M.W., Ito, M., Miyawaki, M., Sato, S., Yoshikawa Y., Wada, S., Maki, H., Nakagawa, I. and Komamine, A. 1997. Plant 21D7 protein, a nuclear antigen associated with cell division is a component of the 26S proteasome. Plant Phys. 113: 281-291.

Smith, R.H. 2000. Plant Tissue Culture. Techniques and Experiments (Second Edition), Academic Press, San Diego, 231 pp.

Smith, R.H. 2013. Plant Tissue Culture: Techniques and Experiments. Elsevier, New Delhi, 188 pp.

Smith, S.M. and Street, H.E. 1974. The decline of embryogenic potential as callus and suspension cultures of carrot (*Daucus carota* L.) are serially subcultured. Am. Bot. 38: 223-241.

Smolders, W. 2005. Plant genetic resources for food and agriculture: Facilitated access or utility plants on plant varieties. IP Strategy Today 13: 1-17.

Snow, R. 1935. Activation of cambial growth by pure hormones. New Phytol. 34: 347-360.

Sommer, H.E., Brown, C.L. and Kermanik, P.P. 1975. Differentiation of plantlets in long leaf pine (*Pinus palustris* Mill.) tissue cultured *in vitro*. Bot. Gaz. 136:196-200.

Sondahl, M.R. and Sharp, W.R. 1977. High frequency induction of somatic embryos in cultured leaf explants of *Coffea arabica* L. Z. Pflanzenphysiol. 81: 395-408.

Sone, T., Nagamori, E., Ikeuchi, T., Mizukami, A., Takura, Y. and Kajiyama, S.i. 2008. A novel gene delivery system in plants with calcium alginate microbeads. J. Biosci. Bioengineer. 94: 87-91.

Spiegel-Roy, P. and Kochba, J. 1980. Role of polyembryony and apomixis in *Citrus* prropagation and breeding. *In*: A. Fletcher (Ed.), Advances in Biochemical Engineering. Springer-Verlag, Berlin, pp. 28-48.

Springer, W. and Bailey, A. 1990. Use of anther culture in plant breeding. *In*: A.B. Bennett and S.D. O'Neill (Eds.), Horticultural Biotechnology. Wiley. Liss, New York, pp. 327-334.

Sree Ramulu, K. 1986. Case histories of genetic variability: Potato. *In*: P.V. Ammirato et al. (Eds.). Cell Cultures and Somatic Cell Genetics of Plant, Vol. 3: Crop Species. Macmillan Publishing Co., N.Y., pp. 449-173.

Sree Ramulu, K., Verhoeven. H.A., Gilissen, L.J.W., Dijkhuis, P. and van der Valk, H.C.P.M. 1987. Induction of micronuclei and their application for transfer of single, specific chromosomes in plant cells. 7th Int. Protoplast Symp., Wageningen, p. 33 (abstr.).

Srivastava, P.S. 1973. Formation of triploid 'plantlets', in endosperm cultures of *Putranjiva roxburghii*. Z. Pflanzenphysiol 69: 270-273.

Stachel, S.E., Timmermann, B. and Zambryski, P. 1987. Activation of *Agrobacterium tumefaciens* Vir gene expression generates multiple single-stranded T-strand molecules from the pTi A6 T-region: requirement for 5' *Vir D* gene products. EMBO J. 4: 857-863.

Staritsky, G. 1997. Backgrounds and principles of *in vitro* conservation of plant genetic resources. *In*: M.K. Razdan and E.C. Cocking (Eds.), Conservation of Plant Genetic Resources *In Vitro*, Vol. 1. Science Publishers Inc., USA., pp. 27-49.

Steiner, A.A. and van Winden, H. 1970. Recipe for ferric salts of ethylene diaminetetra-acetic acid. Plant Physiol. 46: 862.

Stephanopoulos, G. 2007. Challenges in engineering microbes for biofuel production. Science 315: 801-804.

Sterckx, S. 2010. Is the non-patentability of "Essentially Biological Processes" under threat? J. World Intellectual Property. 13: 1-23.

Steward, F.C., Caplin. M and Millar. F.K. 1952. Investigations on growth and metabolism of plant cells. Ann. Bot. (London) 16: 57-77.

Steward, F.C., Kent. A.E. and Mapes, M.O. 1966. The culture of free plant cells and its signification for embryology and morphogenetics. *In*: A.A. Moscona and A. Monroy (Eds.), Current Topics in Developmental Biology. Academic Press, N.Y., pp. 113-154.

Steward, F.C., Mapes, M.O. and Mears, K. 1958. Growth and organized development of cultured cells II. Organization in cultures grown from freely suspended cells. Am. J. Bot. 45: 705-708.

Steward, J.M. and Hsu, C.L. 1978. Hybridization of diploid and tetraploid cottons through *in-ovulo* embryo culture. J. Hered. 69: 404-408.

Stewart Jr., C.N. 2007. Biofuels and biocontainment. Nature Biotechnol. 25: 283-284.

Stewart Jr., C.N. 2008. Plant Biotechnology and Genetics: Principles, Techniques and Applications. John Wiley & Sons Inc., Hobokeu, New Jersey, 377 pp.

Stohs, S.J. and Staba, E.J. (1964). Cited by Purohit and Mathur, 1990, p. 133.

Stone, O.M. 1963. Factors affecting the growth of carnation plants from shoot apices. Ann. Appl. Biol. 52: 199-209.

Street, H.E. 1977. Cell (suspension) cultures—techniques. In: H.E. Street (Ed.), Plant Tissue and Cell. Culture. Univ. California Press, Berkeley, Calif., pp. 61-102.

Street, H.E., McGonagle, M.P. and McGregor, S.M. 1952. Observations on the "staling" of white medium by excised tomato roots. II. Iron availability. Physiol. Plant. 5: 243-276.

Styer, D.J. 1985. Bioreactor technology for plant propagation. In: R.R. Henke et al. (Eds.), Tissue Culture in Forestry and Agriculture. Plenum Press, New York, pp. 117-130.

Sugimoto, K., Gordon, S.P., Meyerowitz, E.M. 2011. Rgeneration in plants and animals: differentiaton, transdifferentiation or just differentiaton. Trends Cell Biol. 21:212-218.

Sun, C.S. and Chu, C.C. 1981. The induction of endosperm plantlets and their ploidy of barley in vitro. Acta Bot. Sin. 23: 265.

Sun, D.Q., Lu, X.H, Liang, G.L., Guo, Q.G., Mo, Y.W. and Xie, J.H. 2011. Production of triploid plants of papaya by endosperm culture. Plant Cell Tiss. Organ Cult. 140:23-29.

Sun, Y., Nie, Y., Guo, X., Huang, C. and Zhang, X. 2006. Somatic hybrids between Gossypium hirsutum L. (4x) and G. davidsonii Kellong (2x) produced by protoplast fusion. Euphytica 151: 393-400.

Sun, Z.X., Zhao, C.Z., Zheng, K.L., Qi, X.F. and Fu, Y.P. 1983. Somaclonal genetics of rice, Oryza sativa L. Theor. Appl. Genet. 67: 67-73.

Sunderberg, E., Landgren, M. and Glimelius, K. 1987. Fertility and chromosome stability in Brassica napus resynthesized by protoplast fusion. Theor. Appl. Genet. 75: 96-104.

Sunderland, N. 1982. Induction of growth in the culture of pollen. In: M.M. Yeoman and D.E.S. Truman (Eds.), Differentiation In Vitro. Cambridge University Press, Cambridge, pp. 1-24.

Sung, Z.R. 1979. Relationship of indole-3-acetic acid and tryptophan concentrations in normal and 5-methyltryptophan resistant cell lines of wild carrots. Planta 145: 339-345.

Suzuki, K., Amino, S., Takeuchi, Y. and Komamine, A. 1990. Differences in the composition of the cell walls of two morphologically different lines of suspension-cultured catharanthus cells. Plant Cell Physiol. 31: 7-14.

Swaminathan, M.S. 1983. Genetic conservation: Microbes to man. Presidential Address, XV. Intl. Congr. Genet., December 12-21, 1983, New Delhi, pp. 3-32.

Tahir, M., Waraich, E.A. and Satsolla, C. 2011. Genetic transformation protocols using zygotic embryos as explants. In: T.A.Thorpe and E.C.Yeung (Eds.), Plant Embryo Culture: Methods and Protocols, Methods in Molecular Biology Series Vol. 710, Humana Press, N.Y., pp. 309-326.

Takebe, I. and Otsuki, Y. 1969. Infection of tobacco mesophyll protoplasts by tobacco mosaic virus. Proc. Natl. Acad. Sci., U.S.A. 64: 843-849.

Takebe, I., Labib, G. and Melchers, G. 1971. Regeneration of whole plants from isolated mesophyll protoplasts of tobacco. Naturwissens-chaften 58: 318-320.

Takebe, I., Otsuki, Y. and Aoki, S. 1968. Isolation of tobacco mesophyll cells in intact and active state. Plant Cell Physiol., Tokyo, 9: 115-124.

Tal, M. 1990. Somaclonal variation for salt resistance. In: Y.P.S. Bajaj (Ed.), Biotechnology in Agriculture and Forestry, Vol. 11: Somaclonal Variation in Crop Improvement I. Springer-Verlag, Berlin, pp. 236-257.

*Tan, M.L., Mc. (1987). Somatic hybridization and cybridization in solanaceous species. Academisch Proefschrift. Vrije Univ to Amsterdam, (cited by Bajaj, Y.P.S. 1989b, p. 24).

Tanaka, H. 1987. Large scale cultivation of plant cells at high density: A review. Process Biochem. 8: 106-113.

Tao, W., Wilkinson, J., Stanbridge, E.J. and Berns, M.W. 1987. Direct gene transfer into human cultured cells facilitated by laser micropuncture of the cell membrane. Proc. Natl. Acad. Sci., U.S.A. 84: 4180-4184.

Taylor, P.W.J. 1997. In Vitro germplasm conservation of sugarcane cultivars, basic sugarcane species and related genera. In: M.K. Razdan and E.C. Cocking (Eds.), Conservation of Plant Genetic Resources In Vitro, Vol. 1. Science Publishers Inc., Enfield, USA, pp. 243-256.

Teixeira da Silva, J.A. and Tanaka, M. 2010. Thin cell layers: the technique In: M.R. Davey and O. Anthony (Eds.), Plant Cell Culture: Essential Methods. John-Wiley, USA, 357 pp.

Tekoah, Y., Shulman, A., Kizhner, T., Ruderfer, I., Fax, L. Nataf, Y., Bartfeld, D., Ariel, T., Gingis-Velit ski, S., Hunania, U. and Shaaltiel, Y. 2015. Large scale production of pharmaceutic proteins in plant cell culture- the protalix experience. Plant Biotechnol. J. 13: 1199-1208.

Tepfer, S.S., Greyson, R.I., Craig, W.R. and Hindman, J.L. 1963. *In vitro* culture of floral buds of *Aquilegia*. Am. J. Bot. 50: 1035-1045.

Terzi, M. and Lo Schiavo, F. 1990. Somatic embryogenesis, *In*: S.S. Bhojwani (Ed.), Plant Tissue Culture: Applications and Limitations. Elsevier, Amsterdam, pp. 1-461.

Thammina, C., He, M., Lu, L., Cao, K., Chen, Y., Tian, L., Chen, J., Mc Avoy, R., Ellis, D., Zhao, D., Wang, Y., Zhang, X. and Li, Y.2011. In vitro regeneration of triploid plants *Euonymus alatus* 'Compactus' (Burning Bush) from endosperm tissues. HortScience 46: 1141-1147.

Thévenin, L. and Doré, C. 1976. *Asparagus* (*Asparagus officinalis* L.) improvement and its chief asset: *in vitro* culture. Ann. Amélior Plantes 26: 655-674.

Thomas, D. 1999. *In Vitro* Production of Haploids and Triploids of *Morus alba* L. Ph.D Thesis, Univ. of Delhi, India, 151 pp.

Thomas, D. and Chaturvedi, R. 2008. Endosperm culture: a novel method of triploid plant production. Plant Cell Tiss. Organ Cult. 93:1-14.

Thomas, D., Bhatnagar, A.K. and Bhojwani, S.S. 2000. Production of triploid plants of mulberry (*Morus alba* L.) by endosperm culture. Plant Cell Rep. 19: 395-399.

Thomas, D., Bhatnagar, A.K., Razdan, M.K. and Bhojwani, S.S. 1999. A reproducible protocol for the production of gynogenic haploids of mulberry, *Morus alba* L. Euphytica 110: 169-173. Thomas, D., Bhatnagar, (1999).

Thomas, E., Miflin, B.J., Bright, S.W.J. and Lancaster, V. 1982. The regeneration of plants from protoplasts of agriculturally important species. *In*: E.D. Earle and Y. Demarly (Eds.), Variability in Plants Regenerated from Tissue Culture. Praeger, N.Y., pp. 58-68.

Thompson, M.R. and Thorpe, T.A. 1997. Analysis of protein patterns during shoot initiation in cultured *pinus radiata* cotyledons. J. Plant Physiol. 51: 724-734.

Thomson, A.J., Gunn, R.E., Jellis, G.I., Boulton, R.E. and Lacey, C.N.D. 1986. The evaluation of potato somaclones. *In*: J. Semal (Ed.), Symposium "Somaclonal Variations and Crop Improvement". Martinus Nijhoff Publishers, Dordrecht/Boston/Lancaster, pp. 236-243.

Thorpe, T.A. (Ed.). 1995. *In Vitro* Embryogenesis in Plants. Kluwer, Dordrecht, pp. 155-205.

Thorpe, T.A. 2007. History of plant tissue culture. Mol. Biotechnol. 37: 169-180.

Thorpe, T.A. and E.C. Yeung (Eds.). 2011. Plant Embryo Culture: Methods and Protocols. Methods in Molecular Biology Series Vol. 710, Humana Press, 377 pp.

Thorpe, T.A. and Stasolla, C. 2001. Somatic embryogenesis. *In*: S.S. Bhojwani and W.Y. Soh (Eds.), Current Trends in the Embryology of Angiosperms. Kluwer, Dordrecht, pp. 219-236.

Tian, L., Ko, Y., Gan, S., Chen, Y., Yang, Z. and Wang, X. 2012. Triploid plant regeneration from mature endosperms of *Sapium sebiferum*. Plant Growth Regul. 68: 319-324.

Tisserat, B. 1981. Date palm tissue culture. Agric. Res. Serv. Adv. Agric. Technol. Western Ser No.17, U.S.D.A., Calif., pp. 1-50.

Tisserat, B. and Murashige. T. 1977. Probable identity of substances in *Citrus* that repress asexual embryogenesis. *In Vitro* 13: 785-789.

Topfer, R., Gronenbom, B., Schell, J. and Steinbiss, H.H. 1989. Uptake and transient expression of chimeric genes in seed-derived embryos. Plant Cell 1: 133-139.

Torres, K.C. 1989. Tissue Culture Techniques for Horticultural Crops. avi-Van Nostrand Reinhold, N.Y., pp. 1-285.

Torrey, J.G. 1975. Tracheary element formation from single isolated cells in culture. Physiol. Plant 38: 158-165.

Toryama, K. and Hinata, K. (1988). Cited by Gleba and Shlumukov. 1990.

Toth, K.F., Wang, Q. and Sjolund, R.D. 1994. Monoclonal antibodies against phloem P-protein from plant tissue cultures. I. Microscopy and Biochemical Analysis. Am. J. Bot. 81: 1370-1377.

*Tramier (1965) C.R. Seanc. Acad. Agric. France 51: 918-922 (cited by Pierik, 1989, p. 340).

Tran Thanh Van, K. 1977. Regulation of morphogenesis. *In*: W. Barz et al. (Eds.), Plant Tissue Culture and Its Biotechnological Application. Springer-Verlag, Berlin, pp. 367-385.

Tran Thanh Van, K. 1980. Control of morphogenesis by inherent and exogenously applied factors in thin cell layers. *In*: I.K. Vasil (Ed.), Perspectives in Plant Cell and Tissue Culture, Suppl. 11 A, Int. Rev. Cytol. Academic Press, N.Y., pp. 175-179.

Tran Thanh Van, K. and Trinh, T.H. 1990. Organogenic differentiation. *In*: S.S. Bhojwani (Ed.), Plant Tissue Culture: Applications and Limitations. Elsevier, Amsterdam, pp. 34-53.

Tran Thanh Van, M., Thi Dien, N. and Chlyah, A. 1974. Regulation of organogenesis in small explants of superficial tissue of *Nicotiana tabacum* L. Planta 119: 149-159.

Trécul, A. 1853. Accroissement des végétaux dicotyledones ligneux (reproduction du bois et de l'ecorce par le bois décortiqué). Ann. Sci. Nat. Dot. Biol. Veg. 29: 157-192.

Trigiano, R.N. and Gray, D.J. (Eds.). 2000. Plant Tissue Culture Concepts and Laboratory Exercises (Second Edition). CRC Press, Boca Raton, 454 pp.

Trigiano, R.N. and Gray, D.J. (Eds.). 2011. Plant Tissue Culture, Development and Biotechnology. CRC Press, Boca Raton, 608 pp.

Trigiano, R.N., May, R.N. and Conger, B.V. 1992. Reduced nitrogen influences somatic embryo quality and plant regeneration from suspension cultures of orchard grass. In Vitro Cell Div. Biol. 28p: 187-191.

Trinick, M.J. 1982. Host-*Rhizobium* associations. *In*: J.M. Vincent (Ed.), Nitrogen Fixation in Legumes. Academic Press London, pp. 111-122.

Tulecke, W. 1953. A tissue derived from the pollen of *Ginkgo biloba*. Science (N.Y.) 117: 599-600.

Tulecke, W., McGranahan, G. and Ahmadi, H. 1988. Regeneration by somatic embryogenesis of triploid plants from endosperm of walnut, *Juglans regia* L. ev. Manregian. Plant Cell Reports 7: 301-304.

Uchimiya, H., Fushimi, T., Hashimoto, H., Harada, H., Syono, K. and Sugarawa, Y. 1986. Expression of a foreign gene in callus derived from DNA-treated protoplasts of rice (*Oryza sativa* L.). Mol. Gen. Genet. 204: 204-207.

Uemura, M. and Sakai, A. 1980. Survival of carnation (*Dianthus caryophyllus*) shoot apices frozen to the temperature of liquid nitrogen. Plant Cell Physiol., Tokyo, 21: 85-94.

UNEP. 1995. Global Biodiversity Assessment. Cambridge Univ. Press, Cambridge, pp. 457-475.

Ushiyama. K., Oda, H. and Miyamoto, Y. 1986. *In*: D.A. Somers et al. (Eds.), Proc. 6th Intnl. Congr. Plant Tissue and Cell Culture. IAPTC, Minneapolis, pp. 252.

Vaek, M., Reynaerts, A. and Höete, H. 1989. Protein engineering in plants: Expression of *Bacillus thuringiensis* insecticidal protein genes. *In*: J. Schell and I.K., Vasil (Eds.), Cell Structure and Somatic Cell Genetics of Plants, Vol. 6: Molecular Biology of Plant Nuclear Genes. Academic Press, Inc., San Diego, pp. 425-440.

Van Harten, A.M., Bouter, H. and Broertjes, C. 1981. *In Vitro* adventitious bud techniques for vegetative propagation and mutation breeding of potato (*Solanum tuberosum* L.). II. Significance for mutation breeding. Euphytica 30: 1-8.

Van Hengel, A.J., Guzzo, F., Van Kammen, A. and de Vries, S.C. 1998. Expression pattern of the carrot EP3 endochitinase genes in suspension cultures and in developing seeds. Plant Physiol. 117: 43-53.

Van Overbeek, J., Conklin, M.E. and Blakslee, A.F. 1941. Factors in coconut milk essential for growth and development of very young *Datura* embryos. Science 94: 350-351.

Varshney, A. and Anis, M. 2014. Synseed conception for short-term storage germplasm exchange and potentialities of regeneration of genetically engineered stable plantlets of desert date tree (*Balanites aegyptica*). Agroforest Syst. 88:321-329.

Vasil, I.K. and Thorpe, T.A. (Eds). 1994. Plant Cell and Tissue Culture. Kluwer, Dordrecht, 362 pp.

Vasil, I.K. and Vasil, V., 1986. Regeneration in cereal and other grass species. *In*: I.K. Vasil (Ed.), Cell Culture and Somatic Cell Genetics of Plants, Vol. 3. Plant Regeneration and Genetic Variability. Academic Press, New York, pp. 121-150.

Vasil, I.K., Vasil, V., Chin-Yi, L., Pegg, O.A., Zsott, H. and Da-Yuan, W. 1982. Somatic embryogenesis in cereals and grasses. *In*: E.D. Earle and Y. Demarly (Eds.), Variability in Plants Regenerated from Tissue Culture. Praeger, N.Y., pp. 3-21.

Vasil, V. and Hildebrandt, A.C. 1965. Differentiation of tobacco plants from single isolated cells in microcultures. Science 150: 889-890.

Vasil, V. and Vasil, I.K. 1980. Isolation and culture of cereal protoplasts. II. Embryogenesis and plantlet formation from protoplasts of *Pennisetum americanum*. Theor. Appl. Genet. 56: 97-99.

*Verhoeven et al. 1987. Cited from Y.P.S. Bajaj (Ed.), 1989b. Biotechnology in Agriculture and Forestry, Vol. 9: Plant Protoplasts and Genetic Engineering II. Springer-Verlag, Berlin, pp. 10.

Verpoorte, R., Van der Heijden, R., Ten Hoopen, H.J.G. and Memclink, J. 1998. Metabolic engineering for the improvement of plant secondary metabolite production. Plant Tissue Culture Biotechnology 4: 3-20.

Verpoorte, R., Van der Heijden, R., Ten Hoopen, H.J.G. and Memelink, J. 1999. Novel approaches to improve secondary metabolite prodection. *In*: T.J. Fu, et al. (Eds), Plant Cell and Tissue Culture for the Production of Food Ingredients. Kluwer, Dordrecht, pp. 85-100.

*Viseur, J. 1989. Cited by Semal, J. and Lepoivre, P. 1990.

*Vöchting, H. 1878. Dber Organbildung im Pflanzenreich. Max Cohen Publ. Bonn (cited by Gautheret 1985).

Von Arnold, S. and Hakman, l. 1986. Effect of sucrose on initiation of embryogenic callus cultures from mature zygotic embryos of *Picea abies* (L.) Karst. (Norway spruce). J. Plant Physiol. 122: 261-265.

Wakasa, K. 1979. Variation in the plants differentiated from tissue culture of pineapple. Jpn. J. Breed. 29: 13-22.

Walbot, V. 1978. Control mechanisms for plant embryogeny. *In*: M.E. Clutter (Ed.), Dormancy and Development Arrest—Experimental Analysis in Plants and Animals. Academic Press, N.Y., pp. 114-116.

Walker, K.A., Wendeln, ML. and Jaworski, E.G. 1979. Organogenesis in callus tissue of *Medicago sativa*. The temporal separation of induction processes from differentation processes. Plant Sci. Lett. 16: 23-30.

Walkey, D.G.A. 1978. *In Vitro* methods for virus elimination. *In*: T.A. Thorpe (Ed.), Frontiers in Plant Tissue Culture, 1978. Univ. Calgary Press, Calgary, Canada, pp. 245-254.

Walkey, D.G.A. and Cooper, J. 1976. Heat inactivation of cucumber mosaic virus in cultured tissues of *Stellaria media*. Ann. Appl. Biol. 84: 425-428.

Wallin, A., Glimelius, K. and Eriksson, T. 1978. Enucleation of plant protoplasts by cytochalasin B. Z. Pflanzenphysiol. 87: 333-340.

Wallin, A., Glimelius, K. and Eriksson, T. 1979. Formation of hybrid cells by transfer of nucleus via fusion of miniprotoplasts from cell lines of nitrate reductase deficient tobacco. Z. Pflanzenphysiol. 91: 89-94.

Wang, A.S. and Phillips, R.L. 1984. Synchronization of suspension culture cells. *In*: I.K. Vasil (Ed.), Cell Culture and Somatic Cell Genetics of Plants, Vol. 1: Laboratory Procedures and Their Applications. Academic Press, Inc., Orlando, pp. 175-181.

Wang, P.J. and Hu, N.Y. 1980. Regeneration of virus-free plants through *in vitro* culture. *In*: A. Flechter (Ed.), Advances in Biochemical Engineering: Plant Cell Culture II. Springer-Verlag, Berlin, pp. 61-99.

Wang, P.J. and Huang, L.C. 1975. Callus cultures from potato tissues and the exclusion of potato virus X from plants regenerated from stem tips. Can. J. Bot. 53: 2565-2567.

Wang, Q., Panis, B., Engelmann, F., Lambardi, M., and Valkonen, J.P.T. 2009. Elimination of plant pathogen by cryotherapy of shoot tips: A technique for pathogen eradication to produce healthy planting material and prepare healthy plant genetic sources for cryopreservation. Ann. Appl. Biol. 154: 351-363.

Warren-Wilson, J., Keys, W.M.S., Warren-Wilson, P.M. and Roberts, L.W. 1994. Effects of auxin on the spatial distribution of cell division and xylogenesis in lettuce pith explants. Protoplasma 183: 162-181.

Watanabe, K. 2000. Effect of post-thaw treatments on viability of cryopreserved plant cells. *In*: M.K. Razdan and E.C. Cocking (Eds.), Conservation of Plant Genetic Resources *In Vitro*. Vol. 2: Applications and Limitations Science Publishers Inc., Enfield, USA, pp. 1-19.

Wedzony, M., Forster, B.P., Zur, I., Golemiec, E., Szechyńska-Hebda, M., Dabas, E. and Gotebiowska, G. 2009. Progress in doubled haploid technology in higher plants. *In*: A. Touraev et al. (Eds.), Advances in Haploid Production in Higher Plants. Springer Sceince+Business Media, pp. 1-34.

Weissbach, A. and Weissback, H. (Eds.). 1986. Methods in Enzymology, Vol. 118. Academic Press, N.Y., 856 pp.

Weller, S.C., Masiunas, J.B. and Gressel, J. 1987. Biotechnologies of obtaining herbicide tolerance in potato. *In*: Y.P.S. Bajaj (Ed.), Biotechnology in Agriculture and Forestry, Vol. 3: Potato. Springer-Verlag, Berlin, N.Y., pp. 280-297.

Went, F.W. 1926. On growth accelerating substances in the coleoptile of *Avena sativa*. Proc. K. Ned. Akad. Wet. Ser. C30: 10.

Went, F.W. 1938. Specific factors other than auxin affecting growth and root formation. Plant Physiol. 13: 55-80.

Wenzel, G., Hoffmann, F. and Thomas, E. 1977. Increased induction and chromosome doubling of androgenic haploid rye. Theor. Appl. Genet. 51: 81-86.

Wetmore, R.H. and Sorokin, S. 1955. On the differentiation of xylem. J. Arnold Arbor. Harv. Univ. 36: 305-317.

Whetten, R. and Sederoff, R. 1995. Lignin Biosynthesis. Plant Cell. 7: 1001-1003.

Whitaker, R.J. and Hashimoto, T. 1986. Production of secondary metabolites. *In*: D.A. Evans et al. (Eds.), Handbook of Plant Cell Culture, Vol. 4: Techniques and Applications. Macmillan Publishing Co., N.Y., pp. 264-286.

White, F.F., Taylor, B.H., Huffman, G.A., Gordon, M.P. and Nester, E.W. 1985. Molecular and genetic analysis of the transferred DNA regions of the hairy root-inducing plasmid of *Agrobacterium rhizogenes*. J. Bact. Biol. 164: 33-44.

White, J.L. 1982. Regeneration of virus-free plants from yellow-green areas and TMV-induced enations of *Nicotians tomentosa*. Phytopathology 72: 866-867.

White, P.R. (Ed.). 1963. The Cultivation of Animal and Plant Cells (2nd ed.). 2 Ronald Press, New York, pp. 1-239.

White, P.R. 1934. Potentially unlimited growth of excised tomato root tips in a liquid medium. Plant Physiol. 9: 585-600.

White, P.R. 1937. Vitamin B_1 in the nutrition of excised tomato roots. Plant Physiol. 12: 803-811.

White, P.R. 1939. Potentially unlimited growth of excised plant callus in artificial nutrient. Am. J. Bot. 26: 59-64.

Widholm, J.M. 1977. Relation between auxin autotrophy and tryptophan accumulation in cultured plant cells. Planta 134: 103-108.

Widholm, J.M. 1987. Selection of mutants which accumulate desirable secondary compounds. *In*: F. Constabel and I.K. Vasil (Eds.), Cell Culture and Somatic Cell Genetics of Plants, Vol. 4; Cell Culture in Phytochemistry. Academic Press, Inc., San Diego, pp. 125-137.

Wiesner, J. 1884. Untersuchungen uber die Wachstunsbewegungen der Wurzeln. *Sitzungsber*, Akad. Wiss. Wien. Math. Naturwiss KI., 1,89: 223-302.

Williams, E. and De Lautour, G. 1980. The use of embryo culture with transplanted nurse endosperm for the production of interspecific hybrids in pasture legumes. Bot. Gaz. (Chicago), 141: 252-257.

Williams, E.G. 1987. Somatic embryogenesis as a tool in plant improvement. *In*: S. Natesh et al. (Eds.), Biotechnology in Agriculture, Oxford & IBH Publishing Co. Pvt. Ltd., New Delhi, pp. 179-184.

Williams, E.G. and Maheswaran, G. 1986. Somatic embryogenesis: factors influencing coordinated behaviour of cells as an embryogenic group. Ann. Bot. 57: 443-462.

Wink, A., Alfermann, A.W., Franke, R., Wetteraner, B., Distl, M., Windhovel, J., Krahn, O., Fuss, E., Garden, H., Mohagzadeh, A., Wildi, E. and Ripplinger, P. 2005. Sustainable bioproduction of phytochemicals by plants in vitro cultures: anticancer agents. Plant Genetic Resour. Charact. Util. 3: 90-100.

Winkler, H. 1908. Besprechung der Arbeit G. Haberlandt's Kultur Versuche mit isolierten Pflanzenzellen. Bot. Z. 60: 262-264.

Withers, L.A. 1979. Freeze-preservation of somatic embryos and clonal plantlets of carrot (*Daucus carota* L.). Plant Physiol. 63: 460-467.

Withers, L.A. 1984. Freeze-preservation of cells. *In*: I.K. Vasil (Ed.), Cell Culture and Somatic Cell Genetics of Plants, Vol. 1: Laboratory Procedures and Applications. Academic Press, Inc., Orlando. pp. 608-620.

Withers, L.A. 1985. Cryopreservation of cultured cells and meristems. *In;* I.K. Vasil (Ed.), Cell Culture and Somatic Cell Genetics of Plants, Vol. 2: Cell Growth, Nutrition, Cytodifferentiation, and Cryopreservation. Academic Press, Inc., Orlando, pp. 253-316.

Withers, L.A. 1988. The current status of *In Vitro* culture for the international movement of plant germplasm. *In*: 1BPGR (1988) Advisory Committee on *In Vitro* Storage Meeting Report, Rome, pp. 47-60.

Withers, L.A. and King, P.J. 1979. Proline: A novel cryoprotectant for freeze-preservation of cultured cells of *Zea mays* L. Plant Physiol. 64: 675-678.

Withers, L.A. and King, P.J. 1980. A simple freezing unit and cryopreservation method for plant cell suspensions. Cryo-Lett. 1: 213-220.

Wu, Y., Ye, D., Installe, P., Jacobs, M. and Negrutiu, J. 1990. Sex determination in the dioecious *Melandrium*. The *In Vitro* culture approach. *In*: Nijkamp, H. J.J. et al. (Eds.), Progress in Plant Cellular and Molecular Biology, Kluwer, Dordrecht, pp. 239-243.

Xu, Y.S., Clark, M.S. and Pehu, E. 1993. Use of RAPD markers to screen somatic hybrids between *Solanum tuberosum* and *S. brevidens*. Plant Cell Rep. 12: 107-109.

Xu, Z.H. and Wei, Z.M. (Eds). 1997. Plant Protoplast Culture and Genetic Transformation. Shanghai Sci. Tech. Publ. (in Chinese), 246 pp.

Xu, Z.H. and Xue, H.W. 1999. Plant regeneration from cultured protoplasts: *In*: W.Y. Soh and S.S. Bhojwani (Eds.), Morphogenesis in Plant Tissue Cultures. Kluwer, Dordrecht, pp. 37-93.

Xynias, I., Kaufalis, A., Gouli- Vavchinoudi, C. and Rupakias, D. 2015. Factors affecting doubled haploid plant production via maize technique in bread wheat. Acta Biol. Cracoviensia Botanica. 56(2): 67-73.

Yadav, V.K. and Mehta, S.L. 1995. *Lathyrus sativus*: a future pulse crop free of neurotoxin. Curr. Sci. 68: 288-292.

Yang, F., Montoya, A.L. and Nester, E.W. 1980. Plant tumor reversal associated with the loss of foreign DNA. *In Vitro* 16: 87-92.

Yang, H.V. and Zhou, C. 1982. *In Vitro* induction of haploid plants from unpollinated ovaries and ovules. Theor. Appl. Genet. 63: 97-104.

Yarrow, S. 1999. Production of cybrids in rapeseed (*Brassica napus*). *In*: R.D. Hall (Ed.), Plant culture Protocol. Humara Press, Totawa, N.J., pp. 211-226.

Ye, X., Al-Babili, S., Klöti, A. Zhang, J., Ducca, P. and Beyer, P. 2000. Engineering provitamin A (β-carotene) biosynthetic pathway into (carotenoid-free) rice endosperm. Science 287: 303-305.

Yeoman, M.M. 1987. Techniques, characteristics, properties, and commercial potential of immobilized plant cells. *In*: F. Constabel and I.K. Vasil (Eds.), Cell Structure and Somatic Cell Genetics of Plants, Vol. 4: Cell Culture in Phytochemistry. Academic Press. Inc., San Diego, pp. 197-215.

Yeung, E.G. and Meinke, D.W. 1993. Embryogenesis in angiosperm development of the suspensor. Plant Cell 5: 1371-1381.

Yeung, E.G. and Sussex, I.M. 1979. Embryology of *Phaseolus coccineus*: The suspensor and the growth of the embryo-proper *in vitro*. Z. Pflanzenphysiol. 91: 423-433.

Yu, X.S., Chu, B.J., Liu, R., Sun, J., Jones, B.J. and Wang, H.Z. 2012. Characteristics of fertile somatic hybrids of *G. hirsutum* L. and *G. trilobium* generated via protoplast fusion. Theor. Appl. Genet. 125: 1503-1516.

Zambryski, P., Herrera-Estrella, L., De Block, M., Van Montagu, M. and Schell, J. 1984. The use of Ti plasmid of *Agrobacterium* to study the transfer and expression of foreign DNA in plant cells: New vectors and methods. *In*: J. Setlow and A. Hollaender (Eds.), Genetic Engineering, Principles and Methods, Vol. 6. Plenum Press, N.Y., pp. 253-278.

Zambryski, P., Joos, H., Genetillo, C., Leemans, J., Van Montagu, M. and Schell. J. 1983. Ti plasmid vector for the introduction of DNA into plant cells without alteration of their normal regeneration capacity. EMBO J. 2: 2143-2150.

Zarsky, V., Rihova, L. and Tupy, J. 1990. Biochemical and cytological changes in young tobacco pollen during *in vitro* starvation in relation to pollen embryogenesis. *In*: H.J.J. Nijkamp, et al. (Eds.), Progress in Plant Cellular and Molecular Biology, Kluwer, Dordrecht, pp. 228-233.

Zelcer, A., Aviv, D. and Galun, E. 1978. Interspecific transfer of cytoplasmic male sterility by fusion between protoplasts of normal *Nicotiana sylvestris* and X-ray irradiated protoplasts of male sterile *N. tabacum*. Z. Pflanzenphysiol. 90: 397-407.

Zeng, J.Z. 1983. Application of anther culture technique to crop improvement in China. *In*: S.K. Sen and K.L. Giles (Eds.), Plant Cell Culture in Crop Improvement. Plenum Press, N.Y., pp. 351-363.

Zenk, M.H., El-Shagi, H. and Schulte, U. 1975. Anthraquinone production by cell suspension cultures of *Morinda citrifolia*. Planta Medica Suppl. 75: 79-81.

Zenk, M.H., El-Shagi, H., Arens, H., Stockigt, J., Weiler, E.W. and Dens, B. 1977. Formation of the indole alkaloids serpentine and ajmalicine in suspension cultures *Catharanthus roseus*. *In*: W. Barz et al. (Eds.), Plant Tissue Culture and Its Biotechnological Application. Springer-Verlag, Berlin, pp. 27-43.

Zenkteler, M. 1980. Intra ovarian and *In Vitro* pollination. Int. Rev. Cytol. Suppl. 11B: 137-156.

Zenkteler, M. 1985. *In Vitro* pollination and fertilization. *In*: I.K. Vasil (Ed.), Cell Culture and Somatic Cell Genetics of Plants, Vol. 1: Laboratory Procedures and Their Applications. Academic Press, Inc., Orlando, pp. 269-282.

Zenkteler, M. and Bagniewska-Zadworna. A. 2001. Distant *In Vitro* Pollination of ovules. Phytomorphology (Golden Jubilee Issue: Trends in Plant Science). 51: 225-235.

Zenkteler, M., Misiura, E. and Guzowska, I. 1975. Studies on obtaining hybrid embryos in test tubes. *In*: H.Y. Mohan Ram et al. (Eds.), Form, Structure and Function in Plants. Sarita Prakashan, Meerut, India, pp. 180-187.

Zhang, H.N., Yang. H., Rech, E.L., Golds, T.J., Davis, A.S., Mulligan, B.J., Cocking, E.C. and Davey, M.R. 1988. Transgenic rice plants produced by electroporated-mediated plasmid uptake into protoplasts. Plant Cell Rep. 7: 379-384.

Zhang, H-X. and Blumwald, E. 2001. Transgenic salt tolerant tomato plants accumulate salt in foliage and not in fruit. Nature Biotech. 19: 693-787.

Zhao, H. 1988. Induction of endosperm plantlets of 'Jinfeng' pear *in vitro* and their ploidy. *In*: Genetic manipulation in crops. Cassell Tycooly, U.K., pp. 123-124.

Zhongyi, L., Tanner, G.J. and Larkin, P.J. 1990. Callus regeneration from *Trifolium subterraneum* protoplasts and enhanced protoplast division by low-voltage treatment and nurse, cells. Plant Cell, Tissue, Organ Cult. 21: 67-73.

Zhou, J.-Y., Guo, F.-X. and Razdan, M.K. 2000. Somatic embryogenesis and germplasm conservation of Plants. *In*: M.K. Razdan and E.C. Cocking (Eds.), Conservation of Plant Genetic Resources *In Vitro*, Vol. 2: Applications and Limitations. Science Publishers, Inc. Enfield, USA, pp. 167-192.

Zhu, Q., Chen, X, Li, W. and Chen, Y. 1988. *In Vitro* regeneration of plantlets from immature endosperm of maize (*Zea mays*). *In*: Genetic Manipulation in Crops. Cassell Tycooly, U.K., pp. 370-371.

Zhu, Z.C. and Wu, H.S. 1979. *In Vitro* induction of haploid plantlets from unpollinated ovaries of *Triticum aestivium* and *Nicotiana tabacum*. Acta. Genet. Sin. 6: 181-183.

Ziebur, N.K. and Brink, R.A. 1951. The stimulative effect of *Hordeum* endosperms on the growth of immature plant embryos *in vitro*. Am. J. Bot. 38: 253-256.

Zimmerman, J.L. 1993. Somatic Embryogenesis: A Model for Early Development in Higher Plants. Plant Cell 5: 1411-1423.

Zimmerman, R.H. 1986. Propagation of fruit, root and vegetable crops—Overview. *In*: R.H. Zimmerman et al. (Eds.), Tissue Culture as a Plant Production System for Horticultural Crops. Martinus Nijhoff, Dordrecht, The Netherlands, pp. 183-200.

Zimmerman, R.H. and Broome, O.C. 1980. Blue-berry micropropagation. *In*: Proceedings of Conference on Nursery Production of Fruit Plants through Tissue Culture—Applications and Feasibility. Agric. Res. Sci. Educ. Admin., U.S.D.A., Beltsville, pp. 44-47.

Zimmerman, R.H. and Broome, O.C. 1981. Phloroglucinol and *in vitro* rooting of apple cultivar cuttings. J. Am. Soc. Hortic. Set 106: 648-652.

Zimmermann, U. and Scheurich, P. 1981. High frequency fusion of plant protoplasts by electrical fields. Planta 151:26-32.

Ziv, M. 1991. Vitrification: morphological and physiological disorders of *in vitro* plants. *In*: P.C. Debergh and R.H. Zimmerman (Eds), Micropropagation: Technology and Applications. Kluwer, Dordrecht, pp. 45-69.

Ziv, M. 1995. *In Vitro* acclimatization. *In*: J, Aitken-Christie et al. (Eds), Automation and Environmental Control in Plant Tissue Culture. Kluwer, Dordrecht, pp. 493-516.

---

* Original not seen.

Subject Index